Eco-efficient Materials for Mitigating Building Cooling Needs

Related titles

Understanding the rheology of concrete
(ISBN 978-0-85709-028-7)

Toxicity of building materials
(ISBN 978-0-85709-122-2)

Building materials in civil engineering
(ISBN 978-1-84569-955-0)

Woodhead Publishing Series in Civil and Structural Engineering: Number 56

Eco-efficient Materials for Mitigating Building Cooling Needs

Design, Properties and Applications

Edited by

F. Pacheco-Torgal, J. A. Labrincha, L. F. Cabeza and C.-G. Granqvist

ELSEVIER

AMSTERDAM • BOSTON • CAMBRIDGE • HEIDELBERG
LONDON • NEW YORK • OXFORD • PARIS • SAN DIEGO
SAN FRANCISCO • SINGAPORE • SYDNEY • TOKYO
Woodhead Publishing is an imprint of Elsevier

WP WOODHEAD PUBLISHING

Woodhead Publishing is an imprint of Elsevier
80 High Street, Sawston, Cambridge, CB22 3HJ, UK
225 Wyman Street, Waltham, MA 02451, USA
Langford Lane, Kidlington, OX5 1GB, UK

British Library Cataloguing-in-Publication Data
A catalogue record for this book is available from the British Library.

Library of Congress Control Number: 2014954884

ISBN 978-1-78242-380-5 (print)
ISBN 978-1-78242-401-7 (online)

For information on all Woodhead Publishing publications
visit our website at http://store.elsevier.com/

Typeset by SPi Global
www.spi-global.com
Printed and bound in the United Kingdom

Contents

List of contributors

N.L. Alchapar CONICET—CCT-Mendoza, CC, Mendoza, Argentina

F. Ascione University of Naples Federico II, Napoli, Italy

C. Barreneche University of Lleida, Lleida, and University of Barcelona, Barcelona, Spain

N. Bianco University of Naples Federico II, Napoli, Italy

K. Boriboonsomsin University of California, Riverside, CA, USA

L.F. Cabeza University of Lleida, Lleida, Spain

A. Carbonari Università Politecnica delle Marche, Ancona, Italy

E.N. Correa CONICET—CCT-Mendoza, CC, Mendoza, Argentina

A. de Gracia University of Lleida, Lleida, Spain

R.F. De Masi University of Naples Federico II, Napoli, Italy

F. de Rossi Univeristy of Sannio, Benevento, Italy

M. D'Orazio Università Politecnica delle Marche, Ancona, Italy

D. Feng Harbin Institute of Technology, Harbin, China

A.I. Fernández University of Barcelona, Barcelona, Spain

F. Frontini University for Applied Sciences and Arts of Southern Switzerland (SUPSI), Canobbio, Switzerland

A.R. Gaspar University of Coimbra, Coimbra, Portugal

C.G. Granqvist Uppsala University, Uppsala, Sweden

S. Kalaiselvam Anna University, Chennai, India

D. Karamanis University of Patras, Agrinio, Greece

H. Li University of California, Davis, CA, USA

B.-S. Lin National Taiwan University, Taipei, Taiwan

Y.-J. Lin National Taiwan University, Taipei, Taiwan

A.G. Martins Institute for Systems Engineering and Computers, and University of Coimbra, Coimbra, Portugal

B. Naticchia Università Politecnica delle Marche, Ancona, Italy

L. Navarro University of Lleida, Lleida, Spain

K. Okada Tokyo Institute of Technology, Tokyo, Japan

F. Pacheco-Torgal C-TAC Research Centre, University of Minho, Guimarães, Portugal

R. Parameshwaran Anna University, Chennai, India

A.L. Pisello CIRIAF – University of Perugia, Perugia, Italy

M. Pittaluga Cagliari University, Cagliari, Italy

F. Reza Minnesota State University, Mankato, MN, USA

A. Speak University of Manchester, Manchester, United Kingdom

I. Susorova Cyclone Energy Group, Chicago, IL, USA

P.F. Tavares Polytechnic Institute of Coimbra, Institute for Systems Engineering and Computers, and University of Coimbra, Coimbra, Portugal

G.P. Vanoli Univeristy of Sannio, Benevento, Italy

H. Wang Harbin Institute of Technology, Harbin, China

N. Xie Harbin Institute of Technology, Harbin, China

Woodhead Publishing Series in Civil and Structural Engineering

1 **Finite element techniques in structural mechanics**
C. T. F. Ross
2 **Finite element programs in structural engineering and continuum mechanics**
C. T. F. Ross
3 **Macro-engineering**
F. P. Davidson, E. G. Frankl and C. L. Meador
4 **Macro-engineering and the earth**
U. W. Kitzinger and E. G. Frankel
5 **Strengthening of reinforced concrete structures**
Edited by L. C. Hollaway and M. Leeming
6 **Analysis of engineering structures**
B. Bedenik and C. B. Besant
7 **Mechanics of solids**
C. T. F. Ross
8 **Plasticity for engineers**
C. R. Calladine
9 **Elastic beams and frames**
J. D. Renton
10 **Introduction to structures**
W. R. Spillers
11 **Applied elasticity**
J. D. Renton
12 **Durability of engineering structures**
J. Bijen
13 **Advanced polymer composites for structural applications in construction**
Edited by L. C. Hollaway
14 **Corrosion in reinforced concrete structures**
Edited by H. Böhni
15 **The deformation and processing of structural materials**
Edited by Z. X. Guo
16 **Inspection and monitoring techniques for bridges and civil structures**
Edited by G. Fu
17 **Advanced civil infrastructure materials**
Edited by H. Wu
18 **Analysis and design of plated structures Volume 1: Stability**
Edited by E. Shanmugam and C. M. Wang
19 **Analysis and design of plated structures Volume 2: Dynamics**
Edited by E. Shanmugam and C. M. Wang

20 **Multiscale materials modelling**
Edited by Z. X. Guo
21 **Durability of concrete and cement composites**
Edited by C. L. Page and M. M. Page
22 **Durability of composites for civil structural applications**
Edited by V. M. Karbhari
23 **Design and optimization of metal structures**
J. Farkas and K. Jarmai
24 **Developments in the formulation and reinforcement of concrete**
Edited by S. Mindess
25 **Strengthening and rehabilitation of civil infrastructures using fibre-reinforced polymer (FRP) composites**
Edited by L. C. Hollaway and J. C. Teng
26 **Condition assessment of aged structures**
Edited by J. K. Paik and R. M. Melchers
27 **Sustainability of construction materials**
J. Khatib
28 **Structural dynamics of earthquake engineering**
S. Rajasekaran
29 **Geopolymers: Structures, processing, properties and industrial applications**
Edited by J. L. Provis and J. S. J. van Deventer
30 **Structural health monitoring of civil infrastructure systems**
Edited by V. M. Karbhari and F. Ansari
31 **Architectural glass to resist seismic and extreme climatic events**
Edited by R. A. Behr
32 **Failure, distress and repair of concrete structures**
Edited by N. Delatte
33 **Blast protection of civil infrastructures and vehicles using composites**
Edited by N. Uddin
34 **Non-destructive evaluation of reinforced concrete structures Volume 1: Deterioration processes**
Edited by C. Maierhofer, H.-W. Reinhardt and G. Dobmann
35 **Non-destructive evaluation of reinforced concrete structures Volume 2: Non-destructive testing methods**
Edited by C. Maierhofer, H.-W. Reinhardt and G. Dobmann
36 **Service life estimation and extension of civil engineering structures**
Edited by V. M. Karbhari and L. S. Lee
37 **Building decorative materials**
Edited by Y. Li and S. Ren
38 **Building materials in civil engineering**
Edited by H. Zhang
39 **Polymer modified bitumen**
Edited by T. McNally
40 **Understanding the rheology of concrete**
Edited by N. Roussel
41 **Toxicity of building materials**
Edited by F. Pacheco-Torgal, S. Jalali and A. Fucic
42 **Eco-efficient concrete**
Edited by F. Pacheco-Torgal, S. Jalali, J. Labrincha and V. M. John

43 **Nanotechnology in eco-efficient construction**
Edited by F. Pacheco-Torgal, M. V. Diamanti, A. Nazari and C. Goran-Granqvist

44 **Handbook of seismic risk analysis and management of civil infrastructure systems**
Edited by F. Tesfamariam and K. Goda

45 **Developments in fiber-reinforced polymer (FRP) composites for civil engineering**
Edited by N. Uddin

46 **Advanced fibre-reinforced polymer (FRP) composites for structural applications**
Edited by J. Bai

47 **Handbook of recycled concrete and demolition waste**
Edited by F. Pacheco-Torgal, V. W. Y. Tam, J. A. Labrincha, Y. Ding and J. de Brito

48 **Understanding the tensile properties of concrete**
Edited by J. Weerheijm

49 **Eco-efficient construction and building materials: Life cycle assessment (LCA), eco-labelling and case studies**
Edited by F. Pacheco-Torgal, L. F. Cabeza, J. Labrincha and A. de Magalhães

50 **Advanced composites in bridge construction and repair**
Edited by Y. J. Kim

51 **Rehabilitation of metallic civil infrastructure using fiber-reinforced polymer (FRP) composites**
Edited by V. Karbhari

52 **Rehabilitation of pipelines using fiber-reinforced polymer (FRP) composites**
Edited by V. Karbhari

53 **Transport properties of concrete: Measurement and applications**
P. A. Claisse

54 **Handbook of alkali-activated cements, mortars and concretes**
Edited by F. Pacheco-Torgal, J. A. Labrincha, C. Leonelli, A. Palomo and P. Chindaprasirt

55 **Eco-efficient masonry bricks and blocks: Design, properties and durability**
Edited by F. Pacheco-Torgal, P. B. Lourenço, J. A. Labrincha, S. Kumar and P. Chindaprasirt

56 **Eco-efficient materials for mitigating building cooling needs: design, properties and applications**
Edited by F. Pacheco-Torgal, J. A. Labrincha, L. F. Cabeza and C.-G. Granqvist

Foreword

It is becoming increasingly evident that human activities are the major cause of climate change and global warming. Human activities have caused major changes in our land, water, and atmosphere leading to permanent modifications to our environment such as removing major areas of forest land on Earth and increasing atmospheric greenhouse gases.

As the threat of climate change becomes more pronounced, a number of scientists have proposed supplementing the full range of mitigation efforts with geo-engineering (manipulation of the Earth's environment) to quickly respond to this threat. Many of these proposed geo-engineering techniques are novel and unproven. One simple technology has been in practice for thousands of years: changing the solar reflectance (albedo) of the built surface. "Cool roofs" and "cool pavements" should be among the first geo-engineering techniques used to combat global warming.

Increasing the solar reflectance of the urban surface reduces its solar heat gain, lowers its temperature, and decreases its outflow of thermal infrared radiation into the atmosphere. This process of "negative radiative forcing" can help counter the effects of global warming. We estimate that resurfacing conventional dark roofs with a cool white material that has a long-term solar reflectance of 0.60 or more increases its solar reflectance by at least 0.40. Our research shows that increasing the albedo of a 1 m^2 area by 0.01 results in a global temperature reduction of 3×10^{-15} K and offsets emission of 7 kg of CO_2. Installing 100 m^2 of cool roof has an effect on radiative forcing equivalent to a one-time offset of 28 tonnes of CO_2. Similarly, the solar reflectance of pavements can be raised on average by about 0.15, the equivalent of a 10 tonne offset in CO_2 per 100 m^2.

In addition, cool roofs and vegetation reduce cooling-energy use in air conditioned buildings and increase comfort in unconditioned buildings. Cool roofs, cool pavements, and urban vegetation mitigate summer urban heat islands and improve outdoor air quality and comfort. Installing cool roofs and cool pavements in cities worldwide is a compelling low-cost, win–win–win (saving cooling energy, reducing urban heat islands, and cooling the globe) activity that can be undertaken immediately, outside of international negotiations to cap CO_2 emissions.

Eco-efficient Materials for Mitigating Building Cooling Needs: Design, Properties and Applications edited by F. Pacheco Torgal, J. A. Labrincha, L. F. Cabeza, C.-G. Granqvist promotes a cool scenario by wide-scale utilization of advanced materials to reduce cooling energy use in buildings. After an introductory chapter, the book is organized in four parts: cool pavements, building facade materials, cool roofing materials, and phase change and chromogenic smart materials, all for reducing cooling demand and mitigating urban heat island effects. This book is a compilation of many recent studies and it is recommended as a good read to both academics and policy makers. Read and enjoy.

Professor Hashem Akbari
Concordia University, Canada

Introduction to eco-efficient materials to mitigate building cooling needs

1

F. Pacheco-Torgal
C-TAC Research Centre, University of Minho, Guimarães, Portugal

1.1 Climate change and urban heat islands (UHIs)

Climate change is one of the most important environmental problems on the planet (Garcia et al., 2014; IPCC, 2007; Rockström et al., 2009; Schellnhuber, 2008).

This is due to the increase of carbon dioxide (CO_{2eq}) in the atmosphere, for which the built environment is a significant contributor, with around one-third of global carbon dioxide emissions. In the early eighteenth century, the concentration level of atmospheric CO_{2eq} was 280 parts per million (ppm). At present it is 450 ppm (VijayaVenkataRaman et al., 2012).

Keeping the current level of emissions (which is unlikely given the high economic growth of less developed countries, with consequent increases in emission rates) will imply a dramatic increase in CO_{2eq} concentration to as much as 731 ppm in the year 2130, leading to a 3.7 °C global warming above pre-industrial temperatures (Valero et al., 2011).

As temperature increases, vector-borne illnesses, which rely upon organisms such as mosquitoes (*Aedes aegypti*, *Aedes albopictus*, *Aedes japonicus*) and other insects that have an active role in the transmission of a pathogen, have been projected to increase both in geographic reach as well as severity (IPCC, 2012; McMichael et al., 2006). Even northern latitudes in Europe will be reached (Githeko et al., 2000). West Nile fever is one that worries Europe (Hubalek and Halouzka, 1999; Sambri et al., 2013). The European Centre for Disease Control (ECDC, 2013) has reported several cases of West Nile fever in Europe and neighboring countries.

In addition to increased mean temperatures, climate change is likely to be responsible for more frequent heat waves, such as the 2003 European heat wave that claimed the lives of more than several thousand people (Table 1.1).

Heat waves also are associated with nonfatal impacts such as heat stroke and heat exhaustion (IPCC, 2007). Heat waves have a much bigger health impact in cities than in surrounding suburban and rural areas because urban areas typically experience higher—and nocturnally sustained—temperatures due to the urban heat island (UHI) effect (Hulley, 2012; IPCC, 2007).

UHIs are originated by heat generated in urban environments that is entrapped by urban structures being aggravated by greenhouse gases and the lack of green spaces. Dark-colored surfaces (like dark asphalt pavements) have low reflecting power

Eco-efficient Materials for Mitigating Building Cooling Needs. http://dx.doi.org/10.1016/B978-1-78242-380-5.00001-7

Table 1.1 Excess mortality attributed to the 2003 heat wave in Europe

Location (date)	Excess mortality (% increase)
England and Wales (Aug 4–13)	2091 deaths (17%)
Italy (Jun 1–Aug 15)	3134 deaths (15%) in all Italian capitals
France (Aug 1–20)	14,802 deaths (60%)
Portugal (Aug)	1854 deaths (40%)
Spain (Jul–Aug)	4151 deaths (11%)
Switzerland (Jun–Sept)	975 deaths (6.9%)
Netherlands (Jun–Sept)	1400–2200 deaths (not reported)
Germany (Aug 1–24)	1410 deaths (not reported)

Haines et al. (2006).

(or low albedo characteristics). As a consequence they absorb more energy, and in summer can reach almost 60 °C, thus contributing to higher UHI effects. Some authors reported a 10 °C temperature increase in the city of Athens due to the UHI effect (Santamouris et al., 2001), and a 8.8 °C increase in London (Kolokotroni and Giridharan, 2008), while a recent three-year investigation in the city of Padua reported an increase up to 6 °C (Busato et al., 2014). According to Li et al. (2014), the waste heat discharged by air conditioners alone was responsible for an increase of almost 2 °C of Beijing's average air temperature in 2005. It's worth noting that since 2005, the population of Beijing has increased from 15 to 21 million people, and by 2020 it will reach 25 million.

According to Santamouris (2013), a renowned expert on energy and solar technologies, UHI is probably the most documented phenomenon of climate change for various geographic areas of the planet. As a result of the urbanization and industrialization of human civilization, UHI has become one of the major problems of the twenty-first century (Rizwan et al., 2008). A recent work published in *Nature/Letter* (Zhao et al., 2014) also addresses this position. If no adaptation measures are taken, the social cost of climate change can be significant. This could mean an additional 26,000 deaths per year from heat waves (and their synergic effects with air pollution) by the 2020s, rising to 89,000 deaths per year by the 2050s and 127,000 deaths per year by the 2080s. These predictions do not even take UHI effects into account.

1.2 Adaptation to climate change and mitigation of UHI effects and of building cooling needs

Even if all the greenhouse gas emissions suddenly ceased, the amount already in the atmosphere would remain there for the next 100 years (Clayton, 2001). In other words, the rise in the sea level, ocean acidification, and extreme atmospheric events will continue.

Hansen et al. (2013) are even more pessimistic, believing that the climate has already been changed in an irreversible manner. This means that adaptation to climate change as well as mitigation of greenhouse gas emissions should be a priority to the built environment (Georgescu et al., 2014; Kwok and Rajkovich, 2010; Olazabal et al., 2014; Reckien et al., 2014).

The rapid increase of river floods and excessive hot days for the next decades is very worrying (EEA, 2012).

That is why the words of the European Commissioner for Climate Action, Connie Hedegaard, on the launch of the EU Strategy on Adaptation to Climate Change in Brussels on 29 April 2013, make a lot of sense: "*Investing now in adaptation will save lives and much greater costs later.*"

According to the COM 216 (2013), "*The minimum cost of not adapting to climate change in the EU is estimated to range from €100 billion a year in 2020 to €250 billion in 2050.*" On 25 June 2014, a study (Ciscar-Martinez et al., 2014) on the effects of climate change in Europe was released, stating that if no further action is taken, climate damages in the EU could amount to at least *€190 billion* and heat-related deaths could reach about 200,000.

The future of the built environment will therefore involve the adaptation towards climate-resilience. Unfortunately, a detailed study that analyzed the climate change adaptation and mitigation plans of more than 200 large and medium-sized cities across 11 European countries concluded that 35% of European cities studied have no dedicated mitigation plan, and 72% have no adaptation plan (Reckien et al., 2014).

Greening of the building envelope by using vegetation is a growing trend both in the adaptation to climate change and in the mitigation of UHIs. Green roofs date back to the fifth century in Babylon, when hanging gardens were implemented, and to ancient Mesopotamia, where they were used in the ziggurats (Berardi et al., 2014).

Though not an innovative technique (first attempts to quantify energy benefits occurred in the 1960s), the greening of the building envelope by using vegetation is a growing trend, and some countries show a remarkable acceptation of such "technology." For instance, Germany has almost 100 million m^2 of green roofs, and the state of Singapore intends to target 0.75 ha of green roofs per 1000 inhabitants (Pacheco-Torgal, 2014).

Green infrastructure is important enough to merit special attention in EU policy. On 6 May 2013, the EU adopted a new strategy for encouraging the use of green infrastructure (IP 404, 2013). The European Commission planned to develop guidance to show how green infrastructures could be integrated into the implementation of these policies from 2014 to 2020 for several areas, including adaptation to climate change. The green building envelope is particularly important, as recent investigations show a strong correlation between the lack of green infrastructures in the urban environment and the increase of allergy-related health problems (Hanski et al., 2012). This is the subject of Chapters 5, 11, 12, and 13 in this book.

An important consequence of climate change is the increase of building cooling needs. In the last two decades, building cooling needs have increased exponentially, from 6 TJ in 1990 to 160 TJ in 2010 (Balaras et al., 2007). According to Crawley (2008), "*the impact of climate change will result in a reduction in building energy use of about 10% for buildings in cold climates, an increase of energy use of up to 20% for buildings in the tropics, and a shift from heating energy to cooling energy for buildings in temperate climates.*" Other authors mention that, depending on the climate zone, cooling loads are likely to increase by 50–90% until the end of the century (Roetzel and Tsangrassoulis, 2012).

The synergistic effect between heat waves and air pollution causes worse outdoor air quality in the summer and will prevent natural ventilation, thus aggravating cooling needs. In the heavily polluted city of Beijing, Li et al. (2014) reported that 28.88% of the total air-conditioning energy consumption is due to the UHI effect.

In Europe, this could complicate the implementation of the ambitious nearly zero-energy building target set by the European Energy Performance of Buildings Directive (EPBD) recast (Directive 2010/31/EU). It could also put at risk the EU strategy to tackle climate change, especially the goal to cut greenhouse gas emissions by 80–95% below 1990 levels by 2050, as well as savings of at least €100 million per year by 2050 (COM (2011) 885/2, Pacheco-Torgal et al., 2013).

Actually, the energy costs represent 12% of the European GDP, and an increase to 14% is expected by 2050 in both scenarios (BAU and decarbonisation) (Figure 1.1). However, the decarbonisation scenario involves a 1.9% increase in infrastructure and energy efficiency investments and a 1.4% reduction of energy imports. This increase in the GDP, despite huge investments, has to do with the increase in energy demand. Of course, a better performance could be obtained if an indicator capable of measuring long-term sustainable use of natural ecosystems and resources was used, like the Genuine Progress Indicator (Constanza et al., 2009; Daly and Cobb, 1989).

Since the urban population will increase dramatically in the next decades, building cooling needs due to UHI effects will also increase dangerously (Allegrini et al., 2012).

According to the World Health Organization (WHO, 2014), around half of all urban dwellers currently live in cities; however, by 2050, cities will hold 70% of

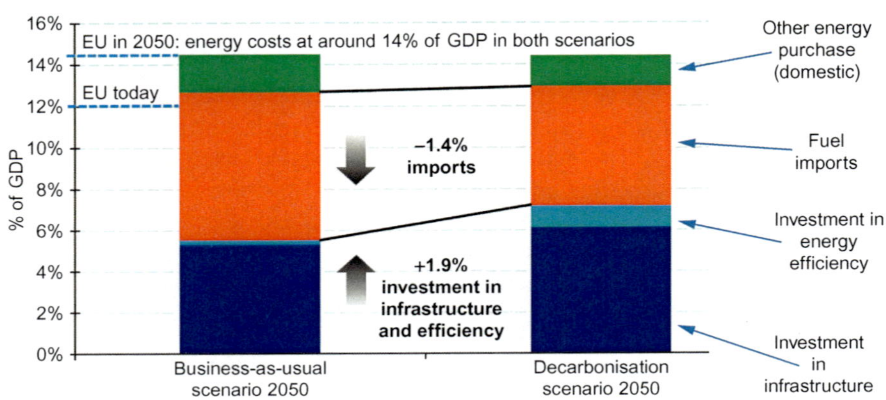

Figure 1.1 Forecasting of energy costs for Europe by 2050.
Barroso (2013).

the world population. This represents an alarming increase of about 3000 million people in the next 40 years (WHO, 2014).

If current trends in population density continue, urban land cover will increase by 1.2 million km^2 by 2030. This is equivalent to an area about the size of South Africa (Seto et al., 2012).

Therefore, it is very important that new eco-efficient materials and technologies capable of mitigating UHI effects and building cooling needs be developed. It is the purpose of this book to make a contribution in this area.

Other books have already been written about mitigation of UHI effects. However, some are short on material for the mitigation of building cooling needs, and others have almost nothing on UHIs. And almost all have nothing on chromogenic smart materials, to which this book dedicates three chapters (16, 17 and 18). Those three chapters, along with the three chapters related to the greening of the building envelope, constitute an important part of this book. With the special contribution of a team of international experts, the book provides an updated state of the art on eco-efficient materials to mitigate UHI effects and building cooling needs.

1.3 Outline of the book

This book provides an updated, state-of-the-art review on the eco-efficiency of materials to mitigate UHI effects and building cooling needs. The first part of the book encompasses an overview on pavements for mitigation of UHI effects (Chapters 2–4). In Chapter 2, different coating overlays of pavements for heat island phenomenon mitigation are introduced. Three types of materials as the main components of the coating materials for pavement to mitigate the heat island effect are provided. Organic polymers applied as the main part of the overlay coating materials, inorganic metal oxide powders used as the solar energy reflective fillers, and fine aggregates used to increase the durability of the reflective coatings are systematically introduced. The advantages and disadvantages of the current coating materials are analyzed. Chapter 3 addresses concrete pavements with high solar reflectance. Materials and methods used and their relative effectiveness for increasing solar reflectance are examined. The basic heat transfer problem relating solar reflectance to pavement surface temperature is explained. Modeling of a hypothetical scenario where high solar reflectance concrete replaces conventional concrete and both concrete and asphalt surfaces in several large cities shows that huge energy savings could be realized. Chapter 4 discusses the albedo and thermal performance of concrete and asphalt-based materials. The performance of different pavement materials, including pavement surface temperature, near-surface air temperature above pavement, thermal impact of pavement on nearby building surfaces, and heat flux from pavement are analyzed.

Façade materials for reducing building cooling needs are the subject of Part Two (Chapters 5–8). Chapter 5 is concerned with green façades. It reviews different types of green façades and living walls and analyzes the individual processes of a vegetated layer that contribute to the cooling effect. It also includes an assessment of energy savings that can be achieved in buildings with vegetated façades and summarizes additional benefits of green walls. Chapter 6 covers the performance of different façade materials. The thermal behavior of 96 façade materials—claddings and coatings evaluated by means of their solar reflectance index (SRI)—is analyzed. A simulation of relative impact of the appropriate selection of materials as a strategy for building and urban rehabilitation is presented. Chapter 7 analyzes innovative new materials for passive evaporative cooling walls. The chapter includes a description of the performance of porous mullite ceramics with unidirectionally oriented pores, prepared by an extrusion method using rayon fibers with various diameters as the pore formers; their capillary rise properties are investigated with a view to developing materials applicable for counteracting UHI effects. Chapter 8 addresses the performance of innovative evaporative cooling walls. A state-of-the-art review is included. The most important approaches currently adopted, namely direct and indirect cooling, are presented. A review of some numerical models relative to the simulation of evaporative cooling components is also presented.

Part Three (Chapters 9–13) deals with roofing materials for reducing building cooling needs. Chapter 9 reviews high albedo coatings for building roof application including its threefold benefits—at building, urban, and global climate scale. Specific characteristics of such coatings are discussed. The physical properties determining their cooling potential, solar reflectance, thermal emittance, and the final overall Solar Reflection Index are presented. Benefits of high albedo coating for roofs through the analysis of recent research demonstrating their effect in reducing building energy requirements for cooling are also presented. Chapter 10 discusses the solar evaporative cooling of roofs with hydrophilic (hydroscopic) porous materials. Urban passive cooling methods of building roofs are briefly presented. The interaction of solar spectrum with the porous materials through reflection and absorption, with or without adsorbed water vapor, is also presented. The concept of solar cooling through the water vapor adsorption–condensation–evaporation–desorption cycle and the experiment for the efficiency of the technique with low-cost materials are analyzed. Chapter 11 looks at cool green roofs. It summarizes three types of green roof systems and gathers materials and properties of each layer, according to maintenance requirements. A green roof design and construction process is presented step-by-step. Design principles for reducing cooling needs are suggested. Chapter 12 is concerned with the influence of damaged vegetation on the reduction of building cooling needs. Suggestions for damage prevention are included. Chapter 13 reviews the technical and economic aspects of green roofs. Economic indicators are described, as well as the physical set of equations for evaluating the energy performance of green roofs. Some case studies are shown, describing the various considered climates, energy and water costs, efficiency of power systems, emissions factors, and all parameters necessary for suitable feasibility studies. Considerations for improving the profitability of green roof by adopting rainwater harvesting systems are presented.

Finally, Part Four concerns phase change materials (PCMs) and chromogenic smart materials (Chapters 14–18). Chapter 14 is concerned with the performance of PCMs. A description of the use of PCMs in passive and active systems is included. Chapter 15 deals with nanomaterials—embedded PCMs for reducing building cooling needs. A review on nanomaterials for thermal energy storage is presented. A simulation on the thermal energy storage properties of nanomaterials is included. Chapter 16 deeply reviews chromogenic glazings based on thermochromic and electrochromic materials for reducing building cooling needs. Chapter 17 covers electrochromic walls, which encompass the innovative use of electrochromic glazing in the opaque partition of the building envelope. Chapter 18 closes Part IV with a case study on the influence of electrochromic windows' impact on the energy performance of buildings located in Mediterranean climates.

References

Allegrini, J., Dorer, V., Carmeliet, J., 2012. Influence of the urban microclimate in street canyons on the energy demand for space cooling and heating of buildings. Energy Build. 55, 823–832.

Balaras, C., Grossman, G., Henning, H., Infante-Ferreira, C., Podesser, E., Wang, L., Wiemken, E., 2007. Solar air conditioning in Europe—an overview. Renew. Sustain. Energy Rev. 11, 299–314.

Barroso, J., 2013. Energy priorities for Europe. Presentation to the European Council in 22 May. European Commission.

Berardi, U., Hoseini, A., Hoseini, A., 2014. State-of-the-art analysis of the environmental benefits of green roofs. Appl. Energy 115, 411–428.

Busato, F., Lazzarin, R., Noro, M., 2014. Three years study of the urban heat island in Padua: experimental results. Sustain. Cities Soc. 10, 251–258.

Ciscar-Martinez, J.C., et al., 2014. Climate impacts in Europe. The JCR PESERAII Project (Projection of Economic impacts of climate change in Sectors of the EU based on bottom-up Analysis). http://publications.jrc.ec.europa.eu/repository/handle/111111111/31757 (accessed 03.07.14.).

Clayton, R., 2001. Editorial: is sustainable development an oxymoron? Trans. Inst. Chem. Eng. 79 (Part B), 327–328.

COM 2011. 885/2. Energy Roadmap 2050.

COM 216, 2013. An EU strategy on adaptation to climate change. Brussels.

Constanza, R., Hart, M., Posner, S., Talberth, J., 2009. Beyond GDP: the need for new measures of progress. The Pardee Papers, N4 Boston University.

Crawley, D.B., 2008. Estimating the impacts of climate change and urbanization on building performance. J. Build. Perform. Simul. 1, 91–115.

Daly, H.E., Cobb Jr., J.B., 1989. For the Common Good: Redirecting the Economy Toward Community, the Environment, and a Sustainable Future. Beacon Press, Boston.

ECDC—European Centre for Disease Control. West Nile fever maps. 2013. http://www.ecdc.europa.eu/en/healthtopics/west_nile_fever/west-nile-fever-maps/pages/index.aspx.

EEA Report 12, 2012. Climate Change, Impacts and Vulnerability in Europe 2012. European Environment Agency, Copenhagen.

European Union. Directive 2010/31/EU of the European Parliament and of the Council of May 19th, 2010 on the energy performance of buildings (recast). Off. J. Eur. Union; 18 June 2010.

Garcia, R., Cabeza, M., Rahbek, C., Araujo, M., 2014. Multiple dimensions of climate change and their implications on biodiversity. Science 2 (3844), 6183.

Georgescu, M., Morefield, P., Bierwagen, B., Weaver, C., 2014. Urban adaptation can roll back warming of emerging megapolitan regions. Proc. Natl. Acad. Sci. U. S. A. 111, 2909–2914.

Githeko, A.K., Lindsay, S.W., Confalonieri, U.E., Patz, J.A., 2000. Climate change and vector-borne diseases: a regional analysis. Bull. World Health Organ. 78, 1136–1147.

Haines, A., Kovats, R., Campbell-Lendrum, D., Corvalan, C., 2006. Climate change and human health: impacts, vulnerability, and mitigation. Lancet 367, 24–30.

Hansen, J., et al., 2013. Assessing "Dangerous Climate Change": required reduction of carbon emissions to protect young people, future generations and nature. PLoS One 8 (12), e81648. http://dx.doi.org/10.1371/journal.pone.0081648. http://www.plosone.org/article/info%3Adoi%2F10.1371%2Fjournal.pone.0081648.

Hanski, I., Von Hertzen, L., Fyhrquist, N., Koskinen, K., Torppa, K., Laatikainen, T., Karisola, P., Auvinen, P., Paulin, L., Mäkelä, M., Vartiainen, E., Kosunen, T., Alenius, H., Haahtela, T., 2012. Environmental biodiversity, human microbiota, and allergy are interrelated. Proc. Natl. Acad. Sci. U. S. A. 109 (21), 8334–8339. http://www.pnas.org/content/early/2012/05/01/1205624109.full.pdf.

Hubalek, Z., Halouzka, J., 1999. West Nile fever—a reemerging mosquito-borne viral disease in Europe. Emerg. Infect. Dis. 5 (5), 643–650.

Hulley, M., 2012. The Urban Heat Island: causes and potential solutions. In: Zeman, F. (Ed.), Metropolitan Sustainability. Understanding the Urban Environment. WoodHead Publishing Limited, Cambridge, pp. 79–98.

IP 404, 2013. Environment: investing in green infrastructure will bring multiple returns to nature, society and people.

IPCC (Intergovernmental Panel on Climate Change), 2007. Climate Change 2007: The physical science basics. In: Contribution of Working Group I to the Fourth Assessment Report of the IPCC, Geneva.

IPCC (Intergovernmental Panel on Climate Change), 2012. Managing the Risks of Extreme Events and Disasters to Advance Climate Change Adaptation Special Report of the Intergovernmental Panel on Climate Change. Cambridge University Press, UK.

Kolokotroni, M., Giridharan, R., 2008. Urban heat island intensity in London: an investigation of the impact of physical characteristics on changes in outdoor air temperature during summer. Sol. Energy 82, 986–998.

Kwok, A., Rajkovich, N., 2010. Addressing climate change in comfort standards. Build. Environ. 45, 18–22.

Li, C., Zhou, J., Cao, Y., Zhong, J., Liu, Y., Kang, C., Tana, Y., 2014. Interaction between urban microclimate and electric air-conditioning energy consumption during high temperature season. Appl. Energy 117, 149–156.

McMichael, A.J., Woodruff, R.E., Hales, S., 2006. Climate change and human health: present and future risks. Lancet 367 (9513), 859–869.

Olazabal, M., De Gregorio Hurtado, S., Olazabal, E., Pietrapertosa, F., Salvia, M., Geneletti, D., D'Alonzo, V., Feliú, E., Di Leo, S., Reckien, D., 2014. How are Italian and Spanish Cities Tackling Climate Change? A Local Comparative Study. BC3 Working Paper Series 2014-03, Basque Centre for Climate Change (BC3), Bilbao, Spain.

Pacheco-Torgal, F., 2014. Eco-efficient construction and building materials research under the EU Framework Programme Horizon 2020. Construct. Build Mater. 51, 151–162.

Pacheco-Torgal, F., Cabeza, L., Mistretta, M., Kaklauskas, A., Granqvist, C.G., 2013. Nearly Zero Energy Building Refurbishment. A Multidisciplinary Approach. Springer Verlag, London, UK.

Reckien, D., Flacke, J., Dawson, R.J., Heidrich, O., Olazabal, M., Foley, A., Hamann, J.J.P., Orru, H., Salvia, M., Hurtado, S.D.G., Geneletti, D., Pietrapertosa, F., 2014. Climate change response in Europe: what's the reality? Analysis of adaptation and mitigation plans from 200 urban areas in 11 countries. Clim. Change 122 (1–2), 331–340.

Rizwan, A., Dennis, L., Liu, C., 2008. A review on the generation, determination and mitigation Urban Heat Island. J. Environ. Sci. 20, 120–128.

Rockström, J., Steffen, W., Noone, K., Persson, Å., Chapin, F.S., III, E. Lambin, Lenton, T.M., Scheffer, M., Folke, C., Schellnhuber, H., Nykvist, B., De Wit, C.A., Hughes, T., van der Leeuw, S., Rodhe, H., Sörlin, S., Snyder, P.K., Costanza, R., Svedin, U., Falkenmark, M., Karlberg, L., Corell, R.W., Fabry, V.J., Hansen, J., Walker, B., Liverman, D., Richardson, K., Crutzen, P., Foley, J., 2009. Planetary boundaries: exploring the safe operating space for humanity. Ecol. Soc. 14 (2), 32. http://www.ecologyandsociety.org/vol14/iss2/art32/.

Roetzel, A., Tsangrassoulis, A., 2012. Impact of climate change on comfort and energy performance in offices. Build. Environ. 57, 349–361.

Sambri, et al., 2013. West Nile virus in Europe: emergence, epidemiology, diagnosis, treatment, and prevention. Clin. Microbiol. Infect. 19, 699–704.

Santamouris, M., 2013. Using cool pavements as a mitigation strategy to fight urban heat island—a review of the actual developments. Renew. Sustain. Energy Rev. 26, 224–240.

Santamouris, M., Papanikolaou, N., Livada, I., Koronakis, I., Georgakis, C., Argiriou, A., Assimakopoulos, D., 2001. On the impact of urban climate on the energy consumption of buildings. Sol. Energy 70, 201–216.

Schellnhuber, H., 2008. Global warming. Stop worrying, start panicking? Proc. Natl. Acad. Sci. U. S. A. 105, 14239–14240.

Seto, K.C., Buneralp, B., Hutyra, L.R., 2012. Global forecasts of urban expansion to 2030 and impacts on biodiversity and carbon pools. Proc. Natl. Acad. Sci. U. S. A. 109 (40), 16083–16088, 17–21.

Valero, A., Agudelo, A., Valero, A., 2011. The crepuscular planet. A model for the exhausted atmosphere and hydrosphere. Energy 36, 3745–3753.

Vijayavenkataraman, S., Iniyan, S., Goic, R., 2012. A review of climate change, mitigation and adaptation. Renew. Sustain. Energy Rev. 16 (1), 878–897.

WHO, 2014. Urban Population Growth. Global Health Observatory. http://www.who.int/gho/urban_health/situation_trends/urban_population_growth_text/en/.

Zhao, L., Lee, X., Smith, R.B., Oleson, K., 2014. Strong contributions of local background climate to urban heat islands. Nature/Letter 511 (7508), 216–219.

Part One

Pavements for mitigating urban heat island effects

Coating materials to increase pavement surface reflectance

2

N. Xie, H. Wang, D. Feng
Harbin Institute of Technology, Harbin, China

2.1 Introduction

The urban heat island (UHI) effect is a common environmental problem occurring in metropolitan areas in which the air temperature is significantly higher than in suburban areas. The UHI effect also leads to a smoggy climate. The UHI effect mainly results from high-density infrastructures, which not only limit air flow but also emit heat stored from solar energy. In addition, the reduction of vegetation and wetlands further weakens the heat-releasing capacity of cities.It is urgent to mitigate the UHI effect, as it has become a huge threat to the environment and human health (Santamouris, 2007; Santamouris et al., 2011; Stathopoulou, 2008; Mirzaei and Haghighat, 2010).

Previous studies indicated that the optical and thermal characteristics of urban structures have a strong relationship with urban climate and temperatures (Chen et al., 2009; White et al., 2010). In many cities, the roofs and pavements comprise about 60% of the total urban area, with roofs contributing about 20–25% and pavements contributing about 40% (Akbari et al., 2003). Cantat indicated that the reflectance of Paris is 16% lower than the surrounding suburban district (Cantat, 1989). Another study claimed that the increase of the solar energy absorption in urban areas is affected by the urban geometrical structures (Aida, 1982; Aida and Gotoh, 1982).

Cool pavement is an effective technology to solve the UHI problem. The primary three types of cool pavements are (1) light color aggregates pavement, (2) permeable pavement, and (3) solar-reflective coating pavement. The albedo of normal asphalt pavement is only about 5%. By using white or light color aggregates, the albedo can be increased to about 30% (Doulos et al., 2004). Recent case study results from Portland State University (Oregon, USA) show that if you increase the albedo of the pavement in a bare courtyard from 37% (black) to 91% (white), the mean radiant temperature will increase 2.9 °C, but the air temperature will decrease 1.3 °C (Taleghani, 2014).

Cotana et al. (2014) believes that using high solar reflective surfaces is an effective way to mitigate the UHI problem and reduce greenhouse gas emission in urban areas. It has been found that albedo coatings can reduce CO_2 emissions. Simulation results indicated that a 115,000 m^2 high-albedo coating area could reduce 16,000 tons of CO_{2eq}, releasing and effectively mitigating global warming if applied for 30 years. Akbari and Matthews (2012) also estimated that using cool roofs and cool pavements could increase urban albedo by about 10%, and reduce CO_2 emissions

Eco-efficient Materials for Mitigating Building Cooling Needs. http://dx.doi.org/10.1016/B978-1-78242-380-5.00002-9

by at least 40–160 Gt. Yu's study claimed a similar result based on a life cycle assessment model (Yu and Lu, 2012). Experimental results indicated that 4500 m^2 cool pavements in an urban parking lot could decrease the ambient temperature about 2 °C, and at the same time decrease the surface temperature by 12 °C during a typical summer day (Santamouris et al., 2012).

Rosenfeld et al. (1998) claimed that cool surface techniques could save at least half a billion dollars if used in the Log Angeles basin, and up to five billion dollars if applied in the entire U.S. by the year 2015. Akbari et al. (2009) estimated that urban albedo could be increased about 10% by using cool roofs and cool pavements, and the increased urban albedo could reduce 44 Gt of CO_2 emissions and save 1.1 trillion dollars worldwide.

To implement the high-reflectance surface coating technology, proper *cool* material selection plays the most important role. In general, the selection rule is governed by two main parameters: solar reflectance and heat emittance (Asaeda et al., 1996; Synnefa et al., 2006). Carnielo and Zinzi (2013) tested the emissivity and solar reflectance of five different albedo samples with colors of green, red, blue, gray, and off-white. It was found that, compared with thermal emittance, the effect of solar reflectance is more evident on mitigating the urban heat island phenomenon. The spectral emissivity values of those samples are similar to the normal asphalt, which are all above 90%, while the solar reflectance of each sample varies a lot, with a maximum difference of about 61% compared to the normal asphalt sample.

The visible light reflectance of the coating is mainly governed by its visible color and surface roughness, which are determined by the inorganic fillers. Uemoto et al. (2010) claimed that the visible light reflectance of the coating will decrease when the color is more dark than light. Other than the visible light reflectance, the near-infrared (NIR) light reflectance is also important for cool pavements. In the NIR reflective region, the colors of the materials are not the dominant factor. High NIR reflectance can be realized in various colors of inorganic fillers. Synnefa et al. (2011) investigated five asphalt pavement overlay coatings with various colors. The results indicated that the reflectance of the coatings with all colors are higher than the asphalt pavement without coatings. The NIR reflectance of these coating ranges from 27% (red and green) to 55%, which could maximally decrease the pavement temperature about 12 °C compared with the noncoated asphalt pavement surface. Kinouchi et al. (2003) developed a new pavement coating with high reflectance and low brightness, which has a relatively low reflectance (20%) in the visible light region (350–750 nm) and high reflectance (83%) in the NIR region (750–2100 nm). The results showed that this coating could decrease the pavement temperature by 15 °C compared with normal asphalt pavement. Wan et al. (2012) also developed a cool pavement coating with high NIR reflectance (81%), low thermal conductivity (0.252 W/m K), and high emissivity (82.8%). The results demonstrated that this pavement could decrease surface temperature about 17 °C compared with conventional asphalt pavement and 5 °C compared with cement concrete pavement.

Unlike cool roofs, which have no traffic safety and traffic load concerns, cool pavement coatings with high reflectance have to satisfy the requirements of a pavement overlay. Therefore, they are generally composed of three main parts: the organic polymer substrate, inorganic fillers, and antiskid aggregates. The polymer substrate is mainly for providing bonding capacity to guarantee the durability of the coatings; the inorganic fillers are mainly used for providing the solar reflectance; and the aggregates are used for transportation safety and wear resistance concerns. The following sections will discuss how these can be used as cooling pavement materials.

2.2 Organic polymers used as coating overlay materials for pavements

In the past decades, various polymers were used as coating overlay materials to protect bridge decks and pavements from chemical attacks and freeze/thaw damage. The bonding or adhesive property is the most important factor to consider before these polymers can be used as coating overlays.

Recently, it was reported that 2724 polymer-based overlays were applied on bridge structures in the United States (Federal Highway Administration, FHWA, 2010). Since 2008, the polymer overlay has increased 24.2%, whereas other surface overlays keep decreasing every year (Young and Durham, 2012).

The requirements of cool pavement are more critical than cool roof because cool pavement must satisfy the requirements of a pavement, including skid resistance, wear resistance, bonding strength, and mechanical properties. Therefore, fewer types of polymer materials can be used as the coating substrate for pavement than can be used for cool roofs. The two main types of materials that are widely used as coating overlays for pavements are epoxy and acrylic polymers. This section presents information relevant to the selection of these two types of polymers for asphalt pavement coatings.

2.2.1 Epoxy resins

Epoxy resins are used as adhesives to join two surfaces from different components by providing a bond between them. Epoxy resins were commercially introduced in 1946 as adhesives and can be applied in many industrial areas. Epoxy resins are easily modified in order to achieve required properties. They are the most widely used adhesive because many materials can be bonded with epoxy resins. In the epoxy industry, the modification is called *formulating* or *compounding*. In the modification process, the curing agent and the reaction mechanisms are the key factors that guarantee the required properties with a reasonable cost; therefore, numerous epoxy adhesive formulations can be obtained. Based on the various curing agents and specific formulations, the final properties of the epoxy resins can be designed in a wide range. Cured epoxies provide magnificent mechanical properties. In addition, they also show good resistance to oil, moisture, and many other solvents.

As demonstrated in Figure 2.1, the epoxy presents a broad group of reactive compounds that are characterized by the presence of an oxirane or epoxy ring. This is a three-member ring containing an oxygen atom bonded with two carbon atoms that

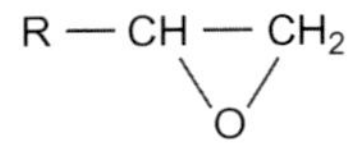

Figure 2.1 The epoxy structure.

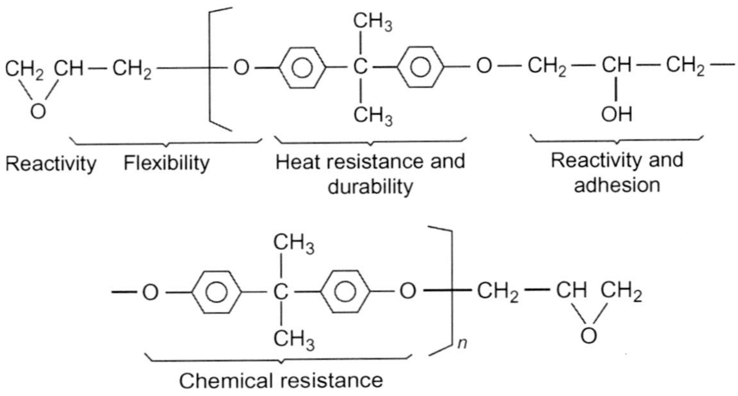

Figure 2.2 The structure and properties of epoxy resins.

have been united with some other elements or groups. More than one epoxy group can be contained in an epoxy resin to form a molecule.

The general formula of an epoxy resin can be expressed as Figure 2.2 (Osumi, 1987). The linear polyether with epoxy groups and the hydroxyl groups along the length of the chain contribute the chemical resistance, heat resistance, durability, and the adhesion, in which the reactivity of the molecules is governed by the terminal of the molecule and the hydroxyl at the midpoint, and the adhesion property of epoxy is governed by the secondary hydroxyl groups located in the molecule chain.

The main mechanism of the epoxy curing is the so-called *ring opening*. In this process, the epoxy group may react in different ways. One is the anionic reaction, and the other is the cationic reaction. Figure 2.3 shows the anionic mechanism of the ring opening (Lee, 1967). In the cationic reaction, the epoxy group will be opened to form a new hydroxyl bond by the introduction of an active hydrogen.

The material that reacts with the epoxy to finally form the epoxy network is called the *curing agent* or *hardener*. The mixing content of the curing agent varies from several percent to as high as fifty percent. A large number of curing agents can react with

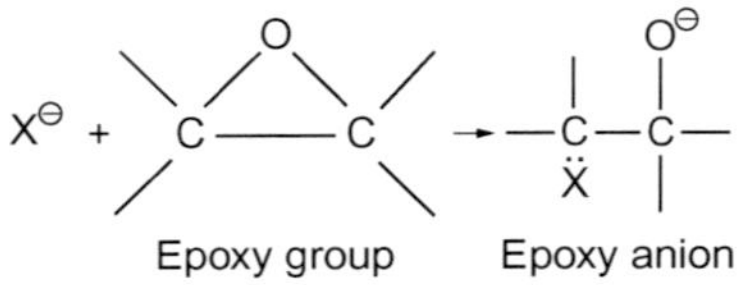

Figure 2.3 Anionic mechanism of the epoxy reaction.

epoxy to form a cross-link structure. The curing agent, in most cases, will react with the available epoxy or hydroxyl groups. Currently, for commercial epoxies, the following materials are widely used as the curing agent.

- Aliphatic amines and modified aliphatic amines
- Polyamides
- Anhydrides
- Polysulfides and mercaptans

The rigidity is one of the shortcomings of epoxy resins when they are used as coating overlay materials for pavement. Therefore, mitigating this problem is the key issue in determining whether or not the epoxy can be applied in the field. With the proper selection of curing agent, the flexibility of inherently rigid epoxy resins can be improved. By controlling stoichiometric mix ratio between the epoxy resin and the curing agent, or by using flexible molecule curing agents, the epoxy formulation can be changed and the flexibility will be adjusted accordingly. For example, by changing the curing agent from hexahydrophthalic anhydride to hexamineethylene-diamine, the impact resistance of a resin system can be doubled and its tensile elongation increased accordingly (Epon Resin Structural Resin Manual, 2001).

There are two types of curing methods for current commercial epoxies. The first is single component epoxies with heat curing, and the second is multiple component epoxies curing at room temperature or at elevated temperatures. In most cases, heat curing will lead to a relatively higher glass transition temperature and a higher cross-linking density than curing at room temperature. As a result, the epoxies cured at high temperatures will provide a higher shear strength and chemical stability. However, due to their rigidity, their toughness and peeling strength will be lower than those cured at room temperature.

Although some epoxies can be cured in minutes at room temperature, most require relatively longer curing time (18–72 h) to obtain full strength. Curing time is temperature sensitive; it can be shortened by increasing curing temperatures. After curing, the epoxy resins are solid materials with outstanding properties. Such properties may include low shrinkage and no volatility during the curing process, compatibility with many types of materials, chemical stability, and durability under a complex environment. In addition, the properties of the epoxy resins can be controlled according to the requirements. Flexible or rigid, high or low modulus, filled or foamed, conductive or insulative—all of these properties can be obtained by changing the components of the epoxies.

Apart from curing time and flexibility, the viscosity of the epoxy resins is another important factor to consider before the material can be applied in the field as a pavement coating It is difficult to utilize epoxy resins with high viscosity. Currently, commercial epoxy resins are available as either liquids or solids. The viscosity of the liquid-state epoxies can range from water-like to crystalline solid. In most cases, the molecular weight decides the viscosity of the epoxy resins. The higher the molecular weight, the higher the viscosity and the melting point. Other than the molecular weight, the viscosity of the epoxy resins is also governed by temperature. Viscosity reduces rapidly with the decrease of temperature.

Epoxy resins were used as pavement coating materials for many years. Most epoxy resin coatings were used as overlays with antiskid or waterproofing functions. Recently, epoxy resin coatings have started to incorporate these functions to anti-icing and wear-resistant surfaces.

Epoxy coatings for concrete pavement have been studied extensively, however, research for epoxy coating on asphalt pavement is very limited due to some bottleneck problems that are still awaiting solutions. First, the inherent brittleness of the cured epoxy resins can easily result in breaks on the top of the asphalt pavement because of the mismatch of the deformation capability. Second, the cost of the materials is relatively high if the materials are adjusted to the required properties. Third, the curing condition, such as relatively long curing time and high curing temperature, is critical to pavement engineering.

2.2.2 Acrylic ester polymers

Acrylic ester polymers were discovered by Otto Rohm when he was conducting his doctoral research in 1901, and they were commercially produced by the Rohm and Haas Co. of Darmstadt, Germany in 1927 (Riddle, 1954). Similar to epoxy resins, acrylic ester polymers can also be used as adhesives to join two surfaces from different components by providing a bond between them.

Figure 2.4 gives the structure of acrylic ester monomers. The R side chain ester group determines the final properties of the formed polymers used in various application fields, from paint to adhesives. In general, acrylic ester monomers are formed from the reaction between acrylic acid and an alcohol as follows:

$$\text{acrylic acid} + \text{alcohol} \rightarrow \text{alkyl acrylate} \tag{2.1}$$

During the copolymerizing process, acrylic ester monomers will randomly incorporate themselves to form polymer chains by the percentage concentration of each monomer. In addition, the acrylic ester monomers can also be copolymerized with methacrylic ester monomers, styrene, acrylonitrile, or vinyl acetate to prepare other products. The final properties of the formed polymers are determined by the molecular weight and the ester side chain of the product. Similar to other polymers, the properties of the acrylic ester polymers can be improved as a function of molecular weight until a threshold is reached. Beyond this threshold, property improvement will cease. In general, for acrylic polymers, the threshold molecular weight value is about 100,000–200,000.

The polymerization of acrylic monomers is based on the chain-growth mechanism. It is realized by the head-to-tail connection of the individual monomer units by breaking the monomer double bond and forming a single bond between the newly incorporated monomer units, shown in Figure 2.5.

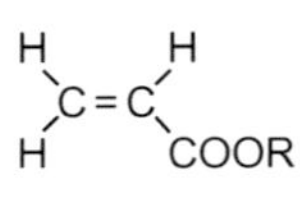

Figure 2.4 The structure of acrylic ester monomers.

Figure 2.5 The polymerization mechanism of acrylic monomers.

Table 2.1 Mechanical properties of acrylic polymers

Polyacrylate	Elongation, %	Tensile strength, kPa
Methyl	750	6895
Ethyl	1800	228
Butyl	2000	21

Stone (2010).

The mechanical properties of acrylic ester polymers are largely counting on the side chain groups of the ester. If the side chain is H, the poly(acrylic acid) is a brittle material, while the poly(methyl acrylate) is a strong and rubbery material at room temperature. The poly(ethyl acrylate), which has a longer chain length, is more rubbery and extensible. The poly(butyl acrylate) is tack at room temperature and is able to be used as the adhesive. Table 2.1 lists the mechanical properties of acrylic polymers (Stone, 2010). The flexibility of acrylic polymers increases with increasing chain length until it reaches the threshold: poly(*n*-nonyl acrylate). At this chain length, the side chains will be crystallized and lead to the stiffening of the polymers.

Acrylic ester polymers are chemically stable at aggressive conditions, even with exposure at 300–400 °C. This chemical stability guarantees the durability of the final product. Acrylic ester polymers are also resistant to oxidation and hydrolysis. However, this resistance is not universal. When exposed to strong acidic or alkaline conditions, acrylic ester polymers will be hydrolysed to poly(acrylic acid) and alcohol. The resistance to hydrolysis is in the order of butyl acrylate > ethyl ethyl acrylate > methyl acrylate.

UV radiation resistance is another important factor that determines the durability of acrylic polymers as a pavement coating material. In general, acrylic polymers have strongUV resistance; however, if they are incorporated with UV-absorbing monomers such as styrene, the UV resistance of the resulting products will be considerably decreased and will deteriorate rapidly. On the other hand, if the incorporating UV-absorbing monomers are bonded in a noncovalent way, the UV resistance of the resulting polymers will not be affected. For example, hydroxybenzophenone will further enhance the UV resistance of acrylic polymers (Burgess, 1952).

In the past decades, acrylic polymers were widely used as marking paint materials in pavement engineering. Recently, they have begun to be used as overlay materials for color pavements. Acrylic polymer coatings have been studied as an antiskid overlay for concrete pavement (Scholer and Forrestel, 1980); however, research on acrylic polymers as coating materials on asphalt pavement is limited. Except for their relatively low adhesive or bonding strength, acrylic polymers have the advantage when compared to epoxy resins. First, the flexibility of cured acrylic polymers is far better than the flexibility of epoxy resins, which makes the former able to match pavement deformation during service time. Second, the chemical and UV stabilities are promising advantages of acrylic polymers when used as overlay materials for pavements. Third, the relatively low cost of acrylic polymers guarantees their wide application in pavement engineering.

2.2.3 Advantages and disadvantages of various polymers

Both epoxy and acrylic polymers provide advantages and disadvantages in their use as coating materials for pavements. *Epoxy polymers* typically bear the benefits of high fatigue strengths, high temperature properties, and chemical stability; however, during implementation, the surfaces that need to be joined have to be carefully cleaned. In addition, the curing time is relatively long and the curing temperature is sometimes high. Apart from the curing problems, the rigidity of the cured epoxy is another problem when used as a coating material for asphalt pavement. *Acrylic ester polymers* typically have good plasticity, which is a merit for asphalt pavement. They are also chemically stable in aggressive environments. However, both epoxy and acrylic polymers are not UV resistant. UV ageing is still the bottleneck problem that needs to be solved.

2.3 Inorganic materials used as polymer fillers to increase reflectance

When an electromagnetic wave irradiates on a substance, the incident wave I_λ can be divided into three categories: absorbed part, transmitted part, and reflected part, which are represented as A_λ, T_λ, and R_λ, and can be expressed as:

$$I_\lambda = \alpha A_\lambda + \beta T_\lambda + \gamma R_\lambda \tag{2.2}$$

where α, β, and γ are the fraction of the absorption, reflection, and transmission, respectively, and the sum of them should be 1.

All solid-state inorganic materials have their own specific electron energy band structures. If the incident light with energy of $E = h\upsilon$ matches the bandgap of a specific material, the electrons in the valence band will be excited and jump to the conduction band to form free electrons and leave holes in the valence band. In this process, light energy will be absorbed and converted to thermal energy. As an electromagnetic wave, sunlight has a wide range of wavelengths, varying from 295 to 2500 nm. Once the sunlight has been absorbed by materials with a selective wavelength, it will be converted into heat energy.

UV light is not visible to the naked eye. Its wavelength ranges from 295 to 400 nm. UV light contributes only about 5% of solar energy; however, it is the main reason for the asphalt binders' or polymer coatings' ageing or degradation.

Visible light, whose wavelength ranges from 400 to 700 nm, combines various colors, and each color corresponds to its own wavelength. Approximately half of sunlight energy comes from this region. Visible light absorption or reflection is governed by the electronic band structures of specific materials. If a material can only reflect light with a specific wavelength, then the observer can only see its corresponding color. For example, if a material shows the color of red, that means the material is selected to reflect red light. If a material shows the color of white, that means the material is selected to reflect white light, which is the combination of all visible light colors. If a material shows the color of black, that means the material is selected to absorb visible light. As a result, color is the dominant factor that determines the visible light reflectivity of materials. In most cases, light colors provide high reflectance of the visual spectrum of radiation. However, to the infrared part, the reflectance is independent of specific colors.

The wavelengths of near-infrared (NIR) light range from 700 to 2500 nm. Around 45% of all solar energy comes from the NIR region. Because NIR light is not visible to the naked eye, the reflectance of NIR light is irrelevant to the colors of the materials. As a result, the coating with high NIR reflectance could be colorful, which is good for pavement coating because white or light color coating of the pavement will introduce glare problems to vehicle drivers. Most NIR reflective materials are metal oxides. Because of their chemical and environmental stability, they are widely used as pigments in various paints.

2.3.1 White color materials for increasing visible light reflectance

Titanium widely exists in earth crust, mostly rutile, anatase, and brookite. It has been found in many places, including rocks, water bodies, and soil. Titanium dioxide has exhibited many outstanding advantages in daily products such as white paint, plastic, paper, ink, food, medicine, makeup, and toothpaste. Meanwhile, it also shows some advanced properties in environmental benefits and energy harvesting, such as photocatalytic water splitting, heterogeneous photocatalysis, charge transfer, Gratzel cells, sensitized solar cells, organic photovoltaic cells, quantum dots, nanorods, nanotubes, photocurrent generation, and electrodes (Kamat, 2012). Thanks to its high refractive index, titanium dioxide is the most widely used nontoxic white pigment at the present time. It has been stated that more than 4.6 million tons of TiO_2 are used as white pigment worldwide, and this number increases annually (Winkler, 2003).

Renz (1921) was possibly the first to discover that TiO_2 is illumination sensitive. It was found that TiO_2 will be reduced with sunlight illumination. The color of white TiO_2 became gray, blue, or black. A similar phenomenon was found in CeO_2, Nb_2O_5, and Ta_2O_5. After that, the famous "chalking" phenomenon, which always happened on TiO_2-based paint, was successfully explained by Goodeve and Kitchener (1938). They proposed an assumed mechanism that TiO_2 will act as a photocatalyst to accelerate the oxidation process, based on an experiment result that TiO_2 powder will help decompose a dye in air under sunlight illumination. In the past 100 years, the properties of TiO_2 have been studied extensively.

Levinson studied the reflectance of rutile TiO_2 in acrylic matrix and compared it with other oxides. It was found that TiO_2 shows similar reflectance curves, with strong backscattering and weak visible and NIR light absorption, as shown in Figure 2.6 (Levinson et al., 2005a,b). This phenomenon has also been proven by Wang's study, as shown in Figure 2.7 (Wang et al., 2013a).

As claimed by Levinson et al. (2005b), the very sharp transition from low absorption to high absorption, which occurs at 400 nm, resulted from the electronic bandgap structure of rutile TiO_2 (3.0 eV of rutile and 3.2 of anatase). This region is the boundary between the visible and ultraviolet regions. With wavelengths shorter than 400 nm

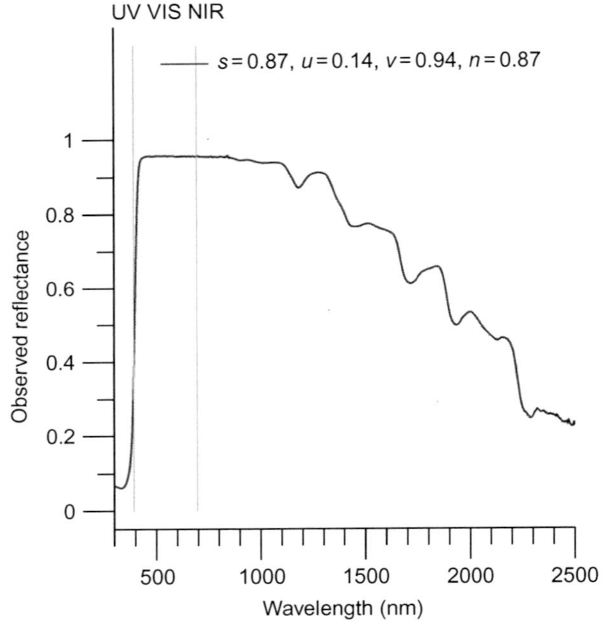

Figure 2.6 Spectral reflectance of 1.5 mm acrylic paint film with rutile TiO_2 as filler (*s*, *u*, *v*, and *n* represent sum, ultra-violet, visible, and near-infrared, respectively).
(Levinson et al., 2005a,b).

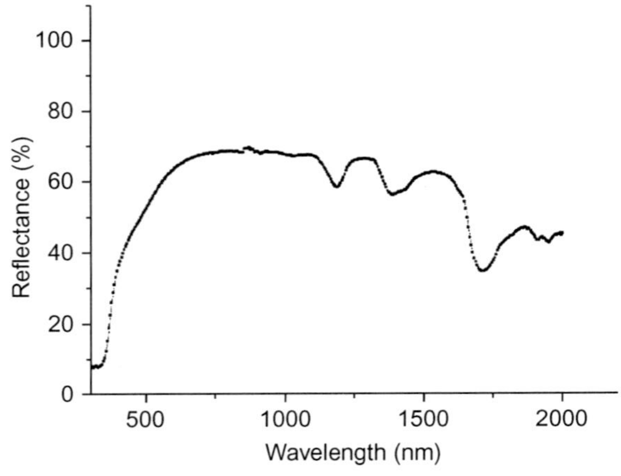

Figure 2.7 Spectral reflectance of acrylic pavement coating with TiO_2 as filler.

(photon energies higher than 3.1 eV), considerable absorption can be observed. With wavelengths longer than 400 nm, absorption starts to decrease, and most of them result from the polymer substrate. Due to the impurities of the TiO_2 and the various polymer substrates, the reflectance curves might have some differences.

Tayca Co. in Japan (http://www.tayca.co.jp/) found that the concentration of rutile TiO_2 in the polymer matrix will influence the reflectance performance. With the increasing filling content, solar reflectance will reach a peak value and then decrease.

Similar to titanium dioxide, zinc oxide (ZnO) is another important white semiconductor, with a bandgap of 3.4 eV. In 1924, the photocatalytic property of ZnO was found by Baur and Perret (1924) at the Swiss Federal Institute of Technology. It was reported that metallic silver can be produced by silver salt combining with zinc oxide. Baur and Perret assumed the oxidation and reduction occurred simultaneously, expressed as:

$$ZnO + h\upsilon \rightarrow h^+ + e^- \tag{2.3}$$

$$4h^+ + 4OH^- \rightarrow O_2 + 2H_2O \tag{2.4}$$

$$e^- + Ag^+ \rightarrow Ag^0 \tag{2.5}$$

After three years, Baur and Neuweiler (1927) found a simultaneous oxidation and reduction process to produce hydrogen peroxide on zinc oxide.

$$2h^+ + 2OH^- + CH_2O \rightarrow CO + 2H_2O \tag{2.6}$$

$$2e^- + 2H^+ + O_2 \rightarrow H_2O_2 \tag{2.7}$$

Frank and Bard (1977) found that some semiconductors can be used to purify water with photocatalytic decomposition of pollutants, and ZnO is the most active one. It was found that the ZnO surface is more hydrophobic than the TiO_2 surface (Sun et al., 2001). In this study, the results show that the ZnO surface can be induced by UV light to generate oxygen vacancies. Figure 2.8 shows that the water contact angle varies from 109° to 5° on the ZnO surface, and from 54° to 0° on the TiO_2 surface after UV irradiation.

The UV/VIS light absorption performance of ZnO was reported by Lin et al. (2005). In this study, the ZnO nanoparticles were prepared with a dc thermal plasma reactor, as in Figure 2.9. The UV/VIS spectra of ZnO nanoparticles were prepared under various environments. It was claimed by the authors that the doping plays an important role in the UV/VIS light absorption performance of ZnO nanoparticles.

Apart from the preparing methods, the effects of the particle sizes and shapes of ZnO on the UV/VIS/NIR reflectance were studied recently (Kiomarsipour et al., 2013). ZnO particles were prepared via hydrothermal approach, and five types of morphology were obtained. As demonstrated by the optical property testing results, the nanorod and microrod particles showed relative low spectral reflectance, scale-like

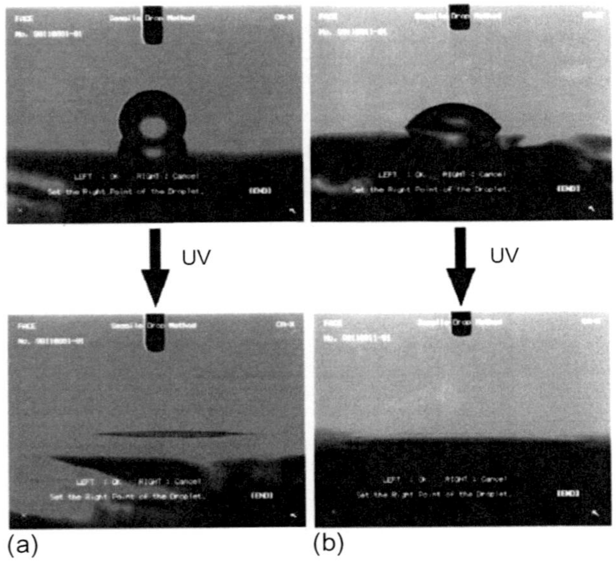

Figure 2.8 Water contact angle of (a) ZnO and (b) TiO_2 surfaces before and after UV illumination.

Figure 2.9 UV/VIS spectra of ZnO nanoparticles prepared under various environments.

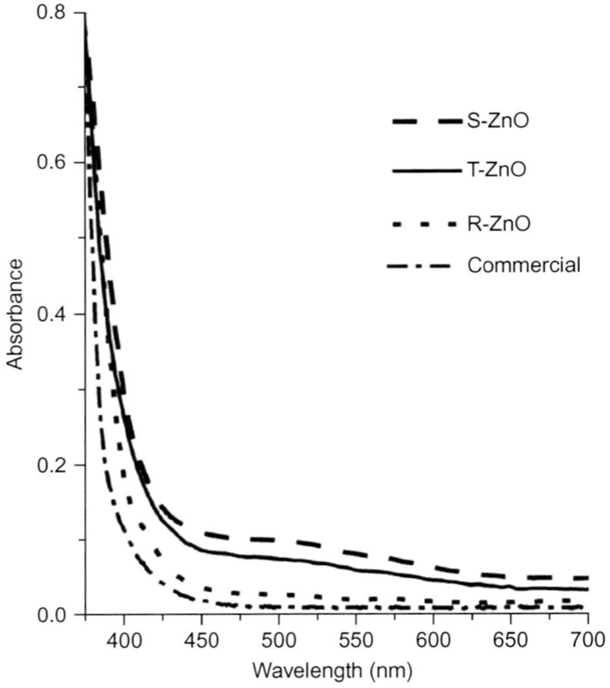

and submicrorod particles showed mean reflectance, and the decorated particle showed the highest reflectance, shown in Figure 2.10.

Li et al. (2006) reported the UV/VIS light absorption performance of ZnO/epoxy nanocomposites. In this study, ZnO was prepared with the homogeneous precipitation method and the ZnO nanoparticles were then calcined at various temperatures. The prepared nanocomposites containing 0.07 wt.% ZnO, with average size of 26.7 nm and calcined at 350 °C, show the highest absorbance of UV/VIS light, shown in Figure 2.11. (In this figure, *pure* represents the ZnO nanoparticles without calcination, and *Z3–Z8* represent the prepared ZnO calcined at various temperatures). Figure 2.12

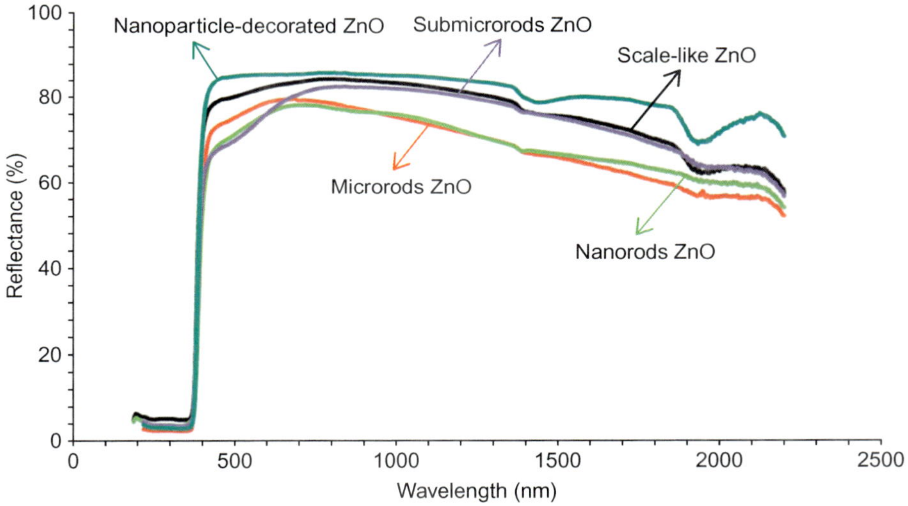

Figure 2.10 UV/VIS/NIR reflectance spectra of various shapes of ZnO particles.

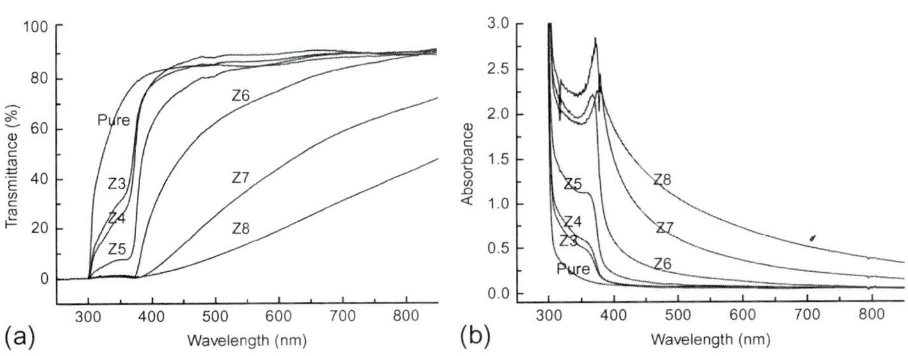

Figure 2.11 UV/VIS spectra of ZnO/epoxy nanocomposites containing 0.07 wt.% ZnO nanoparticles: (a) transmittance, and (b) absorbance.

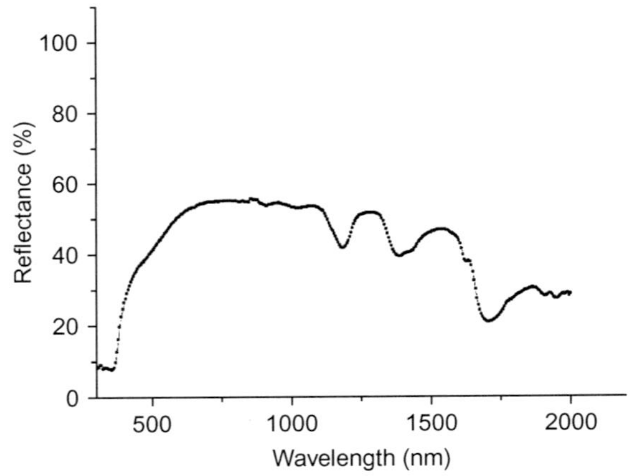

Figure 2.12 Spectral reflectance of acrylic pavement coating with ZnO as filler.

shows the spectral reflectance of acrylic pavement coating with ZnO as filler (Wang et al., 2013a). As seen in this figure, the reflectance curve has a similar trend to TiO_2 but a relatively lower reflectance.

2.3.2 Various color materials for increasing NIR reflectance

Near-infrared reflectance of various metal oxide nanoparticles was tested by Jeevanandam (2007). In this study, five metal oxides, i.e. CeO_2, MgO, TiO_2, Al_2O_3, and ZnO, were investigated. Apart from the types of materials, the size effects on the NIR reflectance were studied as well. Figure 2.13 demonstrates the testing results of the relative NIR reflectance spectra of nanocrystalline oxides and macrocrystalline of various oxides: (a) CeO_2, (b) TiO_2, (c) MgO, (d) Al_2O_3, and (e) ZnO (*the reflectance of PTFE is about 100*). The authors also claimed that the relative reflectance of these metal oxides were influenced by many other factors, such as shape, distribution, porosity, and chemical composition of the nanoparticles.

In most cases, the oxides with high NIR reflectance were prepared by doping various elements in binary metal oxides or composing binary metal oxides to form complex metal oxides. Generally, the doping process can be realized by calcination, plasma sintering, or chemical approaches, including hydrothermal or aqueous co-precipitation methods (Xie et al., 2012). Once the binary metal oxides have been doped or composed with other oxides, their bandgaps will be changed, and as a result, the color and the NIR reflectance performance will be changed accordingly.

The NIR reflectance of the light color yttrium cerate ($Y_2Ce_2O_7$) doped with Mo or Pr was investigated (Sarasamma Vishnu and Lakshmipathi Reddy, 2011). The doping process was realized by calcining Y_2O_3, CeO_2 and $(NH_4)_6Mo_7O_{24}$-$4H_2O$ or Pr_6O_{11} at 1500 °C. After doping with Mo, the bandgap of $Y_2Ce_2O_7$ changed from 3.01 eV to 2.44 eV, and the color changed from ivory white to light yellow. On the other hand, if doped with Pr, the bandgap of $Y_2Ce_2O_7$ changed from 3.01 eV to 1.70 eV, and the

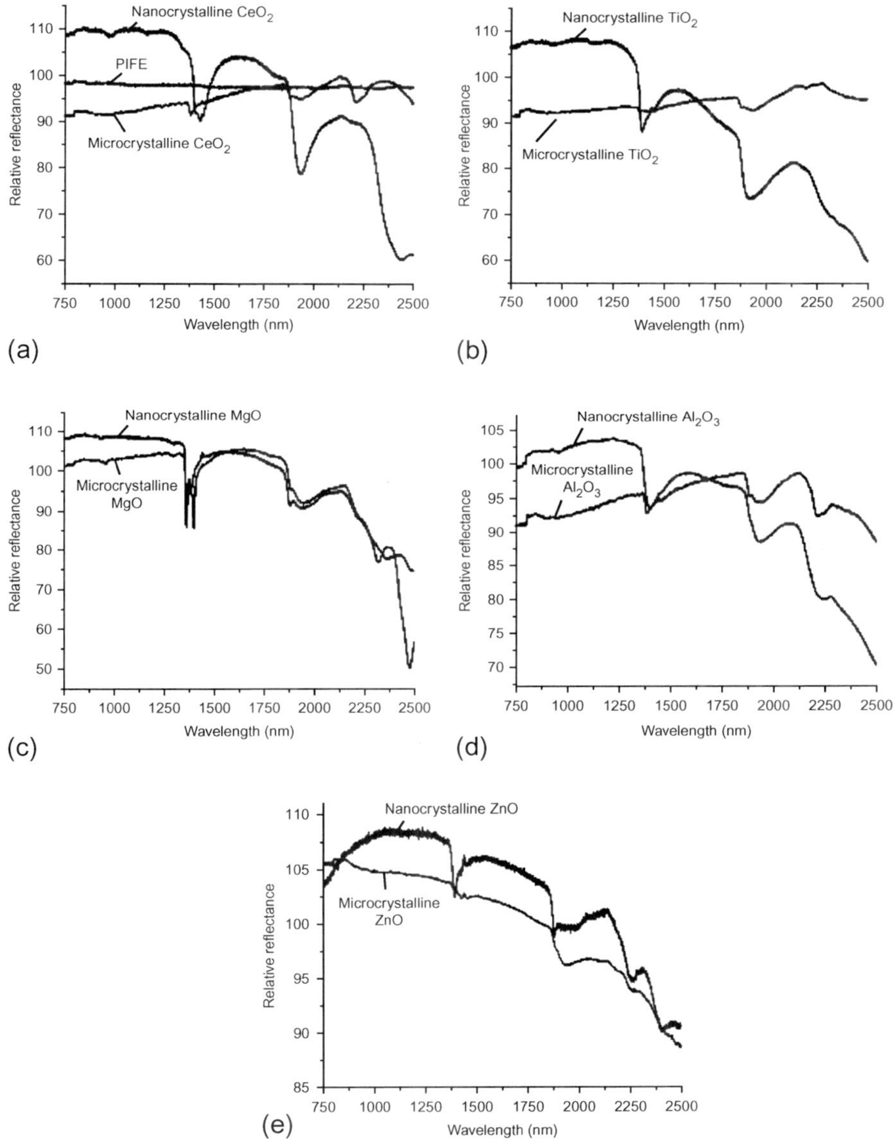

Figure 2.13 Relative NIR reflectance spectra of nanocrystalline oxides and macrocrystalline of various oxides: (a) CeO_2, (b) TiO_2, (c) MgO, (d) Al_2O_3, and (e) ZnO (the reflectance of PTFE is about 100).

color changed from ivory white to dark brown, as shown in Figure 2.14. The NIR reflectance testing results show that NIR reflectance stays relatively high with both Pr and Mo doping (Figure 2.15). Similarly, Zhao et al. (2013) studied the Fe doped $Y_2Ce_2O_7$ and found that this yellow color pigment has a high NIR reflectance (>80.6%).

Most recently, Zou investigated the NIR reflectance of rutile TiO_2 co-doped with Cr and Sb (Zou et al., 2014). The doping process was realized by the aqueous co-precipitate approach followed by calcination. It was found that, after co-doping with Cr and Sb, the color of rutile TiO_2 changed from white to yellow and dark green with different dosage of the doping elements, as shown in Figure 2.16; their reflectance performance is shown in Figure 2.17. Other than the color adjustment, the temperature testing results show that the inner ceramic tile with this coating had a maximum 10 °C lower than the one without the coating.

Similar results were obtained recently (Wang et al., 2013b), as shown in Figure 2.18. $NiTiO_3$ nanoparticles were obtained by a polymer-pyrolysis route. It was found that the high reflection peak occurs at about 580 nm, which gave them the yellow color, and the NIR reflectance of these yellow particles is about 60%.

Figure 2.14 Photographs of $Y_2Ce_2O_7$ doped with Mo and Pr at various dosages.

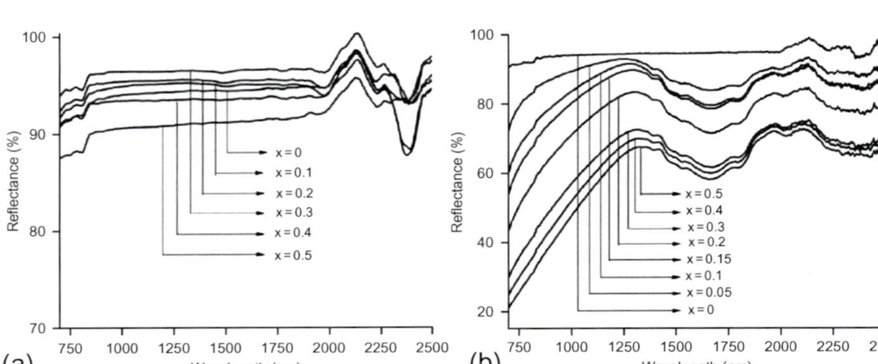

Figure 2.15 NIR reflectance performance of $Y_2Ce_2O_7$ doped with (a) Mo, and (b) Pr.

Figure 2.16 Photographs of Cr and Sb doped rutile TiO_2 pigment.

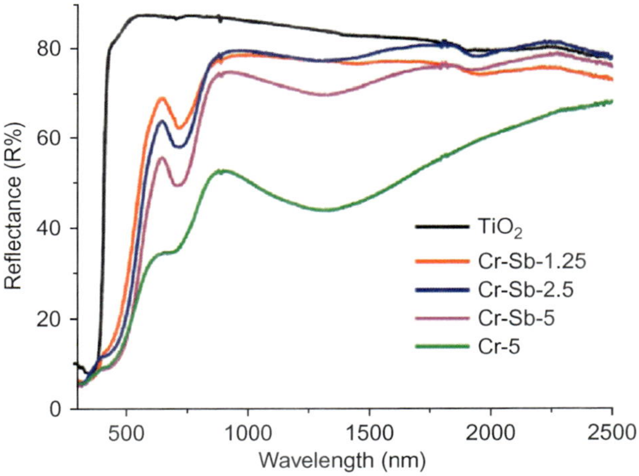

Figure 2.17 UV/VIS/NIR reflectance performance of rutile TiO_2 co-doped with Cr and Sb.

Other than red and yellow, the NIR reflectance of nontoxic blue complex metal oxides with dopants were developed recently (Jose and Reddy, 2013). In this study, the complex metal oxides $Sr_{1-x}La_xCu_{1-y}Li_ySi_4O_{10}$ ($x = y$ ranges from 0 to 0.5) were prepared by calcining the mixture of La_2O_3, $SrCO_3$, CuO, Li_2CO_3, and SiO_2 at 950 °C for 16 h. The obtained blue pigments are shown in Figure 2.19. After the reaction, the bandgap of the light blue $SrCuSi_4O_{10}$ changed from 2.59 eV to 2.68 eV. The NIR reflectance testing results show that the NIR reflectance of the prepared coating on cement concrete and poly(methyl acrylate) (PMMA) surfaces is about 67% and is thermally stable (shown in Figure 2.20).

Most recently, Aceto et al. (2014) summarized the NIR reflectance performance of widely used inorganic pigments with various colors. Nearly 60 types of pigments, which belong to nine categories, were systematically studied. Typical colors of specific materials were reviewed, and their reflectance performances were presented.

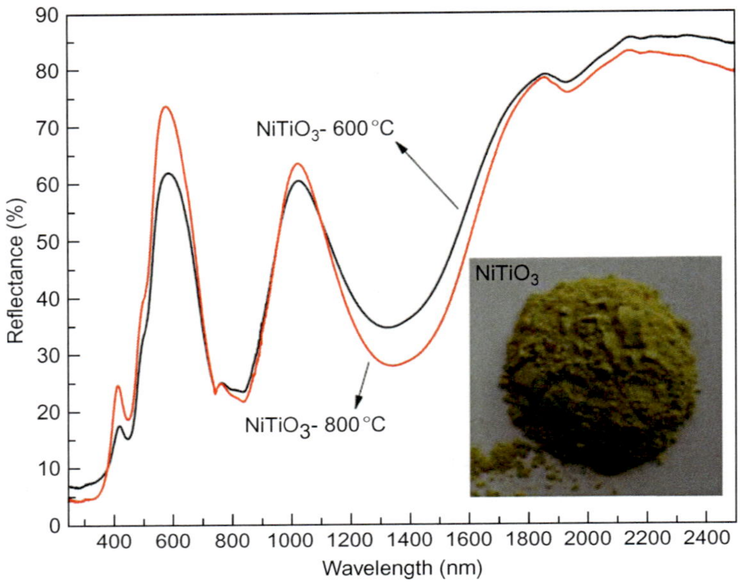

Figure 2.18 Solar radiation reflection of $NiTiO_3$ yellow nanoparticles.

Figure 2.19 Photographs of the $Sr1\text{-}xLa_xCu1\text{-}yLi_ySi_4O_{10}$ ($x=y$ ranges from 0 to 0.5) pigment.

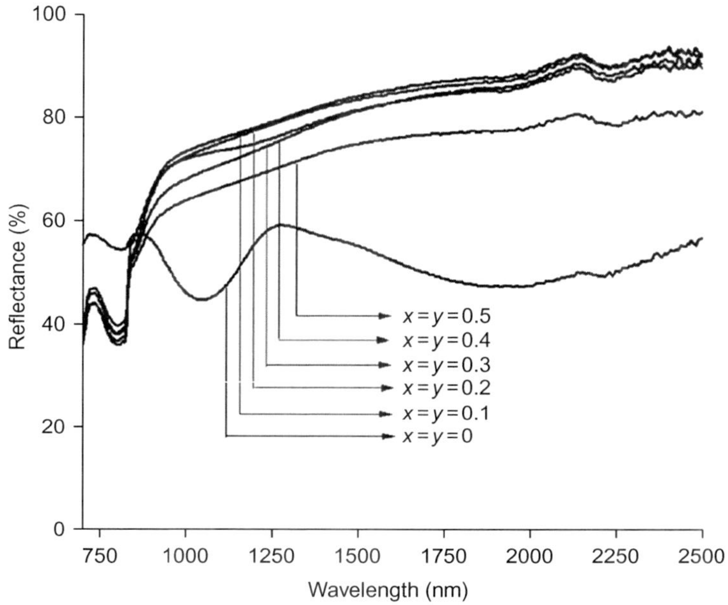

Figure 2.20 NIR reflectance spectra of blue Sr1-xLa$_x$Cu1-yLi$_y$Si$_4$O$_{10}$ ($x=y$ ranges from 0 to 0.5) pigment.

2.4 Aggregate materials with high reflectance

Compared to cool roof, cool pavement coating has to face more critical conditions when it is in service because basic pavement requirements cannot be sacrificed in attempts to increase reflectance. In general, to avoid the skid problem, pavement coatings have to incorporate with fine aggregates to satisfy the skid and wear resistance requirements; therefore, the reflectance of the materials used as fine aggregates has to be considered.

Currently, the most widely used fine aggregates are porcelain and silica sand. Similar to fillers, the solar energy absorption of porcelain and silica sand are not very high. Levinson and Akbari (2002) systematically studied 32 mixes of concrete, in which 2 types of cement, 4 types of sand, and 4 types of rock were used. The reflectance of the sands ranged from 20% to 45%, and the reflectance of the rocks ranged from 17% to 55%. For the sand materials, the dark gray riverbed sand has the lowest reflectance (20%), while the tan beach sand has the highest reflectance (45%). For the rock materials, the dark red volcanic rock has the lowest reflectance (17%), while the gold and white rock has the highest reflectance (55%). Although this study gave brief reflectance data of some aggregates, the research is still very limited and needs to be extensively investigated in future studies.

2.5 Future trends

Future studies on coating materials used for increasing pavement surface reflectance should be focused on three parts: modification of the polymer substrate, development of cost-effective and nontoxic inorganic materials with high UV/VIS/NIR reflectance, and aggregate treatment approaches to enhance reflectance.

For the polymer substrate, new polymers with low viscosity before curing, and with splendid mechanical properties and plasticity, high bonding strength, high wear resistance, and corrosion resistance after curing have to be studied in the future. To realize these targets, the investigation of curing agent development is one of the most important research topics. In addition, whether it be epoxy, acrylic, or another type of polymer, it must be cost-effective. Furthermore, the anti-ageing performance of the polymer substrate under UV irradiation needs to be increased, especially after filled with high reflective pigments.

For inorganic materials as fillers of pavement coatings, in spite of various complex metal oxides being studied systematically, preparing low-cost rear earth metal free complex metal oxides with high reflectance is still a problem because rear earth metals are very expensive and are not allowed to be used for pavement coating on a large scale. In addition, the requirement of a facile doping process is very important because the application of the high reflectance pigment in pavement engineering requires a large quantity of products. A bridge that connects the gap between lab preparation and low-cost industrial production is another important research topic. Other than the development of complex metal oxides, the dispersion approaches, which guarantee the well dispersion of the fillers in polymer substrates and thus determine the final properties of the coatings, have to be extensively investigated. In addition, although many inorganic materials with high UV/VIS/NIR reflectance have been developed, their application as pavement coatings is still very limited. Furthermore, several factors, including measuring system, wind speed, cloudiness, and air temperature have been reported, with evident effects on the diurnal and seasonal changes of the pavements' albedo (Li et al., 2013). Therefore, the relationship between albedo performance and temperature reduction of coated pavements, considering the aforementioned factors, should be systematically studied.

For the aggregates investigation, future studies should be focused on the development of simple surface treatment methods for low reflectance aggregates. The surface-treated aggregates should not only show increased reflectance, but also high durability and chemical stability.

Acknowledgments

This work is financially supported by the ministry of education program for New Century Excellent Talents (NCET-08-0163) and the China Postdoctoral Science Foundation (No. 20110491065).

References

Aceto, M., Agostino, A., Fenoglio, G., Idone, A., Gulmini, M., Picollo, M., Delaney, J.K., 2014. Characterisation of colourants on illuminated manuscripts by portable fibre optic UV-visible-NIR reflectance spectrophotometry. Anal. Methods 6 (5), 1488–1500.

Aida, M., 1982. Urban albedo as a function of the urban structure—a model experiment. Boundary Layer Meteorol. 23 (4), 405–413.

Aida, M., Gotoh, K., 1982. Urban albedo as a function of the urban structure-a two-dimensional numerical simulation. Boundary Layer Meteorol. 23 (4), 415–424.

Akbari, H., Matthews, H.D., 2012. Global cooling updates: reflective roofs and pavements. Energy Build. 55, 2–6.

Akbari, H., Menon, S., Rosenfeld, A., 2009. Global cooling: increasing world-wide urban albedos to offset CO_2. Clim. Change 94 (3–4), 275–286.

Akbari, H., Shea Rose, L., Taha, H., 2003. Analyzing the land cover of an urban environment using high-resolution orthophotos. Landsc. Urban Plan. 63 (1), 1–14.

Asaeda, T., Ca, V.T., Wake, A., 1996. Heat storage of pavement and its effect on the lower atmosphere. Atmos. Environ. 30 (3), 413–427.

Baur, E., Neuweiler, C., 1927. Über photolytische Bildung von Hydroperoxyd. Helv. Chim. Acta 10 (1), 901–907.

Baur, E., Perret, A., 1924. Über die Einwirkung von Licht auf gelöste Silbersalze in Gegenwart von Zinkoxyd. Helv. Chim. Acta 7 (1), 910–915.

Burgess, A.R., 1952. U.S. Patent No. 2,615,860. Washington, DC: U.S. Patent and Trademark Office.

Cantat, O., 1989. Contribution a Study of variation the Balance to energy in region Paris. Ph.D. Thesis, University of Paris Sorbonne.

Carnielo, E., Zinzi, M., 2013. Optical and thermal characterisation of cool asphalts to mitigate urban temperatures and building cooling demand. Build. Environ. 60, 56–65.

Chen, M.Z., Wei, W., Wu, S.P., 2009. On cold materials of pavement and high-temperature performance of asphalt concrete. Mater. Sci. Forum 620, 379–382.

Cotana, F., Rossi, F., Filipponi, M., Coccia, V., Pisello, A.L., Bonamente, E., Cavalaglio, G., 2014. Albedo control as an effective strategy to tackle Global Warming: a case study. Appl. Energy 130, 641–647. http://dx.doi.org/10.1016/j.apenergy.2014.02.065.

Doulos, L., Santamouris, M., Livada, I., 2004. Passive cooling of outdoor urban spaces-the role of materials. Sol. Energy 77 (2), 231–249.

Epon Resin Structural Resin Manual-Additive Selection, 2001. Resolution Performance Polymers. LLC, Houston, Texas.

Federal Highway Administration (FHWA), 2010. Highway bridge by wearing surface. NBI. Inspection and evaluation bridge structures. http://www.fhwa.dot.gov/bridge/wearing.cfm.

Frank, S.N., Bard, A.J., 1977. Heterogeneous photocatalytic oxidation of cyanide and sulfite in aqueous solutions at semiconductor powders. J. Phys. Chem. 81 (15), 1484–1488.

Goodeve, C.F., Kitchener, J.A., 1938. Photosensitisation by titanium dioxide. Trans. Faraday Soc. 34, 570–579.

Jose, S., Reddy, M.L., 2013. Lanthanum–strontium copper silicates as intense blue inorganic pigments with high near-infrared reflectance. Dyes Pigm. 98 (3), 540–546.

Kamat, P.V., 2012. TiO_2 nanostructures: recent physical chemistry advances. J. Phys. Chem. C 116, 11849–11851.

Kinouchi, T., Yoshinaka, T., Fukae, N., Kanda, M., 2003. Development of cool pavement with dark colored high albedo coating. Target 50 (40), 40.

Kiomarsipour, N., Shoja Razavi, R., Ghani, K., Kioumarsipour, M., 2013. Evaluation of shape and size effects on optical properties of ZnO pigment. Appl. Surf. Sci. 270, 33–38.

Lee, H., Neville, K., 1967. Handbook of Epoxy Resins. McGraw-Hill, New York.

Levinson, R., Akbari, H., 2002. Effects of composition and exposure on the solar reflectance of portland cement concrete. Cem. Concr. Res. 32 (11), 1679–1698.

Levinson, R., Berdahl, P., Akbari, H., 2005a. Solar spectral optical properties of pigments—part I: model for deriving scattering and absorption coefficients from transmittance and reflectance measurements. Sol. Energy Mater. Sol. Cells 89 (4), 319–349.

Levinson, R., Berdahl, P., Akbari, H., 2005b. Solar spectral optical properties of pigments—part II: survey of common colorants. Sol. Energy Mater. Sol. Cells 89 (4), 351–389.

Li, H., Harvey, J., Kendall, A., 2013. Field measurement of albedo for different land cover materials and effects on thermal performance. Build. Environ. 59, 536–546.

Li, Y.Q., Fu, S.Y., Mai, Y.W., 2006. Preparation and characterization of transparent ZnO/epoxy nanocomposites with high-UV shielding efficiency. Polymer 47 (6), 2127–2132.

Lin, H.F., Liao, S.C., Hung, S.W., 2005. The dc thermal plasma synthesis of ZnO nanoparticles for visible-light photocatalyst. J. Photochem. Photobiol. A Chem. 174 (1), 82–87.

Mirzaei, P.A., Haghighat, F., 2010. Approaches to study urban heat island—abilities and limitations. Build. Environ. 45 (10), 2192–2201.

Osumi, Y., 1987. One-Part Epoxy Resin. Three Bond Technical News, Three Bond, October, 1.

Renz, C., 1921. Photoreactions of oxides of titanium, cerium and earth acids. Helv. Chim. Acta 4, 961–968.

Riddle, E.H., 1954. Monomeric Acrylic Esters. Reinhold Publishing Corp, New York.

Rosenfeld, A.H., Akbari, H., Romm, J.J., Pomerantz, M., 1998. Cool communities: strategies for heat island mitigation and smog reduction. Energy Build. 28 (1), 51–62.

Santamouris, M., 2007. Heat island research in Europe: the state of the art. Adv. Build. Energy Res. 1 (1), 123–150.

Santamouris, M., Gaitani, N., Spanou, A., Saliari, M., Giannopoulou, K., Vasilakopoulou, K., Kardomateas, T., 2012. Using cool paving materials to improve microclimate of urban areas—design realization and results of the flisvos project. Build. Environ. 53, 128–136.

Santamouris, M., Synnefa, A., Karlessi, T., 2011. Using advanced cool materials in the urban built environment to mitigate heat islands and improve thermal comfort conditions. Sol. Energy 85 (12), 3085–3102.

Sarasamma Vishnu, V., Lakshmipathi Reddy, M., 2011. Near-infrared reflecting inorganic pigments based on molybdenum and praseodymium doped yttrium cerate: synthesis, characterization and optical properties. Sol. Energy Mater. Sol. Cells 95 (9), 2685–2692.

Scholer, C.F., Forrestel, R.R., 1980. Use of a very thin overlay to reestablish the skid resistance of a concrete pavement. Transp. Res. Rec. 752.

Stathopoulou, E., Mihalakakou, G., Santamouris, M., Bagiorgas, H.S., 2008. On the impact of temperature on tropospheric ozone concentration levels in urban environments. J. Earth Syst. Sci. 117 (3), 227–236.

Stone, R.V., 2010. Acrylic ester polymers. In: Encyclopedia of Polymer Science and Technology. John Wiley, New York.

Sun, R.D., Nakajima, A., Fujishima, A., Watanabe, T., Hashimoto, K., 2001. Photoinduced surface wettability conversion of ZnO and TiO_2 thin films. J. Phys. Chem. B 105 (10), 1984–1990.

Synnefa, A., Karlessi, T., Gaitani, N., Santamouris, M., Assimakopoulos, D.N., Papakatsikas, C., 2011. Experimental testing of cool colored thin layer asphalt and estimation of its potential to improve the urban microclimate. Build. Environ. 46 (1), 38–44.

Synnefa, A., Santamouris, M., Livada, I., 2006. A study of the thermal performance of reflective coatings for the urban environment. Sol. Energy 80 (8), 968–981.

Taleghani, M., Sailor, D.J., Tenpierik, M., van den Dobbelsteen, A., 2014. Thermal assessment of heat mitigation strategies: the case of Portland State University, Oregon. USA. Build. Environ. 73, 138–150.

Uemoto, K.L., Sato, N., John, V.M., 2010. Estimating thermal performance of cool colored paints. Energy Build. 42 (1), 17–22.

Wan, W.C., Hien, W.N., Ping, T.P., Aloysius, A.Z.W., 2012. A study on the effectiveness of heat mitigating pavement coatings in Singapore. J. Heat I. Inst. Int. 7, 2.

Wang, J.L., Li, Y.Q., Byon, Y.J., Mei, S.G., Zhang, G.L., 2013a. Synthesis and characterization of $NiTiO_3$ yellow nano pigment with high solar radiation reflection efficiency. Powder Technol. 235, 303–306.

Wang, H., Zhong, J., Feng, D., Meng, J., Xie, N., 2013b. Nanoparticles-modified polymer-based solar-reflective coating as a cooling overlay for asphalt pavement. Int. J. Smart Nano Mater. 4 (2), 102–111.

White, P., Golden, J.S., Biligiri, K.P., Kaloush, K., 2010. Modeling climate change impacts of pavement production and construction. Resour. Conserv. Recycl. 54 (11), 776–782.

Winkler, Jochen, 2003. Titanium Dioxide. Vincentz Network, Hannover.

Xie, N., Feng, D., Li, H., Gong, C., Zhen, L., 2012. Shape-controlled synthesis of zinc phosphate nanostructures by an aqueous solution route at room temperature. Mater. Lett. 82, 26–28.

Young, L.M., Durham, S.A., 2012. Performance of an anti-icing epoxy overlay on asphalt surfaces. J. Perform. Constr. Facil. 27 (6), 836–840.

Yu, B., Lu, Q., 2012. Life cycle assessment of pavement: methodology and case study. Transport. Res. Transport Environ. 17 (5), 380–388.

Zhao, M., Han, A., Ye, M., Wu, T., 2013. Preparation and characterization of Fe^{3+} doped $Y_2Ce_2O_7$ pigments with high near-infrared reflectance. Sol. Energy 97 (2013), 350–355.

Zou, J., Zhang, P., Liu, C., Peng, Y., 2014. Highly dispersed (Cr, Sb)-co-doped rutile pigments of cool color with high near-infrared reflectance. Dyes Pigm. 109, 113–119.

Pavements made of concrete with high solar reflectance

3

F. Reza[1], *K. Boriboonsomsin*[2]
[1]Minnesota State University, Mankato, MN, USA; [2]University of California, Riverside, CA, USA

3.1 Introduction

Pavements play a vital role in the economic and social development of urban areas and therefore unsurprisingly account for a large percentage of the land cover in those areas. Land coverage patterns of four different cities in the United States, namely, Sacramento, Chicago, Salt Lake City, and Houston, were studied via satellite images (Akbari and Rose, 1999, 2001a,b; Rose et al., 2003). The range of values from these four cities found paving to cover 25–26% of land in residential areas and 34–46% in nonresidential areas. The average of the four cities found total impervious surfaces (pavements plus roofs) to be 56%, which could be further broken down into roofs (20%), roads (15%), parking lots (12%), sidewalks (5%), and driveways (3%) (Gartland, 2008).

Compared to asphalt, concrete appears to have the smaller overall share in paving. For example, just considering roads in urban areas in the United States, concrete accounts for only 45,771 km (28,441 miles) out of a total paved surface of 482,924 km (300,075 miles) (USDOT, 2013). This translates to only 9.5%. Similarly, it is estimated that concrete accounts for 15% of parking lots and 20% of pathways and trails, but it does dominate in sidewalks and driveways accounting for 90% and 60%, respectively (Gartland, 2008).

Man-made or built-up surfaces such as concrete and asphalt absorb more heat from sunlight compared to other natural surfaces such as vegetation and earth. This is in part due to their imperviousness that limits the amount of water available for cooling due to evaporation but mainly due to their low solar reflectance. Solar reflectance is often measured in terms of *albedo* (the ratio of the amount of light reflected from a material to the amount of light incident on the material). Albedo is measured on a scale from 0 to 1. An albedo value of 0 indicates a black body cavity that does not reflect any light-while an albedo of 1 indicates a perfectly reflective surface. The stored heat in the pavements subsequently heats the surrounding air thus contributing to the urban heat island (UHI) effect. For this reason, the use of cool pavements is seen as one of the potential mitigation methods for the UHI effect. Concrete has a much higher initial albedo than asphalt because it is light gray in color compared with black, but unfortunately, the albedo of concrete decreases over time because of weathering and the accumulation of dirt. The typical starting value of albedo for new concrete is 0.35–0.40 and 0.25–0.30 for

Eco-efficient Materials for Mitigating Building Cooling Needs. http://dx.doi.org/10.1016/B978-1-78242-380-5.00003-0

Figure 3.1 Newly poured concrete sections appear much lighter than existing aged concrete sections as seen during a highway repair operation.

weathered concrete (Bretz et al., 1998). For comparison, the typical starting value of albedo for asphalt is 0.05–0.10 and this increases over time to about 0.10–0.20 because of the polishing effect of traffic and the exposure of aggregates. The effect of decreasing solar reflectance of concrete with age can be observed clearly in Figure 3.1, which shows a highway repair operation where deteriorated concrete sections are cut out, removed, and then replaced with new cast-in-place concrete. The newly placed concrete sections appear much lighter than the existing aged concrete.

Although concrete starts out with a relatively high solar reflectance and may not be as prevalent as asphalt in pavements, it is still a worthwhile endeavor to attempt to increase the solar reflectance even further since research studies indicate that even a small increase of 0.01 in albedo can have significant environmental benefits (Akbari et al., 2012). Research indicates that concrete can be made with a high solar reflectance in an easy and cost-effective manner using materials that are already widely accepted in the industry. Figure 3.2 shows a simple demonstration where thermocouples measure the surface temperature of two different test plates. The one on the right is made with a conventional concrete mix while the one on the left contains a high amount of slag (70% replacement of cement). The slag concrete is visibly whiter than the conventional concrete. On a warm sunny day with ambient temperature around 28 °C (82 °F), the surface temperature on the conventional concrete is consistently higher, at this instant showing a difference of 2.4 °C (4.3 °F) and a peak difference of around 3 °C (5 °F).

It should be noted at this stage that simple visual appearance of one surface being lighter than another may not necessarily indicate a higher solar reflectance. This is because solar energy is composed of ultraviolet (UV) rays, visible light, and infrared energy. Only about 43% of solar energy is visible light in colors ranging from violet to red. There is about 5% in the UV spectrum and the remaining 52% of solar energy is infrared, felt as heat (USEPA, 2008). For this reason, any measurement method for solar reflectance, for example, ASTM C 1549 (2009), will take measurements at

Figure 3.2 A simple demonstration showing that the surface temperature of a conventional concrete mix test plate (right) is consistently higher than a higher solar reflectance slag concrete mix test plate (left) on a warm sunny day.

different wavelengths all the way from UV to infrared and then weight the responses in proportion to the amount that each wavelength occupies in the total solar radiation.

Although other material properties also affect the surface temperature of concrete pavements such as thermal emittance, thermal conductivity, and heat capacity, there seems to be only a limited opportunity to make a significant impact on the material itself, and there is no strong research thrust along those lines. The dominant factor (not considering permeability) is the solar reflectance, and hence, most research has focused on this area. Within the area of solar reflectance, some research has focused on developing reflective coatings, some in the infrared wavelength, but this topic is addressed in a different chapter of this book. This chapter focuses on what is apparently the easiest and most cost-effective method of improving the solar reflectance of concrete: the careful selection of materials and proportions in the mix design.

Compared to cool roofs, the popularity of cool pavements is not as high with little incentive for owners and contractors to pursue them. One incentive that is being successful in motivating building designers to specify cool pavements is the US Green Building Council's Leadership in Energy and Environmental Design (LEED) certification program (USGBC, 2009). There is no standard definition of what exactly is a cool pavement. One could adopt a LEED guideline of requiring a solar reflectance index (SRI) of at least 29 as a definition of cool pavement; however, at least from the standpoint of concrete, this is a somewhat weak criterion with most conventional concrete mixes being able to achieve this, certainly when considering new concretes. Opportunities are being missed with new construction and repairs of old concrete pavements taking place every year. The process of whitetopping and ultrathin whitetopping is also gaining popularity whereby an existing distressed or rutted asphalt pavement is overlaid with a concrete topping. In this case, the solar reflectance is increased from that of asphalt to that of the concrete overlay. Relevant stakeholders must be provided more education about the urban heat island effect and the need to consider life cycle costs in decision-making that certainly favors concrete pavements in general and high solar reflectance concrete pavements even more.

Inspiration can be drawn from pilot projects that are beginning to take place around the world. For example, under Chicago's Green Alley program and Sustainable Streets Pilot Project, the Chicago Department of Transportation is beginning to repave up to 3058 km (1900 miles) of alleys and 6075 km (3775 miles) of streets, representing roughly 25% of the city's land area (Attarian, 2010). The types of pavements being implemented include pervious asphalt, pervious concrete, and nonpervious high albedo concrete (slag concrete). In Athens, Greece, about 4500 m^2 of existing pavement made of asphalt, concrete, and dark paving materials was removed and replaced with high solar reflectance pale yellow paving blocks in a large city park (Santamouris et al., 2012). After subsequent monitoring, it was found that in the central area of the park, the cool pavements had a noticeable effect: the highest daily air temperatures were reduced by 1.9 °C and the surface temperatures were lowered by 12 °C.

3.2 Materials for high solar reflectance concrete

In this section, a review of the common materials that have been or could be used in the production of high solar reflectance concrete is given. As mentioned before, concrete is already considered to be relatively cool. In fact, if the goal is just to achieve a LEED certification point by having a solar reflectance index (SRI) greater than 0.29, then this can be achieved with many conventional concrete mixes. The unit cost of materials is given in most cases; however, it should be noted that transportation costs may have a significant impact in the event that one material is not locally available and may negate any potential savings in material costs. Fly ash is also mentioned here, although it has been found to decrease the solar reflectance of concrete in one study (Boriboonsomsin and Reza, 2007) and resulted in the worst solar reflectance of all mixes in another study (Marceau and VanGeem, 2007), because it is popularly used in concrete mixtures. It imparts other beneficial properties while at the same time reducing the carbon footprint of concrete production.

3.2.1 *Water*

In its most basic form, concrete is made with water, aggregates, and Portland cement. Water is a key ingredient in the concrete mixture as it is necessary for hydration of the cement. The general rule of thumb is that water safe for drinking is suitable for concrete production. Water may typically occupy 14–21% by volume of freshly made concrete. In an indirect manner, water may affect the solar reflectance of concrete because the water-to-cement (w/c) ratio affects the porosity. Generally, a low w/c ratio will result in low porosity, good strength properties, and high durability. There have been no research projects to quantify the solar reflectance change with either water content or w/c ratio, and in fact, any attempt to affect change in solar reflectance by adjusting water content does not seem to be viable because of the myriad of other beneficial properties that are controlled by water content and w/c ratio.

3.2.2 Aggregates

Aggregates make up roughly 60–75% of the volume of fresh concrete. Aggregates that are 6.5 mm (0.25 in.) or greater are considered coarse aggregates and those less than 6.5 mm (0.25 in.) are referred to as fine aggregates. They are derived for the most part from natural rock (crushed stone or natural gravel). Aggregates not only are an inexpensive filler in the concrete mixture but also significantly reduce the volume instability due to creep and shrinkage. In pavement applications, the gradation, abrasion resistance, freeze–thaw durability, and resistance to alkali-aggregate reaction are important properties. The current average price is around $14/ton ($13/ton) for coarse aggregate and $12/ton ($11/ton) for fine aggregate in the United States.

In Table 3.1, the albedo values of some different types of coarse and fine aggregates are given based on the results of two studies. The study by Levinson and Akbari (2001) appears to be the first research project in the United States to attempt to investigate the effect of composition on the solar reflectance of concrete. Four types of sand and four types of rocks were investigated. Concrete mixtures were made with both gray Portland cement and white Portland cement. The range of albedo values for the coarse aggregates was from 0.17 to 0.55, and for the fine aggregates, from 0.20 to 0.45. One of the drawbacks of the study was that no mechanistic testing was performed; therefore, the practicality or suitability of some of the mixes for pavement applications cannot be gauged. The second study by Marceau and VanGeem (2007) was aimed at

Table 3.1 Solar reflectance of materials used in concrete mixtures

Designation	Description	Albedo
C1	Gray type I-II Portland cement	0.32
C2	White type I Portland cement	0.87
S1	Dark gray riverbed sand (quartz, clay minerals, mica)	0.20
S2	Dark red volcanic sand (basalt)	0.22
S3	Brown sand (quartz, clay minerals)	0.27
S4	Tan beach sand (quartz, clay minerals, mica)	0.45
R1	Dark red volcanic rock (basalt)	0.17
R2	Black and red rock (granite)	0.19
R3	White rock (plagioclase)	0.49
R4	Gold and white rock (chert, iron impurities)	0.55
	Marceau and VanGeem (2007)	
CS	Gray Portland cement	0.47
CW	White Portland cement	0.87
FPB	Pale buff fly ash	0.44
SM	Medium light slag cement	0.75
AE	Eau Claire fine aggregate	0.27
AL	Limestone fine aggregate	0.42
CP	Eau Claire coarse aggregate	0.27
CL	Limestone coarse aggregate	0.42

Levinson and Akbari (2001).

determining if the solar reflectance and hence SRI of concretes made of commonly available materials could achieve the LEED criterion of SRI greater than 29 for cool pavements. About 45 different mixes were developed and tested using 5 types of gray Portland cement, one white Portland cement, 6 types of fly ash, 3 types of slag cement, 4 fine aggregates, and 2 coarse aggregates. Since most gray cements are similar in color, only one is reported in Table 3.1. Similarly, the slag cements albedos were close to each other so only one has been reported.

While there are many different types of coarse aggregates available in the United States, for fine aggregates the most prevalent types of sand tend to be buff to brown in color. Therefore, a most typical mix could be identified in each study and thought of as a control. Other mixes are then benchmarked against the control. For the Levinson and Akbari (2001) study, the gray cement/brown sand/granite rock (black and red) (designated C1:S3:R2) mix was thought of as the control. For the Marceau and VanGeem (2007) study, the typical gray cement/Eau Claire fine aggregate/Eau Claire coarse aggregate (CS:AE:CE) mix was chosen as the control.

Based on the analysis of the available data, it can be concluded that coarse aggregate by itself does not have any effect on the solar reflectance of concrete. Coarse aggregate may have a significant effect in the case of a highly polished concrete, for example, when a pavement is heavily worn down by traffic and where the aggregate has been exposed.

There are some types of sands that are quite light in color and could be referred to as white sands. In this category are eroded limestone, beach sand, and play sand that appear white because iron oxide, crystalline silica, and quartz are often washed away. By benchmarking mix C1:S4:R2 in Levinson and Akbari (2001), mix CS:AL:CP in Marceau and VanGeem (2007), and mix 9 in a third study by Boriboonsomsin and Reza (2007) against their respective controls, it is found that white sand can be slightly effective in increasing the solar reflectance of concrete. This is probably because some fine particles can appear on the finished surface of the concrete.

3.2.3 Portland cement

Portland cement, when hydrated with water, provides the binder in concrete. It occupies about 7–15% of the volume of fresh concrete. It is usually the most expensive component with the current prices in the United States reaching around \$125/ton (\$113/ton). The chemical composition of a typical type I Portland cement is given in Table 3.2. Limestone (calcium carbonate) is the most common raw material that provides the calcium oxide. The second major component is the silicon dioxide, which is usually obtained from clay and silts. Aluminum and iron oxides are added as fluxing agents to lower the fusion temperature to a practical firing temperature (Mindess et al., 2003).

White Portland cement is often employed to provide pleasing architectural finishes such as exposed aggregate and colored concrete. The chemical composition is given in Table 3.2. It is very similar to gray Portland cement, but the white color is achieved by removing the iron oxide. This comes at a penalty of increasing the firing temperature. Iron-free clay must be used and special ball mills are used to prevent iron

Table 3.2 **Chemical composition, percentage of cementitious materials**

Type of cement	Gray Portland type 1	White Portland type 1	Slag cement	Class C fly ash	Class F fly ash
SiO_2	20.5	22.5	32–38	35	52
Al_2O_3	5.4	4.5	8–16	15–25	23
Fe_2O_3	2.6	0.4	<2	<10	11
CaO	63.9	66.3	35–45	20–30	5
MgO	2.1	1.0	5–15		
SO_3	3.0	2.8			
S			1–2		
Blaine fineness, m^2/kg (ft^2/lbm)	369 (1802)	464 (2266)	520 (2539)	420 (2051)	420 (2051)

contamination during grinding. The higher cost of raw materials and changes to the manufacturing process make white cement expensive. At prices around $221/ton ($200/ton), it is roughly twice as expensive as gray cement.

Rather obviously, white cement can be expected to be effective at increasing the solar reflectance of concrete. In fact, a high albedo of 0.77 was achieved using a mix of white cement, white sand, and white rock (Levinson and Akbari, 2001). Similarly, the highest albedo achieved in Marceau and VanGeem (2007) study was 0.69 obtained using white cement, white sand, and normal rock. In order to gain further insight into the effect of individual components on the overall solar reflectance of concrete, one typical mix was chosen as a control mix in each of the studies and then several other mixes where only one variable of interest was changed at a time were chosen. The mixture proportions of the chosen mixes are given in Table 3.3. It should be noted that the actual mixture proportions were not given in Levinson and Akbari (2001) and have been inferred. In the report, the mass proportions of cement/sand/rock/water have been stated to be 1/2.3/2.8/0.6. Using fairly standard specific gravities of 3.15/2.65/2.7/1 and assuming a small amount of entrained air (3%), the volume proportions can be obtained and then actual mixture proportions can be calculated. From this table, it can be seen that the highest increase in solar reflectance over that of the control is achieved with white cement. Next come slag cement and then white sand. White rock is not very effective at increasing solar reflectance and actually decreased the albedo over the control in Levinson and Akbari (2001).

3.2.4 *Slag cement*

The environmental concerns stemming from the high energy use and greenhouse gas emission in cement production spurned interest in cement-replacing materials, often referred to as supplementary cementitious materials (SCMs). Further benefits accrue since the SCMs are often by-products that might have been wasted otherwise and the properties of concrete with SCM are actually enhanced.

Table 3.3 Selected mixture proportions (kg/m^3) (lb/yd^3) and albedos

Designation	Variable of interest	Cement	Slag cement	Water	Coarse aggregate	Fine aggregate	Albedo
C1:S3:R2	Control	343 (578)	0 (0)	206 (347)	962 (1621)	790 (1332)	0.29
C2:S3:R2	White cement	343 (578)	0 (0)	206 (347)	962 (1621)	790 (1332)	0.48
C1:S4:R2	White sand	343 (578)	0 (0)	206 (347)	962 (1621)	790 (1332)	0.44
C1:S3:R3	White rock	343 (578)	0 (0)	206 (347)	962 (1621)	790 (1332)	0.25
CS:AE:CP	Control	335 (565)	0 (0)	134 (225)	1125 (1896)	739 (1245)	0.42
CW:AE:CP	White cement	335 (565)	0 (0)	154 (259)	1120 (1888)	736 (1240)	0.59
CS:AE:CP:SM	Slag cement	155 (261)	126 (213)	135 (228)	1122 (1892)	737 (1242)	0.57
CS:AL:CP	White sand	335 (565)	0 (0)	161 (271)	1081 (1822)	726 (1224)	0.53
CS:AE:CL	White rock	335 (565)	0 (0)	147 (247)	1129 (1903)	739 (1245)	0.46

Source: Levinson and Akbari (2001) and Marceau and VanGeem (2007).

Slag cement or ground-granulated blast furnace slag is a recovered resource obtained during the blast furnace production of iron from ore. Silicate and aluminate impurities from the ore and coke are combined in the blast furnace with a flux that lowers the viscosity of the slag. The flux consists mostly of a mixture of limestone and forsterite. In the blast furnace, the slag floats on top of the iron and can be separated by decanting. The slag is cooled rapidly to achieve a high reactivity. In order to simultaneously cool and fragment the slag into granules, the molten slag is subjected to jet streams of water or air under pressure. The obtained fragments are then ground to a suitable fineness (Snellings et al., 2012). The typical chemical composition of slag cement is given in Table 3.2. The benefits of adding slag cement to concrete include reduced permeability that limits chloride ingress, increased resistance to sulfate attack, and decreased potential for alkali-aggregate reactions. The color of slag cement is whitish. The current price of slag is around \$77/ton (\$70/ton).

A study conducted by Boriboonsomsin and Reza (2007) investigated the effects of the addition of varying amounts of fly ash and slag cement, and single doses of white sand and latex on the solar reflectance of concrete. The mixture proportions used are shown in Table 3.4. In order to ensure that the mixes would be useful for pavement applications, they were all designed to meet or exceed the Ohio Department of Transportation (ODOT) specifications of minimum 28-day compressive strength of 30 MPa (4000 psi) and minimum 7-day modulus of rupture of 8 MPa (600 psi). In addition, rapid chloride permeability tests were performed and showed that all the mixes had low to very low chloride ion penetrability. The control (base) mix was an ODOT class C concrete with no SCM. Two high-performance concrete mixes recognized by ODOT, one with 24% fly ash and one with 30% slag cement, were tested. These standard high-performance mixes were then modified by adding higher amounts of slag cement and fly ash. Some ternary mixes incorporating both fly ash and slag were also investigated. Two additional mixes modified the base mix: one using white sand instead of buff sand and the other using a latex admixture. The albedo results are shown in Figure 3.3. Slag cement was the most effective at increasing the solar reflectance. The albedo values increased with increasing amounts of slag cement, and the highest obtained value was 0.582 at 70% replacement of Portland cement with slag cement. The next best material to increase solar reflectance was white sand followed by latex. Fly ash actually decreased the solar reflectance, and the ternary mixes with both fly ash and slag showed mixed results.

Table 3.5 is a composite table based on the three studies: Levinson and Akbari (2001), Marceau and VanGeem (2007), Boriboonsomsin and Reza (2007). It shows how effective each material was at increasing the albedo of concrete relative to the control mix in each study per unit weight; that is, the percent albedo increase was divided by the actual weight of the material used in that particular concrete mixture. The results show that kilogram for kilogram (pound for pound), slag cement is more effective than white cement at increasing solar reflectance of concrete. When the cost of the materials is taken into account, this favors the slag cement even more.

The use of high-volume replacement of slag cement for Portland cement seems to be gaining momentum. There had been some concerns about decreased salt-scaling resistance, but there has been some research showing that grade 120 (the highest

Table 3.4 **Concrete mixture proportions (kg/m^3 (lb/yd^3) of concrete)**

Mix #	Description	Portland cement	Fly ash	Slag cement	Water	Coarse aggregate	Fine aggregate	w/c ratio
1	ODOT class C concrete	356 (600)	0 (0)	0 (0)	210 (355)	836 (1410)	782 (1320)	0.59
2	ODOT high-performance concrete 1 (HP1)—24% fly ash	314 (530)	101 (170)	0 (0)	156 (263)	877 (1480)	776 (1310)	0.38
3	ODOT high-performance concrete 2 (HP2)—30% slag	290 (490)	0 (0)	124 (210)	156 (264)	886 (1495)	788 (1330)	0.38
4	ODOT HP2 but with 30% fly ash	290 (490)	124 (210)	0 (0)	145 (245)	886 (1495)	788 (1330)	0.35
5	ODOT HP1 but with 60% fly ash	166 (280)	249 (420)	0 (0)	138 (233)	877 (1480)	776 (1310)	0.33
6	ODOT HP2 but with 60% slag	166 (280)	0 (0)	249 (420)	153 (258)	886 (1495)	788 (1330)	0.37
7	ODOT HP2 but with 40% fly ash + 20% slag	166 (280)	166 (280)	83 (140)	138 (233)	886 (1495)	788 (1330)	0.33
8	ODOT HP2 but with 20% fly ash + 40% slag	166 (280)	83 (140)	166 (280)	145 (245)	886 (1495)	788 (1330)	0.35
9	ODOT class C with white sand	356 (600)	0 (0)	0 (0)	210 (355)	836 (1410)	782[b] (1320)[b]	0.59
10[a]	ODOT class C but with latex	362 (611)	0 (0)	0 (0)	77 (131)	836 (1410)	782 (1320)	0.45
11	ODOT HP2 but with 70% slag	124 (210)	0 (0)	290 (490)	152 (256)	886 (1495)	788 (1330)	0.37

[a]This mix includes 100 kg (220 lb) of latex (24% solids content).
[b]White sand.
Source: Boriboonsomsin and Reza (2007).

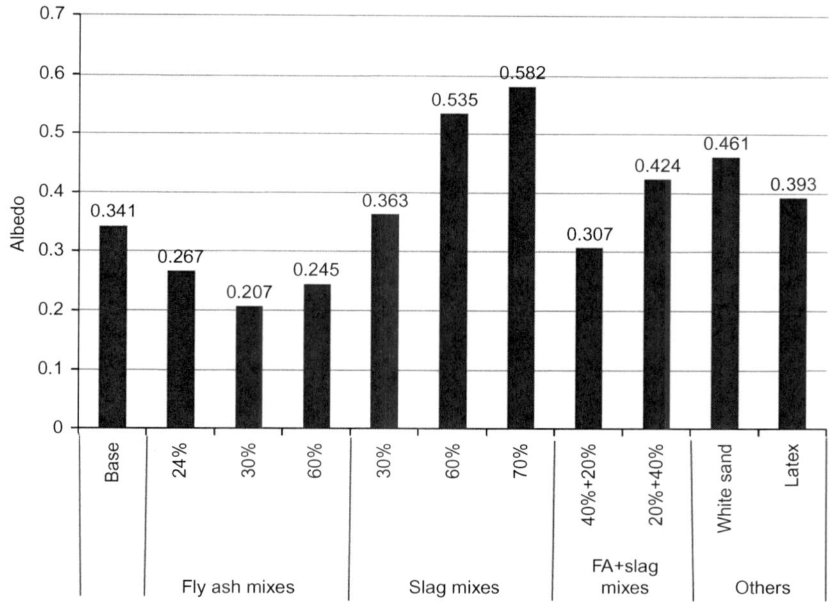

Figure 3.3 Albedo values for different mixes from Boriboonsomsin and Reza (2007).

Table 3.5 Effectiveness of different materials at increasing solar reflectance

Rank	Material of interest	New albedo	% albedo increase w.r.t control	% albedo increase per kg (lb) of material	Mix designation	Data source
1	Slag (45%)	0.57	36	0.370 (0.168)	CS:AE:CP:SM	M&V
2	Slag (70%)	0.582	71	0.318 (0.144)	Mix 11	B&R
3	Slag (60%)	0.535	57	0.299 (0.136)	Mix 6	B&R
4	White cement	0.48	66	0.250 (0.113)	C2:S3:R2	L&A
5	White cement	0.59	40	0.158 (0.072)	CW:AE:CP	M&V
6	Latex	0.393	15	0.153 (0.069)	Mix 10	B&R
7	White sand	0.44	52	0.086 (0.039)	C1:S4:R2	L&A
8	Slag (30%)	0.363	6	0.068 (0.031)	Mix 3	B&R
9	White sand	0.461	35	0.059 (0.027)	Mix 9	B&R
10	White sand	0.53	26	0.047 (0.021)	CS:AL:CP	M&V
11	White rock	0.46	10	0.011 (0.005)	CS:AE:CL	M&V
12	White rock	0.25	−14	−0.02 (−0.01)	C1:S3:R3	L&A

reactivity) slag (at 50% replacement) provides adequate salt-scaling resistance (LaBarca et al., 2007), and Schlorholtz and Hooton (2008) concluded the same for slag at 35% replacement. In Missouri, a 70% slag concrete mix was successfully used in a bridge pier and abutment mass concrete project (Richardson, 2006). The Slag Cement Association (SCA) mentions some notable projects using slag cement that are rewarded by induction into a "Hall of Fame" including (i) 7 World Trade Center, Manhattan, NY (completed in 2006), where the inside concrete core was made with a 40% slag cement mix; (ii) Delaware Route 1 (completed in 2004) where 82 km (51 miles) of road has been paved with slag cement concrete ranging from 35 to 50%; and (iii) Clearview Elementary School, Hanover, PA (completed in 2004), where the insulated concrete from walls incorporates 60% slag cement concrete.

3.2.5 Latex

Polymer latexes when combined with concrete form what is called latex-modified concrete. The primary application is for overlays on bridge decks. Styrene-butadiene rubber (SBR) latex is one of the more common latices. It is an emulsion, that is, a suspension of colloidal polymer beads in water. The solids content can range from about 20 to 55%. The appearance is a milky white liquid. Flocculation of latex particles takes place as water is removed by evaporation, and the particles then coalesce to form a continuous impermeable film. This is quite valuable on bridge decks to prevent chloride ingress from deicing salts that may corrode steel reinforcement. Latex is quite expensive, though in practice, relatively small quantities are required. The price of a 50% solids SBR latex may be as high as \$1654/ton (\$1500/ton). Although, in Table 3.5, it can be seen to be effective in increasing the solar reflectance of concrete behind slag cement and white cement, the high cost of latex would probably make it impractical unless the situation necessitated a highly impermeable surface.

3.2.6 Fly ash

Fly ash is another SCM that is a recovered resource, this time as a by-product from the combustion of coal. It is a pozzolanic material. There are two classes of fly ash. Class F fly ash is produced from bituminous coals found in states east of the Mississippi River, while class C fly ash is produced from lignitic coals found in western states (Mindess et al., 2003). The typical chemical compositions are given in Table 3.2. The benefits of fly ash include improved workability, decreased permeability, and higher ultimate strengths. Typical replacement values do not exceed 25–30%. In some colder climates, fly ash may not be allowed in a concrete mix beyond a certain date because fly ash slows down the setting time. The color of fly ash ranges from tan to dark gray. The current price is around \$66/ton (\$60/ton). The use of fly ash may actually decrease the solar reflectance and hence should be avoided for this application; however, because of the many other benefits that can accrue from its use, it is very popular. In situations where it must be used, a ternary mix where both fly ash and slag cement are used as SCMs may be considered.

3.2.7 Pigments

Color pigments are available in powder form that can be mixed in with concrete to form an integrally colored concrete. The typical applications include countertops, decorative concrete, architectural concretes, and sidewalks. The pigment particles are about an order of magnitude smaller than cement particles. The most common types of pigments are iron oxides, chrome oxides, cobalt blue, and titanium dioxide, which is white and is usually used to neutralize gray cement. Using these pigments, a plethora of colors, not just white, could be used potentially to increase solar reflectance of concrete. The cost of the color pigments may be as high as \$4961/ton (\$4500/ton) but typically no more than 5% by weight of cement is required for color saturation. To date, there does not appear to be any scientific research published to establish the effect of these powder pigments on solar reflectance of concrete.

3.3 Heat transfer in pavements

The role of solar reflectance in reducing surface temperatures of pavements can be understood better by examining the fundamental heat transfer problem in pavements. First, though, besides the solar reflectance, some other relevant material properties must be defined. All hot surfaces give off heat by radiation to the sky. The *thermal emissivity* is defined as the ratio of radiant flux emitted by a surface at a given temperature to that of a black body radiator at the same temperature. Roughness of the surface can also have an effect on the emissivity. A typical value of emissivity for concrete pavements as verified by measurements is 0.90 (Kaloush et al., 2008). The *solar reflectance index* (SRI) is an index that takes into account both solar reflectance and thermal emissivity. It is calculated according to the procedures of ASTM E 1980 (2013). The SRI gives a measure of the surface temperature of a body relative to a standard white (SRI = 100, solar reflectance = 0.8, and thermal emissivity = 0.9) and standard black (SRI = 0, solar reflectance = 0.05, and thermal emissivity = 0.9) under the standard solar and ambient conditions. The SRI is used in the LEED criterion for cool pavements.

Thermal conductivity can be defined as the ratio of heat flux to temperature gradient. The *coefficient of thermal conductivity* represents the uniform flow of heat through a unit thickness of material between two faces of unit area that are subjected to a unit temperature difference. The reference value for coefficients of thermal conductivity of concrete is about 1.4–1.7 W/(m K) (Kaloush et al., 2008).

The *volumetric heat capacity* is a measure of the material's heat storage capability. It represents the amount of energy needed to raise the temperature of a unit volume of the material by one unit. The reference value for volumetric heat capacity of concrete is about 2024 kJ/(m^3 K) (Kaloush et al., 2008). The *thermal diffusivity* is obtained by dividing the coefficient of thermal conductivity by the volumetric heat capacity. It gives an idea of how quickly steady-state conduction will be achieved. The governing equation of transient conduction is given by Equation (3.1)

$$\frac{\partial T}{\partial t} = \alpha \frac{\partial^2 T}{\partial z^2} \tag{3.1}$$

where α is the thermal diffusivity, T is the temperature, z is the distance in the depth direction of the pavement, and t is the time. Analytic models for predicting pavement surface temperatures are available and described in more detail in references such as Kaloush et al. (2008), Li (2012) and Bentz (2000). The model employed in Bentz (2000) is discussed further here.

There are three types of heat transfer mechanisms: radiation, convection, and conduction. The basic heat transfer equation governing the behavior of pavements is given by Equation (3.2)

$$Q_{\text{sun}} = Q_{\text{sky}} + Q_{\text{conv}} + Q_{\text{cond}} \tag{3.2}$$

The first term on the left-hand side of Equation (3.2) represents the portion of incoming solar radiation that is absorbed into the pavement and is given by Equation (3.3)

$$Q_{\text{sun}} = (1 - \hat{a})I \tag{3.3}$$

where $\hat{a}$ is the albedo and I is the solar insolation. The first term on the right hand side is the infrared radiation emitted from the pavement surface to the sky and is given by Equation (3.4)

$$Q_{\text{sky}} = \sigma\varepsilon\left(T_{\text{s}}^4 - T_{\text{sky}}^4\right) \tag{3.4}$$

where σ is the Stefan–Boltzmann constant (5.6669×10^{-8} W/(m^2 K^4) (1.796×10^{-8} Btu/h/(ft^2 K^4))), ε is the thermal emissivity of the concrete, T_{s} is the surface temperature of the concrete in K, and T_{sky} is the sky temperature, also in K. The sky temperature can be estimated from Equation (3.5)

$$T_{\text{sky}} = {\varepsilon_{\text{s}}}^{1/4} \times T_{\text{ambient}} \tag{3.5}$$

where T_{ambient} is the ambient temperature and the sky emissivity ε_{s} is given by Equation (3.6)

$$\varepsilon_{\text{s}} = 0.787 + 0.764 \times \ln\left(\frac{T_{\text{dew}}}{273}\right) \times F_{\text{cloud}} \tag{3.6}$$

where T_{dew} is the dew point temperature in K and the cloud cover factor F_{cloud} is given by Equation (3.7)

$$F_{\text{cloud}} = 1.0 + 0.024C - 0.0035C^2 + 0.00028C^3 \tag{3.7}$$

where C is the "tenths cloud cover" ranging between 0.0 and 1.0. The second term on the right-hand side of Equation (3.2) represents the heat lost by convection from the pavement in warming up the surrounding air and is given by Equation (3.8)

$$Q_{\text{conv}} = h_{\text{conv}} \times (T_{\text{s}} - T_{\text{ambient}}) \tag{3.8}$$

where h_{conv} is the convection coefficient that depends on wind velocity and is given by Equation (3.9)

$$\begin{aligned} h_{\text{conv}} &= 5.6 + 4.0v_{\text{wind}} \quad \text{for } v_{\text{wind}} \leq 5\,\text{m/s} \\ h_{\text{conv}} &= 7.2 \times v_{\text{wind}}^{0.78} \quad \text{for } v_{\text{wind}} > 5\,\text{m/s} \end{aligned} \tag{3.9}$$

The third term in Equation (3.2) represents heat flow due to conduction through the depths of the slab to lower layers such as the base and subgrade soil and is given by Equation (3.10)

$$Q_{\text{cond}} = \pm k(\text{d}T/\text{d}z)|_{z=0} \tag{3.10}$$

where k is the thermal conductivity and z is the depth measured from the top surface. If the surface temperature of the pavement is higher than the temperature at the base, then this term will be positive in Equation (3.2) as expected during the daytime. On the other hand, if the surface temperature is less than that at the base, the term will be negative as may occur at night. If a simplified linear temperature gradient is assumed to exist across the depth of the slab, the heat loss due to conduction can be expressed as in Equation (3.11)

$$Q_{\text{cond}} = k \times \frac{(T_{\text{s}} - T_{\text{base}})}{\Delta x} \tag{3.11}$$

where T_{base} is the temperature at the base of the slab, Δx is the thickness of the slab, and k is the thermal conductivity of the concrete. Solar insolation values as appear in Equation (3.3) can sometimes be obtained from weather stations. In the absence of these data, the solar insolation at any time of the day can be approximated with a simple procedure if the solar insolation peak value for a particular location is known. For example, the solar insolation peak value can be taken as 800–1000 W/m^2 (254–317 Btu/h/ft^2) on a hot sunny day in the Midwestern United States. This peak value occurs at true noon, that is, when the sun is at its highest point or the zenith angle is zero. The values of insolation at other times during the day depend on the zenith angle and are given by Equation (3.12)

$$I = S \cos Z \tag{3.12}$$

where S is the solar insolation value at true noon and Z is the zenith angle. The zenith angle can be calculated from Equation (3.13)

$$Z = \cos^{-1}(\sin\Phi\sin\delta + \cos\Phi\cos\delta\cos H) \tag{3.13}$$

where: Φ = latitude and H = hour angle = 15° × (Time − Time at true noon). Both the time and the time at true noon are entered as the hours from midnight: δ = magnetic declination angle:

Since the eccentricity of the Earth's orbit is quite small, it can be approximated to a circle, and δ is approximately given by the following expression Equation (3.14):

$$\delta = -23.45° \bullet \cos\left[\frac{360°}{365} \bullet (N + 10)\right] \tag{3.14}$$

where N is the number of days spent since January 1.

A source code written in C programming language is provided to perform the computations given in Bentz (2000). Researchers have also used finite-element analysis to model the heat transfer problem in pavements. For example, Li (2012) employed a one-dimensional finite-element model in Abaqus®. Karunarathne et al. (2011) developed a three-dimensional finite-element model in ANSYS® to study temperature variation in a pavement due to solar reflectance. Sensitivity analysis of the parameters in Equation (3.2) will reveal that albedo has the largest impact on surface temperature. The large impact of albedo on reducing surface temperatures of pavements has also been experimentally confirmed by means of thermocouple measurements (Karunarathne et al., 2011; Li et al., 2013). In addition, by means of temperature sensors embedded at various depths in a pavement, Kaloush et al. (2008) had shown the effectiveness of high albedo at reducing the temperature gradient across the pavement thickness.

3.4 Other potential benefits of high solar reflectance concrete

In addition to decreasing the urban heat island effect, other benefits may result from the use of cool concrete pavements. Pomerantz et al. (2000) showed that illumination at night is improved because of increased reflectance of nighttime lighting of the pavements. This helps drivers see traffic signs, pedestrians, and other hazards more clearly, thus potentially reducing accidents. Similarly, high-reflectance concrete provides the option of using fewer or less powerful streetlamps. An increase of albedo by 0.20 was estimated to reduce the lighting requirement by 20%.

In pavements, low solar reflectance can be a significant contributing factor to distress. Low solar reflectance will lead to higher surface temperatures during the daytime. The temperature gradient across the thickness will also increase because of the larger difference between the surface temperature and subgrade temperature. This effect can increase deterioration rates in pavements by causing warping and curling. It is well known that thermal deflections in conjunction with vehicular traffic loading

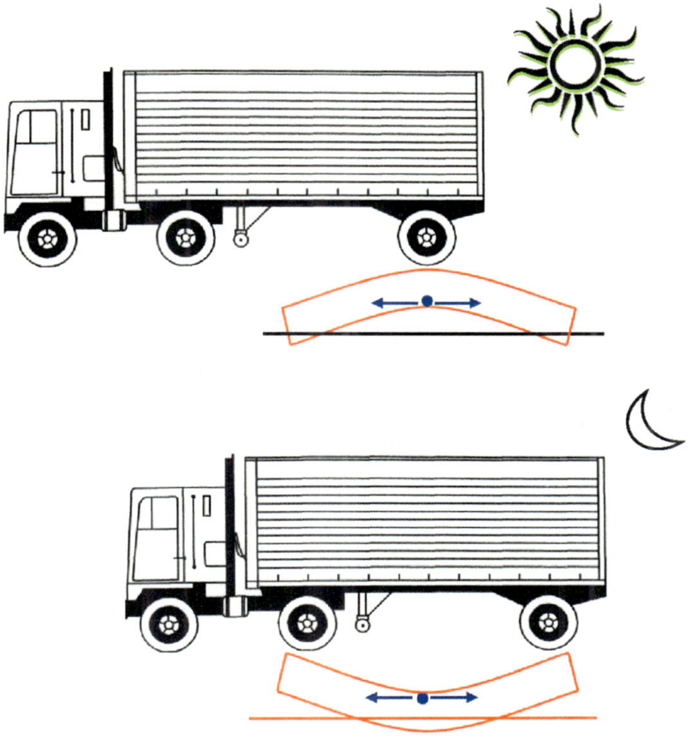

Figure 3.4 Critical combination of thermally induced deflection and truck loading leads to bottom-up cracking during the day-time and top-down cracking during the night-time.

lead to increased cracking in pavements (NCHRP 1-37A, 2004). This phenomenon can be explained with the aid of Figure 3.4. During the daytime, the top surface of a pavement will be hotter than the bottom; thus, the curvature will be concave upward. A void space will be created near the center of the pavement. A critical loading scenario consisting of a concentrated load near the center of the pavement causes bottom-up cracking. At nighttime, the bottom of the pavement will be hotter than the top, and the curvature will be concave downward. A critical loading situation consisting of concentrated loads near the transverse joints of the pavement causes top-down cracking.

The environmental benefits of using SCMs to replace Portland cement have already been discussed. If slag cement is used to increase the solar reflectance, then a life cycle inventory performed in 2006 by Construction Technology Laboratories, Skokie, IL, found that when slag cement replaces 50% of the Portland cement in 52 MPa (7500 psi) concrete, the energy required to produce 0.8 m^3 (1 yd^3) of concrete is reduced by 37%; carbon dioxide emissions are reduced by 46%; and virgin material used is reduced by 15% (Buffenbarger, 2008).

3.5 Modeling the benefits of widespread use of high solar reflectance concrete

The use of high solar reflectance concrete provides both direct and indirect benefits. Since high solar reflectance concrete contains a substantial amount of slag cement in place of Portland cement, which requires a lot of energy and releases a significant amount of harmful emissions during the manufacturing process, it provides direct benefits in terms of reducing energy use and emissions. In addition, high solar reflectance concrete reduces ground surface temperature, and thus, widespread use of such concrete can help mitigate the UHI effect, indirectly resulting in building cooling energy savings and lower levels of air pollution.

3.5.1 Benefits associated with the use of SCMs

The benefits of using SCMs can be determined from a life cycle inventory (LCI) of concrete production. LCI is a compilation of the materials, energy inputs (in the form of electricity, natural gas, diesel fuel, and other fuel oils), and emission outputs associated with the production of one unit of concrete. LCI of concrete production encompasses all the processes from manufacturing cement and replacement materials, aggregate production, transporting materials to the plant, to concrete plant operations.

Table 3.6 compares the energy consumption and emissions between conventional concrete and the high solar reflectance concrete containing 70% slag cement. It is based on data from earlier studies (Marceau and VanGeem, 2002; Prusinski et al., 2004). It is shown that in addition to the 8% savings in total material cost, the use of high solar reflectance concrete helps decrease net energy consumption by 43.5%. It also reduces pollutant emissions in the order of 20–60% where the largest emission reduction is in CO_2. Using an arbitrary highway project as an example, if this high solar reflectance concrete was used in place of a conventional concrete to pave 260 lane km or about 160 lane miles (a 20 miles divided highway segment with 4 lanes in each direction) of a typical 25 cm (10 in.) thick concrete pavement, the benefits would turn out to be approximately 315,000 GJ (228,400 million BTU) of energy savings plus 48,250 tonnes (106 million lbs) of CO_2 reduction. In addition, the utilization of slag cement in high solar reflectance concrete helps minimize the waste that would otherwise take up landfill space.

3.5.2 Benefits as a result of mitigating urban heat island effects

The USEPA provides an urban heat island Mitigation Impact Screening Tool (MIST) to help estimate the expected impacts of urban heat island mitigation strategies on air temperature, air quality, and energy demand, averaged at the city-scale (USEPA, 2005). Under certain assumptions, the tool employs state-of-the-science methods involving meteorology modeling and mitigation simulation to provide qualitatively accurate assessments of the net benefits of albedo modification, vegetation modification, and/or temperature reduction for over 200 US cities (Sailor and Dietsch, 2005).

Table 3.6 Comparison of energy consumption and emissions associated with the production of conventional and high solar reflectance concrete

Attribute	Mix 1	Mix 11	Difference	% Difference
% Portland cement	100%	30%		
% Slag	0%	70%		
Total material cost[a], \$/cu yd concrete	50.33	46.05	−4.28	−8.5
Energy consumption, million BTU/cu yd concrete (1 million BTU/cu yd = 1.379 GJ/m^3)				
Portland cement manufacturing	1.370	0.480	−0.891	−65.0
Slag cement manufacturing	0	0.153	+0.153	n/a
Aggregate production	0.074	0.077	+0.003	+4.2
Transporting materials to plant	0.054	0.059	+0.004	+8.3
Concrete plant operations	0.179	0.179	0	0
Total	1.678	0.948	−0.730	−43.5
Emissions, lbs/cu yd concrete (1 lb/cu yd = 0.5933 kg/m^3)				
Carbon dioxide (CO_2)	585	245	−340	−58.1
Carbon monoxide (CO)	0.722	0.426	−0.296	−41.0
Sulfur dioxide (SO_2)	1.357	0.704	−0.653	−48.1
Oxides of nitrogen (NO_x)	1.763	0.785	−0.978	−55.5
Volatile organic compound (VOC)	0.060	0.047	−0.013	−21.6
Particulate matter (PM)	2.272	1.392	−0.880	−38.7
Methane (CH_4)	0.027	0.014	−0.013	−47.3

[a]Based on the following unit costs: slag \$70/ton; Portland cement \$113/ton, coarse aggregate \$13/ton, and fine aggregate \$11/ton.

To compute the potential albedo change ($\Delta\rho$) of averaged US urban cities due to the implementation of the high solar reflectance concrete, the information regarding the existing surface composition of the cities is needed. Land coverage patterns of four different cities in the United States, namely, Sacramento, Chicago, Salt Lake City, and Houston, are given in Table 3.7(a) (Akbari and Rose, 1999, 2001a,b; Rose et al., 2003).

Over time, the albedo of asphalt can increase from 0.05–0.10 (average = 0.075) at initial installation to 0.15–0.20 (average = 0.175) because of the exposure of aggregates after wearing and the oxidation that causes the binder to fade. On the other hand, the albedo of concrete can decrease from 0.35–0.40 (average = 0.375) when it is new to 0.25–0.30 (average = 0.275) after wearing and weathering. That means the albedo of exposed concrete is about (0.275/0.375) × 100 = 73% of its original value. If the albedo of the high solar reflectance concrete was to drop at the same rate, its albedo would be about 0.582 × 73% = 0.425 after exposure. Thus, the long-term albedo

Table 3.7 Estimates of potential albedo increase due to the implementation of high solar reflectance concrete

Urban area	Roads[a]	Parking lots[b]	Sidewalks[c]	Total
(a) Percentage of each type of pavement structure in a city				
Sacramento, CA	27%	13%	5%	45%
Chicago, IL	19%	16%	3%	37%
Salt Lake City, UT	17%	13%	6%	36%
Houston, TX	10%	17%	2%	29%
Average	*18%*	*15%*	*4%*	*37%*
(b) Potential albedo increase due to paving high solar reflectance concrete in place of conventional concrete and asphalt surfaces				
Sacramento, CA	0.06	0.02	0.01	0.09
Chicago, IL	0.04	0.03	0.00	0.07
Salt Lake City, UT	0.04	0.02	0.01	0.07
Houston, TX	0.02	0.03	0.00	0.06
Average	*0.04*	*0.03*	*0.01*	*0.07*
(c) Potential albedo increase due to paving high solar reflectance concrete in place of conventional concrete				
Sacramento, CA	0.00	0.01	0.01	0.02
Chicago, IL	0.00	0.01	0.00	0.01
Salt Lake City, UT	0.00	0.01	0.01	0.02
Houston, TX	0.00	0.01	0.00	0.01
Average	*0.00*	*0.01*	*0.01*	*0.02*

[a]Assume 3% of roads are concrete surface, 87% are asphalt surface, and the remaining 10% are unpaved.
[b]Assume 45%/45%/10% of concrete/asphalt/other surfaces for parking lots.
[c]Assume 80%/10%/10% of concrete/asphalt/other surfaces for sidewalks.

difference between high solar reflectance concrete and asphalt is $0.425 - 0.175 = +0.25$, and the long-term albedo difference between high solar reflectance concrete and conventional concrete is $0.425 - 0.275 = +0.15$. With this information and the assumptions pertaining to the percentage of each surface type for each pavement facility listed under Table 3.7, the citywide $\Delta\rho$ can be calculated using (3.15).

$$\Delta\rho = \sum_{i,j} \left(\rho_{ij}^{A} f_{ij}^{A} - \rho_{ij}^{B} f_{ij}^{B} \right) \tag{3.15}$$

where ρ_{ij}^{A} and ρ_{ij}^{B} are the albedo of surface i of structure j after and before the change and f_{ij}^{A} and f_{ij}^{B} are the fraction of surface i of structure j after and before the change.

Table 3.7(b) presents the calculated citywide $\Delta\rho$ if the high solar reflectance concrete was paved in place of the conventional concrete and asphalt surfaces. For the four cities analyzed, the $\Delta\rho$ vary from +0.06 to +0.09 with the average value of +0.07. Table 3.7(c) shows the result for the scenario that the high solar reflectance concrete was paved in place of the conventional concrete only. In this scenario, the average value of $\Delta\rho$ is +0.02.

In Table 3.8, the estimated benefits as a result of a citywide albedo increase are shown for some large US cities within 250 miles from slag cement production sources as shown in Figure 3.5 (plotted based on data from USEPA, 2001). According to the results, the estimated mean temperature reduction could vary between 0.1 and 0.5 °C (0.2 and 0.9 °F). The estimated decreases in ozone peak concentrations could range from 0.2 to 2.6 parts per billion (ppb) for 1 h averaged concentration and from 0.2 to 2.1 ppb for 8 h averaged concentration. Lastly, the estimates of energy savings reveal that these cities could be entitled to net energy savings worth millions of dollars annually.

An earlier study demonstrated the simulation results that increasing the albedo of 1250 km^2 (483 $miles^2$) of roadways in Los Angeles by 0.25 would save cooling energy worth $15 million per year and would reduce smog-related medical and lost-work expenses by $76 million per year (Rosenfeld et al., 1998). It is possible that the implementation of high solar reflectance concrete could play a role in achieving such an albedo increase.

3.6 Leadership in Energy and Environmental Design (LEED) credit

One of the most successful programs for assessing and certifying the amount of sustainability incorporated in a building design is the USGBC LEED program. As of May 2013, there were 44,998 registered or certified LEED buildings in North America, mainly in the United States. As a sign of the increasing global popularity, there are also 1704 in Latin America, 1706 in Europe, 1297 in the Middle East and North Africa, and 1995 in East Asia. There are 110 points available to be scored across seven different categories (USGBC, 2009). The lowest base certification requires 40–49 points, silver certification requires 50–59 points, gold certification requires 60–79 points, and platinum requires at least 80 points. When striving for the more prestigious ratings, even 1 or 2 points can be precious and this can be a good incentive for designers to consider cool pavements.

Under the sustainable sites credit 7.1 for urban heat island effect—nonroof—there is one point that can be earned. There are two options. Option 1 is to use any combination of the following strategies for 50% of the site hardscape (including roads, sidewalks, courtyards, and parking lots): (i) provide shade from the existing tree canopy or within 5 years of landscape installation, (ii) provide shade from structures covered by solar panels that produce energy used to offset some nonrenewable resource use, (iii) provide shade from architectural devices or structures that have an SRI of at least 29, (iv) use hardscape materials with an SRI of at least 29, or (v) use an open-grid pavement system (at least 50% pervious). Option 2 is to place a minimum of 50% of parking spaces under cover. Any roof used to shade or cover parking spaces must have an SRI of at least 29, be a vegetated green roof, or be covered by solar panels that produce energy used to offset some nonrenewable resource use.

Table 3.8 **Potential benefits from a citywide albedo increase for large cities within 250 miles from slag production sources**

Rank in 2010	City	2010 Population[a]	Projected average reduction[b]			Projected annual energy savings ($ per 1000 sq ft of roof area)[c]		
			Air temp. (°F)	Max 1-hr O_3 (ppb)	Max 8-hr O_3 (ppb)	Residential	Office	Retail
(a) A citywide albedo increase of 0.07								
1	New York City, NY	8,175,133	0.9	2.6	2.1	5.23	9.35	9.93
3	Chicago, IL	2,695,598	0.9	2.5	2	0.96	4.71	4.75
5	Philadelphia, PA	1,526,006	0.8	1.8	1.4	3.45	6.13	6.51
12	Indianapolis, IN	820,445	0.6	1.3	1	1.02	4.04	4.16
15	Columbus, OH	787,033	0.6	1.3	1	1.18	4.96	5.12
18	Detroit, MI	713,777	0.7	1.9	1.6	0.96	4.73	4.77
20	Memphis, TN	646,889	0.6	0.9	0.7	4.21	5.99	6.20
(b) A citywide albedo increase of 0.02								
1	New York City, NY	8,175,133	0.3	0.7	0.6	1.60	2.73	2.86
3	Chicago, IL	2,695,598	0.3	0.7	0.5	0.33	1.36	1.36
5	Philadelphia, PA	1,526,006	0.2	0.5	0.4	1.05	1.79	1.87
12	Indianapolis, IN	820,445	0.2	0.4	0.3	0.29	1.14	1.20
15	Columbus, OH	787,033	0.2	0.4	0.3	0.34	1.40	1.47
18	Detroit, MI	713,777	0.2	0.5	0.4	0.33	1.36	1.36
20	Memphis, TN	646,889	0.2	0.2	0.2	1.24	1.76	1.76

[a]Based on census data published in Infoplease (2012).
[b]Calculated using data from USEPA (2005).
[c]Estimates are for post-1980 electric-heated buildings; calculated using data from USEPA (2005) and EIA (2004).

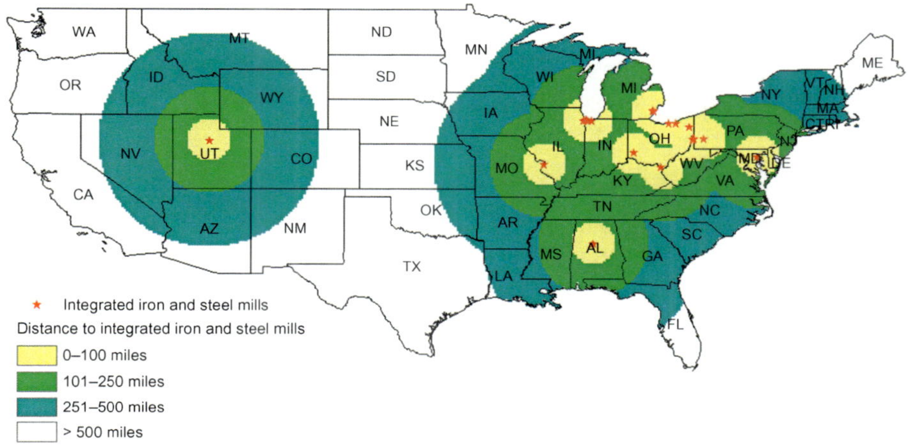

Figure 3.5 Slag production sources in the US (based on data from USEPA (2001)).

Under option 1, a cool nonpervious concrete could be used for 50% of the site hardscape, for example. The cool pavement would need to have an SRI of 29 or greater. As has been shown earlier, this is a somewhat easy target for concrete with most conventional concrete mixes being able to achieve it. However, to make certain to pass the SRI criterion, any of the methods described earlier should be successful at increasing the solar reflectance. It should be noted that if slag cement is used, an additional 1–2 points may potentially be scored under the materials and resources credit 4 for recycled content. One to two points are awarded if 10–20% or more of the total cost of materials contains recycled materials and slag cement is considered as a recycled material in LEED.

3.7 Other resources

In this section, a brief mention of a few other valuable resources for learning about UHI and its mitigation and cool pavements is given. Most of the organizations cited maintain websites that are often updated and can be easily found using any web search engine. The US Environmental Protection Agency (USEPA) has been one of the biggest proponents of UHI mitigation. They suggest a three-pronged approach: the use of cool roofs, the use of cool pavements, and planting more trees. The USEPA sponsored a series of publications related to UHI mitigation named a "Compendium of Strategies," including a compendium of strategies—cool pavements (USEPA, 2008).

One of the research organizations that have been very active in the area of UHI and its mitigation, particularly in cool roofs, is the Lawrence Berkeley National Laboratory and its Heat Island Group. The laboratory is supported by the US Department of Energy and managed and operated by the University of California and is located in Berkeley, CA.

One of the more recent research groups that are studying UHI and cool pavements has been developed at Arizona State University under the National Center of Excellence on SMART Innovations (NCE Sustainable Materials and Renewable Technologies). The center was developed with funding from the USEPA.

3.8 Future trends

In the end, no one single strategy can be expected to address the UHI effect alone; rather, an integrated approach will be needed: for example, on the same street, there may be a pervious pavement mainline, high solar reflectance concrete sidewalks, and shade trees along the side. An integrated research approach is also needed rather than investigating aspects in isolation; for example, can any single concrete mix do it all, that is, meet all of the myriad performance expectations at the same time? For example, it needs to be sustainable, cool, workable and easy to finish, strong, wear-resistant, durable against freeze–thaw, resistant to various types of alkali-aggregate reactions, and resistant to scaling from deicing agents and should provide a good ride quality. As more of these cool pavements are being installed, instrumentation should be incorporated on some test projects for the purpose of long-term performance monitoring. Responsible agencies need to be encouraged to consider life cycle costs and not just initial costs when making decisions about pavements that will favor and therefore promote the use of cool concrete pavements.

Future construction may contain a combination of strategies all in the same pavement. For example, a pavement could have a cool concrete mix, an infrared coating on its top, added latent heat storage materials in the mass of the pavement (Santamouris, 2013), and circulation of cooling fluid in the pavement or its base (Santamouris, 2013).

One interesting recent development is the rating system for sustainability of roads and transportation infrastructure, an example of which is Greenroads. The Greenroads rating system is similar in format to the LEED system for buildings. If the program can enjoy the same success as its counterpart in buildings, then we can expect to see a lot more cool concrete pavements being built in the future.

The Global Cool Cities Alliance was launched in 2010 with five founding partner cities: Athens, Chicago, New York, Singapore, and Taipei. One of the interesting projects of this group is the "100 Cool Cities" program that will attempt to obtain commitments from 100 major cities around the world and major implementation of cool roofs and cool pavements by 2020. If successful, this has the potential to truly make an impact on a global scale in the near future.

References

Akbari, H., Rose, L.S., 1999. Characterizing the Fabric of the Urban Environment: A Case Study of Sacramento, California. Lawrence Berkeley National Laboratory, Berkeley, CA, 48.

Akbari, H., Rose, L.S., 2001a. Characterizing the Fabric of the Urban Environment: A Case Study of Metropolitan Chicago, Illinois. Lawrence Berkeley National Laboratory, Berkeley, CA.

Akbari, H., Rose, L.S., 2001b. Characterizing the Fabric of the Urban Environment: A Case Study of Salt Lake City, Utah. Lawrence Berkeley National Laboratory, Berkeley, CA, 51.

Akbari, H., Matthews, H.D., Seto, D., 2012. The long-term effect of increasing the albedo of urban areas. Environ. Res. Lett. 7 (2), 024004.

ASTM C 1549, 2009. Standard test method for determining solar reflectance near ambient temperature using a portable solar reflectometer. In: Annual Book of ASTM Standards, American Society for Testing and Materials, Philadelphia, PA.

ASTM E 1980, 2013. Standard practice for calculating solar reflectance index of horizontal and low-sloped opaque surfaces. In: Annual Book of ASTM Standards. American Society for Testing and Materials, Philadelphia, PA.

Attarian, J.L., 2010. Greener alleys. Public Roads 73 (6), 26–33.

Bentz, D.P., 2000. A computer model to predict the surface temperature and time of wetness of concrete pavements and bridge decks. In: NIST Report NISTIR 6551. National Institute of Standards and Technology, Washington, DC.

Boriboonsomsin, K., Reza, F., 2007. Mix design and benefit evaluation of high solar reflectance concrete for pavements. Transp. Res. Rec. 2011, 11–20.

Bretz, S., Akbari, H., Rosenfeld, A., 1998. Practical issues for using solar-reflective materials to mitigate urban heat islands. Atmos. Environ. 32 (1), 95–101.

Buffenbarger, J., 2008. Designing for sustainability: cementitious-based building materials contribute to LEED® credits. In: Architectural Record. McGraw-Hill Construction, New York, NY.

EIA, 2004. Electric Sales, Revenue, and Average Price. Energy Information Administration. http://www.eia.doe.gov/cneaf/electricity/esr/esr_sum.html (accessed 30.06.06.).

Gartland, L., 2008. Heat Islands, Understanding and Mitigating Heat in Urban Areas. Earthscan, London.

Infoplease, 2012. Population of the 20 Largest U.S. Cities, 1900–2012. http://www.infoplease.com/ipa/A0922422.html (accessed 30.04.14.).

Kaloush, K.E., Carlson, J.D., Golden, J.S., Phelan, P.E., 2008. The Thermal and Radiative Characteristics of Concrete Pavements in Mitigating Urban Heat Island Effects. SN2969, Portland Cement Association, Skokie, IL.

Karunarathne, A., Mampearachchi, W., Nanayakkara, A., 2011. Modelling of thermal effects due to solar radiation on concrete pavements. In: Proc. of the 7th Annual International Conference of the International Institute for Infrastructure Renewal and Reconstruction, July 2011. Kandalam, Sri Lanka.

LaBarca, I., Foley, R., Cramer, S., 2007. Effect of Ground Granulated Blast Furnace Slag in Portland Cement Concrete – Expanded Study. Report WHRP 07–01, Wisconsin Highway Research Program, Madison, WI.

Levinson, R., Akbari, H., 2001. Effects of Composition and Exposure on the Solar Reflectance of Portland Cement Concrete. Lawrence Berkeley National Laboratory Report LBNL-48334, Lawrence Berkeley National Laboratory, Berkeley, CA.

Li, H., 2012. Evaluation of cool pavement strategies for heat island mitigation. In: Institute of Transportation Studies Report UCD-ITS-RR-12-33. University of California, Davis.

Li, H., Harvey, J.T., Holland, T.J., Kayhanian, M., 2013. The use of reflective and permeable pavements as a potential practice for heat island mitigation and stormwater management. Environ. Res. Lett. 8, 015023.

Marceau, M.L., VanGeem, M.G., 2002. Life Cycle Inventory of Slag Cement Manufacturing Process. http://www.slagcement.org/Sustainability/pdf/Slag%20Cement%20LCI%20Letter%20Report%20August%2011%202003.pdf (accessed 30.04.14.).

Marceau, M.L., VanGeem, M.G., 2007. Solar Reflectance of Concretes for LEED Sustainable Sites Credit: Heat Island Effect. SN2982, Portland Cement Association, Skokie, IL.

Mindess, S., Young, J.F., Darwin, D., 2003. Concrete, second ed. Prentice Hall, Upper Saddle River, NJ.

NCHRP 1-37A, 2004. The Mechanistic Empirical Design Guide for New and Rehabilitated Pavement Structures. National Cooperative Highway Research Program, Washington, DC.

Pomerantz, M., Akbari, H., Harvey, J., 2000. Cooler Reflective Pavements Give Benefits Beyond Energy Savings: Durability and Illumination. Lawrence Berkeley National Laboratory Report, Berkeley, CA.

Prusinski, J.R., Marceau, M.L., VanGeem, M.G., 2004. Life cycle inventory of slag cement concrete. In: 8th CANMET/ACI International Conference on Fly Ash, Silica Fume, Slag and Natural Pozzolans in Concrete, Las Vegas, NV.

Richardson, D.N., 2006. Strength and Durability Characteristics of a 70% Ground Granulated Blast Furnace Slag (GGBFS) Concrete Mix. Missouri Department of Transportation Final Report RI99-035/RI99-035B,Missouri Department of Transportation, Jefferson City, MO.

Rose, L.S., Akbari, H., Taha, H., 2003. Characterizing the Fabric of the Urban Environment: A Case Study of Greater Houston. Lawrence Berkeley National Laboratory, Texas. Berkeley, CA, 61.

Rosenfeld, A.H., Akbari, H., Romm, J.J., Pomerantz, M., 1998. Cool Communities: Strategies for Heat Island Mitigation and Smog Reduction. Energy & Buildings 28 (1), 51–62.

Sailor, D.J., Dietsch, N., 2005. The Urban Heat Island Mitigation Impact Screening Tool (MIST). http://www.heatislandmitigationtool.com/Documents/detailed_help.pdf (accessed 30.04.14.).

Santamouris, M., 2013. Using cool pavements as a mitigation strategy to fight urban heat island-a review of the actual developments. Renew. Sust. Energ. Rev. 26, 224–240.

Santamouris, M., Gaitani, N., Spanou, A., 2012. Using cool paving materials to improve microclimate of urban areas – Design realization and results of the Flisvos project. Build. Environ. 53, 128–136.

Schlorholtz, S., Hooton, R.D., 2008. Deicer scaling resistance of concrete pavements, bridge decks, and other structures containing slag cement, phase 1: site selection and analysis of field cores. In: Federal Highway Administration CTRE Project No. 05–202. National Concrete Pavement Technology Center, Ames, IA.

Snellings, R., Mertens, G., Elsens, J., 2012. Supplementary cementitious materials. Rev. Mineral. Geochem. 74, 211–278.

USDOT, 2013. Highway Statistics 2012. US Department of Transportation, Washington, DC.

USEPA, 2001. Sector Facility Indexing Project. US Environmental Protection Agency, Washington, DC.

USEPA, 2005. US Environmental Protection Agency. Mitigation impact screening tool (MIST). http://www.heatislandmitigationtool.com/ (accessed 30.04.14.).

USEPA, 2008. Reducing Urban Heat Islands: Compendium of Strategies – Cool Pavements. US Environmental Protection Agency, Washington, DC.

USGBC, 2009. LEED 2009 for New Construction and Major Renovations. US Green Building Council, Washington, DC.

A comparison of thermal performance of different pavement materials

4

H. Li
University of California, Davis, CA, USA

4.1 Introduction

This chapter briefly discusses the thermal performance of different pavement types commonly used, including concrete material, asphalt material, and block paver. The chapter first introduces the important material properties such as albedo and thermal properties of different pavement materials. The chapter then discusses the thermal performance of different pavement materials, including pavement surface temperature, near-surface air temperature above pavement, thermal impact of pavement on nearby building surfaces, and heat flux from pavement. Potential impacts of pavements and future trends are also discussed.

4.2 Albedo of pavement materials

The thermal characteristics of paved surfaces (albedo, heat capacity, and thermal conductivity) interacting with solar radiation are among a number of causal factors affecting urban heat islands (UHIs), which also include the population and population density of the urban area, building materials, spacing and height of buildings, waste heat generated by vehicles and building equipment, vegetative cover, and geographic location (Akbari and Matthews, 2012; Asaeda and Ca, 2000; Jo et al., 2010; Navigant Consulting Inc., B. T. P., 2011; Santamouris, 2012; Santamouris et al., 2012; Synnefa et al., 2011). Albedo (or solar reflectivity) is an indicator of the reflecting power of a surface and a key thermal characteristic. It is defined as the ratio of the reflected solar radiation to the incident solar radiation at the surface. Albedo is a dimensionless fraction and is measured on a scale from 0 to 1. An albedo of 0 means no reflecting power of a perfectly black surface (none reflected, all absorbed), while an albedo of 1 means perfect reflection off a perfectly white surface (100% reflected).

The temperatures of pavement surfaces exposed to solar radiation are generally higher than the adjacent air temperatures due to heating by absorbed solar radiation, thus creating a surface heat island. The extent to which solar radiation influences surface temperatures depends on the solar reflectance of the exposed pavement surface. A low solar reflectance material, such as a black stone surface, would result in a very large effect, while a high solar reflectance material, such as fresh snow, would

Eco-efficient Materials for Mitigating Building Cooling Needs. http://dx.doi.org/10.1016/B978-1-78242-380-5.00004-2

result in a small effect on surface temperature. Therefore, solar reflectivity is a factor of great significance for evaluating and modeling the thermal performance of pavements and other land cover types (Akbari and Matthews, 2012; Akbari et al., 2012; Santamouris, 2012).

This section covers: (1) albedo field measurement methods, (2) field measurements of albedo on different pavement materials and other land cover types, (3) a comparison of the albedo for different materials, (4) factors affecting the field measurement of albedo, (5) diurnal and seasonal changes in albedo, and (6) the effect of albedo on pavement temperature.

4.2.1 Albedo field measurement methods and equipment

There are two ASTM standard testing methods for determining solar reflectance of a surface: (1) ASTM C1549 Standard Test Method for Determination of Solar Reflectance Near Ambient Temperature Using a Portable Solar Reflectometer (2009) and (2) ASTM E1918 Standard Test Method for Measuring Solar Reflectance of Horizontal and Low-Sloped Surfaces in the Field (2006).

To improve the convenience and reduce the measurement error of ASTM E1918, a dual-pyranometer can be used to perform the measurement of solar reflectivity (Li et al., 2013c). A dual-pyranometer (also called albedometer) is composed of two star pyranometers. One pyranometer faces upward and the other faces downward. Incident global solar radiation (diffuse and direct solar radiation) with wavelengths from 0.3 to 3 μm is measured by the upward-facing pyranometer, while reflected solar radiation from surfaces is measured by the downward-facing pyranometer (NovaLynx Corporation, 2014). Separate outputs are provided for each pyranometer, which can be read from an indicator or recorded together automatically using a datalogger.

Testing in the research project done by the University of California Pavement Research Center (UCPRC) (see the following description below and references (Li, 2012) and (Li et al., 2013c) for details) was conducted in accordance with ASTM E1918, following the standard method except using a dual-pyranometer. The Model 240-8140 dual-pyranometer was purchased from NovaLynx Corporation (2014), with a calibration certificate and showing the measurement error of less than 1 W/m^2. The data are recorded automatically using a data acquisition system (DAS) composed of a CR10X datalogger from Campbell Scientific, Inc., powered by a battery and connected to a computer. This allows monitoring the solar reflectivity of a surface over long time periods. The whole measurement system is shown in Figure 4.1.

4.2.2 Pavement materials and plan for albedo measurement

4.2.2.1 Pavement materials for albedo measurement

The ongoing study on cool pavements being conducted at the University of California Pavement Research Center (UCPRC) is devoted to investigate the thermal behavior and cooling effect of different pavement types (including asphalt, concrete, and block paver) and different designs (conventional impermeable and novel permeable

Figure 4.1 Albedo measurement system with a dual-pyranometer and DAS.

designs) and use the field measurement data to validate heat transfer modeling and employ the validated model to simulate the thermal behavior and cooling effect of different pavements under various contexts (climates and surroundings) and examine the effect of cool pavements on human thermal comfort (Li, 2012).

The albedo measurements for this study were performed mainly on nine 4 m by 4 m (13 ft by 13 ft) test sections constructed during the summer of 2011 at the University of California Pavement Research Center (UCPRC) test facilities in Davis, California (as shown in Figure 4.2), using the dual-pyranometer test method (modified ASTM E1918). The summary of each test section along with surface mix design and other characteristics is provided in Tables 4.1 and 4.2. These nine test sections include three different pavement surfacing materials, namely, interlocking concrete pavers (surfacing type A), open-graded asphalt concrete (surfacing type B), and portland cement concrete (surfacing type C). For each pavement surface type, one impermeable pavement design (design 1) and two permeable pavement designs were constructed (designs 2 and 3; note that the industry-preferred terms are pervious concrete, porous asphalt, and permeable pavers, and the word "permeable" is used for all surface types for convenience in this chapter). Both of the permeable interlocking concrete paver pavements have the same cross sections; the difference is in the solar reflectivity of the pavers. Both of the permeable asphalt sections have the same surface material; the difference is in the thicknesses of the layers. The two permeable concrete sections have different

Figure 4.2 Nine experimental test sections (4 × 4 m for each) used for albedo measurement (Li et al., 2013c).

thickness and different concrete surface material mix designs, meaning that the aggregate gradations, cement contents, and other ingredient proportions are different. The permeable concrete section C2 is darker than C3, resulting in different albedos.

Besides these nine sections, some extra pavement sections with conventional impermeable asphalt and concrete surfacing were also included in the study for field measurement of albedo. In addition, albedo was measured on some other land cover materials, including gravel, soil, and grass, for comparison. Some of these materials were different ages when the measurement of solar reflectivity was conducted on them. The overall summary of materials used for albedo measurement in this study and dates of measurement are summarized in Table 4.2.

More field albedo measurements were performed around Davis, California, in May 2014 on eight test sites with different materials including slurry seal, fog seal, cape seal, chip seal, and more PCC and AC materials (Table 4.2).

4.2.2.2 Plan for albedo measurement

The dual-pyranometer with DAS was used to measure the albedo of all pavement test sections in the experimental sections (pavers, asphalt, and concrete). For each test section, the albedo measurement was conducted at five different locations: the southeast corner (SE), northeast corner (NE), northwest corner (NW), southwest corner (SW), and center (CT). At least six measurements were performed at each location. A summary of the experimental test plan for the nine test sections is shown in Table 4.1.

The measurements on the nine experimental test sections (A1–3, B1–3, and C1–3) were repeated on four different dates in one year (September 2011 to May 2012) to check the change of albedo over time. The measurements on the other materials were conducted at three to five different locations with at least six measurements for each location. All these measurements were performed at midday (mostly 12:00–3:00 pm). For some materials (mainly for B2), the solar reflectivity was monitored continuously over time. The measurement plan in this study is summarized in Table 4.2.

Table 4.1 **Experimental plan for field albedo measurement on nine test sections**

Section	Pavement layer		Base layer		Test locations[b]	Test method
	Type[a]	Thickness (cm)	Type	Thickness (cm)		
A1	ICP-I	10[c]	AB-I	15	SE, NE, NW, SW, CT	Dual-pyranometer
A2	ICP-P	10[c]	AB-P	15	SE, NE, NW, SW, CT	Dual-pyranometer
A3	ICP-P	10[d]	AB-P	30	SE, NE, NW, SW, CT	Dual-pyranometer
B1	AC-I	10	AB-I	15	SE, NE, NW, SW, CT	Dual-pyranometer
B2	AC-P	10	AB-P	30	SE, NE, NW, SW, CT	Dual-pyranometer
B3	AC-P	20	AB-P	30	SE, NE, NW, SW, CT	Dual-pyranometer
C1	PCC-I	10	AB-I	15	SE, NE, NW, SW, CT	Dual-pyranometer
C2	PCC-P1[e]	10	AB-P	30	SE, NE, NW, SW, CT	Dual-pyranometer
C3	PCC-P2[e]	20	AB-P	30	SE, NE, NW, SW, CT	Dual-pyranometer

[a]ICP, interlocking concrete paver; AC, asphalt concrete; PCC, Portland cement concrete; AB, aggregate base. I, impermeable; P, permeable.
[b]SE, southeast; NE, northeast; NW, northwest; SW, southwest; CT, center.
[c]Includes the thickness of paver (6 cm) and underlying bedding layer (4 cm).
[d]Includes the thickness of paver (8 cm) and underlying bedding layer (2 cm).
[e]The two pervious concrete materials have very different mix designs, cement and aggregate types.
(Li et al., 2013c).

Table 4.2 **Summary of materials and plan for albedo measurement**

Surface category	Permeability	Binder/cement/color type	Code[a]	Measurement date (pavement age at measurement)
Asphalt	Impermeable	Conventional	B1	September 19, 2011, October 13, 2011, February 15, 2012, May 2, 2012[d]
	Permeable	Polymer-modified	B2[b]	September 19, 2011, October 13, 2011, February 15, 2012, May 2, 2012[d]
	Permeable	Polymer-modified	B3	September 19, 2011, October 13, 2011, February 15, 2012, May 2, 2012[d]
	Impermeable	Polymer-modified	PMA	October 13, 2011 (3 years)
	Impermeable	Rubberized	RHMA	October 13, 2011 (3 years)
	Impermeable	Warm mix asphalt	RWMA	October 13, 2011 (2 years)
	Permeable	Polymer-modified	OGFC	October 13, 2011 (1 year)
	Impermeable	Conventional	Aged AC	October 13, 2011 (5 years)
	Impermeable	Conventional (parking lot)	Aged AC	May 21, 2014 (5 years)
	Impermeable	Conventional (highway)	Aged AC	- (2–25 years)
	Impermeable	Conventional (street)	AC	- (0–20 years)
	Impermeable	Reflective (street)	AC with coating	- (0–2 years)
Concrete	Impermeable	Conventional	C1	September 19, 2011, October 13, 2011, February 15, 2012, May 2, 2012 [d]
	Permeable	Conventional	C2[c]	September 19, 2011, October 13, 2011, February 15, 2012, May 2, 2012[d]
	Permeable	White	C3	September 19, 2011, October 13, 2011, February 15, 2012, May 2, 2012[d]

	Impermeable	Conventional	PCC	October 13, 2011 (3 years)
	Impermeable	Conventional (parking lot)	PCC	May 21, 2014 (5 years)
	Impermeable	Conventional (highway)	Aged PCC	- (2–25 years)
	Impermeable	Conventional (street)	PCC	- (0–20 years)
	Impermeable	Reflective (street)	PCC with coating	- (0–2 years)
Interlocking concrete paver	Impermeable	Conventional, orange	A1	September 19, 2011, October 13, 2011, February 15, 2012, May 2, 2012 [d]
	Permeable	Conventional, champagne	A2	September 19, 2011, October 13, 2011, February 15, 2012, May 2, 2012 [d]
	Permeable	Conventional, orange	A3	September 19, 2011, October 13, 2011, February 15, 2012, May 2, 2012 [d]
Treatment	Impermeable	Conventional	Slurry seal	May 21, 2014 (5 years)
	Impermeable	Conventional	Chip seal	May 21, 2014 (5+ years)
	Impermeable	Conventional	Cape seal	May 21, 2014 (3+ years)
	Impermeable	Conventional	Fog seal	May 21, 2014 (0.1 and 3 years)
Gravel	Permeable	–	Gravel, basalt	October 13, 2011 (0.5 year)
Soil	Permeable	–	Soil, yellow clay	October 13, 2011
Grass	Permeable	–	Grass, yellow lawn	October 13, 2011

[a]Code: A1–3, B1–3, and C1–3 are experimental test sections described previously. PMA, polymer-modified asphalt; RHMA, rubberized hot mix asphalt; WMA, warm mix asphalt; OGFC, open-graded friction course; AC, asphalt concrete; PCC, portland cement concrete.
[b]Monitored continuously over time.
[c]Monitored continuously in one day.
[d]Age for A1–3, B1–3, C1–3: September 19, 2011, 1 month; October 13, 2011, 2 months; February 15, 2012, 6 months; May 2, 2012, 9 months
Adapted from Li et al. (2013c).

4.2.3 Albedo values of different pavement materials

4.2.3.1 Albedo of nine test sections without traffic

The albedos of the nine test sections (A1–3, B1–3, and C1–3) were initially measured at five different locations (four corners and one center) at around 1 pm on September 19, 2011. The albedo measured at different locations on one section shows some level of variation; this implies that the pavement surface is not uniform in albedo or color on each section (see reference Li et al. (2013c) for more details). The summary boxplot of albedos at five locations on each section (Figure 4.3) gives an even clearer illustration of the variation of albedo across these nine sections. Sections B2 and C2 are the most uniform, with only small variations across the 4 m wide pavements. Sections C1 and C3 show the largest variation. This suggests that for any pavement, the albedo should be measured at different locations to obtain a representative sample of albedo value, since one point may not be representative of the whole paved area.

The summary statistics of albedo for the nine sections (A1–3, B1–3, and C1–3) measured on September 19, 2011 are listed in Table 4.3. As expected, the asphalt sections, which are black, have lower albedos (0.09 for B1 and 0.08 for B2 and B3). As mentioned previously, sections B2 and B3 have the same surface material (only thicknesses are different) and have the same albedo (0.08) as expected. The three concrete sections (C1–3) have a range of mean albedo of 0.18–0.29. The darker concrete section C2 has a lower albedo of 0.18 compared with the other two concrete sections (0.26 for C3 and 0.29 for C1). The paver sections (A1–3) have albedos close to the more reflective concrete sections (C1 and C3), which are in the range of 0.25–0.28. The relatively low albedo of asphalt pavements will absorb more incident solar radiation and produce

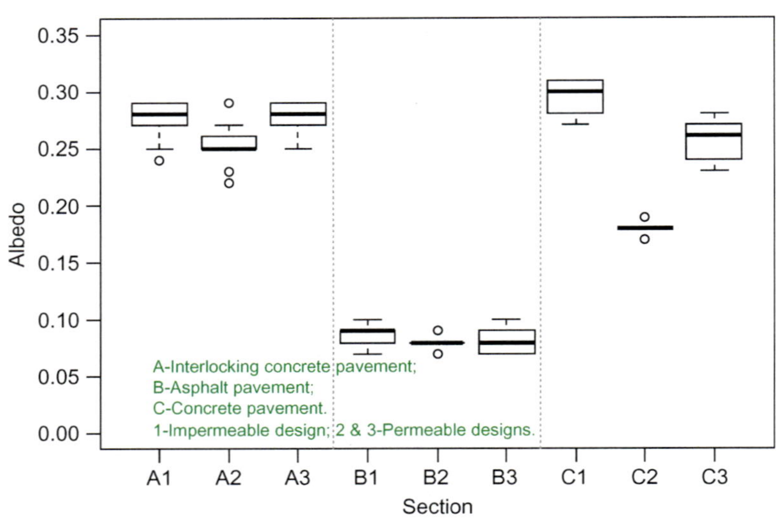

Figure 4.3 Initial albedo of block paver (interlocal concrete pavement), asphalt, and Portland cement concrete on nine test sections from UCPRC (Li et al., 2013c).

Table 4.3 **Summary statistics of initial albedo for different materials on nine test sections from UCPRC**

Section	Surface category	Permeable type	Number of tests	Mean	SD	Median	Min	Max	Max − min
A1	Paver	Impermeable	40	0.28	0.01	0.28	0.24	0.29	0.05
A2	Paver	Permeable	40	0.25	0.02	0.25	0.22	0.29	0.07
A3	Paver	Permeable	33	0.28	0.01	0.28	0.25	0.29	0.04
B1	Asphalt	Impermeable	58	0.09	0.01	0.09	0.07	0.10	0.03
B2	Asphalt	Permeable	58	0.08	0.01	0.08	0.07	0.09	0.02
B3	Asphalt	Permeable	33	0.08	0.01	0.08	0.07	0.10	0.03
C1	Concrete	Impermeable	38	0.29	0.02	0.30	0.27	0.31	0.04
C2	Concrete	Permeable	44	0.18	0.00	0.18	0.17	0.19	0.02
C3	Concrete	Permeable	70	0.26	0.02	0.26	0.23	0.28	0.05

Li et al. (2013c).

a high temperature. In contrast, the concrete and paver pavements generally have a higher albedo, which will reflect more incident solar radiation and produce a lower pavement temperature.

4.2.3.2 Albedo of other different pavement materials from UCPRC testing pavements without traffic

The albedo was also measured on testing pavements with asphalt and concrete surfacing materials other than the above nine experimental sections to include additional pavement surface materials for comparison. These pavements have different mix designs for materials, with different binder or cement and aggregate types than those used in the nine test sections. Moreover, these pavements have ages between 1 and 5 years, while the nine experimental test sections had ages of less than 3 months. Despite their older age, the additional pavement surfaces had not been subjected to much traffic, so aging processes related to traffic were not visible. For example, the asphalt covering the aggregate on the pavement surface had not been worn off. Albedos measured on gravel, bare soil, and grass are also included for comparison. The results are shown in Figure 4.4.

The additional types of asphalt pavement materials tested include a polymer-modified asphalt (PMA, which should not affect the albedo), gap-graded rubberized

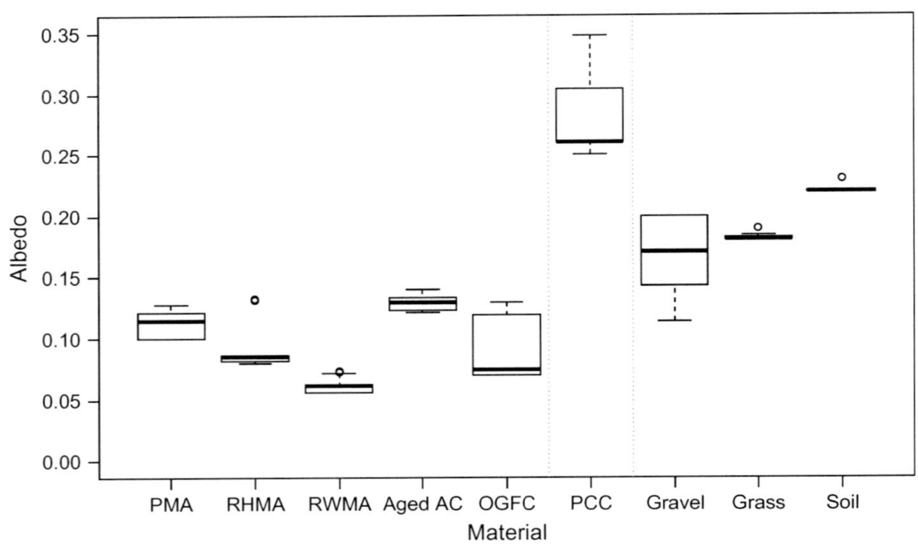

Figure 4.4 Albedo of additional asphalt and Portland cement concrete from UCPRC data with no traffic (Li et al., 2013c). PMA, polymer-modified asphalt; RHMA, rubberized hot mix asphalt; RWMA, rubberized warm mix asphalt; aged AC, aged asphalt concrete; OGFC, open-graded friction course; PCC, Portland cement concrete; gravel, basalt; grass, yellow lawn; soil, yellow clay. See Table 4.2 for pavement ages.

hot mix asphalt (RHMA, which contains a higher asphalt binder content and recycled tire rubber that is expected to affect the albedo), rubberized warm mix asphalt (RWMA, which is a different RHMA mix with a "warm mix" additive that should not affect the albedo), an open-graded friction course (OGFC), and an aged asphalt concrete (AC). The PMA section showed a slightly higher mean albedo of 0.12 compared to 0.08 for the RHMA and 0.06 for the RWMA. The 4-year-old aged asphalt pavement (aged AC) resulted in a higher albedo of 0.14. The newly paved OGFC layer had a low albedo of 0.07. The lower albedos of the RHMA, RWMA, and OGFC were expected because of their higher asphalt contents and inclusion of tire rubber in the binder of the RHMA and RWMA. The higher albedo of the aged asphalt was also expected because of the oxidation of the asphalt.

The extra concrete pavements (PCC) had albedos in the range of 0.25–0.35 with an average of 0.26, which is close to that of concrete pavements in the nine test sections.

The gravel measured in this study is crushed open-graded basalt paved on the yard of the UCPRC facility. It has a maximum aggregate size of 19 mm and is blue/gray in color. This gravel had albedos in the range of 0.12–0.22 with an average of 0.18. The bare soil (native yellow clay between the nine test sections) had an albedo of 0.22. The grass (yellow lawn) had an albedo of 0.19.

4.2.3.3 Albedo of slurry seal, cape seal, fog seal, and chip seal with more data of PCC and AC measured from field pavements with traffic in Davis, California

More field albedo measurements were performed on field pavements around Davis, California, in May 2014, including slurry seal, fog seal, cape seal, chip seal, and more PCC and AC materials. The results are shown in Figure 4.5. The 5-year-old PCC has an average albedo of 0.26, ranging from 0.23 to 0.27. The 5-year-old AC has an average albedo of 0.12, ranging from 0.09 to 0.13. The 5-year-old slurry seal of has an average albedo of 0.08, ranging from 0.07 to 0.1. The 5-year-old or more chip seal has an average albedo of 0.15, ranging from 0.14 to 0.16. The 3-year-old or more cape seal (chip seal with microsurfacing on top) has an average albedo of 0.06, ranging from 0.05 to 0.15. The 0.1–3-year-old fog seal has an average albedo of 0.06, ranging from 0.04 to 0.07.

4.2.4 Main factors affecting field albedo measurement

4.2.4.1 Influence of wind speed and air temperature

Albedo is expected to be constant regardless of wind speed or air temperature. To verify this concept, the albedo of asphalt (B2) pavement on some days with different wind speeds and air temperatures was measured for comparison. As shown in Figure 4.6a, the wind speed and air temperature on February 23, 2012–February 25, 2012 (measured with a nearby mobile weather station) are quite different. The wind speed on February 23, 2012 is much higher than the other two days. The air temperatures on February 23, 2012 and February 24, 2012 are higher than that on February 25,

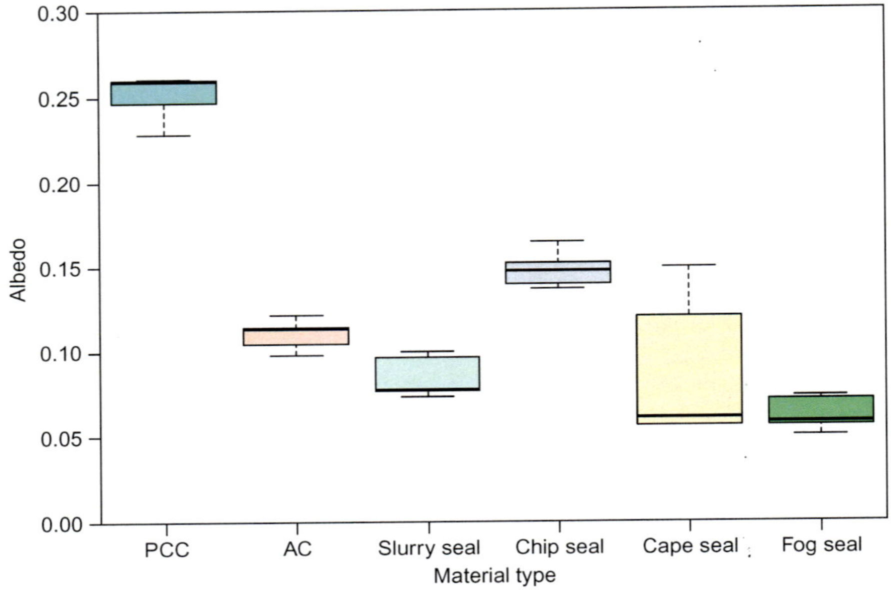

Figure 4.5 Albedo of slurry seal, cape seal, fog seal, and chip seal and asphalt and Portland cement concrete from UCPRC data with traffic. Material age: PCC, 5 years old; AC, 3–5 years old; slurry seal, 5 years old; chip seal, 5 years old or more; cape seal, 3 years old or more; fog seal, 0.1–3 years old.

2012. The albedo and solar radiation on February 23, 2012–February 25, 2012 are plotted in Figure 4.6b. The albedos around noon are all around 0.08 on these three days. No significant variation in midday albedo is observed on this period with quite different wind speeds or air temperatures.

4.2.4.2 Influence of cloudiness

Cloudiness has an important influence on the incident solar radiation, reducing the amount of solar radiation incident to the pavement or other ground surfaces. To examine the influence of clouds on the albedo of pavements, the albedo of asphalt (B2) pavement on days with different cloud levels was measured for comparison. As shown in Figure 4.7, the cloud levels over the days March 9, 2012–March 19, 2012 are different. The cloud levels on March 13, 2012–March 17, 2012 are much higher than all the other days during this period, during which the incident solar radiations are quite low. The midday albedos (~0.05) during the cloudy days and resultant low incident solar radiation are much lower than those (0.08) on days with few clouds or clear days. This implies that the albedo should be measured on a clear day. Otherwise, a lower albedo will be given, even measured at midday.

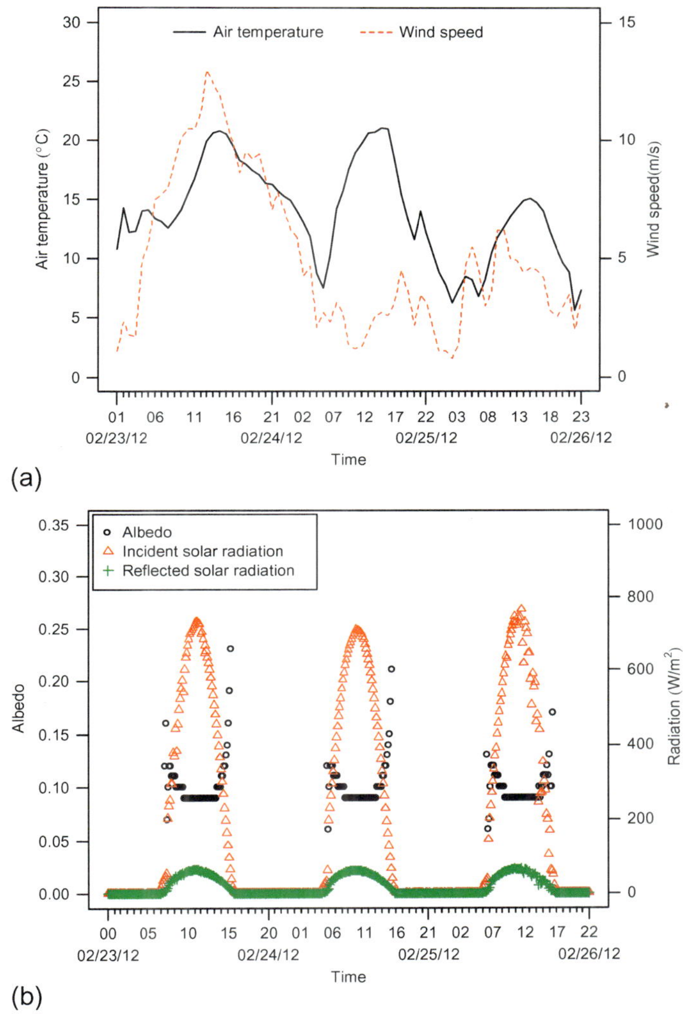

Figure 4.6 Influence of wind speed on solar reflectivity (B2) (Li et al., 2013c). (a) Wind speed and air temperature on February 23, 2012–February 25, 2012. (b) Albedo and solar radiation on February 23, 2012–February 25, 2012.

4.2.5 *Diurnal and seasonal variation of albedo*

4.2.5.1 *Diurnal variation of albedo*

To examine the diurnal variation of albedo, the solar reflectivity was continuously monitored over time on one clear day and on another clear day on an asphalt (B2) pavement and is plotted in Figure 4.8. Measured albedo changes over time during the day: it is high in the early morning and then low and relatively constant around

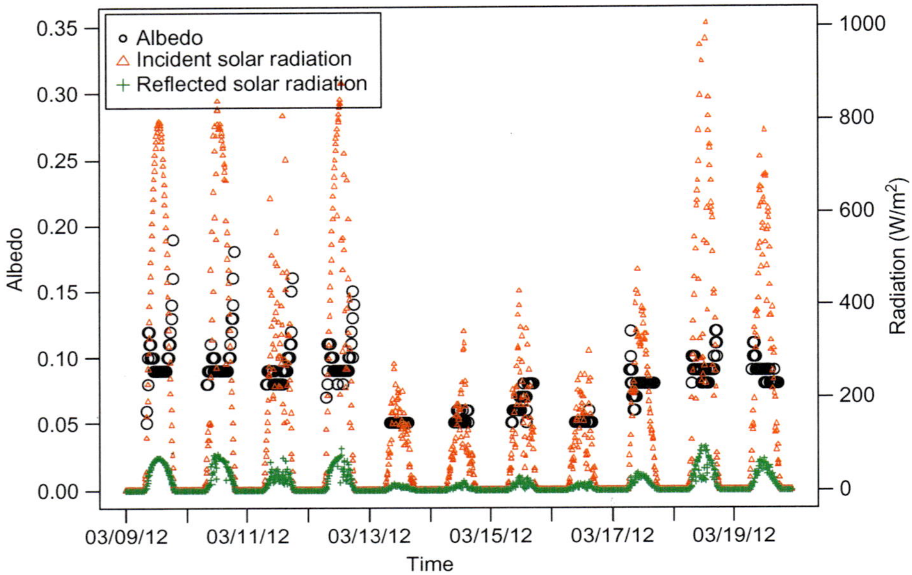

Figure 4.7 Influence of cloud on solar reflectivity (B2) (Li et al., 2013c).

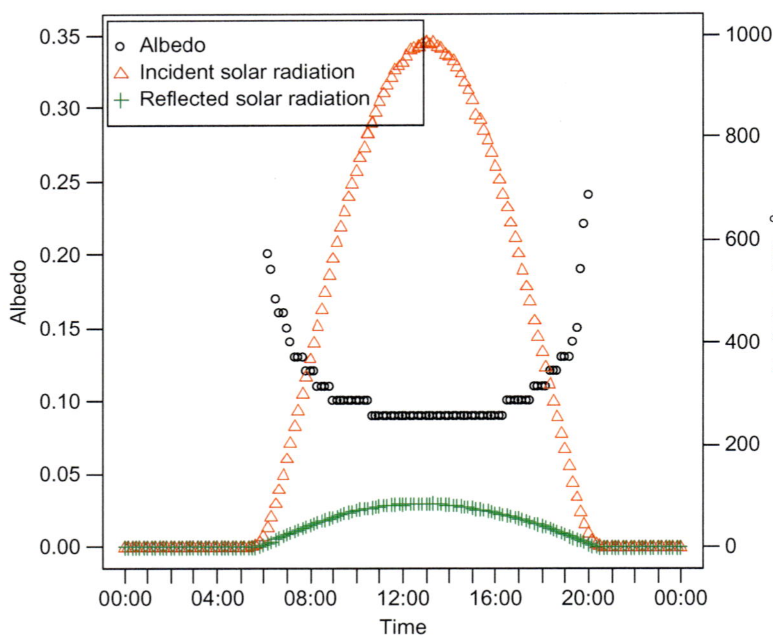

Figure 4.8 Diurnal variation of solar reflectivity in one day (asphalt (B2) pavement) (Li et al., 2013c).

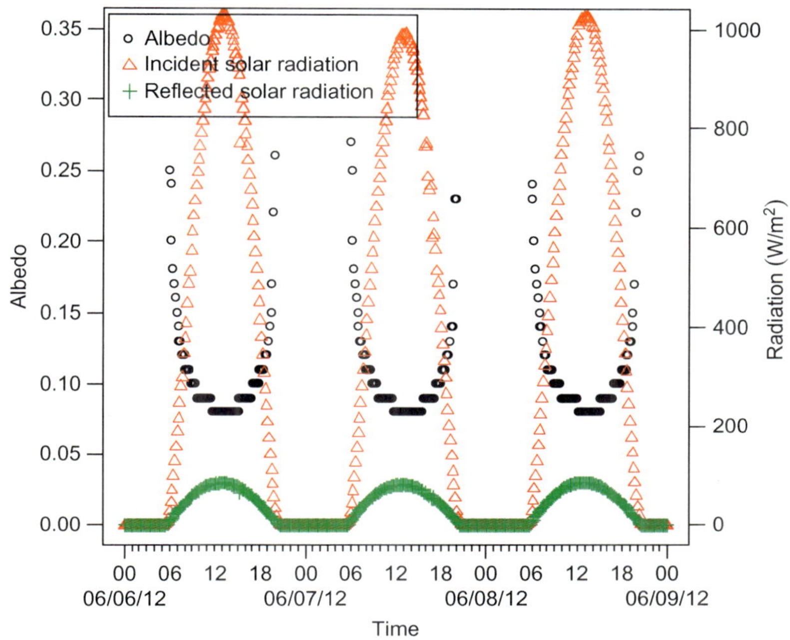

Figure 4.9 Diurnal variation of solar reflectivity in three days (B2) (Li et al., 2013c).

the middle of the day (9:00–15:00) and then low again in the late afternoon when there is a low incident angle of solar radiation as in the early morning. The diurnal variation of solar reflectivity of an asphalt (B2) pavement in three days is plotted in Figure 4.9. The same trend was observed for the concrete pavement. This implies that the albedo should be measured in the middle of the day (9:00–15:00) to obtain a constant and conservative value. Otherwise, if the albedo is measured in the early morning and late afternoon, the value obtained will tend to be larger than that measured in the middle of the day.

4.2.5.2 Seasonal variation of albedo

The solar reflectivity was continuously monitored over time in one year (from the fall of 2011 to the summer of 2012) on a permeable asphalt (B2) pavement. The albedo on three days of each season is plotted in Figure 4.10. As seen from Figure 4.10, there is no significant change of the measured midday albedo over season in a year, except a slightly higher albedo shown in winter (Figure 4.10b) when the angle of the sun is lower.

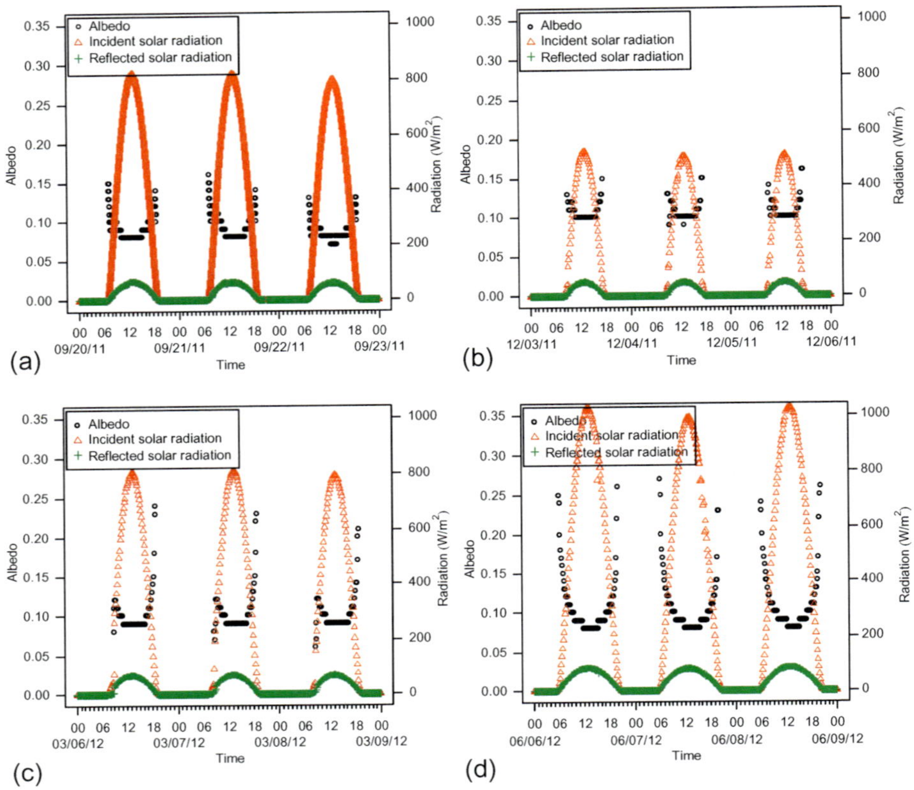

Figure 4.10 Seasonal variation of solar reflectivity (B2; from the fall of 2011 to the summer of 2012) (Li et al., 2013c). (a) Fall. (b) Winter. (c) Spring. (d) Summer.

To verify the impact of season on albedo, the albedos at different times of the day on one clear day for each month in one year are plotted against time in Figure 4.11. As found previously, the albedo measured in early morning and late afternoon is higher than that of midday. The midday albedo (11:00–15:00 in summer and 12:00–15:00 in winter) will be lower and constant over time (falling in one curve in Figure 4.11). As observed from Figure 4.10, it is noted from Figure 4.11 that the midday albedo does not significantly change over season, except slightly higher in wintertime.

4.2.6 Change of albedo over time

The color of the pavement surface tends to change over time due to weathering and traffic. Figure 4.12 shows the albedos of nine test sections at different times after construction. These nine test sections are constructed for testing and were not open to any type of traffic. As noted, the albedos of concrete pavements (C1–3, although albedos

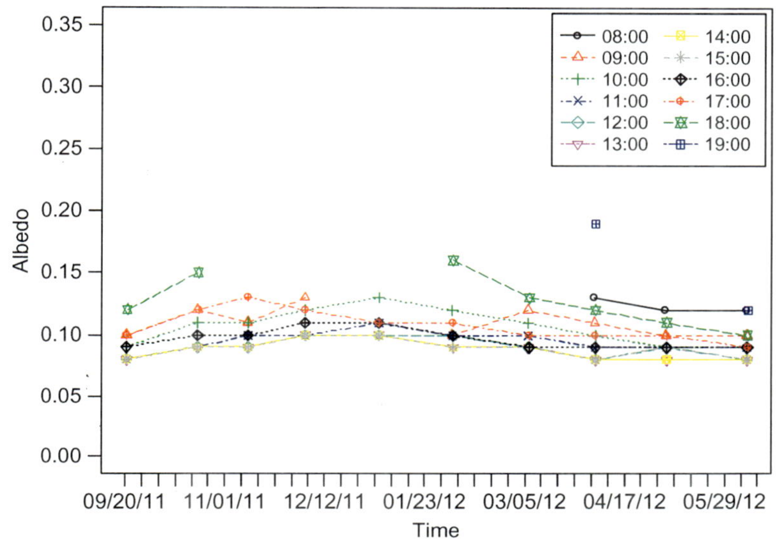

Figure 4.11 Seasonal variation of solar reflectivity at different times of a day (B2) (Li et al., 2013c). (Note: The data with very low solar radiation (<5 W/m^2) but a very high albedo in early morning and late afternoon were neglected.)

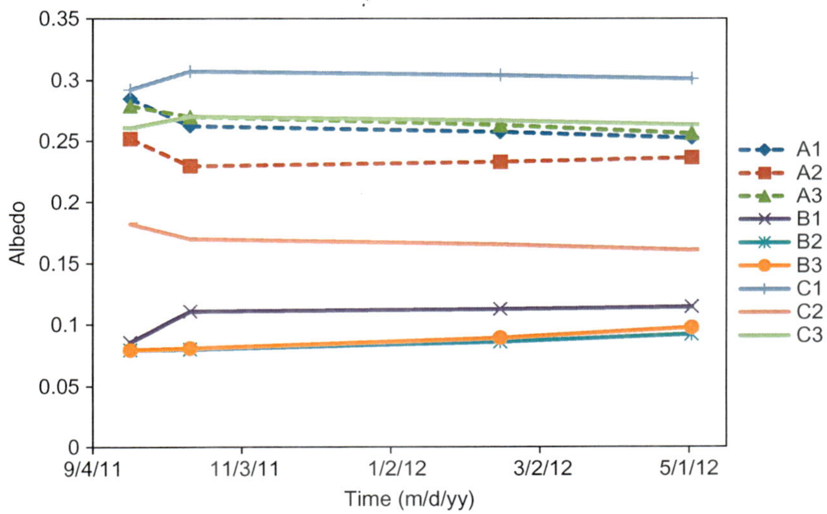

Figure 4.12 Change of albedo over time for materials in test sections with no traffic (Li et al., 2013c). (Nine test sections, only weathered: paver pavements A1–3, asphalt pavements B1–3, and concrete pavements C1–3.)

of C1 and C3 increase slightly in the first month) and interlocking concrete pavers (A1–3) generally tend to decrease over time; in contrast, the albedos of asphalt pavements (B1–3) increase slightly over time. The change in albedo mostly happened in the first month just after the construction due to weathering. Under both continued weathering and trafficking, the change of albedo is expected to be larger, and for concrete and especially asphalt, traffic will wear the binder (cement or asphalt) off of the surface aggregate, which will result in albedo being influenced by the reflectivity of the aggregate.

4.2.7 Summary of albedo of pavement materials

A new albedo measurement system using a dual-pyranometer and automatic data collection system was developed. It can be used to conveniently measure albedo in the field and perform long-term monitoring of albedo when connected to a data acquisition system.

Albedo was measured for commonly used paving materials including asphalt, concrete, and interlocking concrete paver surfacing materials, with different designs. These new data enhance basic knowledge on albedo values for pavements and the other land cover types measured (gravel, soil, and lawn), which can help reduce the uncertainty in understanding, evaluating, and modeling their thermal behavior and their consequences for human thermal comfort and building energy use.

The field albedo measurement of pavement materials is dependent on the solar angle. It is high in the early morning and in the late afternoon and is low and constant over time in the midday. This suggests that the field albedo should be measured in the midday of a clear day to obtain a stable and conservative value. No significant seasonal variation in albedo was found. Pavement albedo will change over time under weathering and trafficking, especially in the first month after construction. Cloud cover will negatively influence the value of albedo measured. No impact of wind speed or air temperature on albedo is observed.

The steady-state albedo values of different pavement materials are summarized in Table 4.4.

Table 4.4 **Summary of albedo of different pavement materials**

Material type	Albedo	
	Range	Average
Asphalt concrete (AC) or overlay	0.05–0.15	0.1
Portland cement concrete (PCC)	0.15–0.35	0.25
Chip seal	0.14–0.16	0.15
Slurry seal	0.07–0.1	0.08
Cape seal	0.05–0.15	0.06
Fog seal	0.04-0.07	0.06

4.3 Thermal properties of pavement materials

4.3.1 Measurement method for thermal properties

Thermal properties (i.e., thermal conductivity and heat capacity) are also important parameters that influence the temperature of building materials and thermal performance of built environment. These properties are required as fundamental inputs for modeling and simulating thermal behavior of built environment (Wong et al., 2010; Tan and Fwa, 1997; Tan and Fwa, 1992; Mahmoud, 2011; Kültür and Türkeri, 2012; Jo et al., 2010; Xu et al., 2012; Santamouris et al., 2012; Mrawra and Luca, 2002; Schindler et al., 2004; Hermansson, 2001; Li and Harvey, 2011; Hermansson, 2000; Gui et al., 2007). Most existing methods for measuring thermal properties of building materials are based on 1-D steady-state heat transfer theory (Carlson et al., 2010; Luca and Mrawira, 2005; Mrawira and Luca, 2006; Mrawra and Luca, 2002). The critical challenge for these methods has been the difficulty in achieving a 1-D heat flow condition for the testing specimen. It is really practically difficult or even impossible for these methods to meet thermally thin slab or thermally long cylinder criteria, epically for the common beam or cylinder specimens prepared in the laboratory or extracted from field in-service pavements. A multidimensional transient method has been developed to reduce the challenge and requirement on the testing specimen size and shape and make it possible to accurately measure all the thermal properties from one single test (Li et al., 2013b).

In the recent study (Li et al., 2013b), a multidimensional transient model and a practical tool were developed to simulate the transient temperature at any location on a beam or cylinder specimen of any size subject to convective heat transfer. Case studies verified that this model can be used to, if the thermal properties are known, simulate the transient temperature at any location for a specimen of various shapes and sizes and predict the time to reach a specified target temperature for mechanical and other testings. The study also developed and validated partly by case studies on both asphalt and concrete materials a procedure for back-calculating the thermal properties of a specimen of various shapes and sizes. Thermal properties of novel building materials (various initiative cool materials such as porous concrete and high thermal resistance materials) can be easily measured with that procedure.

Since it is difficult and even impossible to analytically solve this optimization problem, a numerical method of trail comparison was used instead to find the optimized solution and is implemented through *R* programming. During the numerical method of trail comparison for optimization, all the possible combinations of the two independent parameters α and h/k in the feasible ranges are selected with appropriate step lengths to run the temperature simulations and calculate the corresponding *RMSEs* of the predicted temperatures. The combination that gives a minimal *RMSE* is the optimized solution of the two parameters. The whole procedure of back-calculation of thermal properties based on the new objective function is listed in a flowchart in Figure 4.13. More details can be found in the reference (Li et al., 2013b).

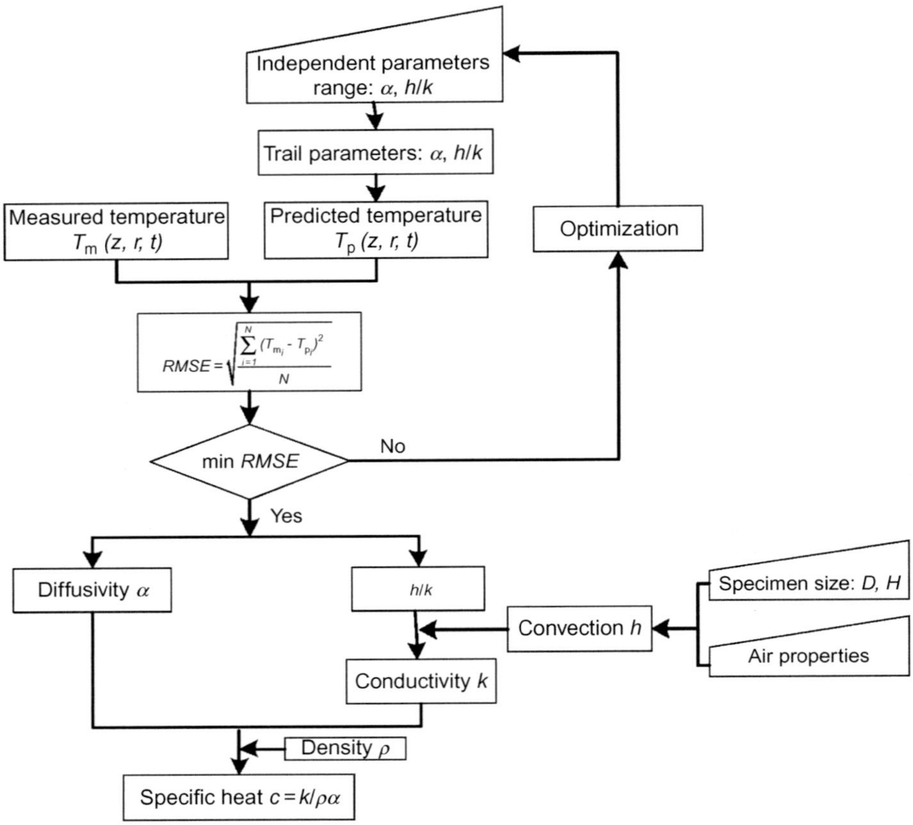

Figure 4.13 Flowcharts for back-calculation of thermal properties (Li et al., 2013b).

4.3.2 Measured thermal properties of pavement surface materials

The developed and validated procedure was used to measure the thermal properties of pavement materials. Cylinder specimens made of some of the materials (mainly concrete and asphalt, both dense-graded and open-graded) were tested at different locations (two specimen sizes for concrete: 4 × 6 in. and 6 × 12 in.). The test setup with example specimens used for testing is shown in Figure 4.14. The results are listed in Table 4.5.

From Table 4.5, it is noted that the dense-graded materials (concrete or asphalt) have a higher thermal conductivity and heat capacity than the open-graded materials. This might give a potential explanation on why the permeable pavement constructed with open-graded materials presents a higher daytime temperature and a lower nighttime temperature under the dry condition than the impermeable pavement constructed with dense-graded materials. The permeable asphalt pavement sections (B2 and B3)

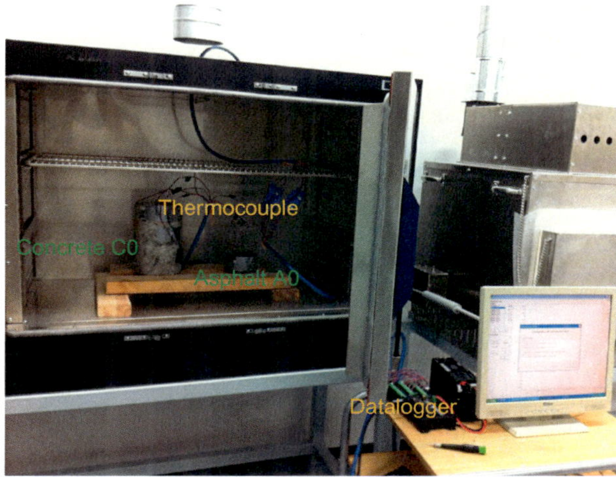

Figure 4.14 Test setup for measurement of thermal properties (Li et al., 2013b).

presented a higher daytime surface temperature than the impermeable pavement (B1) under the dry condition. This might be because the lower thermal conductivity increases the thermal resistance and thus the difficulty in conducting heat into in-depth pavement layers that keeps the heat around the surface. Moreover, because of its lower heat capacity than dense-graded materials, the open-graded materials will be heated up to a higher temperature under the same amount of energy absorbed from solar radiation or surroundings during daytime or hot periods. On the other side, just due to its lower heat capacity, the open-graded materials have less thermal energy (or heat) stored in the solid body around the surface to lose into the cold ambient during night-time or cold periods. Also, because of its lower thermal conductivity compared with the dense-graded materials, it is much more difficult for the open-graded materials to conduct heat to the surface from underlying layers to supply more energy for loss. These two aspects will significantly increase the possibility for the open-graded materials to produce a lower surface temperature when the same amount of heat is lost into ambient air during night-time or cold periods. This theoretically confirms that the permeable pavement composed of open-graded materials can be a potential strategy to counter the night-time heat island effect because of its lower surface temperature and less heat released into the ambient during night. However, as mentioned previously, attention should be given to its potential higher surface temperature during daytime under the dry condition.

In addition, it is noted that concrete materials generally show a slightly higher thermal conductivity and heat capacity than asphalt materials. This implies that when all the others factors and conditions (e.g., albedo, solar radiations, and convection) are identical, the concrete pavements still will give a lower daytime surface temperature compared with the asphalt pavements.

Table 4.5 Summary of average thermal properties of pavement surface materials

Section number[a]	Material	Density[b] (kg/m^3)	Thermal diffusivity[b] (m^2/s)	Thermal conductivity[b] (W/(m °C))	Heat capacity[b] (J/(kg °C))
C1	Dense-graded concrete	2257 (59)	8.12×10^{-7} (2.61×10^{-8})	1.83 (0.07)	1001 (58)
C2	Open-graded concrete	1980 (106)	7.69×10^{-7} (9.25×10^{-8})	1.38 (0.09)	912 (80)
C3	Open-graded concrete	2134 (65)	7.95×10^{-7} (3.74×10^{-8})	1.29 (0.06)	761 (69)
B1	Dense-graded asphalt	2399 (-)	8.46×10^{-7} (1.13×10^{-8})	1.73 (0.09)	852 (42)
B2	Open-graded asphalt	2269 (-)	7.15×10^{-7} (5.13×10^{-8})	1.24 (0.05)	763 (32)
B3	Open-graded asphalt	2239 (-)	7.11×10^{-7} (8.15×10^{-8})	1.23 (0.06)	798 (50)

-, Not available.
[a]See Table 4.2 for details.
[b]Standard deviation listed in parentheses.

4.3.3 Summary of thermal properties of pavement materials

This section is devoted to the discussion of questions related to the back-calculation of the thermal properties of pavement materials with specimens of various shapes and sizes from the measured transient temperatures profile of the specimen based on multidimensional modeling and simulation of transient temperatures at any location on a beam or cylinder specimen of various sizes subject to convective heat transfer:

(1) The multidimensional model and tool developed can accurately predict transient temperature at any location on a beam or cylinder specimen of various sizes subject to convective heat transfer. The case studies presented in this chapter verify that it can also be used, if the thermal properties are known, to simulate the transient temperature and predict the time it takes to reach a specified target testing temperature at any location for specimens of various shapes and sizes, when the specimen is preheated or precooled in the forced convection oven or temperature chamber for mechanical or other temperature-related testing.

(2) Based on the developed multidimensional temperature simulation model, the procedure and tool for back-calculating thermal properties were developed and validated in part by case studies on both asphalt and concrete materials. The partly validated procedure and tool can be used to obtain the thermal properties of a specimen of pavement and building materials from their measured transient temperature profiles, regardless of the shape and size of the specimen. With the fully validated procedure and tool, the thermal properties of novel building materials (various innovative cool materials such as porous concrete and high thermal resistance materials) can be easily measured and then used for evaluating and modeling the thermal performance of built environment composed of these materials.

(3) The developed and validated procedure was employed to measure the thermal properties of pavement materials using cylinder specimens, particularly some surface materials used for the experimental sections in this study. It was found that the dense-graded (i.e., nonporous) materials (concrete or asphalt) have a higher thermal conductivity and heat capacity than the open-graded (i.e., porous) materials. Concrete materials generally show a slightly higher thermal conductivity and heat capacity than asphalt materials.

(4) The open-graded (i.e., porous) materials (e.g., permeable pavement) could present a higher daytime surface temperature than the dense-graded (i.e., nonporous) materials (e.g., impermeable pavement) under the dry condition. This might be because the lower thermal conductivity increases the thermal resistance and thus the difficulty in conducting heat into in-depth pavement layers, which keeps the heat around the surface. Moreover, because of its lower heat capacity than dense-graded materials, the open-graded materials will be heated up to a higher temperature under the same amount of energy absorbed from solar radiation or surroundings during daytime or hot periods. On the other side, just due to its lower heat capacity, the open-graded materials have less thermal energy (or heat) stored in the solid body around the surface to lose into the cold ambient during night-time or cold periods. Also, because of its lower thermal conductivity compared with the dense-graded materials, it is much more difficult for the open-graded materials to conduct heat to the surface from underlying layers to supply more energy for loss. These two aspects will significantly increase the possibility for the open-graded materials to produce a lower surface temperature when the same amount of heat is lost into ambient air during night-time or cold periods. This theoretically confirms that the permeable pavement composed of open-graded materials can be a potential strategy to counter the night-time heat island effect because of its lower surface temperature and less heat released into the ambient during night. However, as mentioned previously, attention should be given to its potential higher surface temperature during daytime under the dry condition.

4.4 Surface temperature of pavement materials

4.4.1 Diurnal variation of surface temperatures of different pavements

The highest and lowest pavement temperatures usually happen at the pavement surface. The critical pavement surface temperatures have a great influence on the near-surface air temperature and consequently affect the human thermal comfort and air quality. To examine the diurnal variation of pavement surface temperature in different seasons, the surface temperatures of three different pavements (concrete (C1), paver (A1), and asphalt (B1)) in one clear sunny day of each season are plotted in Figure 4.15. The weather data (ambient air temperature, wind speed, and solar radiation), which were measured from a nearby weather station at around 2 m high from the ground, are also presented in Figure 4.15 for reference.

It is clearly shown that there are differences in temperature for different pavements, especially around noon with high solar radiation intensity. The black asphalt (B1)

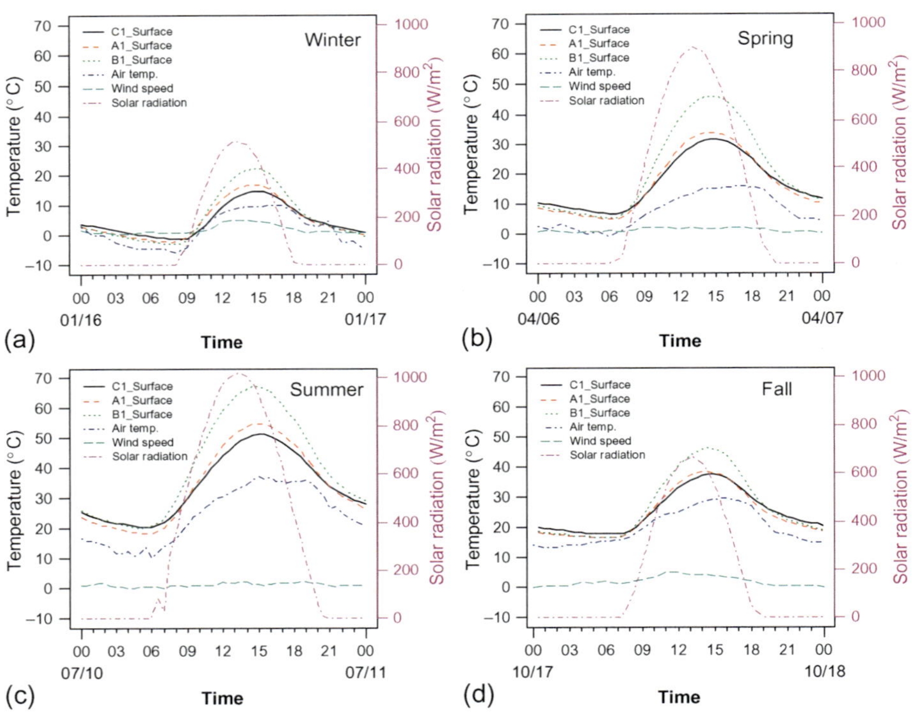

Figure 4.15 Diurnal variation of surface temperatures and weather data in one day of each season. (Weather data (air temperature, wind speed, and solar radiation) were measured from a nearby weather station at around 2 m high from the ground.) (Wind speed shown in the left y axis in m/s.)

pavement has higher surface temperature than concrete (C1) and paver (A1) pavements with light color. The temperature of paver pavement is slightly higher than that of concrete. The surface temperature of the asphalt (B1) pavement reaches up to almost 70 °C (158 °F) in summer (Figure 4.15c), compared with 50 °C (122 °F) of the concrete (C1) pavement. The differences of peak temperatures between asphalt (B1) and concrete (C1) pavements are about 10–20 °C, depending on the weather conditions and seasons. During night-time, the surface temperatures of the three pavements are very close to each other, but still higher than the ambient air temperature by 2–10 °C. The difference in surface temperatures between asphalt and concrete pavements is determined mainly by the color of pavement surfaces (i.e., the solar reflectivity or albedo). The higher the albedo of the pavement, the more the solar radiation is reflected and the less is absorbed by the surface, which will produce a lower surface temperature. The asphalt pavement has a darker color and consequently a lower albedo and a higher surface temperature, compared with the concrete pavement. The difference of surface temperatures between them is higher (~20 vs. ~10 °C) during summer with a high solar radiation (peak intensity of ~1000 W/m^2 in Figure 4.15c) than in winter with a low solar radiation (peak intensity of ~500 W/m^2 in Figure 4.15a). This implies that increasing the albedo is an effective strategy to reduce the surface temperature, especially for the climates and seasons with high solar radiation.

4.4.2 Seasonal variation of surface temperatures of different pavements

To comprehensively evaluate the thermal performance of different pavements, it is of great significance to examine the seasonal change of surface temperatures in one year. The daily maximal and minimal surface temperatures of concrete (C1), paver (A1), and asphalt (B1) pavements over one year are extracted and plotted in Figure 4.16, along with daily maximal and minimal ambient air temperatures for reference. As observed previously, the changes of daily maximal and minimal pavement surface temperatures follow the similar pattern as the ambient air temperature. The pavement surface temperatures are generally higher than the ambient air temperature over the year round, even for the daily minimal temperature during winter (Figure 4.16b). The daily maximal surface temperature of asphalt (B1) pavement is higher than the concrete (C1) and paver (A1) pavements and the ambient air temperature. The temperature differences between them are higher during summer than in winter due to the difference in solar radiation of different seasons. During summer, the daily maximal surface temperature of asphalt (B1) pavement is around 20 °C higher than the concrete (C1) and around 16 °C higher than paver (A1) pavements and about 30 °C higher than the ambient air temperature. During winter, the daily maximal surface temperature of asphalt (B1) pavement is around 10 °C higher than the concrete (C1) and around 8 °C higher than paver (A1) pavements and about 15 °C higher than the ambient air temperature. There are no significant differences in the night-time daily minimal surface

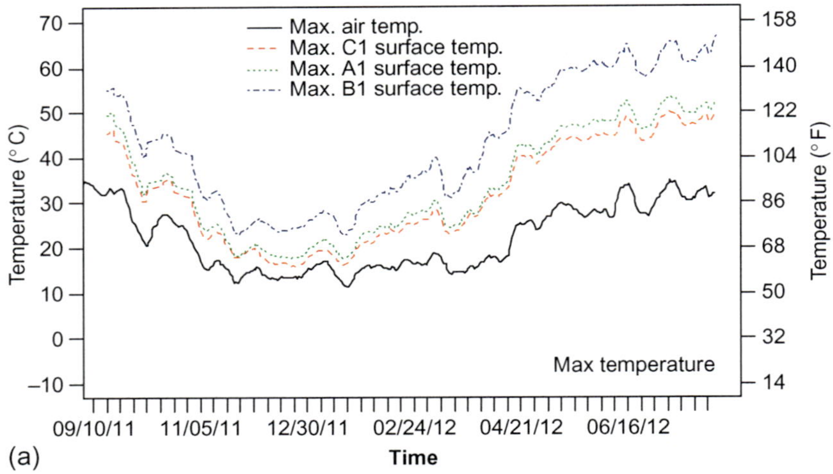

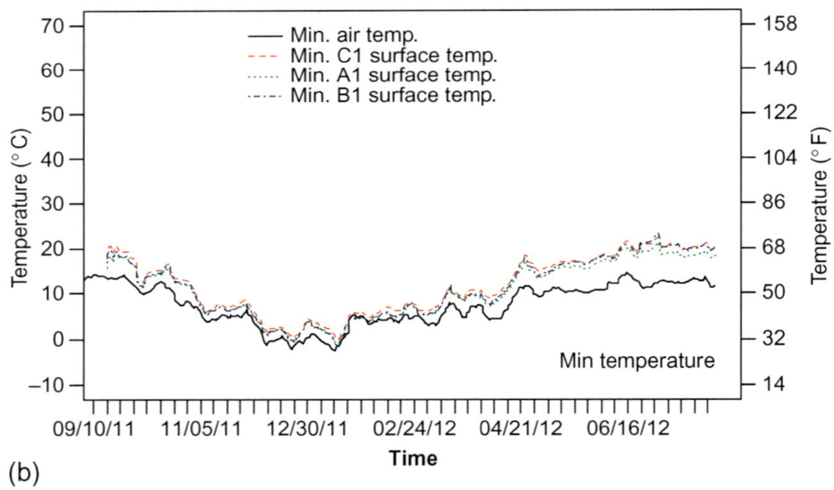

Figure 4.16 Daily maximum and minimum air temperatures and pavement surface temperatures over one year. Concrete, C1; paver, A1; asphalt, B1. Date format: mm/dd/yy.

temperatures between the three pavements for both summer and winter, which are about 10 °C and 5 °C higher than the ambient air temperature for summer and winter, respectively.

4.4.3 Thermal images of experimental pavement sections: comparison of dry and wet conditions

Thermal images were taken during the irrigation experiment at different times for July 9, 2012, July 10, 2012, and July 11, 2012 for all the experimental sections along with the optical images. The water tables were monitored using small monitoring wells for the six permeable experimental sections after irrigation. The weather data were also monitored using a nearby weather station.

From 3:00 pm on July 10, 2012, water was irrigated into the permeable test sections. The irrigation was continued until each permeable pavement section was filled up with water at around 1:00 am on July 11, 2012. After irrigation stopped, the water tables would drop over time due to the nature processes of infiltration and evaporation. The water tables were monitored using small wells for each permeable section and are presented in Figure 4.17.

The weather data during the experiment period were also monitored and are shown in Figure 4.18. It is noted that the air temperature was increasing in the three days. The solar radiation was keeping constant for the period. The wind speed was increasing slightly and the relative humidity was decreasing slightly.

The thermal images on July 10, 2012, are presented in Figure 4.19, along with the optical images. When water was being applied into the permeable pavements (Figure 4.19), the temperatures of the pavement portions with water were as low as ~30 °C, which were much lower than the pavement portion without water. This implies that watering can be an effective way to lower down the pavement temperature in summer.

A comparison of thermal images of six permeable pavements under dry, watering, and wet conditions is presented in Figure 4.20. As observed previously, under the watering condition (July 10, 2012), the pavements had much lower temperatures than under the dry condition (July 9, 2012). Even in the third day (July 11, 2012), without watering but with higher air temperature (see Figure 4.18), the pavements still showed lower temperatures at 4 pm (25 h after watering) by 2–7 °C compared with those in the second day, without considering the higher air temperature on the third day

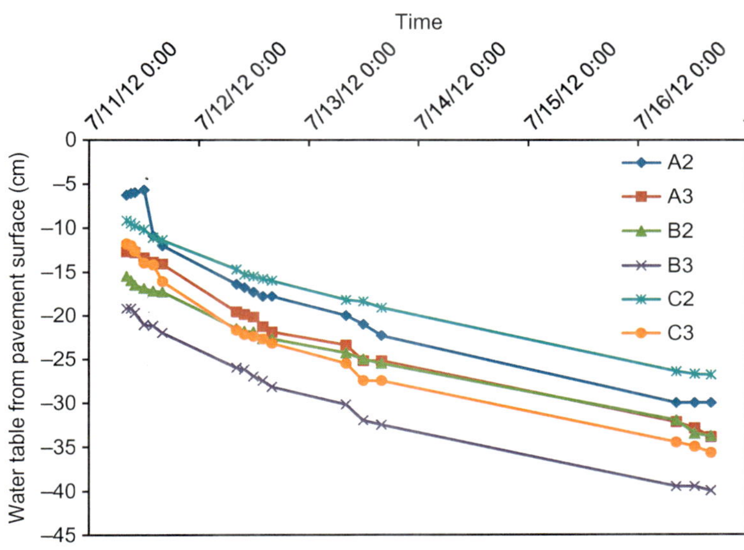

Figure 4.17 Water tables in the monitoring wells for six permeable sections after irrigation.

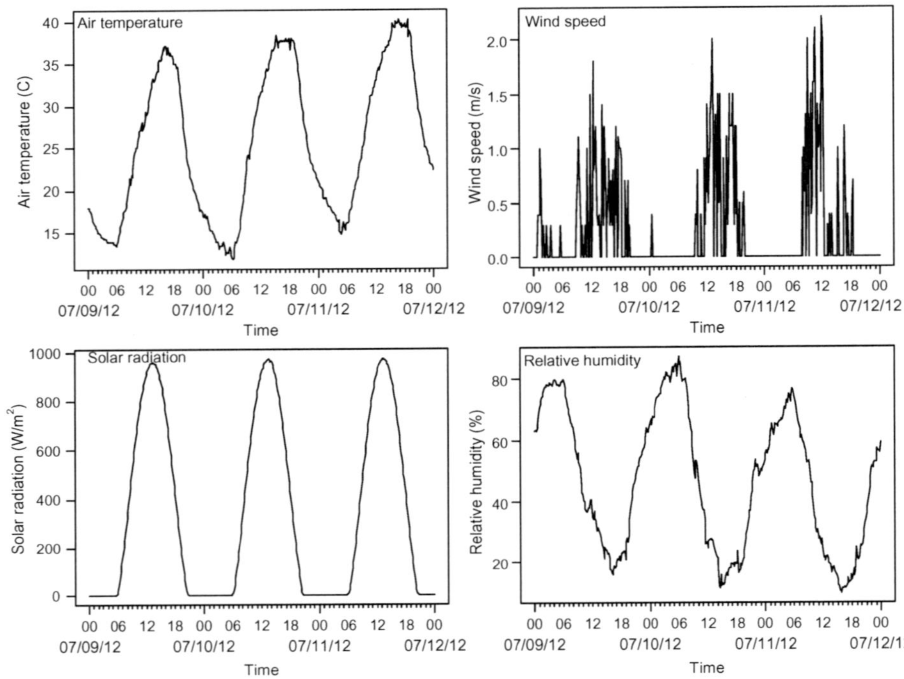

Figure 4.18 Weather data during the experiment period (no rain).

(July 11, 2012; see Figure 4.18) due to evaporative cooling of some moisture existing in the pavements. This implies that watering can effectively lower the pavement surface temperatures using cool water, and evaporation of some moisture existing in pavements also can help produce a low pavement temperature. The specific cooling effect depends on the evaporation rate on the pavement.

4.4.4 Effect of albedo on pavement temperature

As mentioned previously, albedo plays an important role in the pavement temperature. To examine the influence of the albedo on the pavement temperatures, the high (T_{max} at 3 pm) and low (T_{min} at 6 am) temperatures of the nine experimental pavements with different albedos were measured from the fall of 2011 to the summer of 2012 at Davis, California. The temperatures for one day of summer (July 1, 2012) and winter (January 15, 2012) are selected and plotted against their albedos (Figure 4.21). Figure 4.21 illustrates that the albedo has a large influence on the high temperature in both summer and winter. Under the same weather condition in summer, the high temperature of pavement with low albedo (0.08 for the asphalt) is about 15 °C (65 vs. 50 °C) higher than that of pavement with high albedo (0.28 for concrete). In winter, this high temperature difference is ~8 °C (26 vs. 18 °C). An increase in albedo of 0.1 can reduce

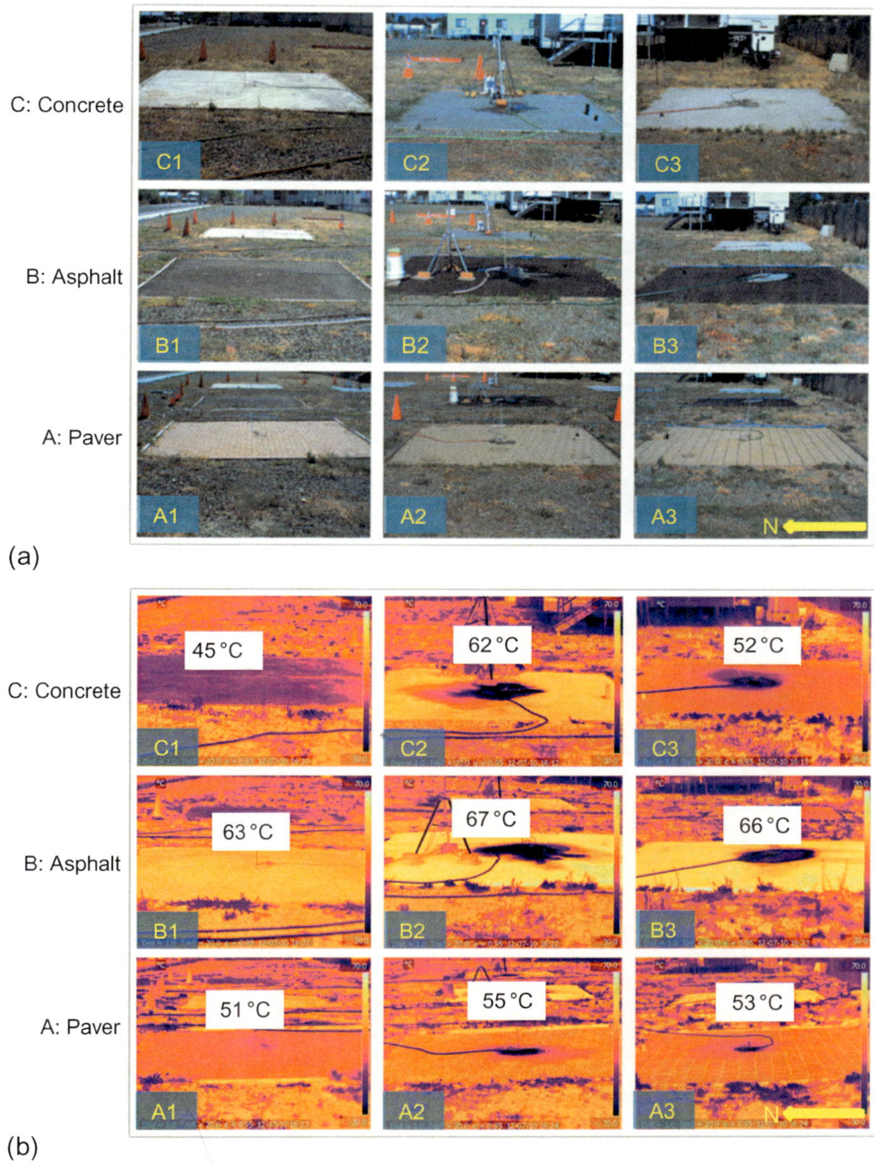

Figure 4.19 (a) Optical images of experimental sections on July 10, 2012. (b) Thermal images of different pavements on July 10, 2012 during watering (16:00) (lighter is hotter).

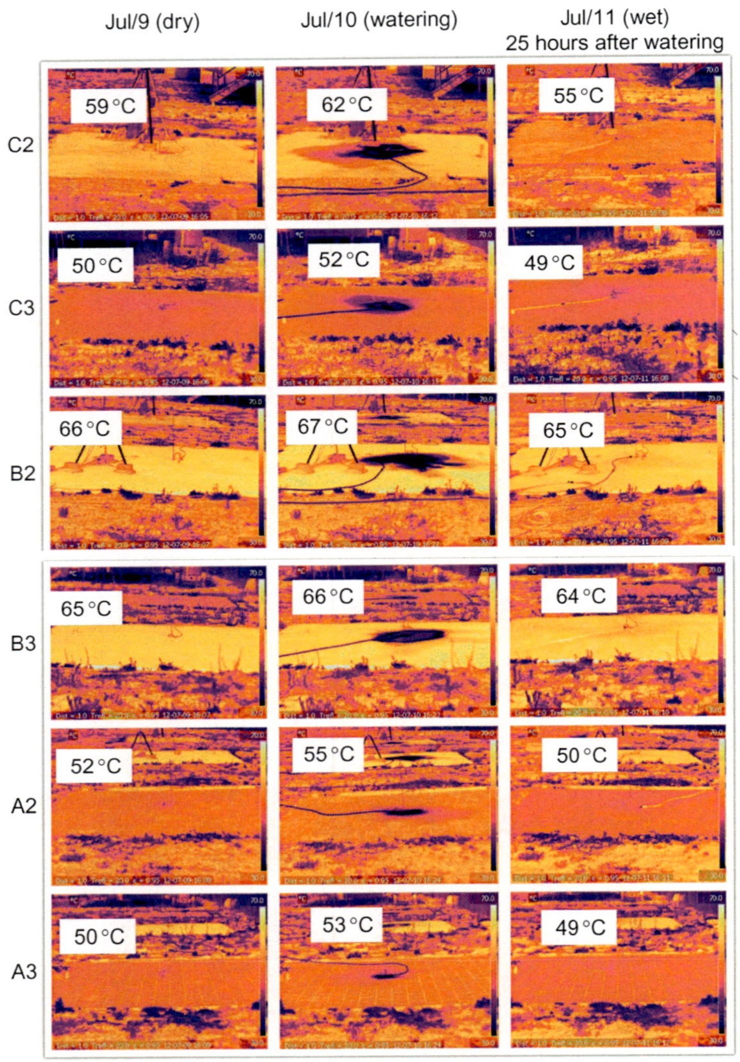

Figure 4.20 Comparison of thermal images of surface temperature of permeable pavements under different conditions (16:00 on July 9–11). Lighter is hotter.

the high surface temperature in summer by ~6 °C (i.e., −6 °C/+0.1 albedo, which is approximately the slope of the fitting line for T_{max}_Summer in Figure 4.21) and by ~3 °C in winter (i.e., −3 °C/+0.1 albedo, which is approximately the slope of the fitting line for T_{max}_Winter in Figure 4.21).

This reveals that albedo has different effects on pavement temperature in summer and winter with different peak solar radiations. The peak solar radiation intensity at around 1:00 pm is ~1000 W/m^2 in summer in Davis, California (see Figure 4.22). It is

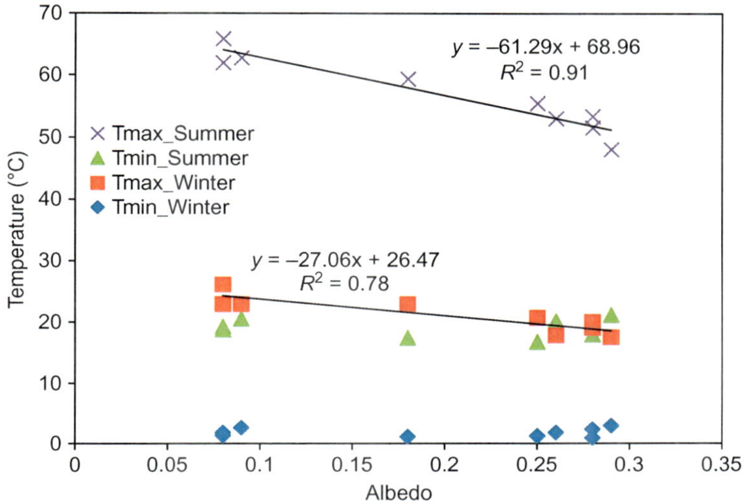

Figure 4.21 Influence of solar reflectivity on pavement surface temperature (Li et al., 2013c). (Summer on July 1, 2012, with daytime peak solar radiation ~1000 W/m^2 and winter on January 15, 2012, with daytime peak solar radiation ~500 W/m^2.)

approximately halved to 500 W/m^2 in winter (see Figure 4.22). The temperature effect of albedo in winter is also approximately half of that in summer (−6 °C/+0.1 albedo in summer and −3 °C/+0.1 albedo in winter). It implies that solar radiation positively influences the cooling effect of increased albedo (CE_{albedo}), and the cooling effect of increased albedo in hot climates with high solar radiation is larger than that in climates with low solar radiation. This also can be seen from the insignificant effect of albedo (~0 °C/+0.1 albedo) on the low temperature during night-time with no solar radiation for both summer and winter.

To further illustrate the influence of solar radiation on the cooling effect of an increased albedo, the cooling effects of increased albedo on high temperatures in each month of a year were calculated from measured temperature data (obtained from the slopes of the fitting lines between surface temperature and surface albedo for each month as shown in Figure 4.21). These calculated cooling effects of increased albedo on high temperatures in each month of a year are presented in Figure 4.22a along with the peak solar radiation intensity in each month. It reveals that the cooling effect of increased albedo does change over seasons and thus changes with the peak solar radiation intensity. The cooling effect is higher in summer with high peak solar radiation intensity and lower during winter with low peak solar radiation. The correlation between cooling effect of increased albedo and solar radiation is shown in Figure 4.22b and with a best-fit linear relation in Equation (1). An increase in peak solar radiation intensity by 100 W/m^2 will produce an increase in cooling effect on surface temperature of albedo of 0.6 °C/0.1 change in albedo:

$$CE_{albedo} = 6q \tag{5.1}$$

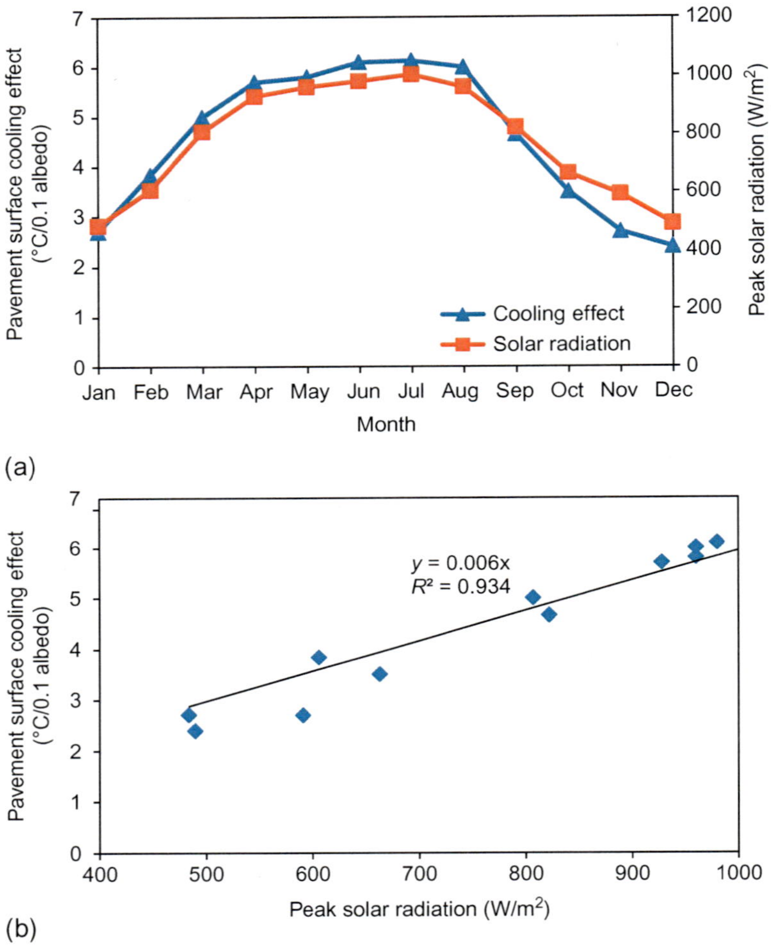

Figure 4.22 Influence of solar radiation on cooling effect of increased albedo (Li et al., 2013c). (a) Cooling effect of increased albedo and peak solar radiation intensity in different months. (b) Correlation between cooling effect of increased albedo and solar radiation.

where q is the peak global solar radiation flux hitting the pavement (kW/m^2) and CE_{albedo} is the cooling effect of albedo increased by 0.1 on pavement temperature (°C/0.1 albedo). This empirical relationship can be used to roughly estimate the cooling effect of increased albedo or heating effect of reduced albedo on pavement for various climates and seasons with different solar radiation.

An increase in albedo indeed can reduce the pavement surface temperature and thus might help mitigate the surface, near-surface, and atmospheric heat island effects and potentially improve outdoor thermal comfort and reduce building cooling energy use during hot periods, depending on the characteristics of the location and the overall urban area. While the urban heat island effect is a large concern for many cities in hot climates, and of course is a great concern in the summer, winter effects of pavement albedo and temperature are also important to consider. The reduced temperature during winter might increase the outdoor thermal discomfort and building energy use for heating for some climates and locations. Building energy use is also highly dependent on whether air conditioners are widely used. There is a chance that this penalty (i.e., colder pavements) in winter might outweigh the benefits gained in summer in some regions. This study looked at pavement surface temperature with focus on the effects of solar radiation and albedo. It must be noted that near-surface temperature is also influenced by wind speed, with higher wind speeds reducing near-surface temperatures. However, compared to solar radiation, the effect of wind speed would be much smaller. The effect of wind speed was assumed identical for each case and not explicitly included in the simple model of cooling effect proposed in this study. For developing a more comprehensive model, however, the wind speed can be included and thermal proprieties of pavement materials (i.e., thermal conductivity and heat capacity; see reference for the measurement method).

One additional factor that was not measured as part of this research is the increased reflected solar radiation caused by high albedo, which might hit and be absorbed by surrounding people or building/vehicle surfaces. This could increase the human thermal discomfort and building cooling energy use during hot periods. Therefore, attention should be given to the complete assessment of both the benefit and penalty in one entire year to ensure a positive net benefit will be obtained. The optimal strategy for changing albedo in a hot climate might be to increase the albedo during summer and reduce or not change it in winter for certain places. This intention is to maximize the benefit in summer and minimize the penalty in winter.

4.4.5 Summary of surface temperature of pavement materials

There are differences in temperature for different pavements, especially around noon with high solar radiation intensity. Increasing the albedo is an effective strategy to reduce the surface temperature, especially for the climates and seasons with high solar radiation. The changes of daily maximal and minimal pavement surface temperatures follow the similar pattern as the ambient air temperature. The pavement surface temperatures are generally higher than the ambient air temperature over the year round, even for the daily minimal temperature during winter. Watering can effectively lower the pavement surface temperatures using cool water in summer, and evaporation of some moisture existing in pavements also can help produce a low pavement temperature. The specific cooling effect depends on the evaporation rate on the pavement.

Albedo has great influence on the pavement surface high temperatures in the daytime and no significant impact on pavement low temperatures in night-time. An empirical relationship between the cooling effect of increased albedo on a pavement high temperature and solar radiation was developed. The cooling effect has a positive correlation with the peak solar radiation intensity in daytime. This simple correlation can help estimate the cooling effect of increased albedo or heating effect of reduced albedo on pavement for various climates and seasons with different solar radiation.

4.5 Near-surface air temperature above pavement

Pavement surfaces could perform heat exchange with the near-surface air through reflected shortwave solar radiation, emitted longwave radiation, and pavement surface convection. These processes of thermal interaction will influence the temperature profile of the near-surface air and consequent human thermal comfort and air quality (e.g., ground-level ozone). Therefore, it is of great importance to investigate the thermal interaction between pavement and near-surface air.

This section investigates the thermal interaction between pavement and near-surface air for different pavement types, measures the temperature profiles of near-surface air, and explores the factors affecting the profiles.

4.5.1 Pavement sections for measurement

Four pavement sections were chosen to conduct the experiments for measuring the temperature profiles of near-surface air above the pavement surfaces. Two of them are asphalt pavements and the other two are concrete pavements. They are of different shapes and sizes. The pavement sections are summarized in Table 4.6, along with the albedos (solar reflectivities) measured by an albedometer.

4.5.2 Measurement method and equipment

Type T thermocouple wires were used to measure the temperatures on pavement surfaces and of near-surface air at different heights above the pavement surfaces. Thermocouple sensor locations are 0, 2, 5, 10, 20, and 40 in height from pavement surfaces (Figure 4.23). The data were recorded at the time interval of 10 min using a CR1000 datalogger. Local weather data were also monitored using an on-site portable weather station.

Table 4.6 **Summary of pavement sections for measurement**

Section	Type	Size	Albedo	Measurement date
PA1	Asphalt	~30 m diameter	0.06	July 2, 2012–July 17, 2012
PC1	Concrete	15 m wide × 45 m long	0.26	July 18, 2012–July 27, 2012
PA2 (B1)[a]	Asphalt	4 m wide × 4 m long	0.08	July 28, 2012–August 20, 2012
PC2 (C1)[a]	Concrete	4 m wide × 4 m long	0.28	July 28, 2012–August 20, 2012

[a]PA2 and PC2 are the experimental test sections B1 and C1 as described previously, respectively.

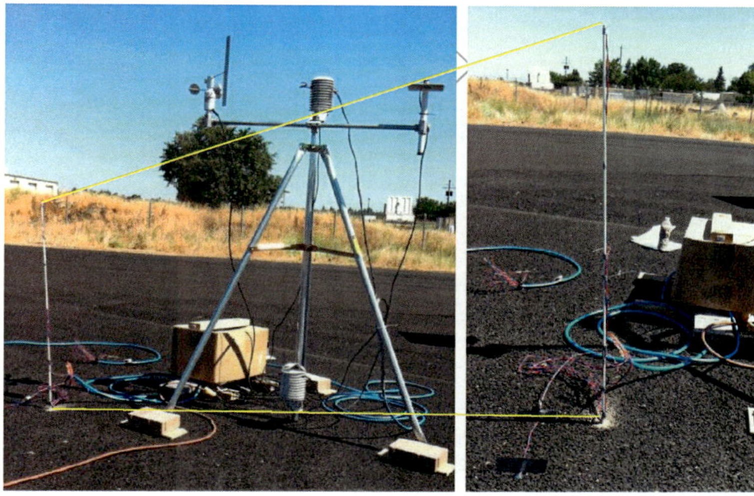

Figure 4.23 Example experimental setup for temperature profile measurement.

4.5.3 Correction for near-surface air temperature measured by thermocouple sensors

To obtain the temperature profile of near-surface air, conveniently conducting multiple measurements of near-surface air temperatures at different heights above ground is critical. Using thermocouple wire sensors is a simple, convenient, and cost-effective choice in performing these multiple measurements, compared to the air temperature sensors with radiation shield used in common weather stations.

However, unlike the air temperature sensors with radiation shield, the thermocouple sensors open in the near-surface air would be influenced by the environmental conditions surrounding the sensors, such as mainly solar radiation and possibly wind speed and ambient air temperature. The near-surface air temperature measured from the thermocouple sensors is not the real air temperature, but representing a combined effect of these environmental factors (several efforts were tried to create a radiation shield small enough for the soft thermocouple wires using light materials such as aluminum foil; however, no good effect was obtained). Usually, a higher (by 1–4 °C) daytime air temperature would be given by the open thermocouple sensors due to the heating effect of solar radiation. Therefore, the near-surface air temperature measured from the thermocouple sensors must be corrected for these environmental factors to reveal the real air temperature, in particular for the daytime air temperature.

To achieve this goal, separate experiments were performed to compare the air temperature measured with an open thermocouple sensor with that measured from an air temperature sensor with radiation shield at the same location and height. A formula for the correction factor was developed against such simultaneous environmental

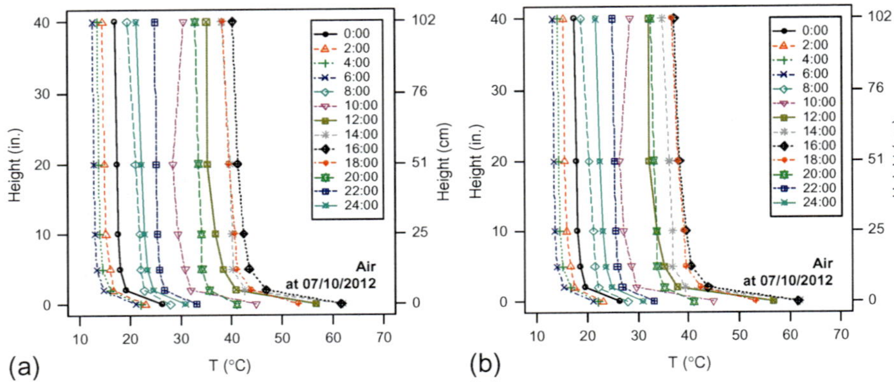

Figure 4.24 Example spatial profile of near-surface air temperature before and after correction (PA1). (a) Before temperature correction. (b) After temperature correction.

conditions as solar radiation, wind speed, and ambient air temperature obtained from a nearby weather station. The correlation for the correction factor is

$$\mathrm{CF}_T = (-1.152) + 0.003 \times \mathrm{SR} + (-0.105 \times \mathrm{WS}) + 0.042 \times T_\mathrm{a} \tag{5.2}$$

where CF_T is the correction factor of near-surface air temperature measured from an open thermocouple sensor, in °C; SR is the solar radiation measured at ~2 m height, in W/m^2, which is the most significant factor for this correction; WS is the wind speed measured at ~2 m height, in m/s; and T_a is the ambient air temperature measured at ~2 m height, in °C. The raw near-surface air temperature measured T_raw from the open thermocouple sensor must be minus the correction factor CF_T to obtain the corrected real values T_cor, that is,

$$T_\mathrm{cor} = T_\mathrm{raw} - \mathrm{CF}_T \tag{5.3}$$

An example spatial profile of near-surface air temperature before and after correction is presented in Figure 4.24. In the clear sunny day with peak solar radiation intensity of ~1000 W/m^2 (July 10, 2012), the daytime temperatures, specifically the peak values at around 16:00, are reduced by ~3 °C after correction. However, the night-time temperatures do not change much due to no solar radiation during night.

4.5.4 Example results on asphalt pavement PA1 and concrete pavement PC1

The near-surface air temperatures on asphalt pavement PA1 and concrete pavement PC1 were monitored at the different periods from July 2, 2012 to July 17, 2012 and from July 18, 2012 to July 27, 2012, respectively. The weather conditions do not change much during this summer period, especially for the sunny days. This makes it comparable to examine the effects of asphalt and concrete pavement on near-surface air temperatures for some clear days. Some results for PA1 and PC1 are presented in Figures 4.25 and 4.26, as well as the example weather conditions during the experiment periods (Figure 4.27).

It is noted that the temporal profiles of near-surface air have a pattern similar to that of ambient air temperature for both asphalt and concrete pavements. For the spatial profiles, the near-surface air temperatures gradually decrease as the distance from pavement surface increases, with greater slope (change rate) in the first 25 cm above the pavement surface. In the range close to the pavement surface, the near-surface air temperatures are much higher than the ambient air temperature (at ~2 m height),

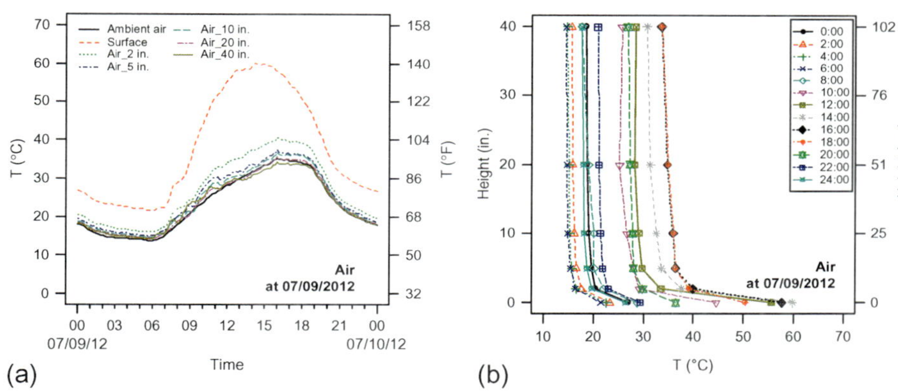

Figure 4.25 Example profiles of near-surface air temperatures on asphalt pavement PA1. (Ambient air is ambient air temperature from a sensor on portable weather station at 67 in. (1.7 m). Surface and Air_x in. are air temperatures at 0 in. and x in. height from pavement surface, respectively.) (a) Temporal profile. (b) Spatial profile.

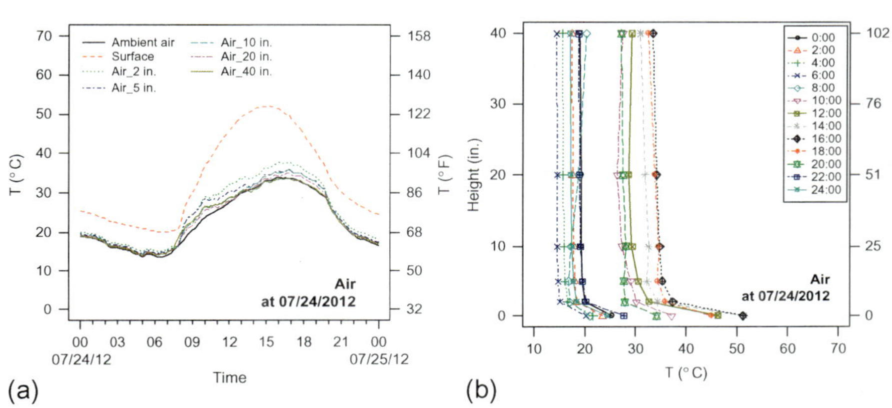

Figure 4.26 Example profiles of near-surface air temperatures on concrete pavement PC1. (a) Temporal profile. (b) Spatial profile.

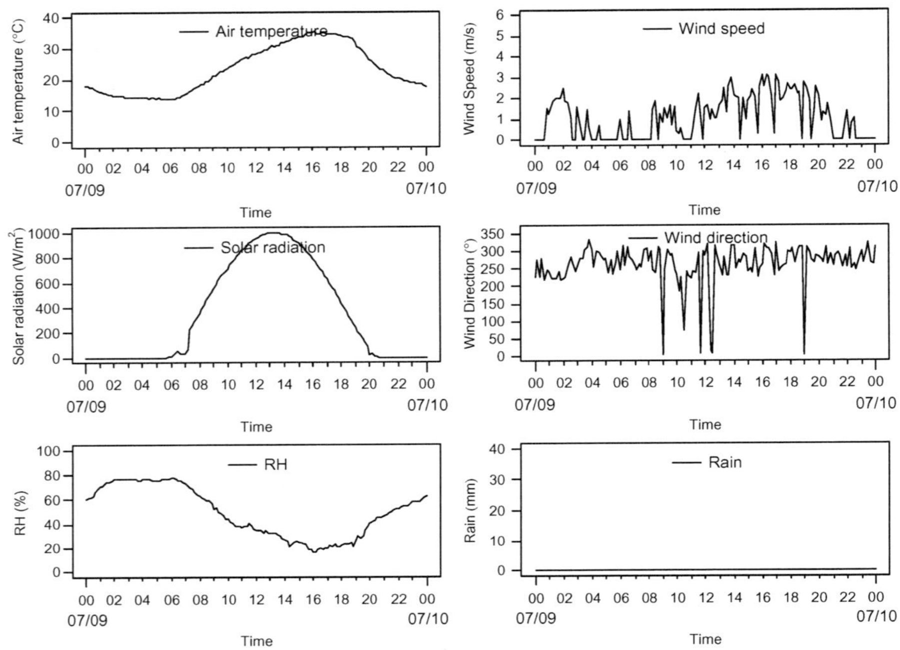

Figure 4.27 Example weather conditions during the experiment period for PA1.

especially for the black asphalt pavement with higher surface temperature compared to high-reflectance concrete pavement. The high near-surface air temperatures would reduce the human thermal comfort, and they are more critical for the babies and children who are shorter and closer to the surface. In addition, the formation of ground-level ozone would be facilitated by these high near-surface air temperatures when smog (VOC and NO_x) is exhausted from vehicle exhaust pipes that are close to the pavement surface on roads or parking lots.

4.5.5 Temporal and spatial variation of near-surface air temperatures on asphalt pavement PA2 and concrete pavement PC2

The near-surface air temperatures on asphalt pavement PA2 and concrete pavement PC2 were monitored at the same time from July 28, 2012 to August 20, 2012. This makes it easy to directly compare the effects of asphalt and concrete pavements on near-surface air temperatures. Some results are presented in Figures 4.29 and 4.30, as well as the example weather conditions during the experiment period (Figure 4.28).

Besides the observations similar to those for PA1 and PC1 presented above, the daytime near-surface air temperatures on asphalt (B1 in Figure 4.29) pavement are much higher than those of concrete (C1 in Figure 4.29) pavements, although they

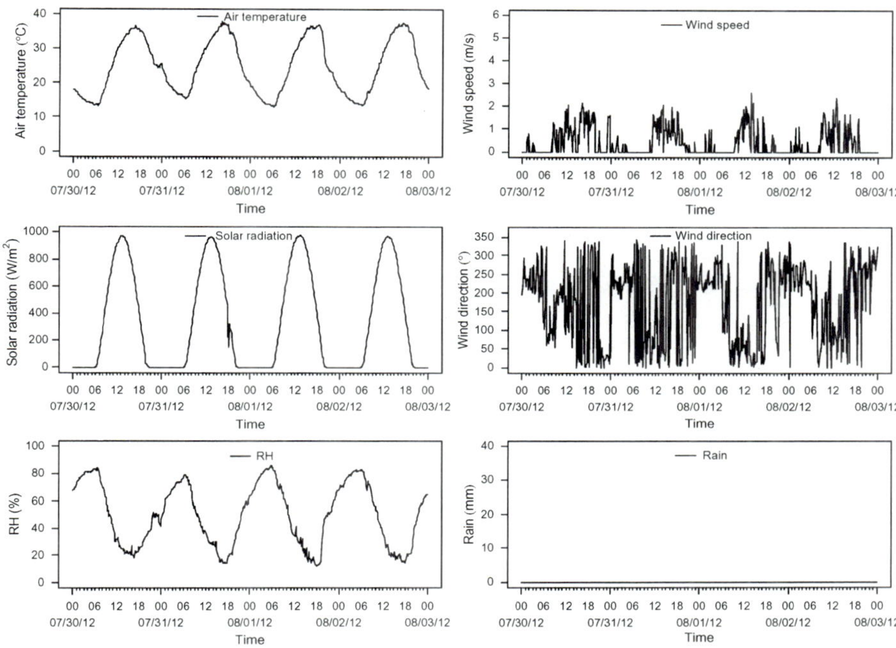

Figure 4.28 Example weather conditions during the experiment period for PA2 and PC2.

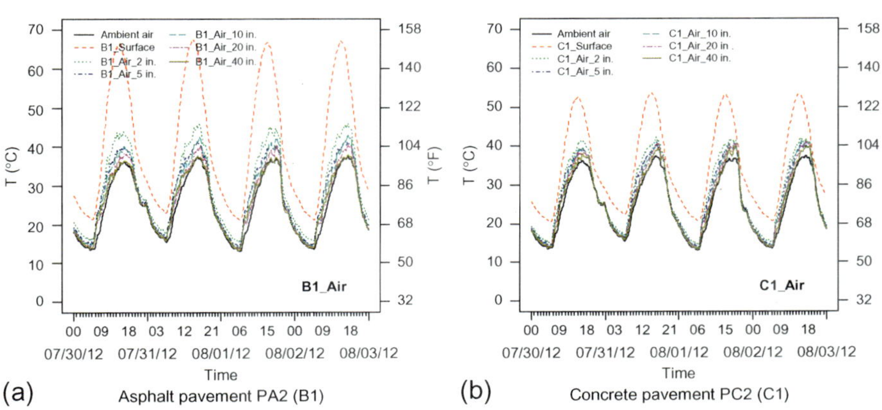

Figure 4.29 Example temporal profiles of near-surface air temperature on asphalt pavement PA2 (a) and concrete pavement PC2 (b). (Ambient air is ambient air temperature from a sensor on portable weather station at 67 in. (1.7 m). Surface and Air_*x* in. are air temperatures at 0 in. and *x* in. height from pavement surface, respectively.)

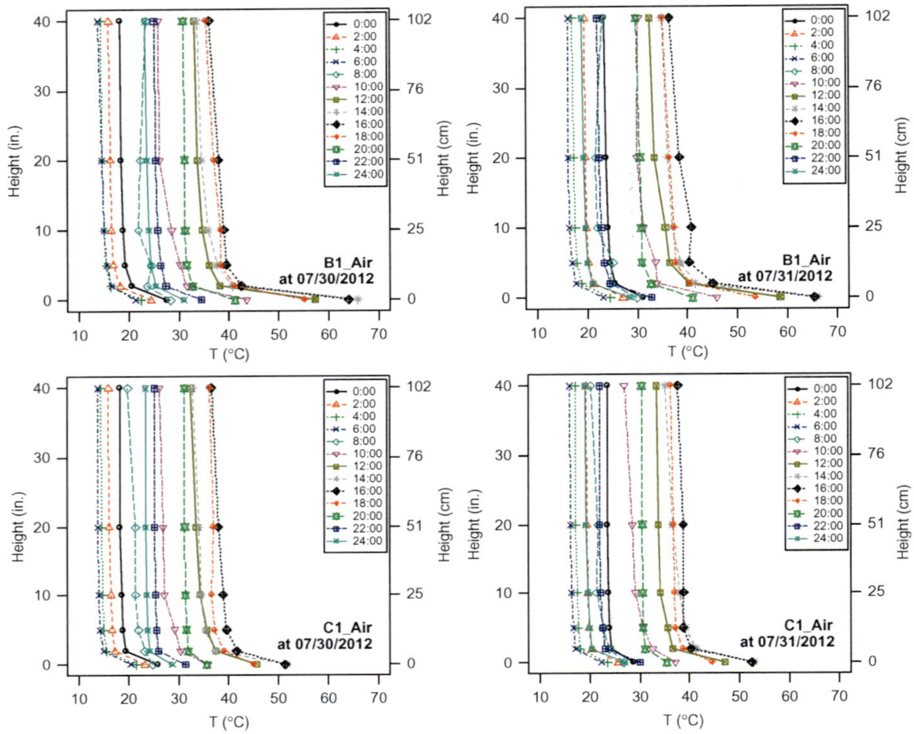

Figure 4.30 Example spatial profiles of near-surface air temperatures on asphalt pavement PA2 (B1) (top) and concrete pavement PC2 (C1) (bottom) on the two same days.

are under the same weather conditions. The higher pavement surface temperature would heat up the near-surface air and produce a higher temperature and a steeper temperature gradient for near-surface air. However, the temperature profiles of near-surface air are quite similar for asphalt and concrete pavements during night-time, and the night-time surface temperatures are close. This implies that pavement surface temperatures play a relatively important role on the near-surface air temperatures and the profiles, which are critical for human thermal comfort and air quality.

4.5.6 Influence of wind speed on near-surface air temperature profile

Wind speed (i.e., airflow velocity) might also significantly influence the near-surface air temperature and the shape of the profiles of near-surface air temperature. A separate experiment was performed on the asphalt pavement PA1 to examine the effect of wind speed.

Like other pavement sections, the asphalt pavement PA1 were also open to the wind in a quite large and open space with few barriers to wind. To break the wind around the temperature sensors, a box served as windbreak was installed around them at the midnight of July 10, 2012 (i.e., 0:00 on July 11, 2012) as shown in Figure 4.31. The temperature profiles of near-surface air were measured and compared for both before and after installing windbreak. The results are presented in Figures 4.32 and 4.33, along with the weather conditions during the experiment period in Figure 4.34. Due to the lack of a handy windmeter, the specific wind speed in the box was not measured.

It is noted from Figure 4.32 that the temporal profiles of near-surface air were changed due to the installation of windbreak. Just after the installation of the windbreak (0:00 on July 11, 2012), the surface temperature and near-surface air temperatures increased to some extent due to the reduced wind speed inside the windbreak. During the daytime, the shape of the temperatures profiles, especially the surface temperature, was also influenced by the shading caused by the windbreak.

Figure 4.33 presents the spatial profiles of near-surface air before and after installing the windbreak. The daytime temperature profiles of near-surface air become

Figure 4.31 Experimental setup for windbreak.

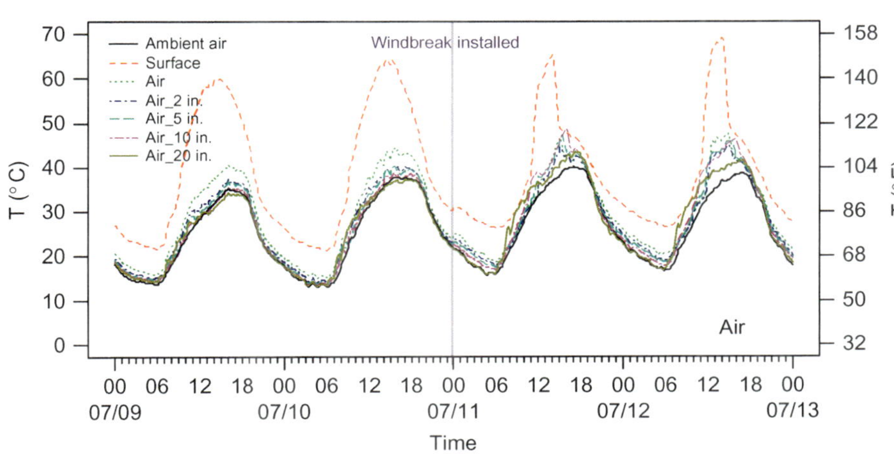

Figure 4.32 Temperature temporal profiles of near-surface air before and after installing windbreak.

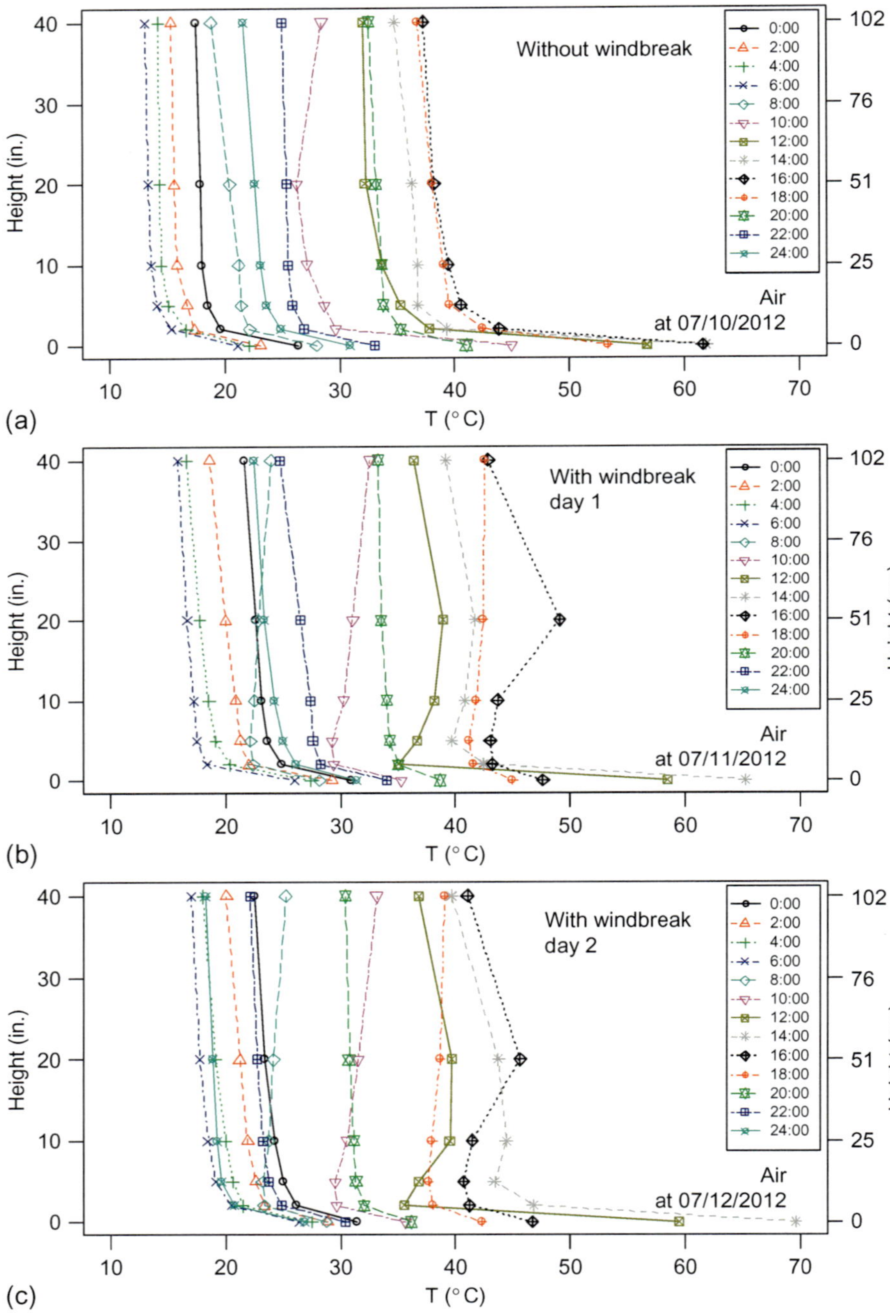

Figure 4.33 Temperature spatial profiles of near-surface air without (a) and with (b, c) windbreak.

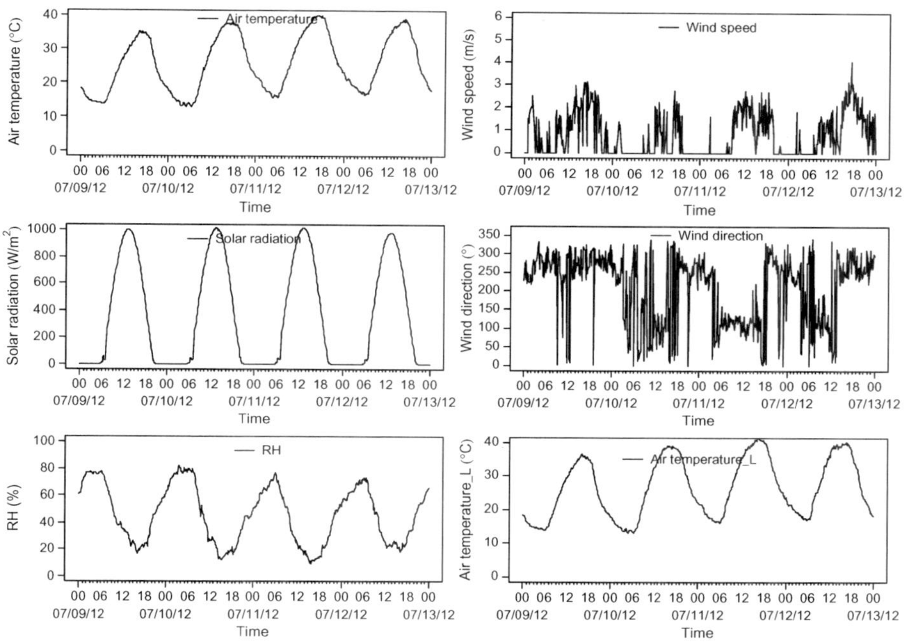

Figure 4.34 Weather conditions during the experiment period for PA1.

steeper and the temperatures around 50 cm (20 in.) height get higher compared to those without windbreak. The daytime temperature profiles of near-surface air do not change significantly.

These findings imply that the wind speed will influence the shape of the temperature profiles of near-surface air. Lower wind speed will make the spatial profiles steeper due to less heat dissipated by wind or airflow. The near-surface air temperatures with higher distances from the surface will be influenced by the pavement heat under lower wind speed. This further implies that the heat effects of pavements will be more critical in the environments with lower wind speed, such as highly developed dense urban areas especially those with high-rise buildings that block prevailing winds.

4.5.7 Modeling of near-surface air temperature profile

It is revealed from the findings above that during hot periods, the near-surface air temperatures diminish gradually as the height above surface increases and the slope of temperature vs. height decreases. The shape of the spatial profiles is influenced by the surface temperature, ambient air temperature, and wind speed.

This section will develop a model to describe the spatial profiles of near-surface air temperatures, which characterize the effect of surface temperature, ambient air temperature, and wind speed.

The assumptions for developing the model include the following:

(1) The spatial variation of near-surface air temperature is considered in the range of up to 1 m from the surface.
(2) The influences of the ambient air temperature and wind speed are characterized with the ambient air temperature and wind speed at ~2 m height, which are usually measured from a nearby weather station.
(3) The size effect of paved area is not explicitly taken into account here.

4.5.7.1 Developing the dimensionless parameters for the model

To simplify the model, the normalized dimensionless parameters are developed for the model as follows:

$$T_n = \frac{T_z - T^+}{T_s - T^+} \tag{5.4}$$

$$z_n = \frac{Z}{Z^+} \tag{5.5}$$

where T_n is the normalized dimensionless temperature, $[-1, 1]$; z_n is the normalized height above the surface, [0, 1]; Z is the height above the pavement surface, in m, $[0, Z^+]$; T_z is the near-surface air temperature at height Z, in °C; Z^+ is the upper bound of height considered in the model, in m, assuming $Z^+ = 1$ m; T^+ is the near-surface air temperature at Z^+, in °C, assuming $T^+ = T_a$ (T_a is the ambient air temperature at ~2 m height, in °C); and T_s is the surface temperature, that is, T_z at $Z = 0$, in °C.

Using the developed normalized dimensionless temperature and height, the normalized spatial profiles of near-surface air temperature can be developed. Some examples of normalized profiles of near-surface air temperature are presented in Figure 4.35, along with the corresponding original profiles.

4.5.7.2 Model development and obtaining coefficient C through regression on the model

According to the trends shown by most of the normalized profiles presented in Figure 4.35, the following simple model is proposed for the normalized profiles of near-surface air temperature:

$$T_n = e^{C \cdot z_n} \tag{5.6}$$

where C is constant.

With the developed model for the normalized profiles of near-surface air temperature, the best-fitting constant C can be obtained through regression on the normalized profiles at different times. Some examples of measured and predicted normalized profiles of near-surface air temperature at different times are presented in Figure 4.36.

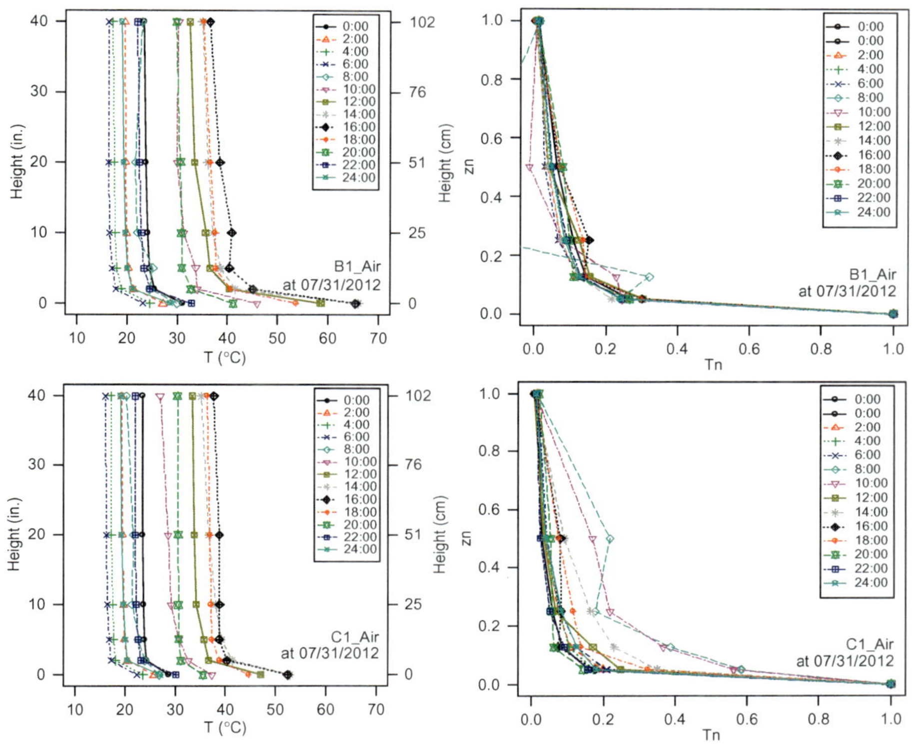

Figure 4.35 Examples of original and normalized profiles of near-surface air temperature (asphalt (B1) and concrete (C1), at different times).

The obtained constant C for each time is also shown in the Figure 4.36. The constant C is different for the different times, which is correlated with the wind speed at that time.

4.5.7.3 Correlation between coefficient C and wind speed

As mentioned previously, the spatial profiles of near-surface air temperature are also influenced by the wind speed besides the ambient air temperature and surface temperature. To quantify the effect of wind speed on the profile, the obtained constant C at different times with various wind speeds is correlated with the wind speed at the corresponding time. The obtained data of constant C and wind speed WS (at ~2 m height) are plotted in Figure 4.37, along with the best-fitting line. The correlation between coefficient C and wind speed is then proposed as follows:

$$C = -5.13 + (-0.57 \times \mathrm{WS}) \tag{5.7}$$

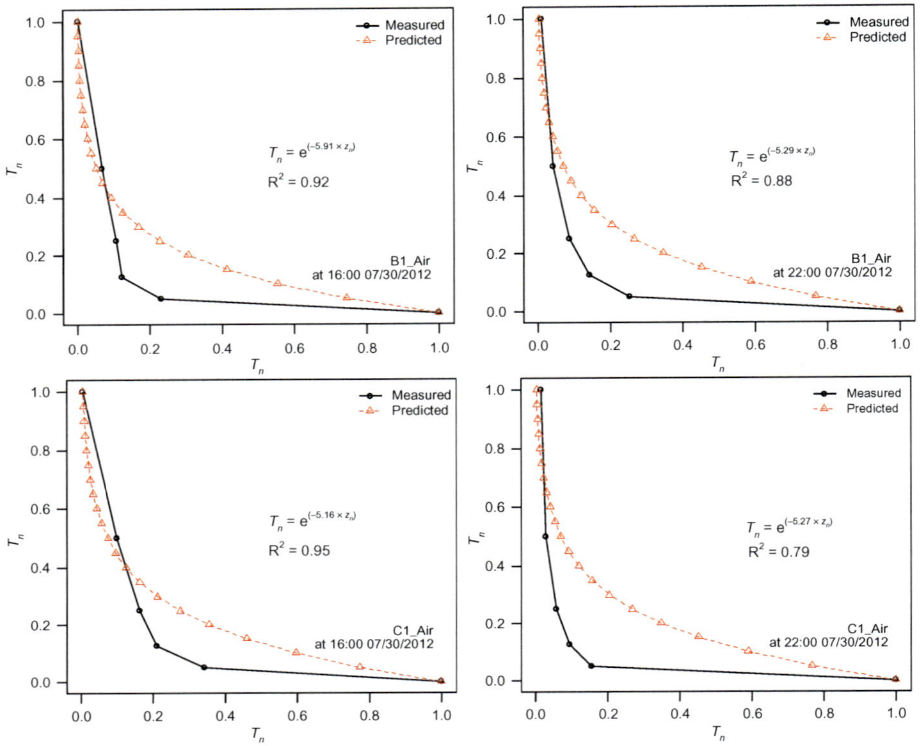

Figure 4.36 Examples of measured and predicted normalized profiles of near-surface air temperature (asphalt (B1) and concrete (C1), at different times).

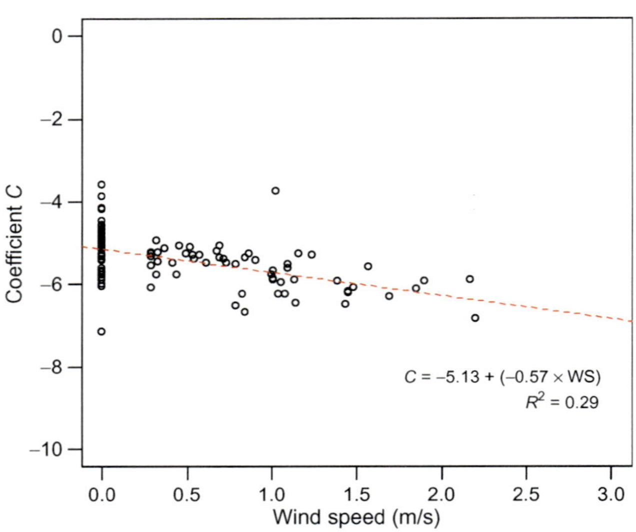

Figure 4.37 Correlation between coefficient C and wind speed at 2 m height for modeling normalized profiles of near-surface air.

4.5.8 Summary of near-surface air temperature above the pavement

This section investigates the thermal interaction between pavement and near-surface air for different pavement types, measures the temperature profiles of near-surface air, and explores the factors affecting the profiles.

The temporal profiles of near-surface air have a pattern similar to that of ambient air temperature for both asphalt and concrete pavements. For the spatial profiles, the near-surface air temperatures gradually decrease as the distance from the pavement surface increases, with greater slope (change rate) within 10 in. (25 cm) from the pavement surface. In the range close to the pavement surface, the near-surface air temperatures are much higher than the ambient air temperature at ~2 m height, especially for the black asphalt pavement with higher surface temperature compared with higher reflectance concrete pavement. The high near-surface air temperatures would reduce the human thermal comfort, and they are more critical for the babies and children who are shorter and closer to the surface. In addition, the formation of ground-level ozone would be facilitated by these high near-surface air temperatures when smog (VOC and NO_x) is exhausted from vehicle exhaust pipes that are close to the pavement surface on roads or parking lots.

The wind speed will influence the shape of the temperature profiles of near-surface air. Lower wind speed will make the spatial profiles steeper due to less heat dissipated by wind or airflow. The near-surface air temperatures with higher distances from the surface will be influenced by the pavement heat under lower wind speed. This further implies that the heat effects of pavements will be more critical in the environments with lower wind speed, such as high-density urban areas especially those with high-rise buildings that block prevailing winds.

It is revealed from the findings that during hot periods, the near-surface air temperatures diminish gradually as the height above the surface increases and the slope (change rate) also diminishes. The shapes of the spatial profiles are influenced by the surface temperature, ambient air temperature, and wind speed.

A simple model was developed to descript the spatial profiles of near-surface air temperatures, which characterizes the effect of surface temperature, ambient air temperature, and wind speed. The spatial profiles of near-surface air temperature could be predicted as long as the surface temperature, ambient air temperature, and wind speed are known.

4.6 Thermal impact of pavement on nearby building wall surfaces

Pavement heat can influence near-surface air temperature profiles and consequently influence the human thermal comfort and air quality in hot periods, as discussed in the previous chapter. Besides the thermal impact on near-surface air, pavement heat potentially would also increase the temperatures of the building surfaces and produce reflection of solar radiation that can be absorbed by building surfaces, consequently increasing the energy use for cooling the building in hot periods.

The objective of this section is to evaluate the thermal interaction between pavement and building walls through experiments and simulations, characterizing the impacts of different pavement types on building surface temperatures.

4.6.1 Experimental materials and methodology

4.6.1.1 Experimental setup

For simplification, two identical building walls (hollow wall box, with a size of $2 \times 4 \times 8$ ft ($0.6 \times 1.2 \times 2.4$ m), made of commonly used half-inch-thick plywood, painted with white paint; Figure 4.38) were developed and put on some of the pavement sections described in Table 4.1 and some other existing pavements to evaluate the thermal interaction between pavement and building walls. Thermocouple wires were used to measure the temperatures on walls at different heights above the pavement surface and temperatures on the pavement surface at different distances from the walls. Thermocouple sensor locations on the walls were 0, 2, 5, 10, 20, 40, and 48 in (0, 5, 12.5, 25, 50, 100, and 122 cm) height above the pavement surface (Figure 4.38). Thermocouple sensor locations on the pavement were 5, 10, 20, 30, 40, 50, 60, and 80 in (12.5, 25, 50, 75, 100, 125, 150, and 200 cm) distance from the walls. Local weather data were also monitored using a portable weather station. Thermal images of walls and pavements were taken for some of the pavement sections at different times in a hot sunny day.

4.6.1.2 Pavement sections and measurement plan

Four pavement sections were chosen to conduct the experiments for measuring the temperature profiles of both walls and pavement surfaces. Two of them are asphalt pavements and the other two are concrete pavements. Two were on small sections and two were in large paved areas. For the first two large pavement sections (PA1 and PC1), the two walls were set up on the same sections in the same way, which

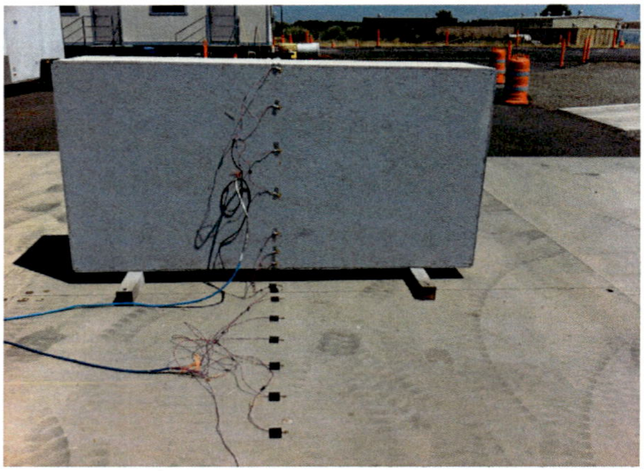

Figure 4.38 Example building wall test setup with temperature sensors.

provided the chance to directly compare the two walls to make sure they are thermally identical when installed under some conditions. For the last two sections (PA2 and PC2), the experiments were performed simultaneously for direct comparison with one wall on each section (Figure 4.39). The pavement sections and measurement dates

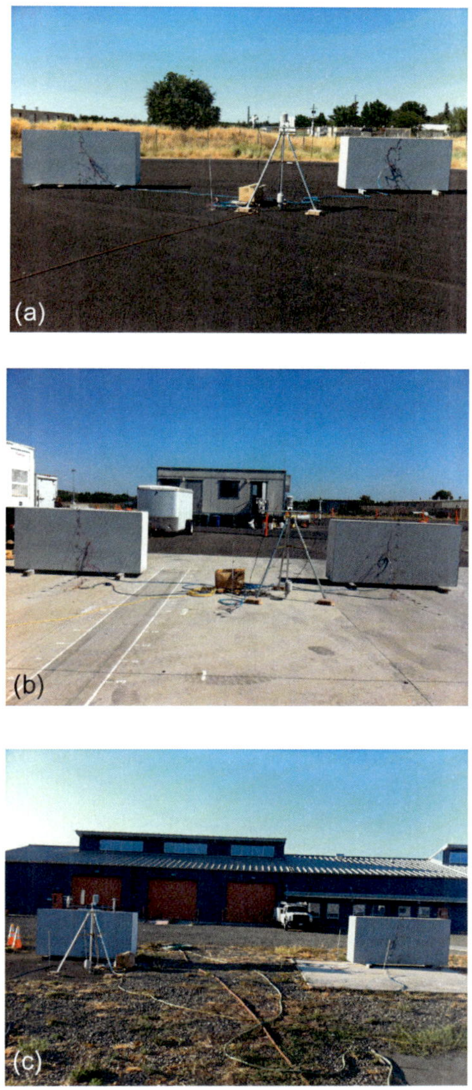

Figure 4.39 Experimental setup for building wall temperature profile measurement. (a) On asphalt pavement PA1. (b) On concrete pavement PC1. (c) On small asphalt (PA2 or B1) and concrete (PC2 or C1) pavements for the same period.

Table 4.7 Summary of pavement sections and walls for experimental measurement

Section	Type	Size	Albedo	Experiment date
PA1	Asphalt	~30 m diameter	0.06	July 2, 2012–July 17, 2012
PC1	Concrete	15 m wide × 45 m long	0.26	July 18, 2012–July 27, 2012
PA2 (B1)[a]	Asphalt	4 m wide × 4 m long	0.08	July 28, 2012–August 20, 2012
PC2 (C1)[a]	Concrete	4 m wide × 4 m long	0.28	July 28, 2012–August 20, 2012
Wall 1	Plywood	0.6 × 1.2 × 2.4 m	0.29	–
Wall 2	Plywood	0.6 × 1.2 × 2.4 m	0.29	–

[a]PA2 and PC2 are the experimental test sections B1 and C1 as described previously, respectively.

are summarized in Table 4.7 along with the albedos (solar reflectivities) of both pavement sections and the two walls measured by an albedometer. The two walls have the same albedo value of 0.29.

4.6.2 Experimental results and discussion

4.6.2.1 Thermal performance comparison of the two walls

Both walls were installed on the big asphalt section (see Figure 4.39a). As shown in Table 4.7, the two walls have the same albedo value of 0.29. Under the small conditions (weather and pavement), these two walls are expected to give the same or close surface temperatures. The temperatures at different heights on both walls in one day were presented in Figure 4.40. The temperatures on both walls at the same height are very close to each other. This implies that both walls are thermally identical when installed under some conditions.

4.6.2.2 Example results on asphalt pavement PA1 and concrete pavement PC1

The wall and pavement temperatures on asphalt pavement PA1 and concrete pavement PC1 were monitored at the different periods from July 2, 2012 to July 17, 2012 and from July 18, 2012 to July 27, 2012, respectively. The weather conditions did not change much during this summer period, especially for the sunny days. This makes it possible to compare the effects of asphalt and concrete pavements on near-surface air temperatures for clear days. Some results for PA1 and PC1 are presented in Figures 4.41–4.44.

It is noted that the temporal profiles (Figures 4.41–4.43) of both wall and pavement temperatures have a pattern similar to that of ambient air temperature for both asphalt

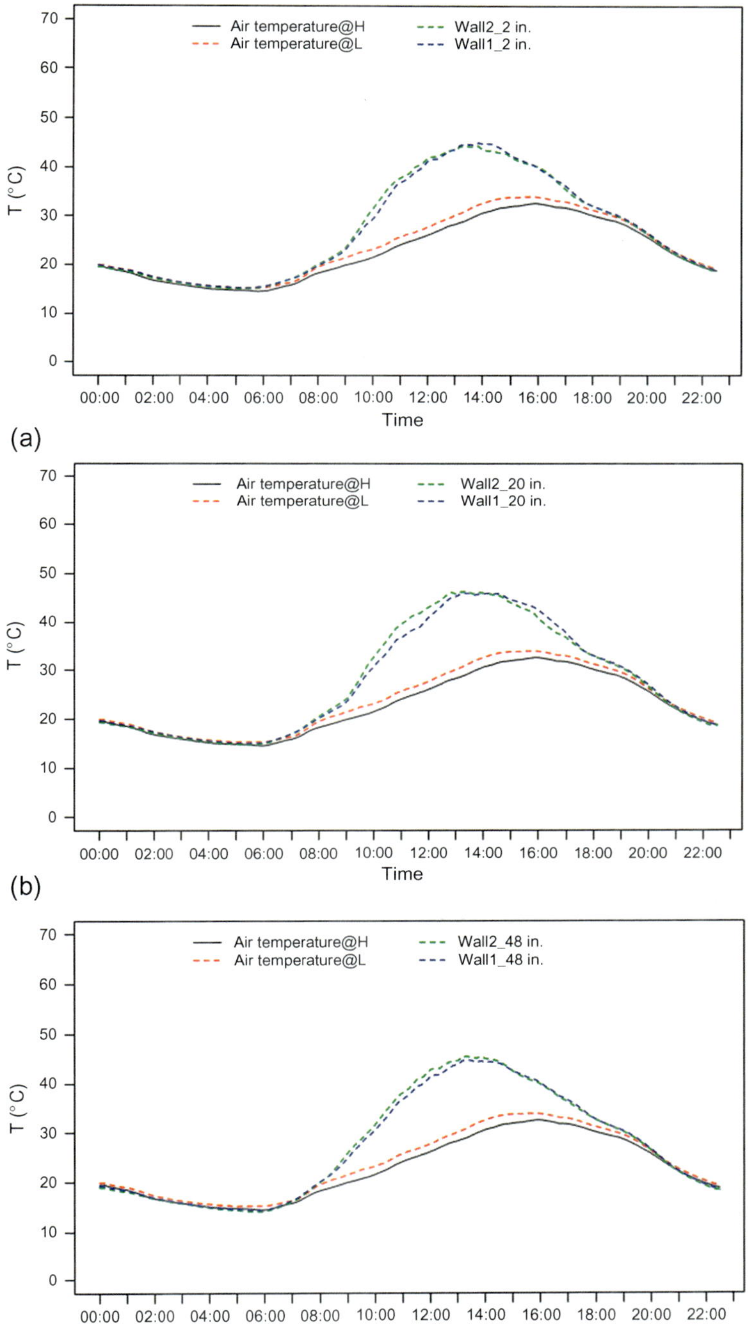

Figure 4.40 Temperatures on both walls under the same conditions. (Air temperature@H and air temperature@L are ambient air temperatures measured at 1.7 m (67 in.) and 0.3 m (11.8 in.), respectively). (a) At 2 in. (5 cm) height. (b) At 20 in. (50 cm) height. (c) At 48 in. (122 cm) height.

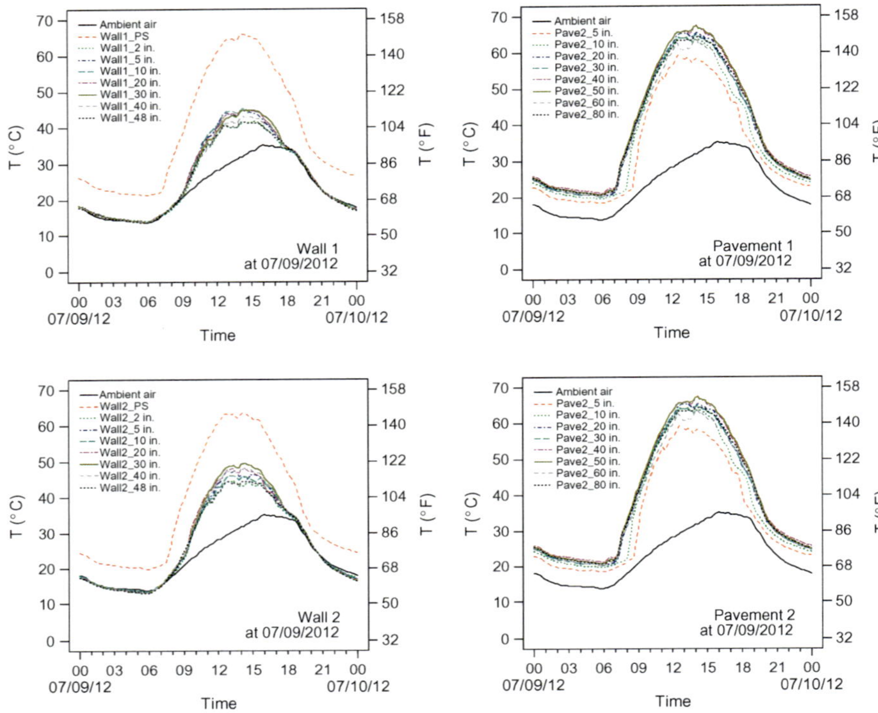

Figure 4.41 Example temporal profiles of wall (left) and pavement (right) temperatures on asphalt pavement PA1. (Ambient air is ambient air temperature from a sensor on portable weather station at 67 in. (1.7 m). Walli_PS and Walli_*x* in. are wall i surface temperatures at 0 in. and *x* in. height from pavement surface, respectively. Pavei_*x* in. are pavement i surface temperatures at *x* in. distance from the wall.)

and concrete pavements. The difference is that the peak temperatures of wall and pavement came earlier than that of ambient air (~14:00 vs. ~16:00). The wall and pavement temperatures around noon are both much higher than the ambient air temperature on the sunny summer day. This might cause more energy use and high peak demand for cooling buildings with air conditioners during hot periods. While the night-time pavement temperature stays higher than the ambient air temperature, the wall temperature drops and stays very close to the ambient air temperature during night-time. This is because the pavement is quite thick (15 cm or thicker for pavement layer), which can absorb and store a quite large amount of heat during daytime and maintain a high temperature with the stored heat during night-time. However, the walls used in the experiments are quite thin (the plywood layer of wall is only ~1.3 cm (0.5 in.) with air in the hollow wall box) and of low heat capacity and thermal conductivity, and consequently, the thin plywood layer cannot absorb and store a large

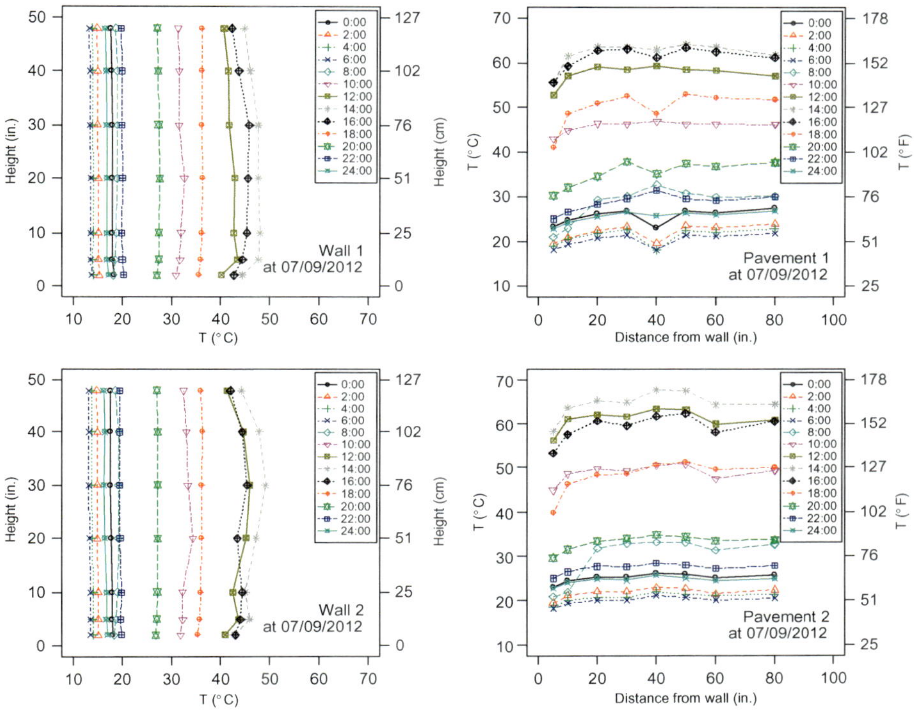

Figure 4.42 Example spatial profiles of wall (left) and pavement (right) temperatures on asphalt pavement PA1.

amount of heat during daytime. During night-time without solar radiation, the wall temperature drops very quickly and stays close to the ambient air temperature. This situation is for simulation and is different from a thick wall or a wall made of materials of high heat capacity and high thermal conductivity, such as aluminum wall surface.

For the spatial profiles (Figures 4.42–4.44), the wall temperatures at various heights on the wall were relatively constant (slightly higher at the bottom close to pavement surface) during night-time. However, during the daytime, around noon, the middle part of the wall presents higher temperatures under the same weather conditions such as solar radiation, ambient air temperature, and wind speed. This indicated that there are some thermal interactions (through reflected solar radiation and emitted radiation) between pavement and building wall. The thermal interactions increase the wall temperature to some extent and potentially will contribute increased energy use for cooling building in hot periods. The nonuniform temperature of the pavement surface at different distances from the wall also confirmed the thermal interactions between pavement and building wall.

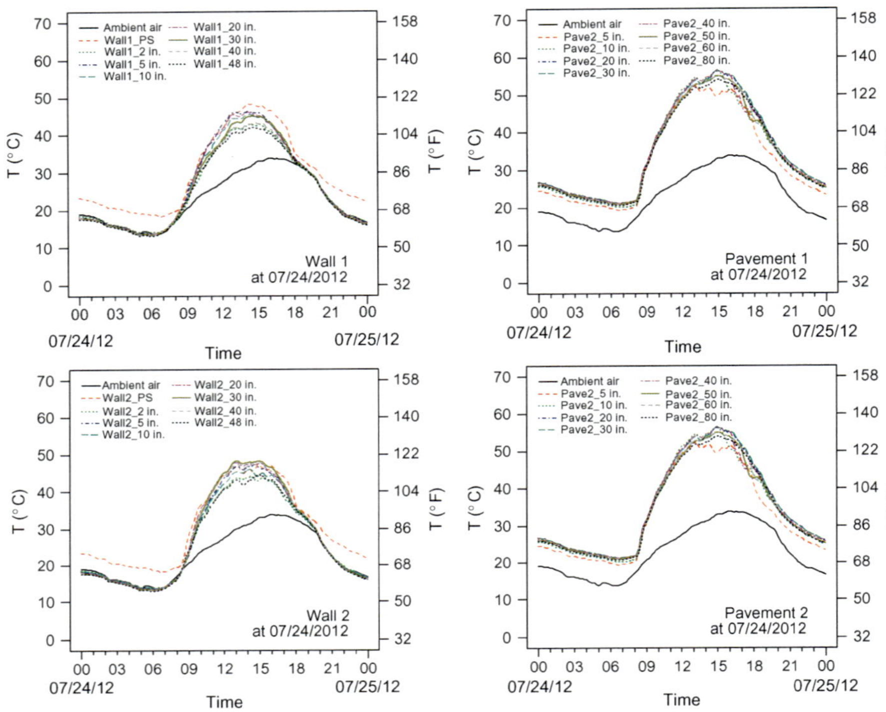

Figure 4.43 Example temporal profiles of wall (left) and pavement (right) temperatures on concrete pavement PC1. (Ambient air is ambient air temperature from a sensor on portable weather station at 67 in. (1.7 m). Walli_PS and Walli_x in. are wall i surface temperatures at 0 in. and x in. height from pavement surface, respectively. Pavei_x in. are pavement i surface temperatures at x in. distance from the wall.)

4.6.2.3 Temporal and spatial profiles of wall temperatures on asphalt pavement PA2 and concrete pavement PC2

The wall and pavement temperatures on asphalt pavement PA2 and concrete pavement PC2 were monitored simultaneously from July 28, 2012 to August 20, 2012. This made it possible to directly compare the effects of asphalt and concrete pavements on wall temperatures. Some results are presented in Figures 4.45 and 4.46.

Besides some observations similar to those for PA1 and PC1 presented above, there are some interesting observations obtained from the direct comparison between asphalt and concrete pavements under the same weather conditions.

The daytime surface temperatures on asphalt (B1 in Figure 4.45) pavement are much higher than those of concrete (C1 Figure 4.45) pavements (70 °C vs. 55 °C), although they are under the same weather conditions. The nonuniform temperature along the wall implies that the thermal interaction between pavement and wall exists for both asphalt (B1 in Figure 4.46) and concrete (C1 in Figure 4.46) pavements,

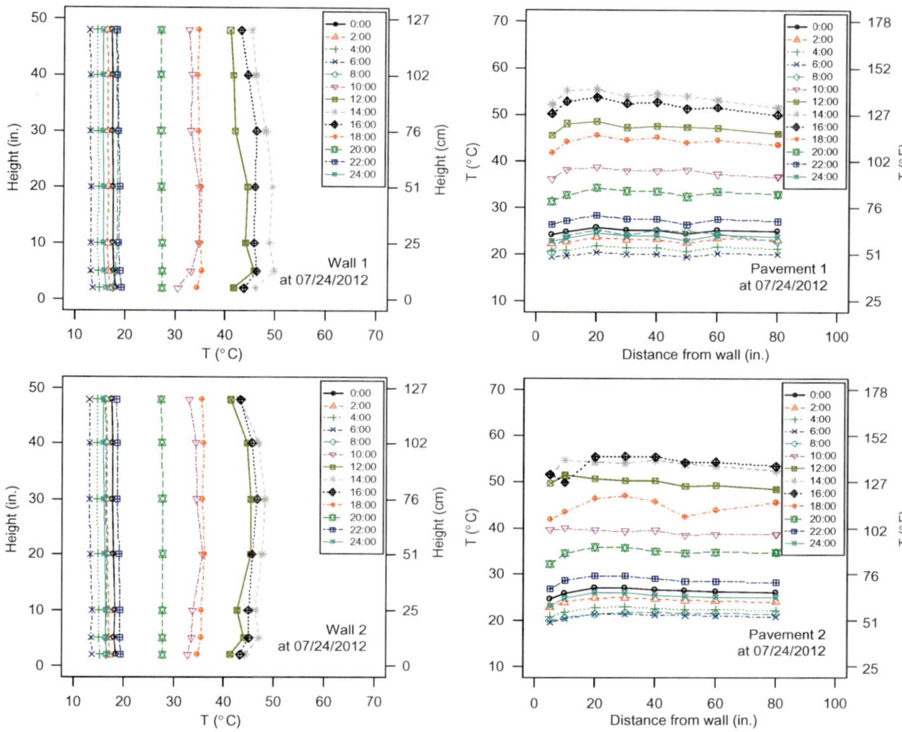

Figure 4.44 Example spatial profiles of wall (left) and pavement (right) temperatures on concrete pavement PC1.

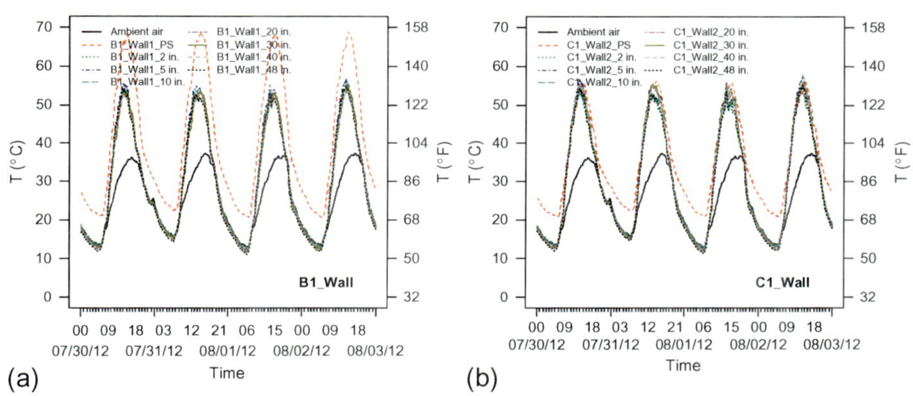

Figure 4.45 Example temporal profiles of wall temperature on (a) asphalt pavement PA2 (B1) and (b) concrete pavement PC2 (C1). (Ambient air is ambient air temperature from a sensor on portable weather station at 67 in. (1.7 m).)

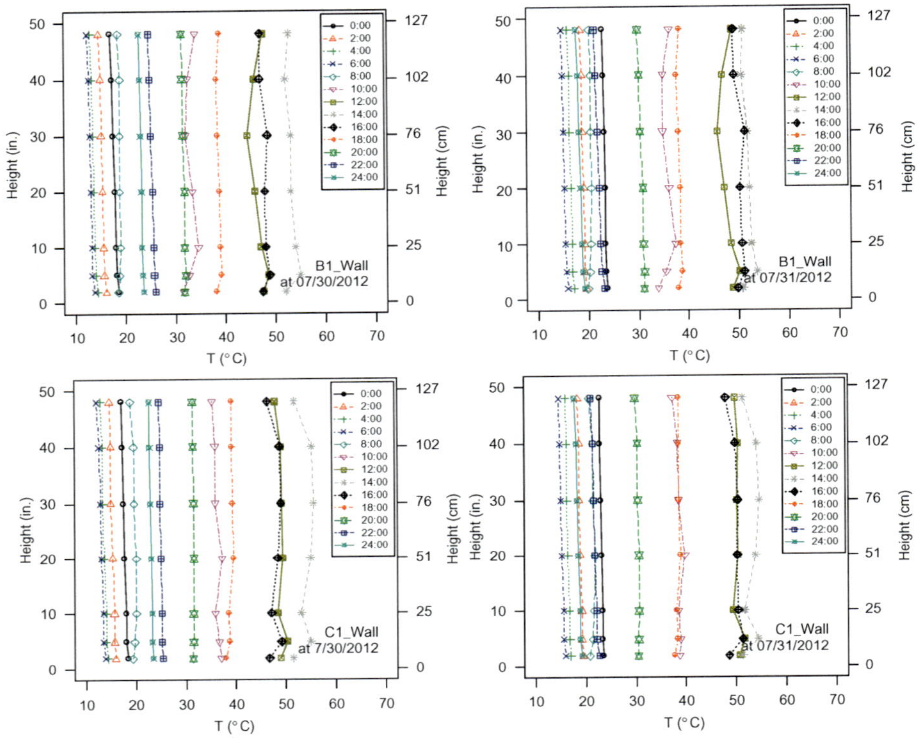

Figure 4.46 Example spatial profiles of wall temperatures on asphalt pavement PA2 (B1) and concrete pavement PC2 (C1) on two days. Concrete, 45 °C. Wall, 52 °C.

especially during the daytime and around noon with high solar radiation. Although the concrete (C1) pavement has lower surface temperature than the asphalt (B1) pavement, the temperatures in the middle part of the wall on the concrete (C1 in Figure 4.46) pavement around noon are higher and more nonuniform than that of the asphalt (B1 in Figure 4.46) pavement. The reason for this is that, compared with the asphalt (B1) pavement, the concrete (C1) pavement has higher solar reflectivity (or albedo) and consequently reflects more solar radiation. Some of the reflected heat will hit and be absorbed by the building wall and heat it up. This implies that increased pavement albedo will reduce the pavement surface temperature; however, the temperature of the building wall will heated up by the reflected energy reflected from the pavement surface and consequently would potentially increase the energy use of cooling building in hot periods. This risk might be not a big issue in open areas but would tend to be more significant for the high-density urban areas. Therefore, special attention should be given to the application of high-reflectance pavement for mitigating the heat island effect, especially in high-density urban areas.

4.6.2.4 Thermal images of walls and pavements on B1 and C1

Thermal images of walls and pavements on B1 and C1 were taken on clear days in summer. Some example thermal images (13:00 on August 15, 2012) are presented in Figure 4.47. This provides a more direct and obvious way to compare the thermal impacts of asphalt and concrete pavements on the walls.

Three thermal images were taken at each time (Figure 4.47b): one for the pavement and wall together (B1/C1, all), one for the pavement only (B1/C1, pavement), and one for the wall only (B1/C1, wall). Besides, the optical images are also presented in Figure 4.47a. The asphalt (B1) pavement presents higher pavement surface temperature than the concrete (C1) pavement (60 vs. 45 °C). However, the average

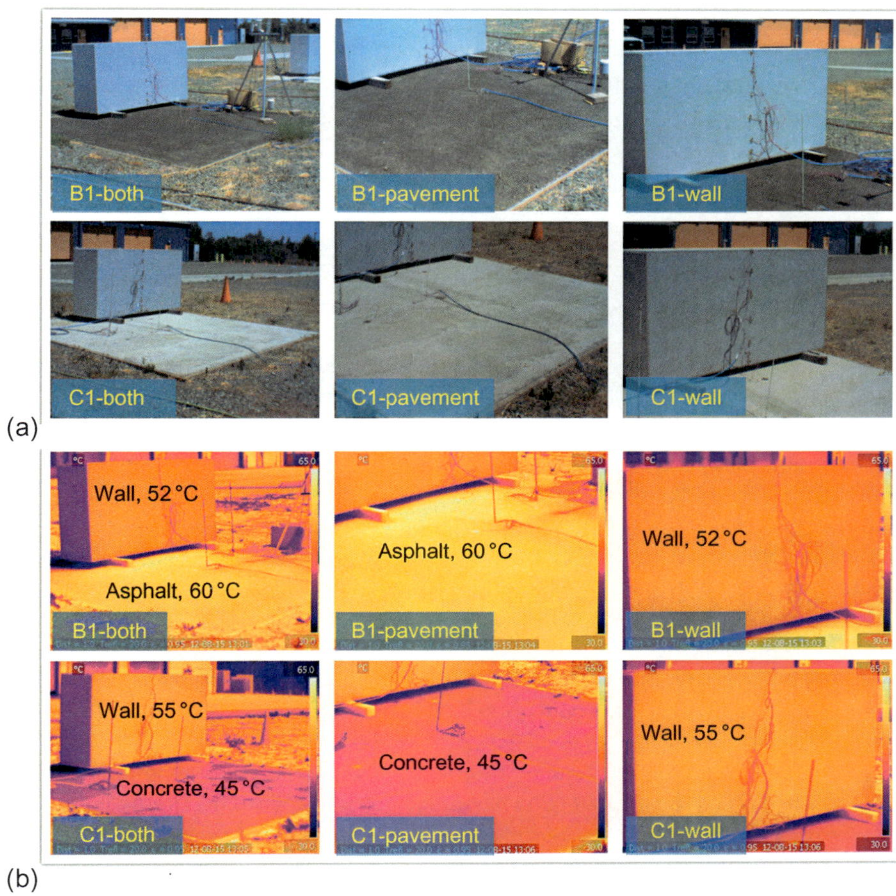

Figure 4.47 Optical (a) and thermal (b) images of walls and pavements on B1 and C1 (average temperatures of walls and pavements are shown in the figures).

temperature of the wall on the concrete (C1) pavement is higher than that of the asphalt (B1) pavement (55 vs. 52 °C). This verified the results obtained from the temperature sensors presented previously. Another more convincing and quantitative way of verifying the thermal impacts of different pavements on the wall temperature is using a heat flux sensor (including both longwave radiation and shortwave radiation) to directly measure the heat flux on the wall surface at different heights and times and to directly compare the thermal impacts. However, such heat flux sensors are quite expensive (over $6000 per set of sensors). With adequate funding available from any source, the direct measurement of heat flux (including both long and short wave radiation) on the wall surface at different heights and times can be performed and would definitely give more convincing and quantitative observations.

4.7 Heat flux from pavement

This section is to comparatively examine the thermal interaction (or heat exchange) between the pavement and near-surface air and other surroundings such as walls and buildings, which is another important aspect to examine besides pavement temperatures. The heat released by the pavement to its surroundings will be the focus of this section.

The heat exchange processes include reflected shortwave solar radiation, emitted longwave radiation, and convective heat. The sum of the reflected shortwave solar radiation and emitted longwave radiation from the pavement surface is called radiosity. It is the total radiation (sensible heat) from the pavement surface that might hit and consequently be absorbed by the surroundings such as building/vehicle surfaces or human bodies. The convective heat is the energy exchanged through convection between pavement and near-surface air.

Based on the theory of heat transfer and radiation (see reference Li (2012) for more details), the heat flux of these heat exchanges of different pavements is calculated and presented in Figure 4.48 for one sunny summer day. The convective heat of the asphalt (B1) pavement is higher than that of other pavements due to its high surface temperature. This means more convective heat will be released into the near-surface air by the asphalt pavement than the concrete (C1 and C2) pavement and the interlocking concrete paver (A1) pavement. In addition, more longwave radiation is emitted by the asphalt pavement with higher surface temperature. However, the concrete pavement with high albedo reflects more shortwave solar radiation than the asphalt pavement. The sum of the reflected shortwave solar radiation and the emitted longwave radiation, that is, radiosity, is higher for the concrete (C1 and C2) pavements than the asphalt (B1) pavement. This implies that, although the concrete pavement produces a lower surface temperature and lower emitted longwave radiation and convective heat due to the low surface temperature led by high albedo, it will increase the reflected shortwave solar radiation and might increase the radiosity, the total sensible energy released by the surface. The increased radiosity of

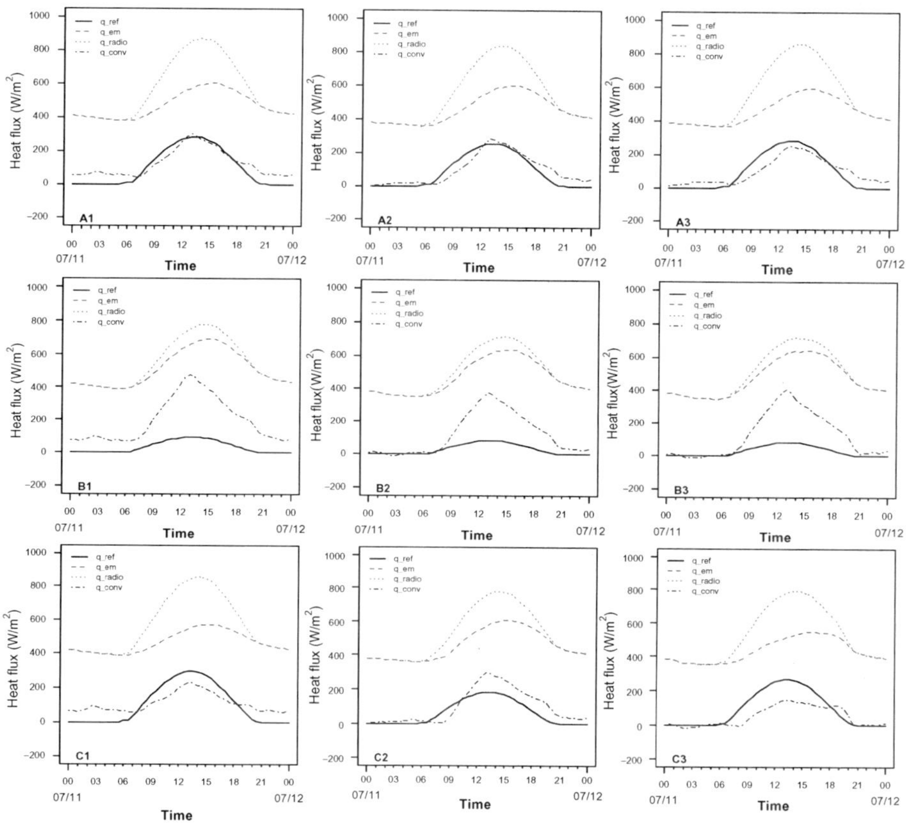

Figure 4.48 Heat flux from different pavement surfaces for one full day during July 2012. (Reflected shortwave solar radiation q_ref $= rI$, where r is the albedo in Table 4.3 and I is the incident solar radiation in W/m^2. Emitted longwave radiation q_em $= \varepsilon\sigma(T_s+273)^4$, where ε is the thermal emissivity and assumed to be 0.9. Stefan–Boltzmann constant $\sigma = 5.670 \times 10^{-8}$ W/(m^2 K^4), where T_s is the pavement surface temperature in °C. Radiosity q_radio = q_ref + q_em. Convective heat q_conv $= h_c(T_s - T_a)$, in W/(m^{2}°C). Convection coefficient $h_c = 6.1 + 3.7v_w$, where v_w and T_a are the wind velocity (m/s) and air temperature (°C) at 2 m above the pavement, respectively. A1, impermeable paver pavement; A2 and A3, permeable paver pavements; B1, impermeable asphalt pavement; B2 and B3, permeable asphalt pavements; C1, impermeable concrete pavement; C2 and C3, permeable concrete pavements.)

concrete pavement might hit and be absorbed by its surroundings such as building surfaces and human bodies, leading to increased building energy use for cooling and reduced human thermal comfort. Therefore, optimal context-sensitive design of the pavement albedo is of great significance to ensure a net benefit in terms of energy use or human thermal comfort or both.

4.8 Potential impacts and future trends of pavements

With extensive development in urban and rural areas, more earth surfaces have been or will be paved by conventional impermeable paving materials. The paved surfaces, over 40% for some urban areas and mostly black, absorb a large amount of solar heat and produce a very high surface temperature, aggravating heat island effect and consequently increasing building cooling energy use, impairing air quality, and compromising human comfort and health. Moreover, the conventional impermeable paving materials substantially alter the permeable nature of original land covers, which causes critical issues associated with stormwater runoff in some regions, including increased water pollution from oil and metal on pavement surfaces to the adjacent public water areas, increased risk of flooding, and reduced groundwater recharge. Cool and permeable pavements made of innovative reflective and pervious materials could be used as a low-impact development (LID) to manage the quantity and improve the quality of stormwater runoff and mitigating the heat island effect in the most effective and cost-effective manner, especially under climate change and extreme events.

A number of research studies on this topic have been conducted by the University of California Pavement Research Center (UCPRC) and some other research groups, including characterization of thermal properties and thermal behavior of different cool pavements, hydraulic and mechanical properties of permeable materials, and hydraulic and structural designs of permeable pavements through both laboratory and field experiment testing and multiscale modeling (e.g., Li, 2012; Li et al., 2012a,b; Li et al., 2013a,b,c,d,e,f; Li et al., in press). It has been preliminarily shown that cool and permeable pavement is a promising strategy to address those urban issues related to heat island effect and stormwater runoff.

4.9 Conclusions

The thermal performance of pavement materials is complex with a number of influencing factors and a number of different impacts in various aspects. Some main findings include the following:

(1) Albedo and thermal properties (including thermal conductivity and heat capacity) play very important roles in the thermal behavior and performance of pavement materials.
(2) Different pavement materials (asphalt vs. concrete; impermeable vs. permeable) have very different material properties (e.g., albedo and thermal properties).
(3) The material properties (albedo and thermal properties) differently influence the pavement surface temperature, near-surface air temperature, and nearby building wall temperature through heat exchange and heat flux with different effects.
(4) Appropriately engineered cool and permeable pavement is a promising strategy to address the urban issues related to heat island effect and stormwater runoff, reducing the associated environmental impacts.
(5) Comprehensive assessment of the specific effects of the material properties on the different aspects (such as building energy use, human thermal comfort, and air quality) should be performed to ensure a net positive benefit from engineered pavement materials.

References

Akbari, H., Matthews, H.D., 2012. Global cooling updates: reflective roofs and pavements. Energy Build. 55, 2–6.

Akbari, H., Matthews, H.D., Donny, S., 2012. The long-term effect of increasing the albedo of urban areas. Environ. Res. Lett. 7, 024004.

Asaeda, T., Ca, V.T., 2000. Characteristics of permeable pavement during hot summer weather and impact on the thermal environment. Build. Environ. 35, 363–375.

ASTM, 2006. ASTM E1918-06 Standard Test Method for Measuring Solar Reflectance of Horizontal and Low-Sloped Surfaces in the Field. American Society for Testing and Materials, Pennsylvania.

ASTM, 2009. ASTM C1549-09 Standard Test Method for Determination of Solar Reflectance Near Ambient Temperature Using a Portable Solar Reflectometer. American Society for Testing and Materials, Pennsylvania.

Carlson, J.D., Bhardwaj, R., Phelan, P.E., Kaloush, K.E., Golden, J.S., 2010. Determining thermal conductivity of paving materials using cylindrical sample geometry. J. Mater. Civ. Eng. 22, 186–195.

Gui, J., Phelan, P.E., Kaloush, K.E., Golden, J.S., 2007. Impact of pavement thermophysical properties on surface temperatures. J. Mater. Civ. Eng. 19, 683–690.

Hermansson, A., 2000. Simulation model for calculating pavement temperatures including maximum temperature. Pavement Manage. Monitor. 1699, 134–141.

Hermansson, A., 2001. Mathematical model for calculation of pavement temperatures - comparison of calculated and measured temperatures. Assess. Eval. Pavements 1764, 180–188.

Jo, J.H., Carlson, J.D., Golden, J.S., Bryan, H., 2010. An integrated empirical and modeling methodology for analyzing solar reflective roof technologies on commercial buildings. Build. Environ. 45, 453–460.

Kültür, S., Türkeri, N., 2012. Assessment of long term solar reflectance performance of roof coverings measured in laboratory and in field. Build. Environ. 48, 164–172.

Li, H. 2012. Evaluation of cool pavement strategies for heat island mitigation. Ph.D., thesis, University of California, Davis.

Li, H., Harvey, J., 2011. Numerical simulation and sensitivity analysis of asphalt pavement temperature and near-surface air temperature using integrated local modeling (11-3125). In: Transportation Research Board 90th Annual Meeting, January 23-27, 2011 Washington, DC.

Li, H., Jones, D., Harvey, J., 2012a. Development of mechanistic-empirical design procedure for fully permeable pavement under heavy traffic. Transp. Res. Rec. J. Transp. Res. Board 2305, 83–94.

Li, H., Jones, D., Harvey, J., 2012b. Material characterization for mechanistic design of fully permeable pavements: low-compaction subgrade and open-graded base materials. In: Transportation Research Board 91st Annual Meeting. Washington, DC.

Li, H., Harvey, J., Jones, D., 2013a. Cooling effect of permeable asphalt pavement under dry and wet conditions. Transp. Res. Rec. J. Transp. Res. Board 2372, 97–107.

Li, H., Harvey, J., Jones, D., 2013b. Multi-dimensional transient temperature simulation and back-calculation for thermal properties of building materials. Build. Environ. 59, 501–516.

Li, H., Harvey, J., Kendall, A., 2013c. Field measurement of albedo for different land cover materials and effects on thermal performance. Build. Environ. 59, 536–546.

Li, H., Harvey, J., Kendall, A., 2013d. Measurement of pavement solar reflectivity and effect on thermal performance. In: Transportation Research Board 92nd Annual Meeting, Jan. 13–17, 2013 Washington, DC.

Li, H., Harvey, J.T., Holland, T.J., Kayhanian, M., 2013e. The use of reflective and permeable pavements as a potential practice for heat island mitigation and stormwater management. Environ. Res. Lett. 8, 14.

Li, H., Kayhanian, M., Harvey, J., 2013f. Comparative field permeability measurement of permeable pavements using ASTM C1701 and NCAT permeameter methods. J. Environ. Manag. 118, 144–152.

Li, H., Harvey, J., Jones, D., in press. Development and preliminary validation of an integrated local microclimate model for numerical evaluation of cool pavement strategies. Transp. Res. Rec. J. Transp. Res. Board. 2444, xx–xx. DOI: 10.3141/2444-17.

Luca, J., Mrawira, D., 2005. New measurement of thermal properties of superpave asphalt concrete. J. Mater. Civ. Eng. 17, 72–79.

Mahmoud, A.H.A., 2011. Analysis of the microclimatic and human comfort conditions in an urban park in hot and arid regions. Build. Environ. 46, 2641–2656.

Mrawira, D.M., Luca, J., 2006. Effect of aggregate type, gradation, and compaction level on thermal properties of hot-mix asphalts. Can. J. Civ. Eng. 33, 1410–1417.

Mrawra, D.M., Luca, J., 2002. Thermal properties and transient temperature response of full-depth asphalt pavements. Des. Rehabil. Pavements 2002, 160–171.

Navigant Consulting Inc., 2011. Building Technologies Program, Assessment of International Urban Heat Island Research - Literature Review of International Studies on Urban Heat Island Countermeasures. Draft Report. Office of Energy Efficiency and Renewable Energy, USDOE, Washington, DC.

NovaLynx Corporation, 2014. 240-8104 Albedometer [Online]. Available from: http://www.novalynx.com/240-8104.html (accessed 10.06.14.).

Santamouris, M., 2012. Cooling the cities – a review of reflective and green roof mitigation technologies to fight heat island and improve comfort in urban environments. Solar Energy 103, 682–703.

Santamouris, M., Gaitani, N., Spanou, A., Saliari, M., Giannopoulou, K., Vasilakopoulou, K., Kardomateas, T., 2012. Using cool paving materials to improve microclimate of urban areas – design realization and results of the flisvos project. Build. Environ. 53, 128–136.

Schindler, A.K., Ruiz, J.M., Rasmussen, R.O., Chang, G.K., Wathne, L.G., 2004. Concrete pavement temperature prediction FHWA HIPERPAV and case studies with the models. Cem. Concr. Compos. 26, 463–471.

Synnefa, A., Karlessi, T., Gaitani, N., Santamouris, M., Assimakopoulos, D.N., Papakatsikas, C., 2011. Experimental testing of cool colored thin layer asphalt and estimation of its potential to improve the urban microclimate. Build. Environ. 46, 38–44.

Tan, S.-A., Fwa, T.-F., 1992. Influence of pavement materials on the thermal environment of outdoor spaces. Build. Environ. 27, 289–295.

Tan, S.A., Fwa, T.F., 1997. Pavement evaluation for thermal/glare comfort during footdrills. Build. Environ. 32, 257–269.

Wong, N.H., Kwang Tan, A.Y., Chen, Y., Sekar, K., Tan, P.Y., Chan, D., Chiang, K., Wong, N. C., 2010. Thermal evaluation of vertical greenery systems for building walls. Build. Environ. 45, 663–672.

Xu, T., Sathaye, J., Akbari, H., Garg, V., Tetali, S., 2012. Quantifying the direct benefits of cool roofs in an urban setting: reduced cooling energy use and lowered greenhouse gas emissions. Build. Environ. 48, 1–6.

Part Two

Facade materials for reducing cooling needs

Green facades and living walls: vertical vegetation as a construction material to reduce building cooling loads

5

I. Susorova
Cyclone Energy Group, Chicago, IL, USA

5.1 Introduction

5.1.1 Background

This chapter examines plants on building facades as a sustainable material that improves exterior wall thermal performance and overall building energy use. The chapter will focus only on the thermal effects of the plant layer in green facades and living walls.

Climbing plants, such as vines and shrubs, naturally grow on building facades by attaching themselves directly to vertical surfaces. Vertical vegetation growing along exterior walls was used for centuries for visual interest, shading, and growing agricultural plants. Recent interest in sustainable materials and methods inspired the development of new ways of integrating plants into building construction through green roofs and vegetated walls (green walls). Green wall typology is very diverse and includes green facades, living walls, vertical gardens, hanging gardens, bioshaders, and biofacades. Perhaps, the best-known examples of green walls include vertical gardens by French botanist and designer Patrick Blanc who conducted intensive research on this topic and invented his own green wall system (Blanc, 2008).

5.1.2 Green wall types

Plants adhere to vertical surfaces through various mechanisms. Self-clinging climbers and self-supporting woody plants can attach themselves directly to the facade surface or grow along the facade without any support. Other plant species, including climbers with aerial roots, suckers, or tendrils, twining climbers, and lax shrubs (ramblers), require the additional support of trellises, netting, or wires attached to the facade surface. In a vegetated facade, or a green wall, plants grow on building facades in a controlled fashion and are regularly maintained.

The main elements of a green wall are plants, soil or growing medium, planting containers, structures that support and attach plants to the facade, and irrigation systems. Depending on what plant species, planting media, and support structures are

Eco-efficient Materials for Mitigating Building Cooling Needs. http://dx.doi.org/10.1016/B978-1-78242-380-5.00005-4

used, there are multiple types of green walls, which are broadly categorized into green facades and living walls (Di, 1999; Grant, 2010).

5.1.2.1 Green facades

A green facade is a system in which plants directly climb the walls (*direct green facade*) (Figure 5.1) or have the additional support of steel, wood, or plastic trellises externally attached to a building facade (*indirect green facade*) (Figure 5.2). Climbing plants and vines used in green facades grow from planters located on the ground or at some intervals along the facade height. Green facades can be two-dimensional,

Figure 5.1 Direct green facade.

Figure 5.2 Indirect green facade.

formed by cables, ropes, and meshes, or three-dimensional, formed by rigid frames and cages.

5.1.2.2 Living walls

A living wall is a system in which vegetation not only is attached to a building facade but also is fully integrated into the facade construction where plants and planting media are both placed on the vertical surface of exterior walls (Figure 5.3). Typically, living walls are separated from the facade surface by a layer of waterproof membrane to protect the facade structure from moisture. There are multiple variations of living walls with the most common systems described below:

- *Vegetated mat living walls* consist of a fabric layer attached to a rigid backup support of steel frame, cast-in-place concrete, or masonry wall. Plants pregrown in nurseries are inserted into holes cut in the fabric layer that serves as a growing medium. Vegetated mats are hydroponic systems without soil in which water and nutrients are delivered to plant roots by irrigation pipes behind fabric layers (Figure 5.4).
- *Hanging pocket living walls* are pocket-shaped fabric containers attached to a rigid backup support. Plants are grown in these containers filled with soil or growing medium.
- *Modular living walls* are rigid rectangular containers filled with soil or growing media that can be attached to an exterior wall or be freestanding (Figures 5.5–5.7). The containers are manufactured from metal or lightweight structural plastic and can be shaped as framed boxes, wire cages, or solid boxes with precut holes. Sometimes, the containers are subdivided into smaller individual cells placed perpendicular or angled to a container's back wall. Modular living walls can also be made of a series of troughs or horizontal miniplanters stacked vertically. Plants are grown directly in containers that are filled with soil, inorganic planting media, or natural fiber.

Different types of green facades and living walls are shown in Figure 5.8.

Figure 5.3 Living wall.

Figure 5.4 Vegetated mat living wall.

Figure 5.5 Framed box modular living wall.

Figure 5.6 Perforated box modular living wall.

Figure 5.7 Wire cage modular living wall.

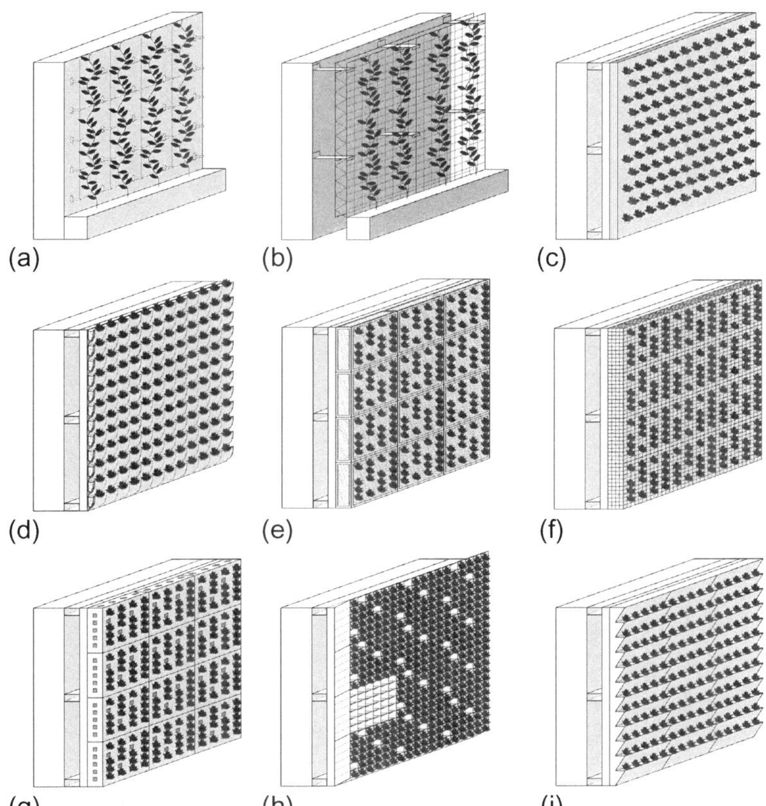

Figure 5.8 Different types of green walls: (a) indirect green facade with a two-dimensional trellis, (b) indirect green facade with a three-dimensional trellis, (c) vegetated mat living wall, (d) hanging pocket living wall, (e) framed box modular living wall, (f) wire cage modular living wall, (g) perforated box modular living wall, (h) slanted cell box modular living wall, and (i) trough planters.

5.1.3 Important aspects of green walls

To use green walls as a cooling strategy in new and existing buildings, the building facade should be designed to account for facade structural requirements, green wall maintenance, and plant selection.

5.1.3.1 Support system

Green walls can be added to new facade construction and to retrofit existing exterior walls. Careful planning is required to ensure that the facade structure has enough strength to support plants. Additionally, exterior wall inspections should be conducted on existing facades to ensure the integrity of wall materials. Direct green facades of self-clinging climbers are best used on exterior walls with rough surfaces, such as brick and concrete. Vine species requiring additional support (indirect green facades) can be grown along any exterior wall materials. It is important to avoid placing vines close to wall elements where plants can climb into holes under wall siding or roof tiles. The type of windows used on the facade is also important; while vines cannot climb along smooth surfaces of glass or metal frames, they can attach themselves to wood window frames and trims that are often featured in old buildings. It is important to ensure that the facade structure can withstand the dead load of an added green wall.

5.1.3.2 Plant selection

Special attention should be given to selecting appropriate plants for green walls. Plants growing on vertical surfaces are more exposed to weather, especially strong winds, than plants on the ground. The best-suited plants for green walls are hardy plant species native to the climate zone, in which the building is located.

The climate zone and length of the cooling season are important criteria for selecting appropriate evergreen or deciduous plant species. Evergreen plants that keep their foliage all year are most suitable when a continuous cooling effect is needed in hot and warm climates. Deciduous plants that shed foliage during cold months are suitable for green walls in temperate and colder climates where they can provide cooling effect in the summer and allow for passive facade heating in winter. Careful consideration should be also given to the plant typical growth rate and biophysiological characteristics, such as maximum height and foliage density.

5.1.3.3 Green wall maintenance

Green facades and living walls require regular irrigation and fertilization. Plants in living wall systems are irrigated automatically through vertical or horizontal pipes installed behind the soil (Dunnett, 2010). Water is treated and pumped to the top of a living wall where it is distributed to plants. Plants in green facade systems are easily irrigated either naturally by rain, manually, or automatically.

Living walls must be periodically examined to ensure the integrity of building materials, the functioning of irrigation systems, and plant health. Depending on wall height, living wall maintenance may require the use of special equipment for maintenance, such as suspended platforms or scissor lifts. Green facades require less maintenance (plant pruning and visual inspection), which can usually be done easily from the ground level (Dunnett, 2010; Grant, 2010).

5.2 Plant cooling mechanisms

5.2.1 Thermal processes in green walls and their effects on energy efficiency

This chapter will focus only on the thermal effects of the plant layer in green facades and living walls. The discussion will not include thermal effects of soil and growing medium that are part of living walls. While soil layer is very beneficial for facade cooling, its thermal properties have been well studied.

Green walls can contribute towards energy efficiency through their ability to affect heat transfer between a building's interior and exterior environments. The main exterior factors affecting heat transfer though a building's facade are solar and thermal radiation from the atmosphere and the ground, air temperature, relative humidity, and wind speed. Plants and some other elements of green walls, such as planting media and support structures, decrease the effects of these weather factors on the exterior wall surface, which causes a reduction in heat transfer through the facade and, as a result, a decrease in heating and cooling energy consumption.

Understanding the energy balance of a vegetated wall and individual thermal physical processes is extremely important in order to evaluate the green wall effects on building thermal performance and potential energy savings. Figure 5.9 shows a schematic diagram of the energy balance and heat flows through a building's facade covered with a plant layer (Campbell, 1998; Gates, 2003; Jones, 1992) (Figure 5.9).

The energy balance of a plant-covered facade accounts for multiple heat flows, including incoming environmental radiation; convection to and from the facade, *C*; evapotranspiration from the plant layer; heat storage in the facade material; and heat conduction through the facade. A typical building facade receives shortwave

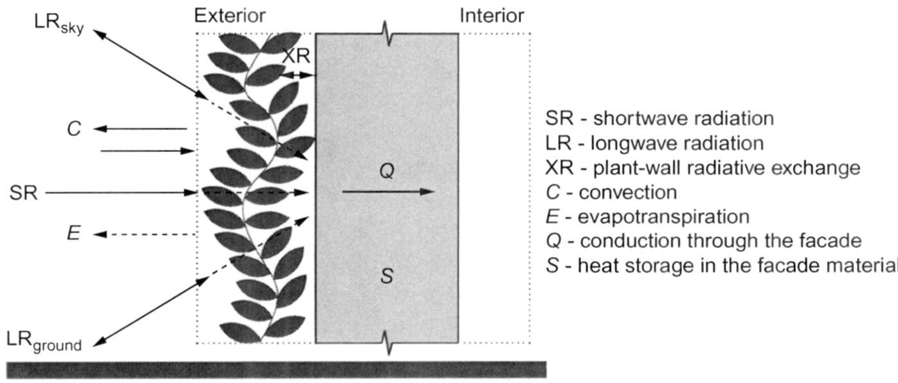

Figure 5.9 Energy balance of a vegetated facade.

radiation, SR, from the sun (direct, scattered from the sky, and reflected from the ground and atmosphere) and also exchanges longwave thermal radiation, LR, between the ground, sky, and surrounding surfaces (Larcher, 2003; Straube, 2005). Radiation absorbed by the facade is either emitted back into the environment; transferred to or away from the wall by convection to the outdoor air, C; or transferred to the building interior by conduction, Q, where energy can also be stored in the wall material, S. At any moment in time, the energy balance of a bare wall, bw, is thus

$$\mathrm{SR}_{\mathrm{bw}} + \mathrm{LR}_{\mathrm{bw}} + C_{\mathrm{bw}} = Q_{\mathrm{bw}} + S_{\mathrm{bw}} \tag{5.1}$$

The energy balance of the vegetated wall, vw, includes an additional term, XR, for the radiative exchange between the leaves of the plant layer and the facade surface, as shown in Equation (5.2):

$$\mathrm{SR}_{\mathrm{vw}} + \mathrm{LR}_{\mathrm{vw}} + \mathrm{XR} + C_{\mathrm{vw}} = Q_{\mathrm{vw}} + S_{\mathrm{vw}} \tag{5.2}$$

The plant layer generally acts as an additional layer of thermal insulation and helps reduce heat gains and losses through a building facade by means of the following mechanisms:

- Shading of the exterior wall from the incoming solar radiation
- Temperature cooling of the air adjacent to the exterior wall through evapotranspiration
- Increasing the exterior wall's thermal insulation value, if the vegetated facade includes a layer of growing medium (soil or inorganic media) placed along the facade as in living walls
- Protection of the exterior wall from wind exposure

This chapter will discuss the first two cooling mechanisms of shading and evapotranspiration because they have the largest impact on reducing heat gains and building cooling loads.

5.2.2 Shading and its cooling effect

A layer of plants placed on a buildings' exterior wall intercepts a fraction of total radiation incident on leaves, reflects some radiation, and transmits the rest to the exterior wall behind it. This shading effect is important in hot months when the facade surface temperature behind the plant layer and the facade surface temperature gradient (the difference in temperatures between the exterior and interior surfaces) of the exterior wall are typically lower than those of a bare facade. During the day, facade exterior surface temperature is reduced by an average of 1–9 °C reaching a maximum of 15–20 °C, as was demonstrated by several previous studies (Chen, 2013; Di, 1999; Evmorfopoulou, 2009; Hoyano, 1988; Olivieri, 2014a,b; Peréz, 2011; Perini, 2011; Price, 2010; Sternberg, 2011; Susorova, 2013a,b, 2014; Wong, 2010a,b). By contrast, during the night, the exterior temperatures of plant-covered facades are 1–2 °C higher than those of bare facades because the vegetative layer prevents nighttime cooling of the wall by longwave radiation to the sky and/or convection to the outdoor air (Hoyano, 1988; Susorova, 2013a,b).

A study by Olivieri et al. found that the presence of a vegetated layer on the facade delays facade peak surface temperatures by approximately two hours (Olivieri, 2014a,b), which can help shift peak cooling load and reduce the associated peak energy charges.

The most significant surface temperature reductions between the bare and the plant-covered exterior walls are recorded during the peak exposure to solar radiation for each facade. In fact, the largest surface temperature reduction is often observed on the east and west facades that are exposed to high-intensity solar radiation in early morning and late evening at low sun angles. This suggests that placing plants on exterior walls of these orientations may be more beneficial for improving facade thermal performance relative to placing plants on north- or south-facing walls, which experience less high-intensity solar radiation.

The amount of solar radiation transmitted to the facade wall surface decreases exponentially with leaf area index (LAI) (Campbell, 1998). The LAI is the total projected area of leaves per unit surface area, which varies with the plant canopy structure, maturity, foliage density, and plant leaf size (Figure 5.10). This ratio varies from <1 for young plants with loose foliage that does not fully cover the wall to 3–5 for mature plants with dense foliage (Yu, 2006; Cameron, 2014). For example, the most common facade vine English ivy (*Hedera helix*) has a LAI of 2.6–7.7 (Ottele, 2011).

The fraction of total solar radiation absorbed by plants increases with a higher plant LAI and the total radiation transmitted through a plant layer decreases exponentially (Jones, 1992; Nobel, 2009). This attenuating effect is expressed through the radiation transmissivity coefficient τ, which is calculated with the widely used relationship developed by Monsi and Saeki in 1953 (Hirose, 2005; Jones, 1992):

$$\tau = \exp(-\kappa \mathrm{LAI}) \tag{5.3}$$

Figure 5.10 Plant layers with different leaf area indexes.

where κ is the radiation attenuation coefficient. The radiation attenuation coefficient is determined by the plant canopy structure and leaf angles. It varies between 0 and 1, where lower values (0.3–0.5) correspond to leaves with leaf angles less than 45° from a vertical surface and higher values (0.7–1) correspond to horizontal leaves with leaf angles more than 45° (Larcher, 2003). When the leaves are perpendicular to the wall, $\kappa = 0$, and when they are parallel to the wall, $\kappa = 1$. Finding precise attenuation coefficient is difficult because of the complexity and irregularity of leave arrangements in canopies and heliotropism when leaf angles change based on the angle of the sun.

An additional factor affecting thermal flux is the temperature gradient, the difference in temperatures between the facade exterior and interior surfaces. The exterior surface temperature fluctuates significantly, while the temperature of the interior surface facing an air-conditioned space is relatively stable. By reducing the facade temperature gradient, the plant layer helps decrease the heat conduction through exterior walls, which leads to reduced building heat gains and consequently cooling load (Di, 1999; Susorova, 2013a,b, 2014).

Plant shading can positively impact facade thermal performance when exterior walls are exposed to sunlight. Large surface temperature reductions between the bare and the plant-covered exterior walls occur during the peak exposure to solar radiation for each facade. Shading plays no role at nighttime, on cloudy days, and when the vegetated facade is permanently shaded by surrounding objects (Evmorfopoulou, 2009; Hoyano, 1988; Susorova, 2014; Wong, 2010a,b). The highest reduction in surface temperature and heat conduction through the facade is observed for facade orientations that have high solar exposure: east, west, and south (facing the equator). In fact, vegetation placed on east and west walls facing high-intensity solar radiation in early morning and late evening hours at low sun angles yields the highest improvements in facade thermal performance (Evmorfopoulou, 2009; Perini, 2011; Susorova, 2014).

5.2.3 Plant evapotranspiration and its cooling effect

Plant foliage absorbs carbon dioxide from the atmosphere and releases oxygen from the plants, while, at the same time, water evaporates into the atmosphere from leaf surfaces. This process is called evapotranspiration.

Evapotranspiration occurs through stomata, microscopic pores on leaf surfaces that control the gas exchange between a plant and its environment. The rate of evapotranspiration is regulated by stomata's aperture, which depends on multiple factors, including outdoor air relative humidity, air temperature, light level, water potential (a tendency of water to flow from one area of the plant to another), and carbon dioxide concentration inside and outside the leaves (Gates, 2003). Stomata are very sensitive to the difference in water potential of the air inside and outside the leaf, and as the ambient air becomes drier, the rate of evapotranspiration reduces significantly to protect plants from drying out. Following the plant's internal rhythm, stomata open during the day and close at night as they react to ambient light level, which means that no evapotranspiration occurs at night.

The rate of water vapor leaving plants through the pores on the leaf surface during transpiration is determined by typical leaf stomatal conductance, one of the physiological characteristics of individual plant species (the reciprocal of stomatal conductance is stomatal resistance). Stomatal conductance depends on the pore size and their number per leaf surface area, which typically amounts to 0.2–2% of the leaf surface. Stomata can be located on both surfaces of a leaf (amphistomatous leaves) or only on the lower surface (hypostomatous leaves) (Nobel, 2009). Typical stomatal conductance values for leaves are 0.002–0.010 m/s (0.08–0.40 mol/m^2 s), and stomatal resistance values are 100–500 s/m (2.5–13 m^2 s/mol) (Gates, 2003; Nobel, 2009).

Generally, the vegetation layer creates a milder local microclimate near buildings as evapotranspiration cools both the plant foliage and the ambient air around plants. The ambient air temperature near exterior walls covered with vines was measured to be 0.8–4 °C lower (depending on orientation) than near bare walls (Cameron, 2014; Hoyano, 1988; Olivieri, 2014a,b; Susorova, 2014; Wong, 2010a,b). The relative humidity is 2–4% higher near facade vegetation. However, even though the plant layer created higher relative humidity near the walls, the absolute humidity ratio (the mass of water vapor per unit volume of air) was similar for both bare and vegetated facades (Sternberg, 2011; Susorova, 2014). This finding disproves a typical misconception about the damaging nature of wall plants that supposedly create high-humidity environment and cause moisture-related problems. A yearlong study monitoring the microclimate conditions near ivy-covered walls in a temperate climate found that vegetated walls experienced smaller daily ranges in outdoor air temperatures and relative humidity (Sternberg, 2011).

The air cooling due to plant evapotranspiration can be beneficial to facade thermal performance because it helps reduce heat conduction through exterior walls and therefore building cooling loads and energy consumption for space cooling. Another potential benefit can be placing the plant layer near mechanical system air intakes as the ambient air precooled by plants will require less energy in order to be conditioned to a desired cooling set point temperature.

Air cooling by evapotranspiration can be used as a method of air cooling all year round in hot and temperate climates and in summertime in cold climates. It is effective at the time when evapotranspiration occurs: during the daytime and during the plant growing season when foliage is present.

Table 5.1 below summarizes the result of different experimental studies on plant cooling effects. Additionally, studies of thermal effects of vegetated facades were analyzed and summarized in a review article by Hunter (2014).

5.2.4 Other thermal effects of vegetated facades

Another effect of plants important for facade thermal performance is a reduction in air velocity near the exterior walls by up to 84% (Perini, 2011). A layer of nearly still air formed behind the facade vegetation serves as an additional layer of thermal

Table 5.1 **Overview of experimental studies on plant cooling effects**

Author and year	Location	Climate zone[a]	Period	Duration	Green wall description	Facade orientation	Exterior facade surface ΔT (decrease)	Ambient air ΔT (decrease)
Hoyano, 1988	Tokyo, Japan	Humid subtropical	Summer	3 days	Vine-covered screen in front of a window	SW	13–15 °C	1–3 °C
					Vine-covered building facade	W	10 °C	
Di, 1999	Beijing, China	Humid continental	Jun	1 month	Vine-covered building facade	W	8.2 °C	
Evmorfopoulou, 2009	Thessaloniki, Greece	Semiarid	Jul–Aug	1 month	Vine-covered building facade	E	1.9–8.3 °C	1–2 °C
Wong, 2010a, 2010b	Singapore	Tropical	Feb, Apr, Jun	3 days	Mock-ups of 9 living wall types	S	1.1–11.6 °C	3.33 °C
Perini, 2011	Delft, the Netherlands	Oceanic	Sep–Oct	1 month	Vine-covered building facade	NW	1.20 °C	0.12 °C
	Rotterdam, the Netherlands				Vine-covered trellis in front of a building facade	NE	2.73 °C	0.17 °C
	Benthuizen, the Netherlands				Planter box living wall	W	3.85 °C	(0.85 °C)
Peréz, 2011	Lleida, Spain	Arid	Apr–Sep	6 months	Vine-covered screen in front of a building facade	NW, SW, SE	5.5 °C	

Sternberg, 2011	Lleida, Spain	Arid	Apr–Sep	6 months	Vine-covered screen in front of a building facade	NW, SW, SE	5.5 °C	
Susorova, 2013a, 2013b	Chicago, the United States	Humid continental	Aug	1 week	Vine-covered building facade	S	1.1–5.7 °C (0.4 °C when cloudy)	
Chen, 2013	Wuhan, China	Humid subtropical	Jul	1 day	Modular box living wall with air gap	W	4.2–20.8 °C	
Mazzali, 2013	Lonigo, Italy	Humid subtropical	Jun–Sep	2.5 months	Vegetated mat living wall	SW	6–20 °C (2–3 °C when cloudy)	
Cameron, 2014	Reading, the United Kingdom	Temperate oceanic	Aug–Sep	1 month	Potted plants in front of a facade mock-up	S	9.9 °C	3 °C
Susorova, 2014	Chicago, the United States	Humid continental	Jul	10 days	Vine-covered building facade	N, E, S, W	0.7–12.6 °C	0.8–2.1 °C

[a]Köppen climate classification.

insulation. Reducing air speed and temperatures near facades can help reduce air infiltration and exfiltration through exterior walls, which is driven by differences in wind pressures and air temperatures between the interior and the exterior of a building.

Air infiltration is one of the highest contributors to building heating loads and heating energy use in cold weather. On the other hand, the opposite of air infiltration, air exfiltration from the air-conditioned space towards the exterior, contributes to building cooling loads. A recent study estimated that a 20 cm thick layer of Boston ivy (*Parthenocissus tricuspidata*) placed on exterior masonry walls of four-story buildings in Chicago could reduce air infiltration by an average of 4–12% and by a maximum of 42–97%, depending on a selected air infiltration model (Susorova, 2014). At the same time, reduction in wind speed near exterior walls might have a negative impact on facade thermal performance in hot weather as facade surface cooling by convection might be impeded by vegetation. The effect of facade vegetation on air infiltration and exfiltration has not been quantified exactly and merits further study.

A recent study examined plant cooling mechanisms by monitoring six different plant species in a controlled environment facility (Cameron, 2014). Potted plants of various species were placed in front of bare wall inside of growth cabinets. The plants were allowed to grow until their foliage covered the controlled walls in a uniform way. To assess the contribution of the shading effect versus the evapotranspiration effect, some leaves were sealed to prevent gas exchange between the plants and the environment. The findings showed that *Hedera helix*, *Lonicera*, and *Jasminum officinale* cooled the air and wall surface through shading, while *Fuchsia* cooled by evapotranspiration. The *Prunus laurocerasus* and *Stachys byzantina* cooling effect could be equally attributed to shading and evapotranspiration. Thus, the exact cooling mechanism depends on multiple factors, such as plant maturity, LAI, canopy structure, leaf size, stomata size, density, stomatal pore conductance, and dynamic weather conditions.

5.3 Effective thermal resistance of a plant layer

Reduced surface and air temperatures of exterior walls covered by plants lead to reduced heat flux through the opaque facade. Several studies estimated that mean instantaneous heat flux reduction due to the plant layer can vary and reach 10% (Susorova, 2014), 75% (Hoyano, 1988), and 70–80% (Mazzali, 2013), depending on the experiments' weather conditions, solar radiation intensity on the wall, and plant layer characteristics.

To evaluate the effect of the plant layer on heat conduction through facades and to compare results with those of other insulating materials, one can utilize a metric of "effective thermal resistance." The effective R-value of bare and vegetated facades is calculated as the temperature gradient between the indoor and the outdoor surfaces divided by the dynamic heat flux through the wall. The reduction in heat flux through the facade caused by the plant layer, due to shading and evapotranspiration, increases thermal resistance of the exterior wall. If instantaneous surface temperatures and heat

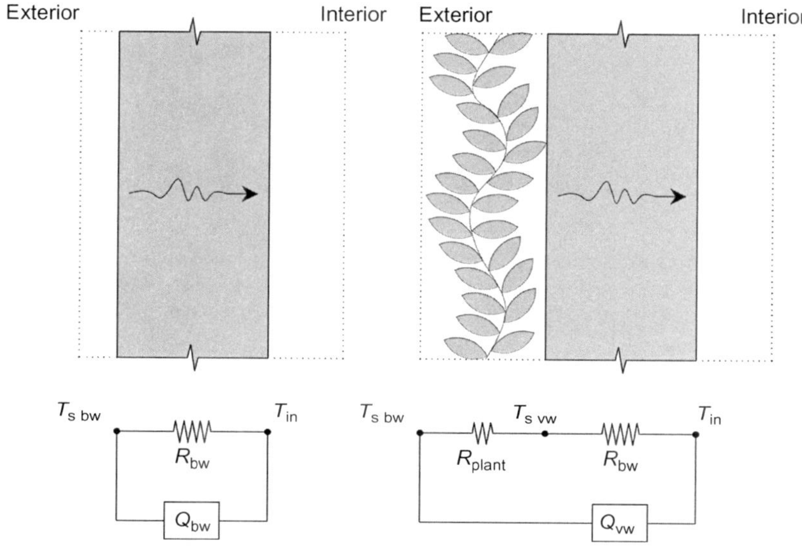

Figure 5.11 Heat flux through the bare facade (left) and vegetated facade (right) expressed as an electrical circuit diagram.

flux through the exterior wall are known, effective R-values of a plant layer can then be calculated for any weather conditions, plant characteristics, and facade material properties. This concept can be illustrated through the analogy of an electric circuit diagram where thermal resistances are presented as resistors in series, temperature gradient is analogous to voltage, and heat flux is analogous to current (Figure 5.11).

From the heat flux through the facade, one can find the instantaneous effective thermal resistances of the bare facade R_{bw}:

$$R_{bw} = \frac{T_{s\ bw} - T_{in}}{Q_{bw}} \tag{5.4}$$

where $T_{s\ bw}$ is the bare facade exterior surface temperature (K), T_{in} is the indoor surface temperature, and Q_{bw} is the heat flux through the exterior wall (W/m^2). The instantaneous effective thermal resistance of the vegetated facade R_{vw}, which includes the bare wall and a layer of plants, is calculated as

$$R_{vw} = R_{bw}\frac{Q_{bw}}{Q_{vw}} = \frac{T_{s\ bw} - T_{in}}{Q_{vw}} \tag{5.5}$$

where Q_{vw} is the heat flux through the wall with the plant layer (W/m^2). Knowing the effective thermal resistances of the bare and vegetated facades, the instantaneous effective thermal resistance of the plant layer R_{plant} can be estimated as

$$R_{plant} = R_{vw} - R_{bw} \tag{5.6}$$

A study by Ottele, which measured thermal properties of English ivy on a wall in a controlled environment using a hot-box apparatus, found that the plant thermal resistance is approximately 0.18 m^2 K/W (Ottele, 2011). However, experiments conducted in controlled environments are very different from field measurements where weather conditions constantly change throughout a day. These findings were confirmed with a study by Susorova that calculated the effective thermal resistance of a plant layer using a model of a vegetated facade that dynamically accounts for changing weather conditions (Susorova, 2013a,b). Plant thermal resistance calculated using weeklong experimental data for Chicago varied throughout the day and reached a maximum of 0.14 m^2 K/W. It suggests that the plant layer reduced conductive heat transfer by as much as an additional 10 cm layer of brick or 0.5 cm layer of expanded polystyrene insulation for this climate and for these conditions (ASHRAE Handbook Fundamentals, 2009; Straube, 2005).

Plant effective thermal resistance is the highest when the plant layer covers exterior walls made of materials with high solar absorptivity (e.g., dark brick walls) because such material surfaces can heat to higher temperatures due to a larger fraction of absorbed solar radiation and when exterior wall assembly has little or no thermal insulation. The plant thermal insulation effect is less pronounced for reflective and lightly colored facade materials with low solar absorptivity and for well-insulated exterior walls.

Note: Although plant elements, such as leaves and stems, have some thermal value (e.g., thermal conductivity of fresh leaves is 0.37–0.41 W/m K for *Citrus limon*, 0.36–0.38 W/m K for *Arbutus menziesii*, 0.32–0.36 W/m K for *Eucalyptus globulus*, and 0.55–0.56 W/m K for *Peperomia obtusifolia*), it is negligible in comparison with the thermal insulation of the whole plant layer (Hays, 1975).

5.4 Building energy savings with vegetated facades

While the positive effect of the plant layer on facade thermal performance is widely studied, it is not yet well quantified, especially the effect of the plant layer on building energy consumption, which is an open field of study. Several studies examined the effects of vegetation (trees and climbing plants) on heating and cooling loads and energy consumption in buildings, primarily using the method of energy simulations. The most prominent studies are reviewed below.

A study by Huang (1987) researched the effect of trees on annual energy consumption, peak cooling loads, and energy cost by simulating energy use in a typical air-conditioned, one-story wood-frame house located on sites with different amounts of tree area coverage in four different hot climates. The simulation results showed that annual energy cooling requirements were significantly reduced with an 11–18% energy use reduction for 10% tree-covered sites, a 27–43% reduction for 25% tree-covered sites, and a 33–53% reduction for 30% tree-covered sites.

Another study by McPherson (1988) examined the shading and wind reduction effects of vegetation on building cooling and heating energy cost. The researchers simulated energy use in a one-story square building for four different climate zones (Wisconsin, Utah, Arizona, and Florida) testing multiple parameters, such as building orientation, plant shading factors, and wind speeds. Tree shading effects reduced the space cooling cost by 53–61% and peak cooling load by 32–49% in temperate and hot climates. However, in cold climates, with predominant heating requirements, reducing building solar exposure with trees in winter increased the energy cost by 21–24%. On the other hand, the wind reduction effect allowed for lower building air infiltration and decreased heating loads by 9–11% in cold climates but increased cooling loads by 17–23% in hot climates.

Few studies have evaluated energy savings specifically in buildings with green walls. For example, Di (1999) measured a 28% reduction in the peak cooling load through a west-facing wall of a building covered with thick ivy on a clear summer day. Similarly, a study by Price (2010), who developed a mathematical model based on his earlier experimental findings to evaluate energy savings in low-rise buildings with plant-covered trellises, found that building cooling load could be reduced by a maximum of 28% in a temperate climate. Another study by Kontoleon (2010) using a one-dimensional thermal resistance model to analyze the influence of plant cover on energy performance in a one-story brick building in Thessaloniki, Greece, found that the vegetation layer can decrease building cooling load by 5% for the north-facing wall, 18% for the east-facing wall, 8% for the south-facing wall, and 20% for the west-facing wall.

A parametric study by Wong (2009), which simulated the energy performance of a hypothetical 10-story office building in Singapore with various degrees of plant facade coverage, showed a significant cooling load reduction of 74% in a building with a 100% opaque facade fully covered with vegetation, a cooling load reduction of 10% in a building with a 50% opaque facade fully covered with vegetation and 50% glazing, and a cooling reduction of 32% in a fully glazed building with a curtain wall facade fully covered with vegetation.

A more comprehensive way to evaluate thermal performance of vegetate facades is to use mathematical models rather than only experiments. Such models should be able to dynamically simulate the effect of plants on the facade thermal performance for variable plant characteristics, facade properties, building orientations, and weather conditions.

Several mathematical models of ground vegetation canopy were developed in the field of climatology (Deardorff, 1978; Zhang, 1997) and were later adapted to simulate the thermal effects of plants in green roofs (Alexandri, 2007; Djedjig, 2012; Sailor, 2008). The model of vegetated facades developed by Susorova (2013a,b) offers an improvement over previous models by computing plant physiological processes, including evapotranspiration and radiative and convective heat exchange between the plant layer, the facade, the surrounding environment, and the ground and using individual plant characteristics inputs (e.g., leaf absorptivity, typical leaf dimension, LAI, radiation attenuation coefficient, and leaf stomatal conductance) and weather data to simulate the impacts of vegetated walls on facade thermal performance. A few new thermal models of green walls were recently developed by Malys (2014) and Olivieri (2014a,b).

The vegetated facade mathematical model (without soil) was used to evaluate the impact of wall plants on building energy consumption (Susorova, 2013a,b). The study's objectives were to evaluate the energy use reduction in buildings with plant-covered facades and to examine the effects of various parameters on building energy efficiency, including weather conditions, climate zones, facade orientation, wall assembly types, and a variety of plant characteristics. Annual energy simulations were performed to find energy use reduction due to the plant layer. The author compared the annual energy use in two air-conditioned thermal zones—the first with a bare and the second with a vegetated facade—located in the hot climate of Phoenix. A thermal zone in this context is an indoor room with a constant temperature set point maintained at the thermal comfort level, which is achieved through mechanical heating, cooling, and ventilation (Figure 5.12).

Each thermal zone was 6 m wide, 6 m deep, and 2.75 m high and had an adiabatic floor, ceiling, and three interior walls. Heat transfer between the defined thermal zone and the exterior occurred only through the single exterior wall, which was left bare in the first model and covered with a plant layer in the second model. Energy use in the tested buildings was analyzed for a range of parameters, including building occupancy type (office and residential), zone geometry (one and two exterior walls), exterior wall opaqueness (100% opaque and 60% opaque + 40% glazing), facade orientation (north, east, south, and west), exterior wall assembly type (brick wall with and without insulation and metal stud wall with and without insulation), and plant LAI (1, 2, and 3). The results of energy simulations are discussed in detail below and summarized in Tables 5.2 and 5.3:

- Annual cooling energy use was reduced by an average of 1–33% for LAI 1, 2–55% for LAI 2, and 2–66% for LAI 3. Total energy consumption was reduced by an average of 0.3–2.7% for LAI 1, 0.6–4.4% for LAI 2, and 0.2–5.4% for LAI 3. The highest total energy reduction of 6.2% corresponding to an annual cooling energy reduction of 34.6% was achieved in a residential thermal zone with one east-facing opaque exterior wall made of uninsulated brick material, covered with a dense plant layer (LAI 3).
- Among two studied building types, the energy use reduction in the residential thermal zones was higher than that of the office zones. This variation is explained by the different nature of office and residential buildings in their schedule of operation, people occupancy, and lighting and equipment loads. Office buildings are typically dominated by internal heat gains generated by office equipment, lighting, and high occupant density, while residential

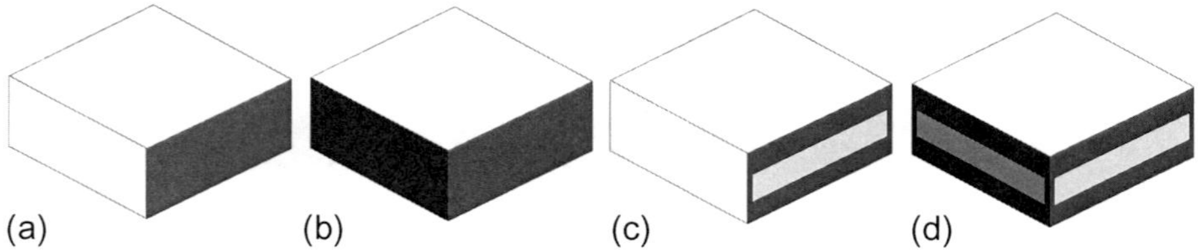

Figure 5.12 Thermal zones used for energy savings calculations: (a) zone with a single exterior wall, (b) zone with two exterior walls, (c) zone with a single exterior wall and 40% glazing, and (d) zone with two exterior walls and 40% glazing.

Table 5.2 **Total annual energy savings in the tested thermal zone for various parameter combinations (without fenestration)**

	Office thermal zone (100% opaque wall)						Residential thermal zone (100% opaque wall)					
	1 ext. wall				2 ext. walls		1 ext. wall				2 ext. walls	
Facade wall type	N	E	S	W	N+E	S+W	N	E	S	W	N+E	S+W
Cooling annual energy use reduction (%)												
Brick wall (ins.)	0.2	5.1	4.8	2.9	1.7	7.4	1.3	16.8	15.6	9.6	5.5	21.2
Brick wall (unins.)	1.1	14.0	12.7	8.5	5.2	16.8	4.4	34.6	32.0	20.9	11.0	35.5
Metal stud wall (ins.)	(0.0)	1.4	1.0	0.4	(0.1)	1.5	(0.1)	4.9	3.2	1.2	(0.5)	4.6
Metal stud wall (unins.)	0.1	4.5	3.3	1.6	0.3	4.5	0.2	12.1	8.9	4.4	0.6	10.4
Total annual energy use reduction (%)												
Brick wall (ins.)	0.0	1.4	0.3	0.7	0.4	0.4	0.1	2.1	0.4	1.1	0.6	0.6
Brick wall (unins.)	0.3	4.8	1.6	2.7	1.9	2.3	0.6	6.1	1.9	3.4	2.0	2.5
Metal stud wall (ins.)	—	0.3	(0.1)	0.1	(0.0)	(0.2)	(0.0)	0.5	(0.2)	0.1	(0.0)	(0.3)
Metal stud wall (unins.)	0.1	1.4	0.1	0.6	0.3	0.2	0.1	1.9	0.2	0.8	0.4	0.2

Note: N, north; E, east; S, south; and W-west facade orientations; ins., insulated; unins., uninsulated.

Table 5.3 Total annual energy savings in the tested thermal zone for various parameter combinations (with fenestration)

	Office thermal zone (60% opaque wall+40% glazing)						Residential thermal zone (60% opaque wall+40% glazing)					
	1 ext. wall				2 ext. walls		1 ext. wall				2 ext. walls	
Facade wall type	N	E	S	W	N+E	S+W	N	E	S	W	N+E	S+W
Cooling annual energy use reduction (%)												
Brick wall (ins.)	0.3	2.8	2.6	1.6	0.8	3.0	0.6	5.5	5.0	3.0	1.3	5.0
Brick wall (unins.)	1.1	8.7	8.0	5.3	3.0	9.2	2.2	15.9	14.9	9.3	4.3	14.5
Metal stud wall (ins.)	(0.0)	0.8	0.5	0.2	(0.1)	0.6	(0.1)	1.5	1.0	0.4	(0.1)	1.0
Metal stud wall (unins.)	0.1	2.7	2.0	1.0	0.2	2.3	0.1	5.1	3.8	1.8	0.2	3.7
Total annual energy use reduction (%)												
Brick wall (ins.)	0.1	1.0	0.2	0.5	0.3	0.2	0.1	0.9	0.2	0.4	0.2	0.2
Brick wall (unins.)	0.3	3.5	1.2	2.0	1.3	1.5	0.3	3.2	1.0	1.7	0.9	1.1
Metal stud wall (ins.)	—	0.2	(0.1)	0.0	(0.0)	(0.1)	—	0.2	(0.1)	0.0	(0.0)	(0.1)
Metal stud wall (unins.)	0.1	1.0	0.1	0.4	0.2	0.1	0.1	0.9	0.1	0.4	0.2	0.1

Note: N, north; E, east; S, south; and W, west facade orientations; ins., insulated; unins., uninsulated.

buildings are dominated by heat gains and losses from exterior envelope. Heat conduction through exterior walls is a major contributor to cooling loads in residential buildings, and any thermal improvements to exterior envelopes produce noticeable building cooling energy reduction.

- The improvements in energy consumption depend directly on the fraction of the building facade covered with the plant layer. The maximum energy improvements occurred in thermal zones that had two opaque exterior walls covered with vegetation. The energy savings effect of the vegetated facade decreased moderately for thermal zones whose exterior walls contained glazing that reduced the fraction of the plant cover to 60%.
- The facade orientation plays a significant role in achieving energy savings, which is explained by the facade receiving a different amount of solar radiation depending on the latitude, solar angle, and elevation of the building site. The highest energy reduction, due to the plant layer on the facade in Phoenix, occurred for east and south orientations.
- The materials used in exterior walls behind the plant layer played an important role in energy use as well. Among the thermal zones with four different exterior wall assemblies, the highest energy use reduction was achieved in the zones with uninsulated exterior brick walls. The energy improvements were high when the thermal resistance of the original exterior walls was low and occurred mainly for poorly insulated walls. The improvements to well-insulated exterior walls were marginal.
- The highest energy savings occurred in the thermal zones whose exterior walls were covered with a very dense plant layer that had a LAI of 3.

To conclude, green walls can contribute towards cooling energy savings in buildings, but the reduction in energy use greatly varies depending on multiple environmental factors and building geometry and materials. One cannot expect green walls to be the sole solution to reduce building energy consumption. However, the greatest reduction in cooling loads and energy consumption can be achieved when the use of green walls in buildings is combined with other energy-efficiency measures.

5.5 Additional benefits of vegetated facades

5.5.1 *Improving urban air quality*

Cities have a lower level of oxygen than the countryside due to the scarcity of oxygen-producing plants. Urban air quality is further degraded by pollutants and greenhouse gas emissions from vehicles, buildings, and industries.

As a by-product of photosynthesis, plants generate oxygen, which is essential to sustain life on this planet. A mature beech tree with a leaf surface area of 1600 m^2 can generate 1.71 kg of oxygen hourly while using 2.4 kg of carbon dioxide, which is equivalent to a typical hourly oxygen consumption of 10 people (Minke, 1985). Plants in green walls can help increase oxygen levels in urban environments. For example, 1000 m^2 of English ivy can produce 1712 kg of oxygen and consume 2351 kg of carbon dioxide (Ottele, 2011).

Green wall plants can also help mitigate urban air quality problems by reducing the amount of greenhouse gases in the atmosphere and by adsorbing air pollutants. Plant foliage adsorbs carbon dioxide (CO_2) and nitrogen oxide (NO_2) gases, volatile organic compounds (VOC), and particulate matter during gas exchange between stomatal pores on the leaves and the atmosphere (Ottele, 2011). A study by Pugh (2012) shows that green walls applied across large wall areas of street canyons can reduce concentrations of nitrogen oxide and particular matter with particle diameters less that 10 μm by 15–23% (these are common air pollutants from vehicle emissions, a common cause of pulmonary and cardiac diseases).

Leaves also have the ability to adsorb pollutant particles of heavy metals from the atmosphere, including cadmium, copper, lead, and zinc, and can reduce fine dust concentration in the air by 20% (Minke, 1985; Ottele, 2011). A German study demonstrated that the air pollution level in a street without trees was 10,000–20,000 dirt particles per liter, as opposed to 3000 dirt particles per liter in a tree-lined street (Minke, 1985).

Air pollutants are present not only in the atmosphere but also inside buildings where various building materials (adhesives, carpets, electronic equipment, and cleaning fluids) emit volatile organic compounds (VOCs), chemical compounds that can negatively affect human health. Air filtration with plants in indoor green walls can be a natural alternative to energy-consuming artificial ventilation. For example, Nedlaw Living Walls, a proprietary biofilter that consists of pollutant-degrading plants, could remove up to 80% of the formaldehyde, 50% of the toluene, and 10% of the trichloroethylene with a single pass of the air through the 5 cm thick living wall (Nedlaw Living Walls, 2008).

5.5.2 Health benefits

Plants are known to have positive effects on psychological and physiological health. Results of multiple studies demonstrated that when inside buildings, people prefer to have a visual contact with the exterior vegetation behind windows (Beatley, 2010; Peck, 1999; White, 2011). Additionally, the air filtering and oxygenating abilities of plants can greatly benefit people suffering from breathing diseases caused by urban pollution, such as asthma and allergies (Peck, 1999). Overall, skillful use of vegetation in green facades and living walls can reduce stress, a common cause of psychological disorders and cardiovascular diseases (Browning et al., 2012).

5.5.3 Envelope protection

Facade vegetation protects wall construction behind the plant layer from multiple external factors that can cause material deterioration, such as excessive solar and ultraviolet radiation and wide temperature fluctuations. By decreasing daily temperature fluctuations, plants help reduce internal stress in building materials, which can lead to material cracking and deterioration. On extreme days, the exposed facade temperature can vary between −10 °C and 60 °C, while the temperature of a plant-covered facade fluctuates only between 5 °C and 30 °C (Minke, 1985).

The external plant layer protects building envelop from driving rain (moisture under pressure entering an exterior wall assembly), which is the main source of moisture in a typical bare wall assembly. Measurements by Bartfelder and Kohler showed that the wall surface behind English ivy was drier that the bare walls (Bartfelder, 1987). However, if moisture is trapped behind the vegetation layer, it dries out more slowly due to reduced air flow and sun exposure.

Overall, wall materials that are protected from external factors do not require as much maintenance, have an increased life span, and have a lower life cycle cost.

5.5.4 Noise reduction

A layer of vegetation in green walls helps reduce noise and can be a good sound insulator (Van Renterghem, 2013). A study by Wong et al. examining sound absorption of several green wall types found that a plant layer can reduce sound by 5–10 dB in the low and middle frequency range (200–1000 Hz) that is clearly noticeable by human ear (Wong, 2010a,b).

5.6 Future trends

5.6.1 Further research

Man-made environment and structures consume a great deal of limited natural resources through their extensive use of energy and materials. Further, urban development poses multiple problems, including environmental pollution, reduction in biodiversity, and disappearance of the natural environment. Green walls can help mitigate the problems cause by urban development. The vegetation component of green walls can lower energy use in buildings, reduce urban heat island effect, improve air quality, lower the risk of some diseases, and increase biodiversity of plants and animals in cities.

The effects of vegetation on building thermal and energy performance are a relatively new area of study. To fully explore the potential of plants as an insulating material, further investigations should be conducted on the following topics:

- Study the effect of wind speed reduction due to the plant layer and its role on air infiltration through experiments and building monitoring.
- Quantify normalized energy savings in buildings with plant-covered envelopes (energy use reduction per unit area of the vegetated facade) using simulation methods and energy consumption monitoring in actual buildings.
- Explore how vegetated facades affect the microclimate and urban heat island effect.
- Analyze the life cycle cost of vegetated facade systems.
- Evaluate vegetated facade thermal performance and potential energy savings in cold climates.

Green wall researchers advocate standardization of metrics for experimental and simulation-based studies of vegetated facade thermal performance (Hunter, 2014). Future research studies on thermal effects of vegetated facades should take into consideration the following basic metrics and variables, which can be used to estimate exterior wall energy balance, cooling loads, and energy consumption:

(1) Weather conditions (a) near the bare facade and (b) facade behind the vegetation layer:
- Outdoor air temperature
- Relative humidity
- Wind speed
- Direct solar radiation and solar radiation incident at the facade
- Precipitation

(2) Building characteristics:
- Orientation
- Building occupancy type (indoor thermal set points and schedules of operation)
- Amount of opaque walls and fenestration

(3) Thermal properties of the exterior wall assembly:
- Thermal conductance (U-value) or resistance (R-value)
- Thermal capacity of wall materials
- Solar absorptivity coefficient of the exterior wall
- Surface temperature of (a) the bare facade and (b) facade behind the vegetation layer
- Heat flux through (a) the bare facade and (b) vegetated facade

(4) Physiological characteristics of individual plant species used in green walls:
- Leaf area index (foliage density)
- Leaf typical stomatal conductance
- Leaf reflectivity
- Plant species growing rate and pattern

5.7 Sources of further information and advice

Further information on green walls can be found on websites of professional organizations and green wall manufacturers.

5.7.1 Professional organizations

- Green Roofs for Healthy Cities, http://www.greenroofs.org/
- International Green Roof Association, IGRA, http://igra-world.com/
- World Green Infrastructure Network, http://www.worldgreenroof.org/

5.7.2 Green facade manufacturers

- Carl Stahl DecorCable, http://www.decorcable.com/
- Greenscreen, http://www.greenscreen.com/
- GSky Plants Systems, Inc., http://gsky.com/
- Jacob UK, http://www.jakob.com/
- Helix Plant Systems, http://www.helix-pflanzensysteme.de/

5.7.3 Living wall manufacturers

- Biotecture, http://www.biotecture.uk.com/
- Elmich VGM Green Wall modules, http://www.elmich.com.au/VGM/
- ELT Easy Green, http://www.eltlivingwalls.com/

- G-O2, http://www.myplantconnection.com/
- Green Living Technologies, http://agreenroof.com/
- Green over Grey, http://www.greenovergrey.com/
- Greenwall, http://www.greenwall.fr/
- Greenwall Biosistemas Urbanas, http://greenwall.com.br/
- Greenwalls Vertical Planting Systems, http://greenwalls.com/
- Gruenwand, http://www.gruenwand.at/
- Nedlaw Living Walls, http://naturaire.com/
- Plants On Walls, http://www.plantsonwalls.com/
- Sage Vertical Garden Systems, http://www.sageverticalgardens.com/
- Sempergreen, http://www.sempergreen.com/us
- The Greenwall Company, http://www.greenwall.com.au/
- The Living Wall, http://www.thelivingwallco.com/
- Tournesol Siteworks, http://www.tournesolsiteworks.com/category-Greenwall%20and%20Roof.html
- Vertical Garden Patrick Blanc, http://www.verticalgardenpatrickblanc.com/
- Woolly Pocket, http://www.woollypocket.com/
- 90DEGREEN, http://www.90degreen.com/

References

Alexandri, E.J., 2007. Developing a one-dimensional heat and mass transfer algorithm for describing the effect of green roofs on the built environment: comparison with experimental results. Build. Environ. 32, 2835–2849.

ASHRAE Handbook Fundamentals, 2009. Atlanta: American Society of Heating, Refrigerating and Air-Conditioning Engineers.

Bartfelder, F.M., 1987. Experimentelle Untersuchungen zur Function von Fassadenbegrünungen. Dissertation TU Berlin 612S.

Beatley, T., 2010. Biophilic Cities: Integrating Nature into Urban Design and Planning. Island Press, Washington.

Blanc, P., 2008. The Vertical Garden: From Nature to the City. W. W. Norton & Company, New York.

Browning, B., Garvin, C., Ryan, C., Kallianpurkar, N., Labruto, L., Watson, S., Knop, T., 2012. The Economics of Biophilia. Terrapin Bright Green LLC, New York.

Cameron, R.W., 2014. What's 'cool' in the world of green facades? How plant choice influence the cooling properties of green walls. Build. Environ. 73, 198–207.

Campbell, G.S., 1998. An Introduction to Environmental Biophysics. Springer, New York.

Chen, Q.B., 2013. An experimental evaluation of the living wall system in hot and humid climate. Energy Build. 61, 298–307.

Deardorff, J.W., 1978. Efficient prediction of ground surface temperature and moisture, with inclusion of a layer of vegetation. J. Geophys. Res. 83, 1889–1902.

Di, H.F., 1999. Cooling Effect of Ivy on A Wall. Exp. Heat Transfer 12 (3), 235–245.

Djedjig, R.O., 2012. Development and validation of a coupled heat and mass transfer model for green roofs. Int. Commun. Heat Mass Transfer 39, 752–761.

Dunnett, N.a., 2010. Planting Green Roofs and Living Walls. Timber Press, Portland.

Evmorfopoulou, E.A., 2009. Experimental approach to the contribution of plant-covered walls to the thermal behaviour of building envelopes. Build. Environ. 44 (5), 1024–1038.

Gates, D.M., 2003. Biophysical Ecology. Dover Publications, Inc., New York.

Grant, G., 2010. Green Roofs and Facades. IHS–BRE Press, Bracknell.

Hays, R.L., 1975. The thermal conductivity of leaves. Planta 125, 281–287.

Hirose, T., 2005. Development of the Monsi–Saeki theory on canopy structure and function. Anal. Bot. 95, 483–494.

Hoyano, A., 1988. Climatological uses of plants for solar control and the effects on the thermal environment of a building. Energy Build. 11 (1–3), 181–199.

Huang, Y.J., 1987. The potential of vegetation in reducing summer cooling loads in residential buildings. J. Clim. Appl. Meteorol. 26, 1103–1116.

Hunter, A.M., 2014. Quantifying the thermal performance of green facades. A critical review. Ecol. Eng. 63, 102–113.

Jones, H.G., 1992. Plants and microclimate. A quantitative approach to environmental plant physiology. Cambridge University Press, Cambridge.

Kontoleon, K.J., 2010. The effect of the orientation and proportion of a plant-covered wall layer on the thermal performance of a building zone. Build. Environ. 45, 1287–1303.

Larcher, W., 2003. Physiological Plant Ecology, fourth ed. Springer, New York.

Malys, L.M., 2014. A hydrothermal model to assess the impact of green walls on urban microclimate and building energy consumption. Build. Environ. 73, 187–197.

Mazzali, U.F., 2013. Experimental investigation on the energy performance of living walls in a temperate climate. Build. Environ. 64, 57–66.

McPherson, E.G., 1988. Impacts of vegetation on residential heating and cooling. Energy Build. 12, 41–51.

Minke, G.a., 1985. Häuser Mit Grünem Pelz. Ein Handbuch zur Hausbegrünung. Edition Fricke, Köln.

NEDLAW Living Walls, 2008. Retrieved February 28, 2011, from http://www.naturaire.com/index.php.

Nobel, P.S., 2009. Physicochemical and Environmental Plant Physiology. Academic Press, Burlington.

Olivieri, F.O., 2014a. Experimental study of the thermal-energy performance of an insulated vegetal façade under summer conditions in a continental mediterranean climate. Build. Environ. 77, 61–76.

Olivieri, F.R., 2014b. Experimental characterization and implementation of an integrated autoregressive model to predict the thermal performance of vegetal facades. Energy Build. 72, 309–321.

Ottele, M., 2011. The Green Building Envelope Vertical Gardening. Doctoral Thesis, Delft Technical University.

Peck, S.W., 1999. Greenbacks from Green Roofs: Forging a New Industry in Canada. Canada Mortgage and Housing Corporation, Ottawa.

Peréz, G.L., 2011. Green vertical systems for buildings as passive systems for energy savings. Appl. Energy 88, 4854–4859.

Perini, K.M., 2011. Vertical gardening systems and the effect on air flow and temperature on building envelope. Build. Environ. 46, 2287–2294.

Price, J., 2010. Green Facade Energetics. M.S. Thesis, University of Maryland.

Pugh, T.A., 2012. Effectiveness of green infrastructure for improvement of air quality in urban street canyons. Environ. Sci. Technol. 46, 7692–7699.

Sailor, D.J., 2008. A green roof model for building energy simulation programs. Energy Build. 40, 1466–1478.

Sternberg, T.V., 2011. Evaluating the role of ivy (Hedera helix) in moderating wall surface microclimates and contributing to the bioprotection of historic buildings. Build. Environ. 46, 293–297.

Straube, J.a., 2005. Building Science for Building Enclosures. Building Science Press Inc., Westford.

Susorova, I., 2013a. Evaluation of the effects of vegetation and green walls on building thermal performance and energy consumption. Illinois Institute of Technology, Chicago.

Susorova, I.M., 2013b. A model of vegetated exterior facades for evaluation of wall thermal performance. Build. Environ. 67, 1–13.

Susorova, I.P., 2014. The effects of climbing vegetation on the local microclimate, thermal performance, and air infiltration of four building facade orientations. Build. Environ. 76, 113–124.

Van Renterghem, T.M., 2013. The potential of building envelope greening to achieve quietness. Build. Environ. 61, 34–44.

White, E.V., 2011. Greenery on residential buildings: does it affect preferences and perceptions of beauty? J. Environ. Psychol. 31, 89–98.

Wong, N.H., 2009. Energy simulation of vertical greenery systems. Energy Build. 33, 1401–1408.

Wong, N.H., 2010a. Acoustics evaluation of vertical greenery systems for building walls. Build. Environ. 45, 411–420.

Wong, N.H., 2010b. Thermal evaluation of vertical greenery systems for building walls. Build. Environ. 45 (3), 663–672.

Yu, C., 2006. The Intervention of Plants in the Conflicts between Buildings and Climate—A Case Study in Singapore. Ph.D. Thesis for Department of Building, National University of Singapore.

Zhang, J.Q., 1997. A heat balance model for partially vegetated surfaces. Infrared Phys. Technol. 38, 287–294.

Comparison of the performance of different facade materials for reducing building cooling needs

6

N.L. Alchapar, E.N. Correa
CONICET—CCT-Mendoza, CC, Mendoza, Argentina

6.1 Introduction

The thermal balance of a city is defined by the optical characteristics of the materials that compose the urban envelope. The building cover is responsible for the most significant loads that affect cooling energy use and the quality of the outdoor environment. The materials in urban areas absorb solar and infrared radiation, and the accumulated heat is transferred into the buildings or is dissipated to the atmosphere. This means that the urban envelope plays a key role in the reduction of heat gains and the city overheating. Therefore, improving the optical properties (albedo and thermal emissivity) of urban surfaces could reduce cooling energy consumption, due to the reduction of the air and surface temperatures.

Solar reflectance and emissivity are more important optical parameters of materials. Numerous works have studied the thermal performance and impact over energy consumption of the materials used for the resolution of building envelopes as a strategy of low-cost refurbishment and great factibility (Akbari et al., 1997; Bretz and Akbari, 1997; Kai et al., 2010; Levinson et al., 2005, 2007; Revel et al., 2014).

Surfaces with low solar reflectance or albedo absorb a great amount of incident solar energy. The extended application of materials with high albedo and thermal emittance known as "cool materials" can reduce the urban temperature on a summer day between 1 and 2 °C (Akbari and Konopacki, 1998; Akbari et al., 1999; Pomerantz et al., 1999; Young, 1998). On a building scale, the energy consumption for refrigeration can be significantly reduced by means of the limitation in the gain of solar heat of the envelope, which depends on the intensity of the impact of solar radiation and the capacity of absorption (albedo) of the external surface (Cheng et al., 2005). It has been shown in numerous experimental studies and simulations that the use of a high dotation albedo can reduce the temperature of indoor air considerably and, at a medium scale, it can also diminish the urban environmental temperatures, thus mitigating the effect of the urban heat island (Akbari et al., 1997; Givoni, 1998; Taha et al., 1988).

Several studies about urban radiative cooling attempt to develop special materials in order to maximize cooling derived from this source (Cheng et al., 2005; Dornelles

Eco-efficient Materials for Mitigating Building Cooling Needs. http://dx.doi.org/10.1016/B978-1-78242-380-5.00006-6

et al., 2011; Doya et al., 2012; Kai et al., 2010; Karlessi et al., 2009; Synnefa et al., 2005). Usually, the component of the building that is recommended for use as a radiating surface is the roof, as it turns out to be the most advantageous orientation to "see" the sky (Givoni, 1998). The vertical components of the building (walls) have low viewing angles of the sky and therefore when they are exposed to sunlight have lower thermal load (Radhi et al., 2014). However, the area covered by the walls is significant in relation to the total surface area of the urban volumetry. From this derives the hypothesis that optimizing its thermal behavior will improve comfort conditions in outdoor spaces. In addition, there is less development in this field compared with the scientific research over ceiling materials; so in this research, we have focused on the thermal behavior of claddings and paintings of facades.

This chapter reports the main results from an investigation conducted by the Human Environment and Housing Laboratory, CONICET—CCT-Mendoza. In this research project, the ability of materials to reduce urban temperatures has been quantified by means of the value of their solar reflectance index (SRI). SRI quantifies the radiation that would be absorbed in a material relative to a white and black pattern surface, under standard environmental conditions. This method is used for surfaces with emissivities higher than 0.01 and superficial temperatures lower than 423° K (ASTM Standard E1980-11, 2011). The index enables a direct comparison between diverse materials of the urban facades with different optical properties (solar reflectance and emissivity) in order to determine the efficiency of each of them to mitigate the urban heat island. Particularly, the thermal behavior of 96 facade envelope materials—vertical claddings and façade coatings—which present different composition, finish, texture, color, and shape and are widely used in the region, is evaluated.

Finally, even though internationally the SRI is used to measure the material's ability to reduce urban temperatures, it is a dynamic property due to the aging of materials. The solar reflectance of the building envelope tends to change with time due to outdoor weather conditions and dirt accumulation. The effect of aging was quantified by measuring the percentage modification of the SRI of each material throughout a period of 3 years. The results obtained have been analyzed according to the influence of different characteristics of the claddings and paintings over the thermal behavior of the material throughout time.

6.2 Selection of sample unit

The study was made in the city of Mendoza, Argentina—an urban conglomerate located in a seismic and semiarid zone with a population of close to a million people. It presents an open urban geometry, formed by wide streets and a heavily forested building structure. The city has a pyramidal shape: the high building density is concentrated in the center of town, whereas the low building density is diluted to the periphery.

Conceptually, the building's designs respond to its climate adaptation in some cases. They result from the application of traditional elements available regionally and technologies associated with the seismic character of the site—bare brickwork and reinforced concrete. In other cases, the buildings respond to formal trends resulting from the use of new technologies—steel and glass. Roofs are flat in the buildings but tilted—with different materials and colors—in residential areas of low density. The vertical facades are composed of bare brickwork and rustic or smooth cladding.

The urban areas present the same appearance as other current cities: asphalt or concrete pavement and sidewalks with a wide range of materials—concrete and stone, with clear colors in the past and a strong tendency to use dark colors nowadays. Into the city, pavements represent 40%, sidewalks 15%, and roofs 22% of the urban envelopes (Alchapar et al., 2014).

6.2.1 Material selection and classification

The study samples are 80 vertical claddings (acrylic and concrete) and 16 facade paintings. The vertical coatings—blocks of 30 × 30 cm—were formed by three layers: support (10 cm thick high-density expanded polystyrene), an extended 3 mm mortar, enabling a better adherence of the cladding to the polystyrene surface, and, lastly, the layer of textured and colored cladding.

Sixteen (16) samples of latex paint for walls were applied over a concrete base of 40 × 20 cm. This set is representative of the most used paints in the local market. According to the manufacturer's specification, before painting, the surface was prepared by putting on a layer of external fixator and one layer of sealer. Then, the surface was sanded, and after this, three layers of each of the latex paints were arranged in each sample to study. The samples were positioned on a horizontal surface of expanded polystyrene (7 cm thick). The latex paints for facades were selected based on three compositions available in the local market: acrylic (ac), waterproof (im), and athermic (at), in diverse costs available in the regional market. These samples were placed in the Regional Center for Scientific and Technical Research, located at the west area of the city (32°53′ latitude, 68°51′ western longitude).

The vertical textured claddings (quoted according to the manufacturer's labeling) are differentiated according to their composition into acrylic (SIP) and concrete (CW). The first ones—SIP—are composed of acrylic polymers, mineral loads, and inorganic pigments with high UV resistance and chemical additives. The concrete claddings (CW) are composed of white concrete, spur stone, lime, pigments, organic and inorganic additives, mineral loads, synthetic resins, fungicides, and algaecides in powder.

In this stage, six textures of different particle sizing and finish were selected according to the demand and frequency of use in the local market: In the acrylic claddings—SIP: travertine rulato (both fine (fr) and rustic (rr)), llaneado (both rustic (fl) and fine (rl)), and medium granitex (mg); in the concrete claddings—CW: rulato textured (mr), travertine textured (mt), salpicrete (both salpic (ss) and ironed (is)), and medium granitex (mg) (in the 8 most-demanded colors: white, ivory, paris stone, ochre, terracotta, clear gray, jade green, and dark gray).

	Composition	Finish	Texture	Color	CIELab system
Vertical claddings	Acrylic (SIP)[1]	Rulato travertine	Fine rustic	White	98.618/0.005/–0.010
		Llaneado	Fine rustic	Ivory	94.320/–1.381/22.189
		Granitex	Medium	Paris stone	67.857/6.516/23.710
	Concrete (CW)[1]	Rulato textured	Medium	Ochre	58.607/13.716/58.966
		Travertine textured	Medium	Terracotta	32.882/42.305/25.852
		Salpicrate	Salpic ironed	Light gray	76.309/–1.156/–2.14
		Granitex	Medium	Green	58.426/–5.622/7.638
				Dark gray	50.070/2.358/–2.886
Paints	Acrylic	Matte		White	
	Waterproof	Gloss		Terracotta	
	Athermic	Matte		Black	

Figure 6.1 Facade materials classification (vertical cladding and paints) according to composition, finish and texture, color, and shape.
[1]Designation according to manufacturer Weber-Saint-Gobain Argentina S.A.

The acrylic latex paint (ac) is a styrene acrylic emulsion composed of resins in water dispersion, titanium dioxide pigments, and toners. The impervious latex paint (im) is an elastomeric acrylic emulsion that provides flexibility and strength to the coating to prevent cracking caused by changes in temperature and humidity. The athermic paint (at) is a highly durable coating formulated with acrylic polymers and thermally reflective components designed to protect from sunlight and UV rays. Figure 6.1 shows the features that were taken into account for the classification and study of claddings and paints.

6.3 Test and instrumentation

Optic and thermal characterization of materials were evaluated: albedo, emissivity, surface temperature, and SRI.

6.3.1 Solar reflectance index (SRI)

In this research project, the ability of all evaluated materials to reduce urban temperatures has been quantified by means of the value of their SRI. SRI is the fraction of the incident energy solar which is reflected by the surface in relation to a white and black pattern under standard environmental conditions. This method is used for surfaces with emissivities higher than 0.01 and superficial temperatures lower than 423°K (ASTM E1980-11). The index enables a direct comparison between diverse materials used in the urban envelopes with different optical properties (solar reflectance and emissivity) in order to determine the efficiency of each of them to mitigate the urban heat island. It is calculated by using equations based on values of solar reflectance and IR emittance. It is expressed as a value (0.0 to 1.0) or as a percentage (0% to 100%). For a surface exposed to the sun and isolated underneath, the superficial temperature of equilibrium (T_s) is obtained as from ASTM E1980-11:

$$T_s = 309.07 + \frac{(1066.07\alpha - 31.98\varepsilon)}{(6.78\varepsilon + h_c)} - \frac{(890.94\alpha^2 + 2153.8\alpha\varepsilon)}{(6.78\varepsilon + h_c)} \tag{6.1}$$

where α = solar absorptance = 1 (solar reflectance); I = solar flux, W/m^2; ε = thermal emissivity; σ = Stefan Boltzmann constant, 5.66961×10^{-8}, W/m^2 °K^{-4}; T_s = steady-state surface temperature, °K; T_{sky} = sky temperature, °K; h_c = convective coefficient, W/m^2 °K^{-1}; T_a = air temperature, °K.

Therefore, the solar reflectance index is defined as

$$\mathrm{SRI} = 100\frac{T_b - T_s}{T_b - T_w} \tag{6.2}$$

where T_b = black pattern temperature, °K; T_w = white pattern temperature, °K; T_s = steady-state surface temperature, °K. By means of Equation (6.1), the superficial temperatures of the patterns are calculated under local environmental conditions, according to regulations.

6.3.2 Emissivity and solar reflectance of materials

The thermal performance of the materials is determined mainly by their optical and thermal properties. The solar reflectance and the thermal emissivity are the most important factors (Chimkla et al., 2004; Doulos et al., 2001). To classify materials according to their SRI, previous knowledge of their optical properties is needed (solar reflectance and emissivity).

The value of the emissivity is obtained—according to the ASTM 1933-06 when the temperature registered by a T-type thermocouple and the superficial temperature measured by IR thermometer Fluke 568 are similar (ASTM, 2006). The radiant energy registered by the infrared chamber to determine the material's surface temperature depends not only on the object temperature but also on its emissivity (ε). Then, the value of emissivity obtained for each sample was set into the IR camera for the measurements of the surface temperature.

The solar reflectance measurement according to ASTM E1918 may only be used when the target is at least 10 m^2 in area and roughly square or circular in shape. The difficulty of transporting such large samples for the natural exposure makes the sample-size requirements of E1918 costly and inconvenient. Therefore, in this study, a method (variant to ASTM E1918) proposed by Akbari et al. (2008) has been used. This method measures the solar reflectance of materials samples of about 1 m^2 using a pyranometer and a pair of black and white masks (Akbari et al., 2008; Sailor et al., 2006).

The use of a single pyranometer for solar reflectance measurements is economical but introduces additional uncertainties associated with the inversion of the instrument. For these reasons, we have opted to use a two-pyranometer system to determine the solar reflectance value ($\hat{a}$), where both pyranometers are Kipp & Zonen CM3 type; the set is named albedometer. The albedometer registers the solar radiation, which is received over the horizontal surface, and the reflected solar radiation. The solar reflectance of each material is determined by the difference between the received and reflected radiation (Kipp & Zonen, 2012). The radiative spectrum covered by the instrument is 285 to 2800 nm and a maximum solar irradiance of 4000 W/m^2. Their nominal sensitivities are in the range 1.5×10^{-6} V/W/m^2. The stated response time (95%) is 18 s.

The thermal conductivity has not been considered because all samples have similar thicknesses. Besides, the layer of expanded polystyrene works as an adiabatic limit regarding the grounding conductivity.

6.3.3 Test method

The solar reflectance, emissivity, and superficial temperature were registered during the summers of 2011 and 2014. Measurements of the solar reflectance and emissivity were recorded during the second and third weeks of January 2011. Superficial temperatures were measured during the summers of 2011, 2012, 2013, and 2014, being days in standard environmental conditions (Table 6.1). The values were recorded at 13:00 hours and were used to calculate the SRI following the ASTME1980-11.

The results derived from this research correspond to the analysis of a set of weather data selected from the full series of measurements. They were selected because they represent typical weather conditions on summer days in Mendoza: sunny, dry, clear sky, no clouds, and light winds. Tables 6.2 and 6.3 detail the values set to 13:00 hours used to calculate the SRI according to the reference standard. These measurements were taken on the site by means of Onset's HOBO® Weather Station Data Logger—H21-001. Its operative range is between 253 and 323 °K.

As was said previously, in this experiment for surface temperature measurements, the IR chamber Fluke Ti 55 was used, which detects the infrared radiation of long

Table 6.1 Schedule of measurements made

	2011		2012		2013	2014
Material	**Summer**	**Winter**	**Summer**	**Winter**	**Summer**	**Summer**
Vertical cladding	10-March	24, 27, 28-June	07, 08-February; 22-March and 28-December			
Facade paint			9, 17, 18, 19, 24, 25-March and 13 to 25-December	12-July	02, 03, 07-January and 23, 24, 28, 29, 30-December	03, 04, 05, 07, 08-June

Table 6.2 **Weather data corresponding to selected typical days for the SRI calculation of facade claddings**

Weather data at 13:00	Vertical cladding		
	First year	**Second year**	**Third year**
Date	10/03/2011	08/02/2012	28/12/2012
Solar radiation (W/m^2)	846	938	959
Air temperature (°C)	33	27	28
RH (%)	36	32	25
Wind speed (m/s)	1.2	1.3	1.9
h_c (W/m^2 °C)	6.4	6.7	8.4

wave in the range of 7.5 to 14 microns within the electromagnetic spectrum. In thermographic pictures, each pixel contains a certain temperature value (Fluke, 2010).The software SmartView 2.1, through algorithms, assigns a specific color corresponding to a temperature value in the x–y coordinate of the image. The values of surface temperatures obtained were averaged for each material analyzed. Besides, a set of data logger and T-type thermocouples, model Akribis TPD8356, were used to verify the values of surface temperatures obtained from the IR camera.

6.4 Materials thermal behavior: their impacts on building design decisions and energy consumption

The thermal behaviors of materials of the urban envelope are evaluated using the solar reflectance index as an indicator established by regulation ASTM E1980-11. SRI is influenced by two main factors: material composition and surface finish. The optical properties of the building envelope materials are dynamic, that is, they tend to change with time due to the elements, wearing, and accumulation of dirt. The SRI changes as a result of wear and discoloration. Therefore, during a period of three years, albedo, emissivity, and surface temperature were monitored in all materials.

In this chapter, the SRI values of each material are shown in their initial stage and with the variation produced during three years of aging. In the exterior latex paint, due to its washable condition, aging has been analyzed in two stages: dirty paint and clean paint. In order to quantify the thermal behavior response of the material to the passing of time, the following formulas were used:

$$\mathrm{SRI}_1 - \mathrm{SRI}_2 = \Delta\mathrm{SRI}_1 \tag{6.3}$$

$$\mathrm{SRI}_1 - \mathrm{SRI}_3 = \Delta\mathrm{SRI}_{\mathrm{total}} \tag{6.4}$$

where SRI_1 = initial solar reflectance index (at first year), SRI_2 = aged initial solar reflectance (at second year), SRI_3 = aged initial solar reflectance (at third year), and $\Delta\mathrm{SRI}_{\mathrm{total}}$ = difference between SRI_1 and SRI_3 values.

Table 6.3 Weather data corresponding to selected typical days for the SRI calculation of facade paintings

Weather data at 13:00	Facade painting								
	First year	Second year				Third year			
Date	18/03/2012	28/12/2012	29/12/2012	30/12/2012	03/12/2013	23/12/2012	28/12/2013	04/01/2014	08/01/2014
Solar radiation (W/m^2)	849	954	961	951	959	973	988	1017	1018
Air temperature (°C)	31	31	33	35	32	36	35	36	35
HR (%)	36	24	30	27	28	31	28	23	18
Wind speed (m/s)	1.4	1.1	0.9	1.1	0.8	0.9	0.7	1.3	1.5
h_c (W/m^2 °C)	7.0	6.1	5.4	6.1	5.3	5.6	5.0	6.7	7.2

The established criterion for analysis was to consider the thermal behavior of aged material as stable when the differences between initial and aged SRI values (ΔSRI_{total}) during the three years are below 20%. Ranges of ΔSRI_{total} between 20% and 30% show a moderate degradation of the material, and values higher than 30% show strong degradation between periods, that is, $\Delta SRI \leq 20\%$ = stable; $\Delta SRI > 20 \leq 30\%$ = moderate degradation; $\Delta SRI > 30\%$ = strong degradation. Vertical cladding is considered efficient when it has an SRI value higher than 50% after three years of exposure and for white exterior paint the index must exceed 85%.

6.4.1 Values of albedo, emissivity, superficial temperature, and the SRI for vertical cladding and facade coating

Tables 6.4 and 6.5 describe the surface temperatures (T_s) in degrees Celsius, albedo ($\hat{a}$), SRI of the three years of aging ($SRI_1 - SRI_2 - SRI_3$), and the difference between the SRI values of the first and last periods (ΔSRI_{total}) for the studied materials according to the nomenclature used to distinguish them. The surface temperatures were registered at 13:00 hours according to the established reference regulation with a solar radiation between 850 and 1000 W/m^2 and wind speed between 0.8 and 2 m/s (Tables 6.2 and 6.3).

6.4.2 Thermal behavior and the SRI for each material type

Vertical cladding is analyzed according to the comparison between extreme cases and differences by composition, colors, and finish and texture. Facade painting is analyzed according to the comparison between extreme cases and differences by composition.

6.4.2.1 Vertical cladding

Initial behavior (SRI_1)

By analyzing all samples, the acrylic composition is observed to register ranges of SRI between 100% and 33% and its surface temperature between 35 and 72 °C. SRI differences equal to 77% (see SIP 02 vs SIP 32). In claddings of cement composition, the SRI ranges oscillate between 100% and 41% and between 33 and 67.5 °C in terms of their surface temperature, with SRI differences of up to 59% (see CW 49 vs CW 64). Diversity in the thermal behavior of materials is observed depending on color and finish, which is even more accentuated for claddings of acrylic composition (Figures 6.2 and 6.3).

Compositions (acrylic claddings (SIP) vs concrete claddings (CW)): 43% of acrylic claddings and 28% of the concrete claddings register SRI = 80–100%; 7.5% of acrylic claddings and 30% of the concrete cladding register SRI = 60–80%. It is observed that the 32.5% of acrylic claddings and 40% of the concrete claddings evaluated are set in

Table 6.4 Albedo ($\hat{a}$), surface temperature (T_s °C), and solar reflectance index (%) of textured claddings analyzed during the first year (SRI_1), second year (SRI_2), and third year (SRI_3) and their differences (ΔSRI_{total})

Vertical cladding		1st year			2nd year			3rd year			
		$\hat{a}$	T_s	SRI_1	$\hat{a}$	T_s	SRI_2	$\hat{a}$	T_s	SRI_3	ΔSRI_{total}
SIP 01	Acrylic fine rulato travertine white	0.86	39	97	0.65	46	82	0.5	55	61	−37
SIP 02	Acrylic fine rulato travertine ivory	0.9	35	100	0.68	45.5	84	0.51	55.5	60	−40
SIP 03	Acrylic fine rulato travertine paris stone	0.81	41	92	0.58	52.5	70	0.47	58	55	−37
SIP 04	Acrylic fine rulato travertine ochre	0.51	59	57	0.43	61	53	0.35	64	41	−16
SIP 05	Acrylic fine rulato travertine terracotta	0.45	63	50	0.39	63	50	0.31	66	37	−13
SIP 06	Acrylic fine rulato travertine light gray	0.6	55.5	64	0.49	57	61	0.39	62	46.5	−18
SIP 07	Acrylic fine rulato travertine green	0.44	63	49	0.45	60	56.5	0.4	61	47	−2
SIP 08	Acrylic fine rulato travertine dark gray	0.34	69	37	0.4	63	50	0.32	65	38	1
SIP 09	Acrylic rustic rulato travertine white	0.79	43	89	0.61	50	75	0.45	59	53	−36
SIP 10	Acrylic rustic rulato travertine ivory	0.86	38	99	0.61	50	75	0.46	58	54	−45
SIP 11	Acrylic rustic rulato travertine paris stone	0.82	40	94	0.48	59	58	0.29	69	31	−63
SIP 12	Acrylic rustic rulato travertine ochre	0.47	62	51.5	0.33	67	42	0.23	71	26	−25
SIP 13	Acrylic rustic rulato travertine terracotta	0.35	69	38	0.37	65	47	0.26	69	30.5	−8
SIP 14	Acrylic rustic rulato travertine light gray	0.51	60	56	0.49	57	61	0.34	64	40	−16
SIP 15	Acrylic rustic rulato travertine green	0.42	65	46	0.33	67	42	0.26	68	31	−15
SIP 16	Acrylic rustic rulato travertine dark gray	0.34	69	38	0.36	66	45	0.29	67.5	33	−5
SIP 17	Acrylic fine llaneado white	0.82	42	91	0.61	50	74.5	0.35	65	39	−52
SIP 18	Acrylic fine llaneado ivory	0.53	60	55	0.58	53	70	0.33	66	36	−19
SIP 19	Acrylic fine llaneado paris stone	0.9	35	100	0.53	55	66	0.33	65	39	−61
SIP 20	Acrylic fine llaneado ochre	0.83	40	95	0.5	58	60	0.34	66	37	−58
SIP 21	Acrylic fine llaneado terracotta	0.43	64	48	0.32	67	42	0.28	68	33	−15
SIP 22	Acrylic fine llaneado light gray	0.77	44	87	0.56	54	68	0.36	64	40	−46
SIP 23	Acrylic fine llaneado green	0.41	65	45	0.41	63	51	0.24	70	28	−17
SIP 24	Acrylic fine llaneado dark gray	0.26	74	29	0.43	62.5	51	0.23	72	24	−5
SIP 25	Acrylic rustic llaneado white	0.83	40	94	0.79	38	98	0.45	60	50.5	−43

SIP 26	Acrylic rustic llaneado ivory	0.75	46	83	0.74	42	91	0.45	60	50	−33
SIP 27	Acrylic rustic llaneado paris stone	0.58	57	61	0.58	52	71	0.34	65	38	−23
SIP 28	Acrylic rustic llaneado ochre	0.74	46	83	0.39	64	48	0.24	70	28	−55
SIP 29	Acrylic rustic llaneado terracotta	0.5	60	56	0.43	61	54.5	0.35	63	42	−13
SIP 30	Acrylic rustic llaneado light gray	0.74	46	83	0.5	57	62	0.4	61	48	−36
SIP 31	Acrylic rustic llaneado green	0.36	68	39	0.36	65.5	45	0.28	68	33	−6
SIP 32	Acrylic rustic llaneado dark gray	0.3	72	33	0.39	65	46	0.25	71	26	−6
SIP 33	Acrylic medium granitex white	0.86	38.5	98	0.73	41	92	0.46	58	54	−44
SIP 34	Acrylic medium granitex ivory	0.77	44.5	86	0.67	46	84	0.45	59	52.5	−33
SIP 35	Acrylic medium granitex paris stone	0.67	51	73	0.48	58	60	0.38	62	46	−28
SIP 36	Acrylic medium granitex ochre	0.72	47	80.5	0.42	62	52.5	0.28	68	33	−48
SIP 37	Acrylic medium granitex terracotta	0.44	63	49	0.4	63	50	0.32	65	38	−11
SIP 38	Acrylic medium granitex light gray	0.5	61	54	0.48	58	60	0.35	64	41	−13
SIP 39	Acrylic medium granitex green	0.39	66	43	0.39	64	49	0.28	68	32	−12
SIP 40	Acrylic medium granitex dark gray	0.32	70	35	0.31	68.5	40	0.23	70.5	27	−9
CW 41	Concrete medium rulato white	0.85	38	98	0.79	38	99	0.53	54	63	−35
CW 42	Concrete medium rulato textured ivory	0.78	43.5	88	0.71	43	89	0.49	56	58	−30
CW 43	Concrete medium rulato textured paris stone	0.72	49	77	0.68	46	84	0.41	61	47	−30
CW 44	Concrete medium rulato textured ochre	0.55	58.5	58	0.5	57	62	0.37	62.5	44	−14
CW 45	Concrete medium rulato textured terracotta	0.53	58.5	58	0.46	59	57	0.33	65	39	−19
CW 46	Concrete medium rulato textured light gray	0.64	53.5	68	0.58	52.5	70	0.38	63	43.5	−25
CW 47	Concrete medium rulato textured green	0.47	62	51.5	0.55	54	67	0.32	67	35	−17
CW 48	Concrete medium rulato textured dark gray	0.37	67	41	0.45	60	56	0.3	67	35	−6
CW 49	Concrete medium travertine textured white	0.94	33	100	0.83	35	100	0.54	55	62	−38
CW 50	Concrete medium travertine textured ivory	0.77	44.5	86	0.75	41	92	0.47	59	52	−34
CW 51	Concrete medium travertine textured paris stone	0.71	48	79	0.69	44.5	86	0.45	59	52	−27
CW 52	Concrete medium travertine textured ochre	0.63	53	69	0.54	54	67	0.36	63.5	42	−27
CW 53	Concrete medium travertine textured terracotta	0.48	61	53	0.56	53	70	0.36	63.5	42	−11
CW 54	Concrete medium travertine textured light gray	0.7	49	78	0.62	50	75	0.37	65	39	−39
CW 55	Concrete medium travertine textured green	0.54	59	58	0.5	57	62	0.38	62	45	−13
CW 56	Concrete medium travertine textured dark gray	0.5	61	54	0.4	63	50.5	0.27	68.5	31	−22
CW 57	Concrete salpic salpicrete white	0.81	42	91	0.72	43	89	0.45	60	49	−41

Continued

Table 6.4 Continued

Vertical cladding		1st year			2nd year			3rd year			
		$\hat{a}$	T_s	SRI_1	$\hat{a}$	T_s	SRI_2	$\hat{a}$	T_s	SRI_3	ΔSRI_{total}
CW 58	Concrete salpic salpicrete ivory	0.8	42	91	0.7	44	88	0.5	56	59	−32
CW 59	Concrete salpic salpicrete paris stone	0.68	49.5	76	0.6	51	73	0.44	60	50	−26
CW 60	Concrete salpic salpicrete ochre	0.58	55	65	0.58	53	70	0.36	64	40	−25
CW 61	Concrete salpic salpicrete terracotta	0.6	54.5	66	0.45	60	56.5	0.3	67	35	−31
CW 62	Concrete salpic salpicrete light gray	0.72	48	79	0.63	49	78	0.4	62	46	−33
CW 63	Concrete salpic salpicrete green	0.52	60	56	0.44	62	52	0.28	69	29	−27
CW 64	Concrete salpic salpicrete dark gray	0.39	67.5	41	0.32	68	40	0.23	71	26	−15
CW 65	Concrete ironed salpicrete white	0.84	40	94	0.76	40	94	0.45	60	50	−44
CW 66	Concrete ironed salpicrete ivory	0.75	46	83	0.66	46	82	0.4	62	45	−38
CW 67	Concrete ironed salpicrete paris stone	0.63	54	68	0.59	52	72	0.35	65	39	−29
CW 68	Concrete ironed salpicrete ochre	0.52	59	58	0.51	56	64	0.38	62	45	−12
CW 69	Concrete ironed salpicrete terracotta	0.48	61	54	0.49	57	61	0.3	67	35	−19
CW 70	Concrete ironed salpicrete light gray	0.67	51	73	0.62	49	78	0.43	59	51	−22
CW 71	Concrete ironed salpicrete green	0.5	61	54	0.49	57.5	61	0.32	66	37	−17
CW 72	Concrete ironed salpicrete dark gray	0.36	68	40	0.4	63	50	0.28	68	33	−7
CW 73	Concrete medium granitex white	0.82	41	92	0.7	44	88	0.48	57	57	−36
CW 74	Concrete medium granitex ivory	0.77	45	85	0.67	46	82.5	0.44	59	51	−34
CW 75	Concrete medium granitex paris stone	0.73	47	82	0.61	50	75	0.39	63	44	−38
CW 76	Concrete medium granitex ochre	0.53	59	57	0.53	56	64	0.36	64	40	−17
CW 77	Concrete medium granitex terracotta	0.53	59	57	0.48	59	58	0.33	66	36	−21
CW 78	Concrete medium granitex light gray	0.72	48	78	0.61	50.5	74	0.41	61.5	46.5	−32
CW 79	Concrete medium granitex green	0.49	61	53	0.44	61.5	53	0.27	70	28	−25
CW 80	Concrete medium granitex dark gray	0.71	47	80	0.4	64	48	0.24	71	25	−55

Table 6.5 Albedo ($\hat{a}$), surface temperature (T_s °C), and solar reflectance index (%) of dirty and clean painting analyzed during the first year (SRI_1), second year (SRI_2), and third year (SRI_3) and their differences (ΔSRI_{total})

		First year			Second year			Third year			
		$\hat{a}$	T_s	SRI_1	$\hat{a}$	T_s	SRI_2	$\hat{a}$	T_s	SRI_3	ΔSRI_{total}
Facade painting dirty											
L01	Athermic matte white	0.82	41.5	92	0.84act	40	96	0.78	46	82.5	9
L02	Acrylic matte black	0.4	70	36.5	0.58	58	62	0.52	63	49	−12
L03	Waterproof gloss white	0.83	41	93	0.86	38	98	0.77	46.5	81.5	11.5
L04	Waterproof gloss white	0.82	41	92	0.82	41	93	0.77	46.5	81.5	10
L05	Waterproof gloss white	0.84	40	94	0.75	46	83	0.76	49	76.5	18
L06	Waterproof gloss white	0.84	40	94	0.79	44	87.5	0.79	47	80.5	14
L07	Waterproof gloss white	0.96	31	100	0.79	44	86	0.8	47.5	82	18
L08	Waterproof gloss white	0.84	40	95	0.75	46	83	0.74	50.5	73.5	21
L09	Waterproof gloss white	0.84	40	94	0.8	43	91	0.82	46	84	10
L10	Waterproof gloss white	0.81	42	90	0.82	41	94	0.82	45.5	85	5
L11	Acrylic matte white	0.87	40	94	0.81	42	92	0.82	46	84	11
L12	Acrylic matte white	0.82	44.5	86	0.82	42	92	0.81	46.5	83.5	3
L13	Acrylic matte white	0.82	44.5	86	0.84	40.5	94	0.72	52	70.5	16
L14	Waterproof gloss white	0.83	43	88	0.83	41	93	0.8	46	82	6
L15	Waterproof gloss white	0.83	43	89	0.86	38	98	0.8	47.5	82	7
L16	Waterproof gloss terracotta	0.61	59.5	58	0.68	51	73	0.58	63.5	51	7

Continued

Table 6.5 Continued

		First year			Second year			Third year			
		$\hat{a}$	T_s	SRI_1	$\hat{a}$	T_s	SRI_2	$\hat{a}$	T_s	SRI_3	ΔSRI_{total}
Facade painting clean											
L01	Athermic matte white	0.82	41.5	92	0.85	38	99	0.8	44.5	85.5	6
L02	Acrylic matte black	0.4	70	36.5	0.51	61	57	0.55	61	53	−16
L03	Waterproof gloss white	0.83	41	93	0.83	39	98	0.77	46.5	81.5	11.5
L04	Waterproof gloss white	0.82	41	92	0.8	41	93	0.76	47	80.5	11
L05	Waterproof gloss white	0.84	40	94	0.73	46.5	82	0.75	49.5	75.5	19
L06	Waterproof gloss white	0.84	40	94	0.76	44	86	0.8	46.5	81.5	13
L07	Waterproof gloss white	0.96	31	100	0.76	44.5	86	0.83	45.5	85	15
L08	Waterproof gloss white	0.84	40	95	0.73	46	83	0.74	50	74.5	20
L09	Waterproof gloss white	0.84	40	94	0.78	42.5	91	0.83	45.5	85	9
L10	Waterproof gloss white	0.81	42	90	0.8	41	93.5	0.84	44	88	2
L11	Acrylic matte white	0.87	40	94	0.79	42	92	0.81	46.5	83	12
L12	Acrylic matte white	0.82	44.5	86	0.8	42	92	0.81	46.5	83.5	3
L13	Acrylic matte white	0.82	44.5	86	0.85	38	99	0.76	49.5	78	8
L14	Waterproof gloss white	0.83	43	88	0.84	38.5	98	0.81	45	84	4
L15	Waterproof gloss white	0.83	43	89	0.83	39	97	0.84	44.5	87	2
L16	Waterproof gloss terracotta	0.61	60	58	0.71	48	79.5	0.64	59.5	59	−1

Figure 6.2 Best and worst acrylic samples.

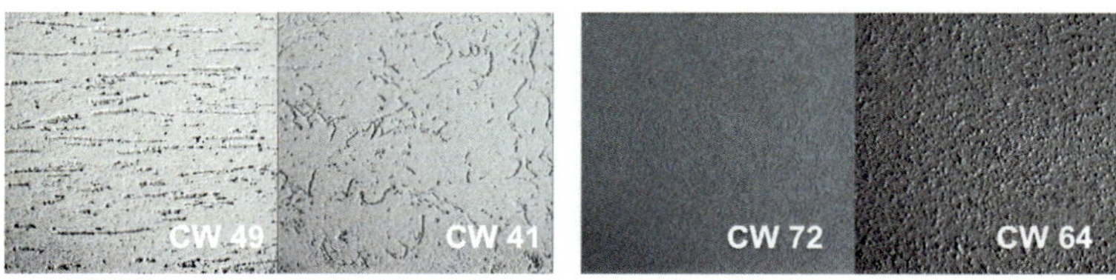

Figure 6.3 Best and worst concrete samples.

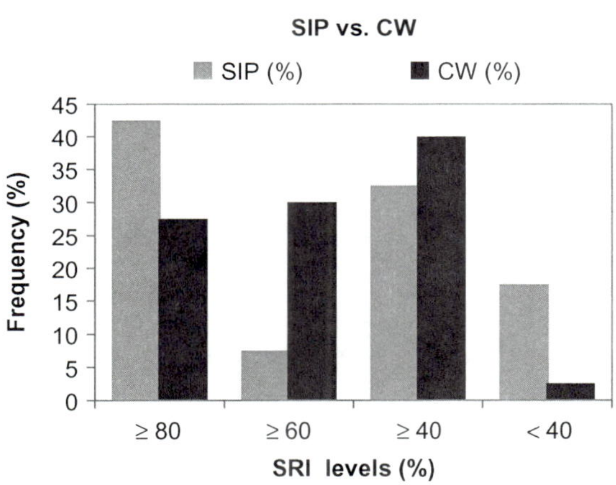

Figure 6.4 SRI levels of textured claddings according to composition.

the range of 60–40% of SRI, while the other 17.5% of acrylic and 2.5% of the concrete have less than 40% of SRI (Figure 6.4).

In order to evaluate the effect of the cladding composition, the case of acrylic medium granitex vs concrete medium granitex is analyzed comparatively. In light colors (white, ivory, and ochre), the acrylic cladding (SIP) shows better thermal performance than the concrete cladding (CW). For example, the surface temperature of acrylic medium granitex ochre SIP 36 is 12 °C lower than concrete medium granitex ochre CW 76, and its SRI is 23.5% higher. In paris stone, terracotta, light gray, green, and dark gray colors, which have lower albedos, this trend is inverted. That is, in these colors, the concrete cladding has better performance than acrylic to mitigate the effects of the UHI. SRI in dark gray concrete medium granitex (CW 80) is 80%, while in dark gray acrylic medium granitex (SIP 40), it is 35%.

Finish and texture: they have different impacts on the thermal performance of cladding depending on its color. This impact could be classified in three ranges:

Range 1: the SRI varies between the different textures and finishes evaluated around 40% for the same color.
Range 2: the SRI varies between the different textures and finishes evaluated around 20% for the same color.
Range 3: the SRI varies between the different textures and finishes evaluated around 10% for the same color.

The acrylic claddings (SIP) are found in different colors: in range 1, ivory, paris stone, ochre, and light gray; in range 2, terracotta; and in range 3, white, green, and dark gray for the five finishes. These ranges show that thermal behaviors of light and dark claddings colors (white and dark gray) are scarcely influenced by the texture and finish, while intermediate colors have significant differences associated with these variables. For example, in the case of SIP 19 vs SIP 27, the surface temperature difference is 22 °C and the SRI difference 38.5%.

Fine rulato travertine finish has better thermal performance associated with fine texture for all evaluated colors, while in the fine llaneado finish, the effect of texture does not show a dominant tendency and the thermal behavior is commanded by the color. For example, acrylic rustic llaneado ivory (SIP 26) shows an SRI = 83%, whereas acrylic fine llaneado ivory (SIP 18) SRI = 55%; on the contrary, the acrylic rustic llaneado paris stone (SIP 27) shows an SRI = 62%, whereas acrylic fine llaneado paris stone (SIP 19) SRI = 100%.

In range 1, concrete claddings (CW) are white and light gray; in range 2, they are terracotta, paris stone, and ochre; and in range 3, they are ivory, light gray, and green for the five finishes. These ranges show that thermal behaviors of light and dark claddings colors (white and dark gray) are strongly influenced by the finish and texture, while intermediate colors have little differences associated to these variables.

For example, in range 1 (concrete ironed salpicrete dark gray (CW 72) vs concrete medium granitex dark gray (CW 80)), the surface temperature difference is 21 °C and solar reflectance 40%.

Color: the differences in color are evaluated maintaining fixed variables of texture and finish. In general, in all compositions, finishes, and textures, white and ivory colors remain colder. They reach an average solar reflection index of 95% and 85%, respectively.

Color is the most important composition in acrylic claddings since it determines the thermal behavior, while in the case of cement claddings, the effect derives from both color and texture, for example, the fine travertine rulato texture: white color SIP 01 reaches a temperature of 39 °C and SRI of 97%, while the hottest alternative, fine travertine rulato-dark gray-SIP 08 reaches up to 69 °C and the SRI to 37%. That is, the surface temperature difference is 30 °C and the SRI difference is 60%.

In the case of concrete claddings, the selection of color among different textures and finishes can improve its thermal behavior to 50% with respect to the SRI and decrease its surface temperature to 21 °C (i.e., CW 72 vs CW 80). Dark gray color is less influenced by finish and texture, while intermediate colors (ivory, light gray, paris stone, terracotta, green, and ochre) present significant differences associated to these variables, where their SRI varies between 20% and 40%.

Finally, in the lighter tonalities (white, ivory, and ochre), the acrylic claddings show better thermal behavior than the concrete claddings (i.e., SIP 36 vs CW 76). This tendency is inverted in most colors with lower albedos: paris stone, terracotta, light gray, dark gray, and green cement. That is, in the case of these colors, the cement cladding has better performance in diminishing thermal charges (i.e., CW 80 vs SIP 40).

Aging behavior ($SRI_2 - SRI_3$)

Figure 6.5 shows the tendency of vertical claddings to decrease the levels of albedo and increase their surface temperature; this leads to a decrease in the SRI values as the time of exposure increases. 80% of the vertical claddings are within the range of SRI between 50% and 100%, both in the initial condition (SRI_1) and in the first year of aging (SRI_2). However, its reflective ability in the third period of aging (SRI_3) decreases significantly. In this period, the vertical claddings are grouped in an SRI

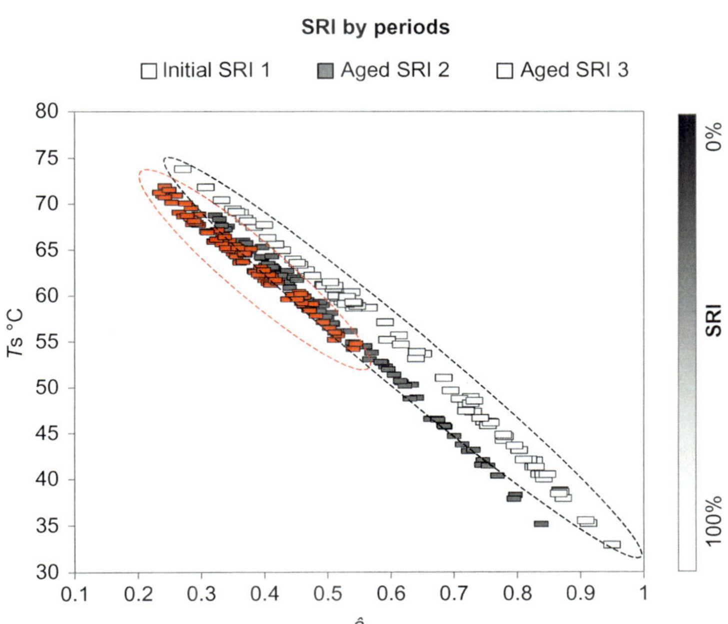

Figure 6.5 Solar reflective index of vertical claddings: initial, second, and third year of aging.

range that oscillates between 63% to 24%. This indicates that 80% of the initially efficient materials have a worsened SRI value of 36% and increased surface temperature between 1° and 29 °C.

SRI aging according to compositions. Acrylic (SIP) vs concrete (CW): 50% of acrylic materials kept a stable thermal behavior and 30% of the concrete materials integrated this range. The materials that moderately diminished their capacity are found within 7.5% of the acrylics and 32.5% of the concrete materials evaluated. 42.5% of the acrylic materials and 37.5% of concrete claddings showed a strong degradation, decreasing more than 30% their SRI value. Therefore, by analyzing the two groups of materials (acrylic and concrete) according to composition in general terms, acrylic claddings are observed to have a more extreme behavior, since half of these register a high thermal behavior and practically the total of the remaining acrylics are inefficient. In contrast, the cement materials show a more balanced distribution between the degradation ranges (Figure 6.6).

SRI aging according to finish and texture: by analyzing the claddings separately according to their finish and texture, it is observed that the most efficient finish regarding passing of time, weather, and dirt is fine rulato travertine. That is, 62.5% of materials with this finish have remained stable, diminishing their reflective capacity in less than 20%. It should be noted that the finish fine rulato travertine is very resistant to wear in the dark tonalities—dark gray, cement green, and terracotta (see SIP 08-07-05)—with decreasing SRI values inferior to 16% (Figure 6.7).

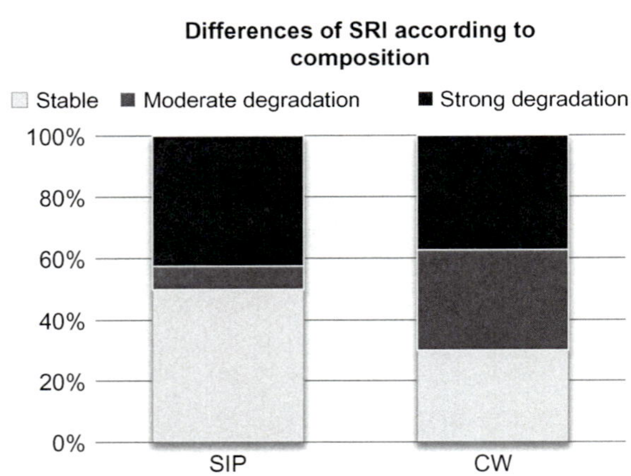

Figure 6.6 Differences in SRI of vertical claddings (ΔSRI) according to their composition.

Figure 6.7 Textures more resistant to wear: fine rulato travertine.

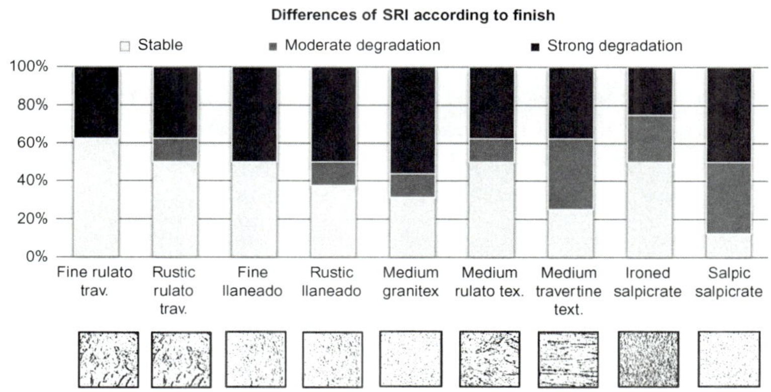

Figure 6.8 Differences in SRI of vertical claddings (ΔSRI) according to finish and texture.

The finish medium granitex has shown strong degradation in its thermal behavior. 56% of the materials with this finish reduced their SRI in the range between 32% and 55%. In Figure 6.8, degradation of claddings is shown in relation to finish and texture.

SRI aging according to color: by evaluating the color variable for both compositions in isolation, we detected that 80% of the materials of green cement color, terracotta, and dark gray have remained stable. Within the colors with moderate degradation in their SRI levels are the paris stone and ochre tonalities. 100% of white claddings, 80% of ivory, and 50% of light gray claddings have strong degradation. It should be noted that within the light tonalities, the color that wears the least is the ochre tonality with an SRI aging between 12% and 58% (Figure 6.9).

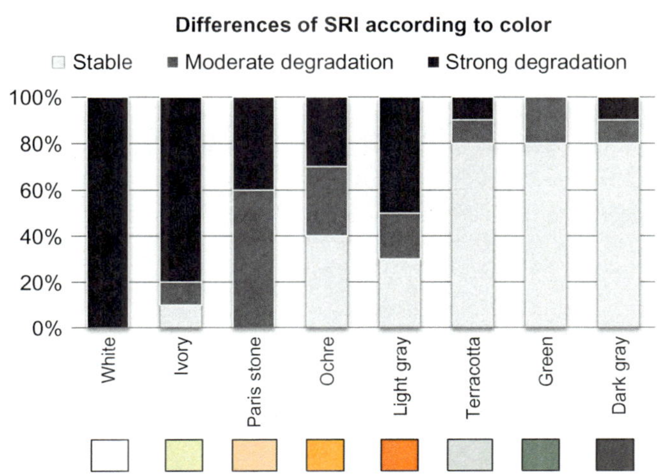

Figure 6.9 Differences in SRI of vertical claddings (ΔSRI) according to their color.

6.4.2.2 *Facade painting*

Initial behavior (SRI_1)

From the total of the studied samples, waterproof gloss white—L07—paint is the material that registers the most efficient behavior with a 100% SRI. The least favorable paint is acrylic matte black (L02), and within the white tonalities, the acrylic matte paints (L12 and L13) register the lowest SRI levels.

Compositions: waterproof gloss white registers SRI ranges between 100% to 88%, and its surface temperature goes from 31 °C to 43 °C, that is, SRI differences equal 12% (see L07 vs L14). For the acrylic matte composition, the SRI ranges obtained vary between 94% and 86% and the surface temperature between 40 °C and 44.5 °C, reaching between the extreme cases analyzed differences in SRI of 8% (see L11–L12). These results evidence that the white paint of waterproof gloss presents a wider range of behavior according to the manufacturer in comparison to acrylic matte.

In Figure 6.10, thermal behavior and optical characteristics of the total of studied paints are represented. In detail D1, information is expanded due to the fact that the great concentration of white paints oscillates between the thermal ranges of 31 °C and 44.5 °C and the albedo values between 0.82 and 0.96.

It should be noted that the athermic matte white—L01—paint does not show efficient behavior in relation to the rest of the claddings. It stays in fourth place, with an SRI = 92% and thermal differences of 10.5 °C with respect to the coldest paint—L07.

Aging behavior ($SRI_2 - SRI_3$)

In Figure 6.11, the SRI values of paints are shown to study the modification of the thermal behavior of each material associated with wear in detail, during the three periods. If we consider that the white paint reaches a high degree of efficiency when it registers levels of solar reflectivity higher than 85%, it is then observed that initially, 100% of them correspond to this range (SRI ≥ 85%). In the second period of aging,

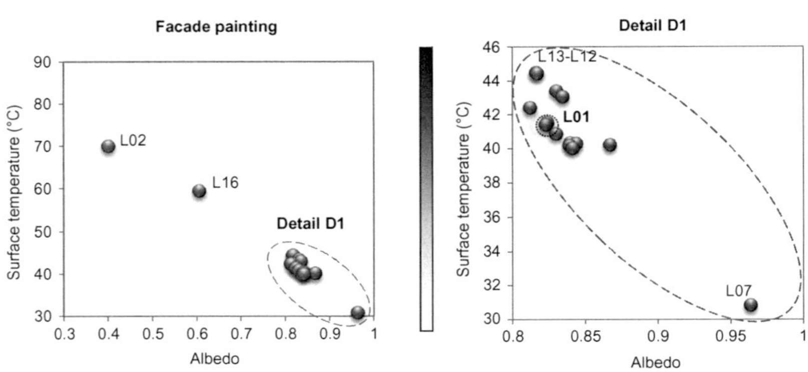

Figure 6.10 Solar reflective index (SRI), surface temperature (°C), and albedo in facade paintings.

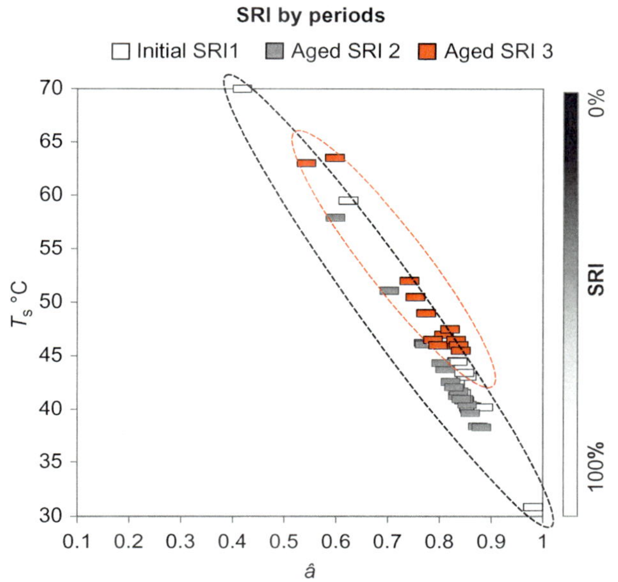

Figure 6.11 Solar reflective index of facade painting. (a) Initial (SRI_1), white boxes. (b) Aging for second year of exposure (SRI_2), light tinted boxes. (c) Aging for third year of exposure (SRI_3), dark tinted boxes – according to surface temperature and albedo.

75% of white paints occupied this place. Meanwhile, in the third period, only 6% manage to maintain this level of efficiency. In the dark tonalities, SRI oscillates between 73% and 51% for the case of terracotta (L06) and between 62% and 36.5% for black (L02) paint.

SRI aging according to compositions (acrylic matte vs waterproof gloss vs athermic): after exposing the paints for three years, the SRI levels registered differences for the athermic composition in contrast to the waterproof and acrylic ones. 100% of athermic paint, 50% of waterproof materials, and 33% of acrylics remained stable. 20% of waterproof materials and 33% of acrylics registered a moderate degradation in their SRI levels. Among the materials with strong degradation, 30% of the waterproof and 33% of the acrylic materials were evaluated. On average, the SRI values of white paints decreased to 10% in the case of athermic composition, 13% in the case of waterproof materials, and 11% in the case of acrylic materials (Figure 6.12).

SRI aging according to cleaning: due to the characteristics of washable latex exterior paints, SRI changes were studied after being submitted to cleaning.

In Figure 6.13, it is shown that the second year of aging does not show notable differences between the SRI levels of clean and dirty alternatives; however, in the third year, the exposure to a clean condition increases the SRI average in 3% in all alternatives. Terracotta (L16) and black (L02) paints improve their thermal behavior

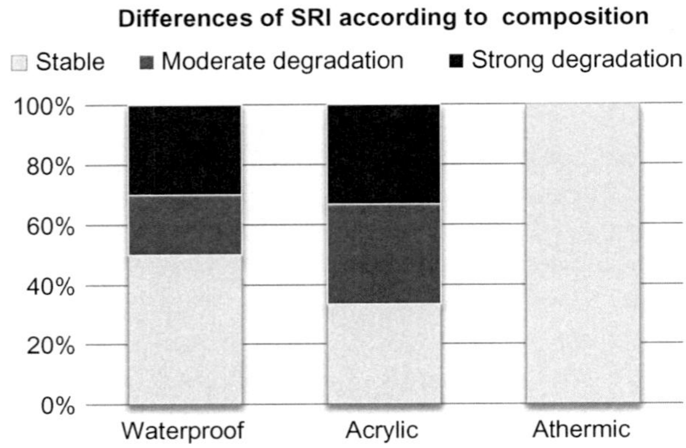

Figure 6.12 SRI differences on white facade paint (ΔSRI) according to its composition.

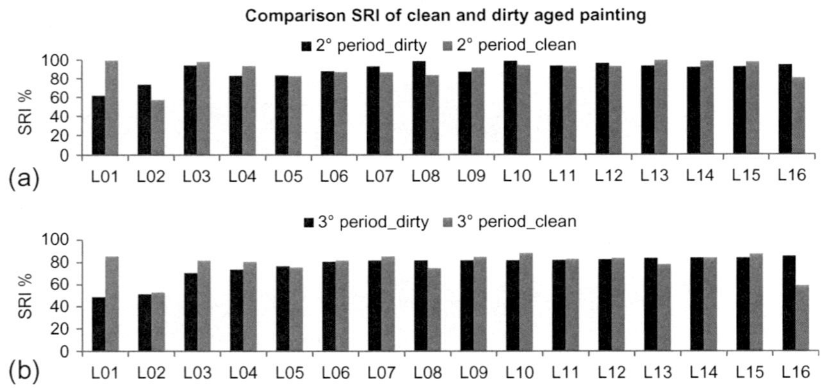

Figure 6.13 SRI of clean and dirty facade paintings. (a) At second period of aging. (b) At third period of aging.

at 11% and 15%, respectively. Meanwhile, in the white alternatives, clean surfaces show higher impact in acrylic white matte (L13) with an increase of SRI = 13%.

6.4.3 Cool materials used in building and urban scale: evaluation of thermal performance

As was referred to previously, to increase the solar reflectance (albedo) of urban surfaces it is necessary to modify the thermal conditions in both indoor and outdoor spaces. The application of materials with high-albedo "cool materials" on facades is a viable and low-cost strategy. To assess the relative impact of the appropriate selection of materials as a strategy for building and urban rehabilitation, we have carried out simulations on two scales of intervention.

6.4.3.1 Building scale

Two typologies of social welfare houses (square and linear) built in Mendoza were simulated in order to evaluate the effect of modifying the albedo of the envelope material over indoor housing temperatures, in both summertime and wintertime.

The thermal simulations of the housing typologies were carried out with SIMEDIF for Windows, a code developed at the Non-Conventional Energy Research Institute (INENCO, Argentina). This software is a tool for the design and simulation of transitory thermal behavior in buildings. SIMEDIF enables the determination of the thermal behavior of the analyzed prototype and the detection of comfort problems, caused by either overheating or low temperatures (Casermeiro and Saravia, 1984). The first version was developed for DOS by Casermeiro and Saravia (1984), and the code has been largely validated since then through years of experimental work in Argentina and in the world (Filippin et al., 2002; Larsen and Lesino, 2000, 2001a, 2001b; Larsen et al., 2008).

Software SIMEDIF: basic aspect

The SIMEDIF software needs the building to be divided into thermal zones. A zone is a building space that can be considered isothermal (i.e., a zone can be a group of various rooms in the building or a part of a room). Thus, each zone has a unique temperature whose temporal evolution is determined by the code, using the building data, materials, location, orientation, ambient temperature, and solar irradiance on a horizontal surface in the simulation period. Those zones are connected to each other and with the environment by elements that have any of the following thermal characteristics:

- Storage and heat transfer by conduction, such as brick walls, adobe walls, and concrete walls. These elements are called massive walls.
- Storage into a uniform temperature mass, such as water walls where the water is supposed to convect and has a uniform temperature.
- Heat transfer by conduction without storage, such as wood and expanded polystyrene. In SIMEDIF, these elements are called lightweight walls.
- Heat transfer by conduction with two heat transfer coefficients (day and night), such as windows with night insulation. These elements are called windows, and they do not describe any solar gain or ventilation.
- Heat transfer by convection, such as openings in the walls that allow a convective air exchange. These elements are doors and vents in SIMEDIF.

Input data

The following configuration data were entered into the simulator (see Table 6.6). Interior gains are not considered to be a study variable due to the fact that the objective of the work is to determine the impact of the strategy over the different morphological configurations and not over its use. In order to evaluate the degree of impact of the reflective materials over the indoor temperatures, the selected typologies were simulated with three ranges of albedo in walls: range 1 (average albedo = 0.37), range 2

Table 6.6 Basic input parameters to the model

Parameters	Input data
Georeference area	32°54′8″S 68°50′51″W. *Area*: low density and residential use
Date and time of simulation	*Summer*: 17th to 23rd of January 2010; 7 days *Winter*: 25th of July to 31st of July: 7 days
Geometric variables of environment	A medium albedo value of 0.3 was applied for surfaces of neighboring areas Orientation of main facade: north
Zones	The houses were divided into four rooms each (living area, bathroom, bedroom, and master bedroom)
Thermal variables for building components	*Floors*: made up of three layers: 1 m of soil with thermal conductivity (k) of 0.6 W/m^2 °C, density (δ) of 1600 kg/m^3, and heat capacity (C_e) of 1130 J/kg °C; the second layer is formed by 10 cm of underlayment (k = 1.73 W/m^2 °C, δ = 1893 kg/m^3, and C_p = 794 J/kg °C); and finally the calcareous tile (k = 1.38 W/m^2 °C, δ = 1500 kg/m^3, and C_p = 1000 J/kg °C) *External and internal walls*: made of brick, k = 0.72 W/m^2 °C, δ = 1797 kg/m^3, and C_p = 919 J/kg °C); and cement plaster on both faces (k = 1.43 W/m^2 °C, δ = 1184 kg/m^3, and C_p = 836 J/kg°C) *Roofs*: conduction coefficient was considered (k) = 0.71 W/m^2 °C. Composed of French tile (k = 0.76 W/m^2 °C), glass wool insulation (k = 0.043 W/m^2 °C), poplar girders and vapor barrier, pine bridle joints 1/2″ (k = 0.199 W/m^2 °C), and purlins of each 8 × 4″ 50 cm (k = 0.138 W/m^2 °C) *External doors*: k = 5.7 W/m^2 °C. *Windows:* transmittance = 5.8 W/m^2 °C (day and night)
Coefficients of transference by convection (h_c)	*Internal walls*: h_c = 7 W/m^2 °C; *external walls* h_c = 16 W/m^2 °C; *internal floor* h_c = 5 W/m^2 °C; *internal roofs* h_c = 9 W/m^2 °C; and *external roofs* h_c = 16 W/m^2 °C

(average albedo = 0.57), and range 3 (average albedo = 0.73). The determination of albedo ranges corresponds to the statistical calculation of the average albedo levels of facade materials registered through experimental tests.

Description case studies

Given their morphological configurations, two typologies of welfare housing were selected, built in the province of Mendoza in the year 2007, under the Federal Housing Plan (Mitchell and Acosta, 2009): a typology of square plant known as "ALAS" (covered area = 55 m^2) and another house with lineal plan "ALGARROBO" (covered area = 68 m^2).

According to compactness index (CI), both square typology and lineal typology are compact with different levels IC = 89% and 77%, respectively.

The functional design of the square housing plan is made up of a living area and a main bedroom oriented to the north. The service areas and the second bedroom are located to the south. The gabled roof is light and inclined north–south with a 20% slope. The lineal plan has four rooms: a living area, a service area, and two bedrooms. All areas open to the north. The living/kitchen area also has a window to the west. The roof is inclined to the north with a 36% slope. Both typologies are built with a traditional system: structure of reinforced concrete and 20 cm thick walls. Gabled wooden roof has an insulation of glass wool of 5 cm and French ceramic tile. The windows are metallic with simple glass and low infiltration (one air renovation per hour) (see Figure 6.14).

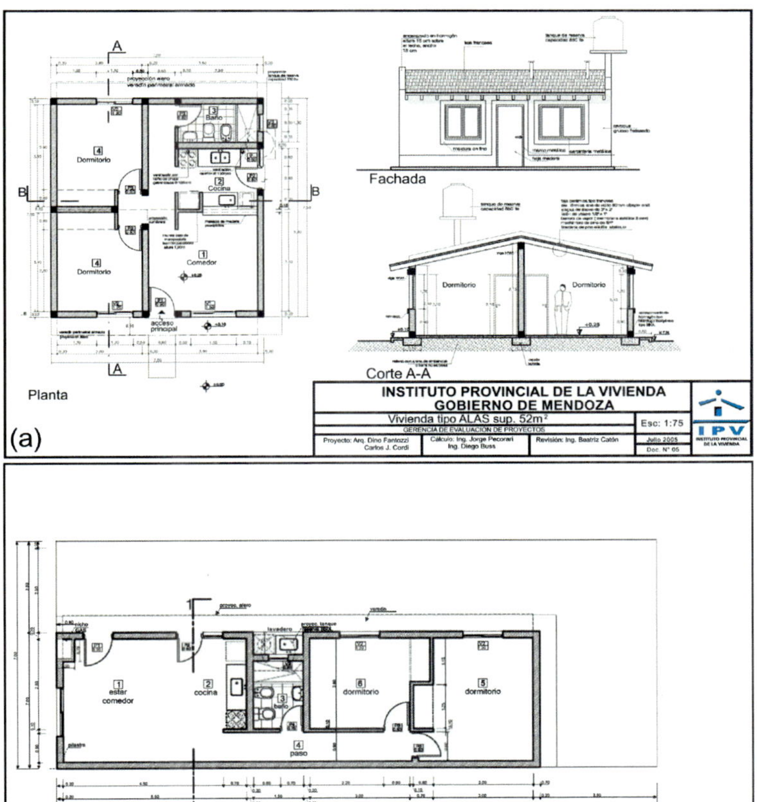

Figure 6.14 Plans of social housing studied. (a) Square typology. (b) Lineal typology.

Results of simulation

The thermal behavior of the typologies was simulated according to the three selected ranges for the variation of the albedo of their envelopes during summertime and wintertime.

Thermal behavior according to albedo ranges: the resulting graphs of the simulation model (Figure 6.15) describe the following behavior according to ranges and periods:

Summer period: considering the exterior average temperature of 29 °C (min: 23 °C and max: 36.5°), the behavior of the houses was analyzed according to their configuration:

- Square typology: in range 1, with an average albedo of 0.33, the typology reaches an average interior temperature of 35 °C. By increasing the amount of albedo to 0.57 (range 2), the resulting interior temperature is 33 °C. In range 3, the temperature decreases to 31 °C.
- Lineal typology: in range 1, this configuration reaches an interior temperature (Ti) of 34 °C. Range 2 reaches Ti = 32 °C. In range 3, the value is Ti = 31 °C.

Winter period: the medium exterior temperature is 9.8 °C (min: 4.5 °C and max: 15.5°).

- Square typology: in winter, as might be expected, more efficiency is registered in the rooms located towards the north (living area and bedroom), being more evident in the alternative of $\hat{a} = 0.33$. In range 1, the rooms reach an average temperature of 14.5 °C.

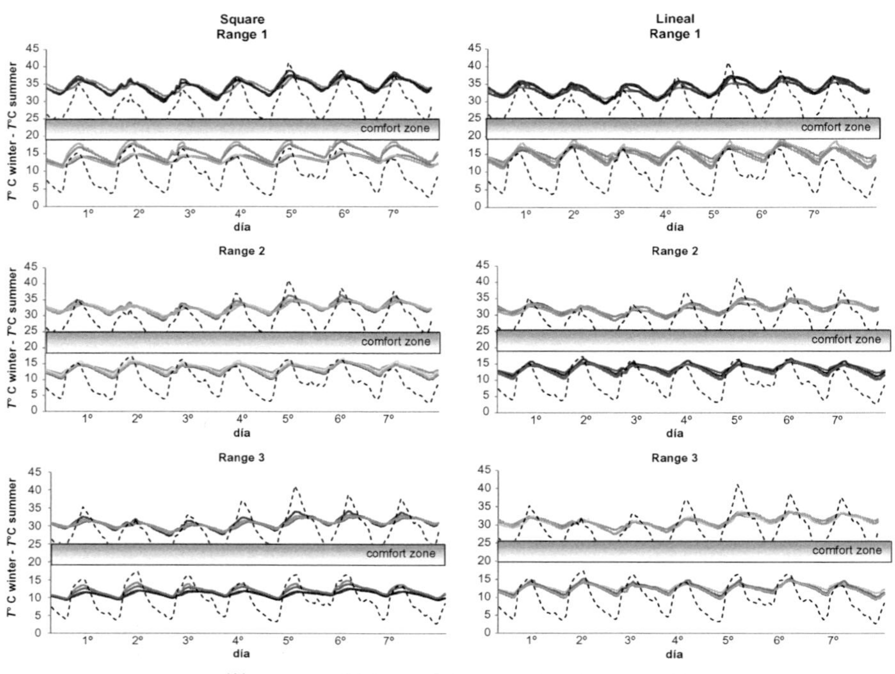

Figure 6.15 Interior temperatures °C of the four areas for summer and winter periods studied in the square (left) and the lineal (right) typologies.

In range 2 ($\hat{a}=0.57$), temperature decreases to 13 °C. In range 3 ($\hat{a}=0.73$), the registered temperature was 11.5 °C.

- Lineal typology: in range 1, a Ti = 15.5 °C was registered. In range 2, Ti = 14 °C. In range 3, the value was Ti = 12.5 °C.

During summer, even though a priori it could be expected that the typology with the most availability of direct solar gain (lineal house) would show higher interior temperatures, this does not occur due to its building technology, which is not very conservative, in addition to the higher envelope area (166 m^2) and a large portion of this surface being oriented to the south. However, the square typology—with the same technology—has less south wall area and a larger east–west wall area, envelopes which in the summer work as collectors. If we analyze that the collector areas of the north windows are similar in both typologies (difference of 0.9 m^2), we see why the square typology is hotter (interior average temperature = 35 °C), while the average temperature in the lineal typology is of 34 °C in the summer.

This behavior inverts during winter, where the lineal house presents higher interior temperatures due to the larger surface of gain to the north (average interior temperature = 14.5 °C in the square typology and 15.5 °C in the lineal).

Inner temperatures and comfort: both typologies stay completely out of the ranges of comfort of 18 °C in winter and 25 °C in summer. The consumption of auxiliary energy in the house depends on the difference between the indoor air and that necessary to reach comfortable conditions, for both heating and refreshing.

For the evaluated cases (two typologies and three ranges of albedo) in winter, it can be observed in all alternatives that the distance between the interior temperature and the comfort temperature is shorter than for summer. The interior temperatures of both houses in winter oscillate between 11.5 °C and 15.5 °C and in summer between 31 °C and 35 °C (see Table 6.7). If we add to these values, the internal gains interior temperatures would increase in approximately 2 °C. This fact favors the winter conditions and worsens the behavior of houses for the summer period. It is thus deduced that the technology used for building this type of social housing is more apt to the weather requirements of winter within the urban context analyzed than for the summer periods.

Average thermal difference according to the ranges of albedo: summer period: by modifying the levels of albedo of the building envelope by 57% (delta between ranges 1 and 3), interior thermal differences were registered. The square typology decreased its interior temperature by 4 °C in this period, which is the most demanding, whereas

Table 6.7 Difference between the interior temperature (with internal gain) and the comfort temperature for the square and lineal typologies according to the three ranges of albedo

	Diff. T° summer (°C)		Diff. T° winter (°C)	
Range	**Square**	**Lineal**	**Square**	**Lineal**
1 ($\hat{a}=0.37$)	12	11	−1.5	−0.5
2 ($\hat{a}=0.57$)	10	9	−3	−2
3 ($\hat{a}=0.73$)	8	8	−4.5	−3.5

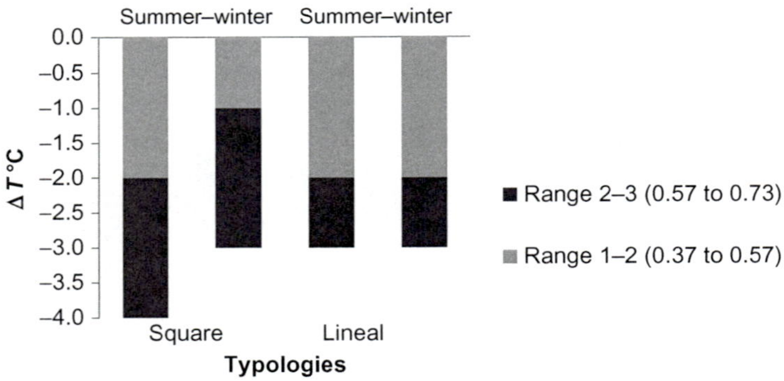

Figure 6.16 Thermal differences (°C) by housing typology and periods according to modifications in the range of albedo.

the lineal typology reduced its temperature by 3 °C. Winter period: a decrease of 3 °C was registered both in the square and in the lineal typologies (see Figure 6.16).

Comparison of strategies of facade (albedo increase vs insulation): to compare the efficiency of the application of this strategy—by modifying solar reflectance—of envelopes with respect to other forms traditionally used for the house energy conservation like the insulation of walls, the effect of insulating the house with 5 cm of expanded polystyrene was compared to the square typology, because it offers more sensitivity to the albedo modifications.

The results of the simulation showed that in summer, the benefit of wall insulation is equivalent to the effect of modifying the wall albedo level from 0.37 to 0.57. This shows that modifying the optical properties of the envelope is a strategy of cost-effective summer thermal refurbishment, since it does not involve practically any investment in relation to the original budget. Besides, the painting cost is necessary, even when it has been decided to insulate the house (see Figure 6.17).

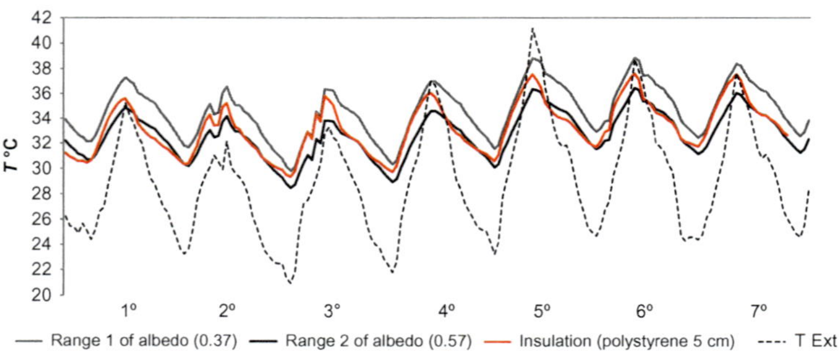

Figure 6.17 Comparison of interior thermal behavior (°C) (in summer) of square typology range 1, square typology range 2, and square typology modified with 5 cm of wall insulation.

Influence of orientation over the thermal performance according to the housing typology: in order to study the effect of modifying of albedo levels according to house orientation, a set of simulations were carried out in which the housing typologies were rotated towards the other three orientations (E, W, and S) and for each selected range of albedo.

Figure 6.18 charts the thermal behavior of the housing typologies—square and lineal—in the three albedo ranges according to orientation:

Summer period:

- Square typology: due to its symmetrical morphology, this typology does not present thermal differences modified by orientation. There is a thermal difference of 4 °C between ranges 1 and 3 for the four orientations (N, S, E, and W).
- Lineal typology: the east and west orientations register a thermal difference of 4 °C between ranges, while the orientation north–south reaches a thermal difference of 3 °C.

Winter period:

- Square typology: like in the summer period, this typology does not show a different thermal behavior by modifying its orientation, with registers of 3 °C of thermal difference.
- Lineal typology: it reaches a thermal difference of 3 °C between ranges in the north, south, and east orientations and 2 °C to the west (Figure 6.18 and Table 6.6).

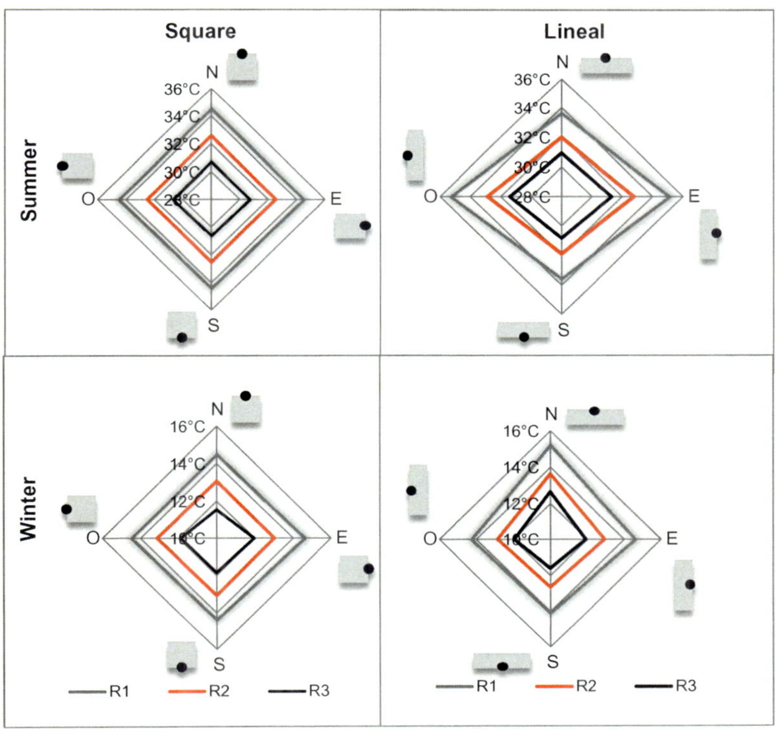

Figure 6.18 Comparison of interior thermal behavior (°C) of square (left) and lineal (right) typology according to the three ranges of albedo for orientations (N, S, E, and W) in summer (top) and winter (bottom).

6.4.3.2 Urban scale

A portion of the metropolitan area of Mendoza was simulated with the ENVI-met (Bruse, 1999) software. The objective was to determine the more efficient levels of albedo on facades to reduce the urban temperatures and its interface with building density of urban plot.

The ENVI-met software is able to simulate microscale interactions between urban surfaces, vegetation, and the atmosphere. ENVI-met proved to be a good tool for the prognosis of microclimatic modifications due to urban environments and for assessing the thermal comfort of pedestrians (Arnfield, 2003). The use of this tool is widely internationally valid (Acero, 2010; Bruse, 2009; Chowa and Brazel, 2012; Han et al., 2007; Huttner et al., 2009; Krüger et al., 2011; Perini and Magliocco, 2014; Samaali et al., 2007; Taleghani et al., in press.

Software ENVI-met and LEONARDO: basic aspects of the environmental model

ENVI-met is a three-dimensional nonhydrostatic microclimate model and is able to calculate and simulate climate in urban areas with a typical grid resolution of 0.5 to 10 meters in space and 10 seconds in time. It calculates the dynamics of microclimate during a diurnal cycle (24 to 48 hours) using fundamental laws of fluid dynamics and thermodynamics. Main prognostic variables of the program are wind speed and direction, air temperature and humidity, turbulence, radiative fluxes, bioclimatology, and gas and particle dispersion.

The modeling program has four user interfaces. The first one is editing the input of prepared digital maps of layout of the domain. Therefore, one has to generate baseline data from other geographic information systems and/or has to generate the data herself/himself in ENVI-met's own cartographic format. This stage can be quite complex depending on the environmental domain that is chosen to work. The high-resolution aspects of the program enable the user to go into finer details in smaller scales (0.5) and to be more coarse (and less detailed) in lower scales (e.g., 10 m in a neighborhood study of 1 km by 1 km). The program consists of modules of working areas such as 130 m by 130 m maximum, so theoretically one can go to a study area of approximately 1300 m by 1300 m.

The second interface is the configuration editor, where databases for soil types, humidity, temperature, temporal input, etc. are entered. The third one is the modeling area, where additional parameters are present and the modeling process takes place. The output data can be interpreted and visualized in LEONARDO. It is also possible to edit the data in other programs because the program structure is public.

Input data

ENVI-met was used to simulate the microclimate of the Mendoza metropolitan area. The model input parameters used are shown in Table 6.8.

Table 6.8 **Basic input parameters to the model**

Parameters	Input date
Date and time of simulation	14th January; 96 hours
Initial wind	3 m/s at 10 m from 150 deg. (south–west)
Boundary conditions	Temperature (at 2500 m) = 300 °K Specific humidity (at 2500 m) = 2.8 g/kg
Grid size	70 × 70 × 30 m. X-Y- spacing: 3 m
Plants	Tree species: Morus alba. Height: 15 m
Surface/soil profiles	Concrete road profile and pedestrian pavement
Soil initial conditions	300 °K

Description of case studies

The ENVI-met simulation was carried out in the west area in the city of Mendoza, Argentina. It is a low-density residential area inserted into a potential place that provides the basis for a future vertical growth. The sector study covers an area of 4.4 ha (210 × 210 m), in Godoy Cruz, Mendoza (32°55′ S; 68°50′ W). It has an orthogonal configuration, and its building height is between 1 and 3 levels, with an aspect ratio of $H/W = 0.3$ (Figure 6.19). In order to validate the simulated curve with the numerical model, we proceeded to compare the simulated results with the measured reference day ($R^2 = 0.98$, standard deviation = 0.7, and mean standard deviation = 0.025) (Figure 6.20).

Results of simulation

Albedo range in facade according to building density: to determine the most efficient albedo level over building height profile, two simulations were run with a base case of low density ($H/W = 0.3$) and the other with high density ($H/W = 1.5$) for albedo ranging from 0.3 to 0.8. Figure 6.21 shows that when the level of albedo on walls increases, the air temperature increases too. This fact becomes more evident at high density, because the vertical area is larger. That is, in the high-density alternative, each 10% increase in the level of albedo increases the temperature of the surrounding air from 0.4 to 0.5 °C, while the temperature modification in low density is minimal ($\leq$0.07 °C).

Optimal albedo level in facade according to building height: building height level was gradually increased in two scenarios: one with high albedo and the other with low albedo. Table 6.9 and Figure 6.22 show that in the high-density scenario, the facade with low albedo is more efficient, while in the low-density scenario, we recommend raising the levels of albedo over the facade. If we analyze the average air temperature as a variable, we note that in up to 12 m of building height, the high albedo of urban facades generates lower temperatures. At 15 m of height, it is advisable to apply low albedo over walls. This means that the inflection point occurs between the fourth and fifth levels of building height. To evaluate the minimum temperature, it is recommended to increase the levels of albedo up to 9 m in height (3rd level). In the case of maximum temperatures, the strategy begins to be detrimental from 18 m of building height.

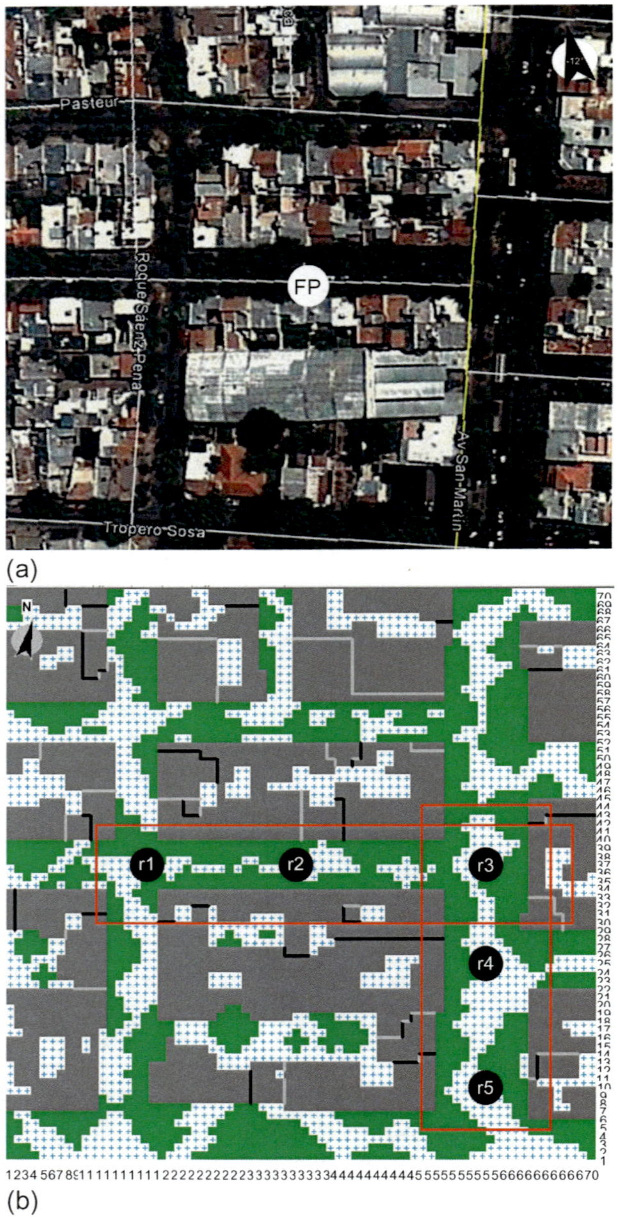

Figure 6.19 Aerial view of the study area; location of the fixed point (FP); and the reference receptors (r1, 2, 3, 4, 5).

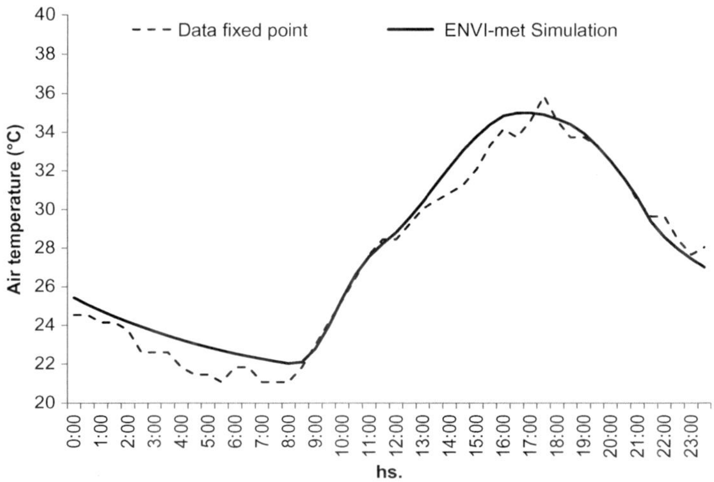

Figure 6.20 Adjustment of air temperature curve (°C) between the numerical models and measured.

In order to decrease urban temperatures, Figure 6.23 represents the scenarios with albedo levels and building height extremes to show that increasing albedo is an efficient strategy in low-density scenarios and conversely it is detrimental in high-density scenarios.

6.4.4 Labeling of vertical material

The results of this study confirm the importance of classifying the thermal behavior of the materials available at the regional level to resolve the envelopes. To use this knowledge by the developers of the habitat and the general public, it is necessary to develop effective communication strategies (see Figures 6.24 and 6.25). With this purpose, we preliminarily developed a label to classify facade materials according to their SRI levels and resistance to aging.

6.5 Conclusions and future trends

The results show that surface temperatures of materials for walls are derived from the combined effect of composition, texture, and color, but the color is the most important variable that determines their behavior. In the initial stage, the acrylic composition claddings and acrylic paints in bright colors are more efficient; in contrast to the application of darker colors over facade, the selection of concrete composition claddings is recommended, since they reach higher levels of SRI.

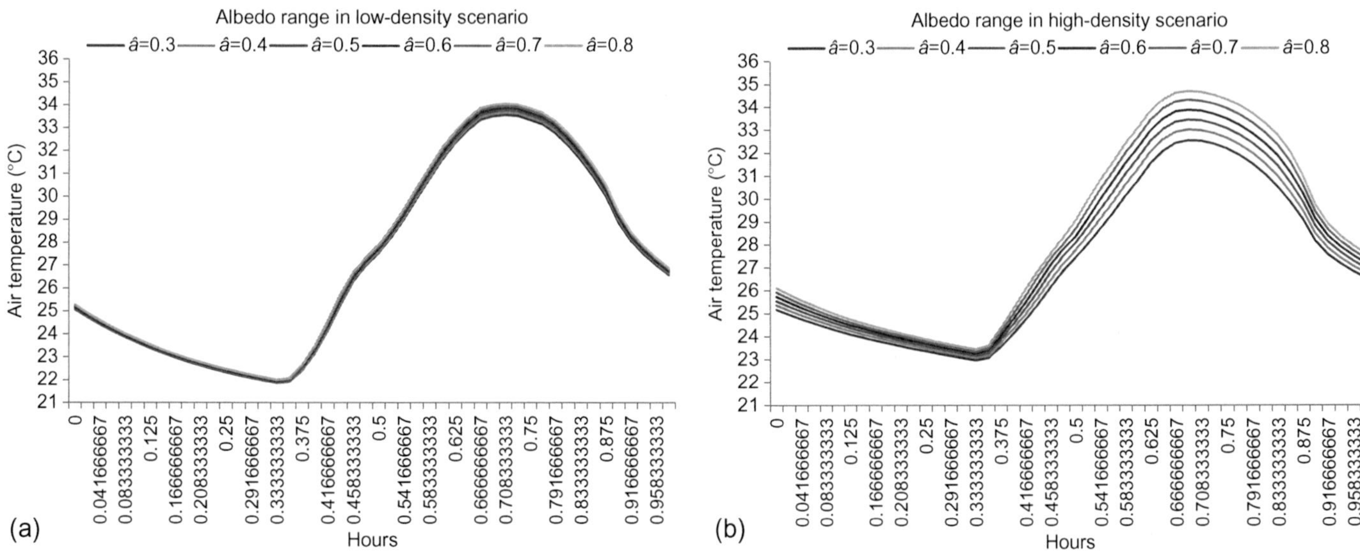

Figure 6.21 Thermal behavior curve of scenarios of low (a) and high (b) density, according to albedo graduation on facades.

Table 6.9 Thermal behavior (°C) of scenario according to albedo and building height

Variable		Air temperature (°C)		
Albedo ($\hat{a}$)	**Building height (m)**	**Min T**	**Max T**	**Average T**
$\hat{a}$ high (0.75)	6	22.3	33.8	27.5
	9	22.5	33.6	27.5
	12	22.7	33.8	27.6
	15	22.8	33.9	27.8
	18	23.0	34.1	28.0
	21	23.1	34.2	28.1
	24	23.3	34.2	28.2
	27	23.4	34.3	28.2
$\hat{a}$ low (0.3)	6	22.3	34.6	27.8
	9	22.5	34.1	27.6
	12	22.6	34.0	27.6
	15	22.7	34.1	27.7
	18	22.8	34.0	27.7
	21	22.9	33.9	27.7
	24	23.0	33.8	27.7
	27	23.1	33.7	27.6

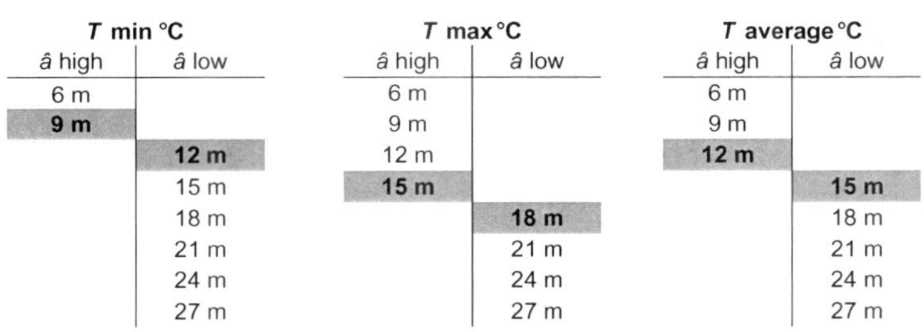

Figure 6.22 Optimal albedo according to building height.

The solar reflectance of textured claddings and paints tends to change over time due to wear caused by weather conditions and dirt accumulation. After three years of exposure, the materials more resistant to aging are those of acrylic composition, because 50% of them are stable. Unlike its initial performance, the acrylic cladding and acrylic paints in dark colors are the least aging. In general, the value of SRI of paints and claddings in light colors is the most affected by aging.

On a *building scale*, the results show that as the levels of albedo increase in the building envelope, the inner temperatures decrease. The inner temperature of the square typology is more susceptible to modifications as a consequence of the variation of the optical characteristics of the materials of its envelope (thermal differences of 4 °C in summer and 3 °C in winter). On the other hand, the lineal typology—north orientation—registers a thermal difference in winter and summer of 3 °C.

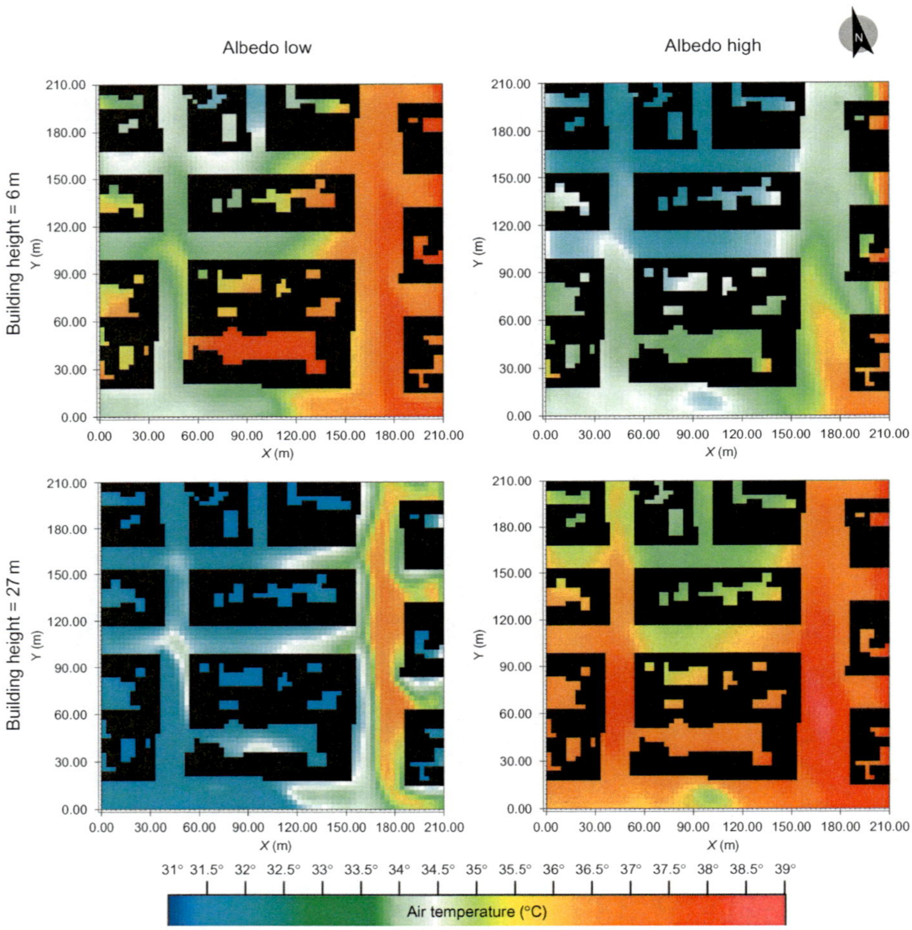

Figure 6.23 Thermographic images of scenarios during maximum temperature (17:00 hours). Air temperature difference (°C) according to albedo level and building height.

If we compare the results of the albedo variations to other refurbishment strategies, such as insulation, we see that the impact of applying 5 cm of expanded polystyrene on walls decreases the inner temperature by 2 °C, an equivalent value to modifying the albedo level from 0.37 to 0.57. Analyzing the effect of albedo modification according to the building orientation, we observed that the lineal housing typology is more susceptible because the exhibition area of facades is not symmetrical.

In *urban scale*, to determine the most efficient level of albedo according to high of building profile, the application of materials with high albedo is recommended when the building height is less than 12 meters.

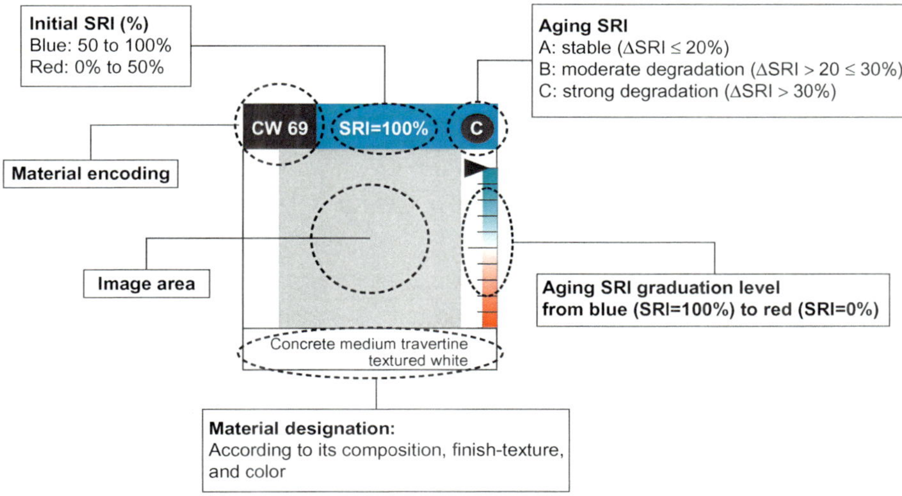

Figure 6.24 Preliminary format of material label.

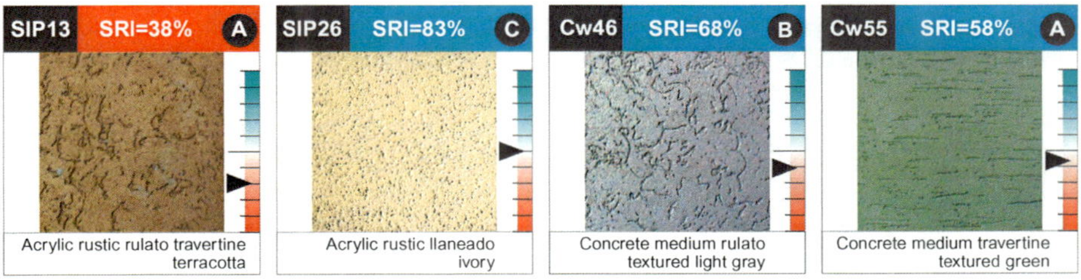

Figure 6.25 Example of vertical cladding label.

In order to spread this knowledge to the developers of the habitat, it is necessary to develop effective communication strategies. With this purpose, we preliminarily developed a label to classify facade materials according to their SRI levels and resistance to aging. The results show that the role of the vertical urban envelope is critical for reducing heat gains provided by multiple reflections within an urban canyon, affecting not only the indoor temperature of the building but also the comfort of pedestrians by increased urban temperatures.

The purpose of this investigation was to assess the resulting energy budget from the use of diverse strategies for improving the surface temperature of vertical urban envelopes with the final objective to mitigate the adverse effects of the city over the urban climate, reduce the consumption of fossil energy in the buildings, and improve the thermal comfort in the urban space.

Acknowledgments

The Agencia Nacional de Promoción Científica y Tecnológica—ANPCYT (National Agency for Scientific and Technological Promotion) provided funds for the research. Also, the Consejo Nacional de Investigaciones Científicas y Técnicas—CONICET (National Council of Scientific and Technical Researches) partially financed the staff involved in this work. The authors would like to thank Weber Saint-Gobain Building Products Inc., Argentina, for providing the vertical cladding samples evaluated.

References

Acero, J., 2010. Influencia de la Vegetación en la Calidad del aire y el Clima Urbano. In: CONAMA10, 1–13.

Akbari, H., Konopacki, S.J., 1998. The Impact of Reflectivity and Emissivity of Roofs on Building Cooling and Heating Energy Use. In: Proceedings of the Thermal Performance of the Exterior Envelopes of Building VII. December 6–10. Clearwater Beach, FL.

Akbari, H., Sarah, B., Dan, M., James, H., 1997. Peak power and cooling energy savings of high-albedo roofs. Energy Build. 25, 117–126. http://dx.doi.org/10.1016/S0378-7788(96)01001-8.

Akbari, H., Konopacki, S., Pomerantz, M., 1999. Cooling energy saving potential for reflective roofs for residential and commercial buildings in the United States. Energy 24, 391–407. http://dx.doi.org/10.1016/S0360-5442(98)00105-4.

Akbari, H., Levinson, R., Stern, S., 2008. Procedure for measuring the solar reflectance of flat or curved roofing assemblies. Sol. Energy 82, 648–655. http://dx.doi.org/10.1016/j.solener.2008.01.001.

Alchapar, N., Correa, E., Cantón, M., 2014. Classification of building materials used in the urban envelopes according to their capacity for mitigation of the urban heat island in semiarid zones. Energy Build. 69, 22–32. http://dx.doi.org/10.1016/j.enbuild.2013.10.012.

Arnfield, J., 2003. Two decades of urban climate research. A review of turbulence, exchanges of energy and water, and the urban heat island. Int. J. Climatol. 23, 1–26. http://dx.doi.org/10.1002/joc.859.

ASTM Standard E1933-06, 2006. Standard test methods for measuring and compensating for emissivity using infrared imaging radiometers.

ASTM Standard E1980-11, 2011. Standard practice for calculating solar reflectance index of horizontal and low-sloped opaque surfaces.

Bretz, S., Akbari, H., 1997. Long-term performance of high albedo roof coatings. Energy Build. 25 (2), 159–167. http://dx.doi.org/10.1016/S0378-7788(96)01005-5.

Bruse, M., 1999. The influences of local environmental design on microclimate. Ph.D. Thesis. University of Bochum, Bochum (in German).

Bruse, M., 2009. Stadtlandschaft im Klimawandel—Zu den Grundlagen des städtischen Mikroklimas. In: Collinet, H.D., Pesch, F. (Eds.), Stadt und Landschaft, Klartext Verlag Essen.

Casermeiro, M., Saravia, L., 1984. Cálculo Térmico Horario de Edificios Solares Pasivos, Actas de 98 Reunión de ASADES. In: Asociación Argentina de Energía Solar, San Juan, Argentina. 39–45.

Cheng, V., Ng, E., Givoni, B., 2005. Effect of envelope color and thermal mass on indoor temperatures in hot humid climate. Sol. Energy 78, 528–534. http://dx.doi.org/10.1016/j.solener.2004.05.005.

Chimkla, P., Hagishima, A., Tanimoto, J., 2004. A computer system to support albedo calculation in urban areas. Build. Environ. 39 (10), 1213–1221. http://dx.doi.org/10.1016/j.buildenv.2004.02.006.

Chowa, W., Brazel, A., 2012. Assessing xeriscaping as a sustainable heat island mitigation approach for a desert city. Build. Environ. 47, 170–181. http://dx.doi.org/10.1016/j.buildenv.2011.07.027.

Dornelles, K., Roriz, M., Roriz, V., Caram, R., 2011. Thermal performance of white solar reflective paints for cool roofs and the influence on the thermal comfort and building energy. Use in hot climates. In: Solar World Congress. ISES. International Solar Energy Society.

Doulos, L., Santamouris, M., Livada, I., 2001. Passive cooling of outdoor urban spaces. The role of materials. Sol. Energy 77, 231–249. http://dx.doi.org/10.1016/j.solener.2004.04.005.

Doya, M., Bozonnet, E., Allard, F., 2012. Experimental measurement of cool facades performance in a dense urban environment. Energy Build. 55, 42–50. http://dx.doi.org/10.1016/j.enbuild.2011.11.001.

Filippin, C., Larsen, S.F., Lesino, G., 2002. Simulación térmica de verano de un sector del bloque de residencias estudiantiles bioclimáticas en Santa Rosa, La Pampa. Renew. Energy Environ. Programme. 6 (1), 53–58.

Fluke Ti 55-IR chamber 2010.

Givoni, B., 1998. Climate Considerations in Building and Urban Design. John Wiley & Sons, Inc., New York.

Han, S., Mun, S.-H., Huh, J.-H., 2007. Changes of the micro-climate and building cooling load due to the green effect of a restored stream in Seoul, Korea. In: Proceedings: Building Simulation, 1131–1138.

Huttner, S., Bruse, M., Dostal, P., Katzschner, A., 2009. Strategies for Mitigating Thermal Heat Stress in Central European Cities. In: The project KLIMES; Seventh International Conference on Urban Climate ICUC-7, 29 June–3 July 2009, Yokohama, Japan (Poster).

Kai, L., Neide, M.N., Vanderley, M., 2010. Estimating thermal performance of cool colored paints. Energy Build. 42, 17–22. http://dx.doi.org/10.1016/j.enbuild.2009.07.026.

Karlessi, T., Santamouris, M., Apostolakis, K., Synnefa, A., Livada, I., 2009. Development and testing of thermo chromic coatings for buildings and urban structures. Sol. Energy 83, 538–551. http://dx.doi.org/10.1016/j.solener.2008.10.005.

Kipp & Zonen, 2012. Product catalogue. http://www.kippzonen.com (accessed: 1.03.12.).

Krüger, E., Minella, F., Rasia, F., 2011. Impact of urban geometry on outdoor thermal comfort and air quality from field measurements in Curitiba, Brazil. Build. Environ. 46, 621–634. http://dx.doi.org/10.1016/j.buildenv.2010.09.006.

Larsen, S.F., Lesino, G., 2000. SIMEDIF 2000: Nueva versión del programa de diseño y cálculo de edificios. Renew. Energy Environ. Programme. 4 (2), 53–58.

Larsen, S.F., Lesino, G., 2001a. A New Code for the Hour-by-Hour Thermal Behavior Simulation of Buildings. In: Proceedings of VII International Building Simulation Congress, 75–82, Brazil.

Larsen, S.F., Lesino, G., 2001b. Modelo térmico del programa SIMEDIF de simulación de edificios. Renew. Energy Environ. Programme. 9, 15–24.

Larsen, S.F., Filippin, C., Beascochea, A., Lesinoa, G., 2008. An experience on integrating monitoring and simulation tools in the design of energy-saving buildings. Energy Build. 40, 987–997. http://dx.doi.org/10.1016/j.enbuild.2007.08.004.

Levinson, R., Berdahl, P., Berhe, A., Akbari, H., 2005. Effects of soiling and cleaning on the reflectance and solar heat gain of a light colored roofing membrane. Atmos. Environ. 39, 7807–7824. http://dx.doi.org/10.1016/j.atmosenv.2005.08.037.

Levinson, R., Akbari, H., Reilly, J., 2007. Cooler tile-roofed buildings with near infrared-reflective non-white coatings. Build. Environ. 42, 2591–2605. http://dx.doi.org/10.1016/j.buildenv.2006.06.005.

Mitchell, J., Acosta, P., 2009. Evaluación comparativa de tipologías de viviendas sociales en la provincia de Mendoza. Renew. Energy Environ. Programme. 13, 161–167.

Perini, K., Magliocco, A., 2014. Effects of vegetation, urban density, building height, and atmospheric conditions on local temperatures and thermal comfort. Urban For. Urban Green. 13 (3), 495–506. http://dx.doi.org/10.1016/j.ufug.2014.03.003.

Pomerantz, M., Akbari, H., Berdahl, P., Konopacki, S.J., Taha, H., Rosenfeld, A.H., 1999. Reflective surfaces for cooler buildings and cities. Philos. Mag. B79, 9457–9476. http://dx.doi.org/10.1080/13642819908216984.

Radhi, H., Assem, E., Sharples, S., 2014. On the colours and properties of building surface materials to mitigate urban heat islands in highly productive solar regions. Build. Environ. 72, 162–172. http://dx.doi.org/10.1016/j.buildenv.2013.11.005.

Revel, G., Martarelli, M., Emiliani, M., Gozalbo, A., Orts, M., Bengochea, M., Delgado, L., Gaki, A., Katsiapi, A., Taxiarchou, M., Arabatzis, I., Fasaki, I., Hermanns, S., 2014. Cool products for building envelope—part I: development and lab scale testing. Sol. Energy 105, 770–779. http://dx.doi.org/10.1016/j.solener.2014.03.029.

Sailor, D., Resh, K., Segura, D., 2006. Field measurement of albedo for limited extent test surfaces. Sol. Energy 80, 589–599. http://dx.doi.org/10.1016/j.solener.2005.03.012.

Samaali, M., Courault, D., Bruse, M., Olioso, A., Occelli, R., 2007. Analysis of a 3D boundary layer model at local scale. Validation on soybean surface radiative measurements. Atmos. Res. 85, 183–198. http://dx.doi.org/10.1016/j.atmosres.2006.12.005.

Synnefa, A., Santamouris, M., Livada, I., 2005. A study of the thermal performance of reflective coatings for the urban environment. Sol. Energy 80, 968–981. http://dx.doi.org/10.1016/j.solener.2005.08.005.

Taha, H., Akbari, H., Rosenfeld, A., Huang, J., 1988. Residential cooling loads and the urban heat island: the effects of albedo. Build. Environ. 23, 271–283. http://dx.doi.org/10.1016/0360-1323(88)90033-9.

Taleghani, M., Kleerekoper, M., Tenpierik, M., Dobbelsteen, A., in press. Outdoor thermal comfort within five different urban forms in the Netherlands. Build. Environ. http://dx.doi.org/10.1016/j.buildenv.2014.03.014, Continuous in press.

Young, R., 1998. Cool roofs: light-colored coverings reflect energy savings and environmental benefits. Build. Des. Constr. 39 (2), 62–64.

Lotus ceramics for counteracting urban heat island effects

7

K. Okada
Tokyo Institute of Technology, Tokyo, Japan

7.1 Introduction

Most fields of science and technology have developed remarkably from the middle of 20th century. It could therefore be said that the 20th century was the "Science and Technology" century. Thus, various social activities and population increased significantly. As a result, consumption of energy by the burning of fossil fuels and also emission of greenhouse gases such as carbon dioxide, methane, and nitrous oxide increased dramatically from the middle of the 20th century. These factors are thought to strongly relate to "global warming." Actually, the clear tendency of increased temperature is much more significant in big cities due to factors such as higher density of population, number of large buildings, congested housing, and massive volume of logistics.

Figures 7.1 and 7.2 show examples in Tokyo for changes of average temperature, number of tropical nights (minimum temperature >25 °C throughout the day), and number of days of winter sun (minimum temperature <0 °C) over a hundred years (Japan Meteorological Agency, 2013). On the change of average temperature in Figure 7.1, a clear increase of average temperature was observed from the beginning of the 1910s, and the increase reached about 3 °C in 100 years. Since the average temperature change of the world from 1981 to 2010 is reported to be 0.69 °C over 100 years (Japan Meteorological Agency, 2013), the change of 3 °C over 100 years in Tokyo is remarkably higher than in the rest of the world. This is one of the examples that shows the large increase of temperature in big cities. With such an increase of the average temperature, numbers of tropical nights increased steeply while those of winter suns decreased, as clearly observed in Figure 7.2. The latter effect (milder winter) is a positive change, but the former effect (more severe summer) is negative for Tokyo residents. Basically, summer in Tokyo is hot and humid like a tropical city due to the effect of strong high pressure from the Pacific Ocean. Most of the surfaces in Tokyo are covered by artificial materials with low thermal conductivity, so after the surfaces are warmed by sunshine in daytime, the temperature decreases very slowly after dark, and retaining residual heat even up to midnight. Thus, people tend to use air conditioners even at night, which makes more heat exhaust in the atmosphere, resulting in an increased number of tropical nights.

Apparent warming of big cities in summer nights has been well-known in Japan for years; such phenomena are called urban heat island effects (Yoshino, 1991). Similar phenomena had already been recognized in the world, but they were generally considered localized issues and not as serious as they are in Japan. Local governments of big

Eco-efficient Materials for Mitigating Building Cooling Needs. http://dx.doi.org/10.1016/B978-1-78242-380-5.00007-8

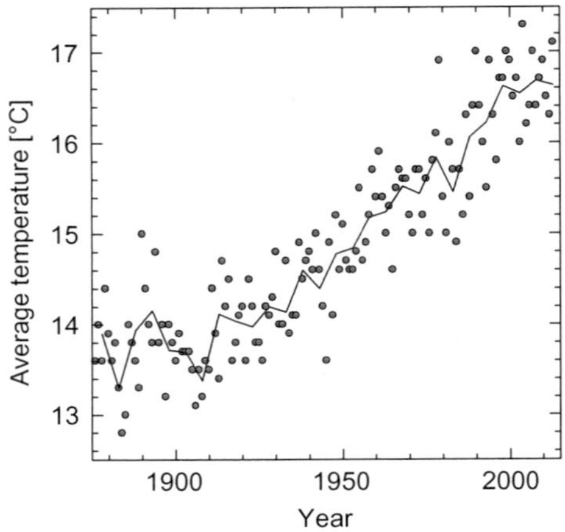

Figure 7.1 Change of average temperature in Tokyo. The blue line graph shows five year average data.
Japan Meteorological Agency (2013).

cities in Japan have tackled this problem and proposed a number of measures to suppress this phenomenon. The major proposed measures include greening; making wind paths and/or water paths; developing energy-efficient electronics; surface cooling by TiO_2 photocatalysts, utilizing their superhydrophilicity (Guo et al., 2009); watering by many citizens; covering the surfaces of buildings and pavements with heat-resistant paint; and using water-retentive materials for buildings, houses, and pavements.

We believe it is necessary to develop highly functional porous ceramics for water-retentive materials and pavements. One of the important and effective properties of this application is capillary rise ability. A good example is trees and plants, which have excellent capillary rise abilities and great cooling effects by water vapor evaporation. Thus, we decided to develop porous ceramics mimicking the microstructures of xylems in trees with unidirectionally aligned cylindrical pores. We call the porous ceramics with such microstructures "lotus ceramics."

7.2 Porous ceramics with a similar microstructure to the root of the lotus

7.2.1 Various prior preparation methods for lotus-type ceramics

Porous ceramics are widely used for various applications and are required to have a variety of porous properties, depending on their applications. To satisfy such a variety of requirements, various preparation methods (Saggio-Woyansky et al., 1992) have been developed, such as partial sintering (Lam et al., 1994) and foaming (Peng et al., 2000). The most common preparation method is the partial sintering of a

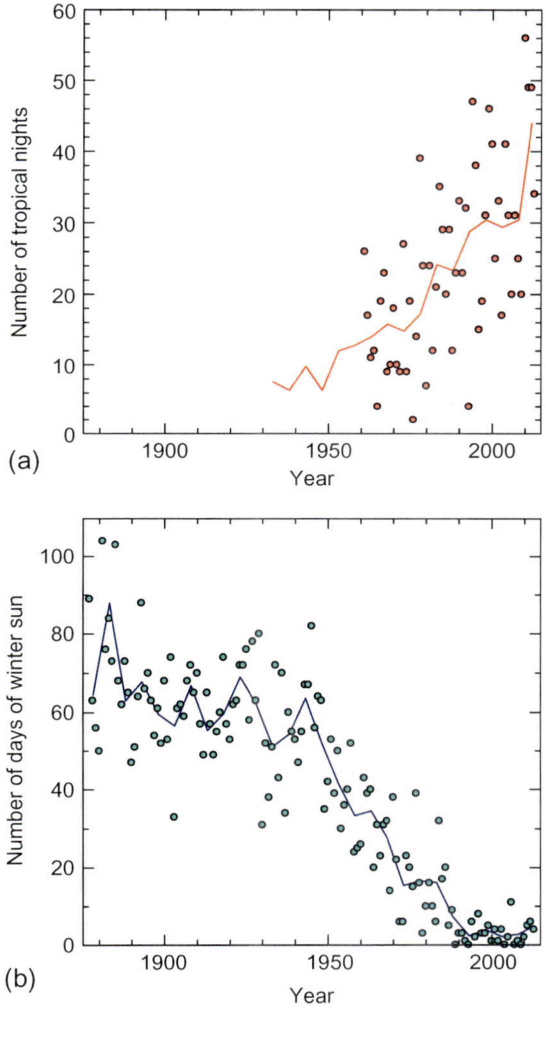

Figure 7.2 Change of numbers of tropical nights (a) and days of winter sun (b) in Tokyo. Red and blue line graphs show five year average data.
Japan Meteorological Agency, 2013.

ceramic powder using organic matter as a pore former, but it is difficult to produce products with a well-controlled microstructure from this method.

On the other hand, many works have already been reported on the preparation of porous ceramics with a similar microstructure to the root of the lotus, i.e., well-ordered and unidirectionally oriented cylindrical through holes (one-dimensional (1D) porous structure). The reported typical preparation methods were anodic oxidation, templating, unidirectional solidification, extrusion, and other miscellaneous methods. Some of the examples of 1D porous ceramics utilizing the aforementioned methods are introduced in more detail in the following sections.

7.2.1.1 Anodic oxidation method

Surfaces of metals are more or less oxidized in an ambient atmosphere. Some kinds of metals were known to be treated to form an oxidized layer to protect the metals from oxidation. One of the typical examples is aluminum metal. Its surface is generally electrochemically treated and is called an "almite or alumite," which was invented in RIKEN, Japan in 1924. The surface of almite is coming to be known to have unique porous microstructures, well-ordered hexagonal pore arrays perpendicular to the surface. Therefore, this unique porous microstructure has attracted great interest as a new type of porous material (Therese and Kamath, 2000). The electrochemical treatment to form such a porous microstructure is performed in a concentrated acid solution, applying high direct voltages of typically 10–200 V. The resulting microstructures are comprised of cylindrical mesopores in a hexagonal array, with interpore distances of several tens to several hundreds of nm (Li et al., 1998). The well ordered hexagonal pore arrays are formed under certain preparation conditions, especially under a longer anodization time ($\geq$160 h) and low temperature (0 °C) (Masuda and Fukuda, 1995). Due to the well-ordered unique porous microstructure, porous aluminas prepared by the anodic oxidation process have been widely studied to fabricate various nano-sized functional materials. This process has been investigated not only for aluminum but also for silicon and titanium. Porous silicons are mostly prepared to study properties related to photoluminescence (Brandt et al., 1992), while porous titanias are investigated for their photocatalytic (Gong et al., 2001) and bioactive performances (Chen et al., 2012). This process is, however, unsuitable for the fabrication of large-sized products.

7.2.1.2 Templating method

As previously mentioned, we wanted to develop lotus ceramics by mimicking wood-like unidirectionally aligned cylindrical pores. Thus, preparation of porous ceramics using the wood templating method is a direct approach for the purpose because the porous microstructures of wood can be transferred or replicated by this method. In the templating, the simplest method is just carbonizing wood (Gao et al., 2010). Activated carbon can be also prepared by low temperature physical activation of wood and corrugated cardboard, maintaining the original microtextures (Okada et al., 2005). Greil (2001) summarized the approaches to form porous ceramics by the templating method as (i) pyrolytic decomposition resulting in a porous carbon replica (template), which may subsequently be reacted to form carbide phases or may be infiltrated with nonreacting sols or salts that can be further processed to yield oxide reaction products; and (ii) infiltration of chemically preprocessed native or technical lignocellulosic products with gaseous or liquid organometallic and metal-organic precursors and subsequent oxidation to remove the free carbon phase. Although various biomorphic porous ceramics have been prepared using those replicating techniques, their disadvantages generally include poor mechanical strength or low porosity, multistep processes for fabrication, and the difficulty of fabricating large-sized products.

7.2.1.3 Unidirectional solidification method

Two processes were reported for unidirectional solidification methods: freeze casting based methods and unidirectional solidification of melts. The former method involves using the formation of frost pillars in the freeze drying of water solvent in the ceramic slurry (Fu et al., 2008). Arrangement of well-dispersed and stable ceramic slurry is important to obtain uniform unidirectionally oriented porous microstructures. By using a highly evaporative organic solvent such as alcohol instead of water, a gradient in pore size can be intentionally introduced by acceleration of evaporation of the solvent from the top opened surface of the slurry (Hu et al., 2010). Formation of unique unidirectional pores with hexagonal cross sections was reported by using tert-butyl alcohol (Hu et al., 2010). Ceramic matrix tends to be porous and forms a three-dimensional pore network as well as a unidirectionally connected porous structure by using organic solvents. The resulting porous ceramics are, however, generally poor in mechanical properties and it is also difficult to fabricate large-sized products.

Nakajima (2005) developed porous metals with a unidirectional porous structure by unidirectional solidification of melts and called these materials lotus-type porous materials. In this method, metals were melted under H_2 in a pressurized atmosphere to dissolve H_2 into the melt. Since the H_2 that was dissolved in the melt evolved to form bubbles by the solidification due to the lower solubility in the solid phase, this formed unidirectional cylindrical pores in the solidified products. This technique was applied for preparation of porous ceramics such as Al_2O_3 (Ueno et al., 2008), mullite (Yang et al., 2010), and $MgAl_2O_4$ (Ueno et al., 2009). The pores formed in their microstructures ranged from 100 μm to mm size, but it is rather difficult to prepare smaller pores by this method.

7.2.1.4 Extrusion method

Extrusion is a very common and widely used process in the ceramic industry. It is therefore easy to fabricate large-sized products. Lee and Sarkar (2009) proposed a multipass extrusion method to prepare lotus-type porous ceramics. In the preparation, a core/shell rod structure was prepared from a rod formed by a mixture of carbon and ethylene vinyl acetate (EVA), and a tube formed by a mixture of alumina and EVA at the first stage. EVA was mixed to both the core rod and the shell tube as the binder and pore former. The core/shell rod was loaded into the warmed extrusion die and extruded to form a filament. Then, these filaments were stacked and reloaded into the extruded die. The resulting product was comprised of bundles of the elongated and thinned core/shell filaments. A porous texture with unidirectionally aligned cylindrical pores was formed after the firing. The pore size obtained ranged from several tens to several hundreds of μm in diameter. The products showed good mechanical strength because of the well-controlled unidirectional porous microstructures and densified matrix during repeated extrusion passes. This technique is suitable for fabricating well-controlled unidirectional porous ceramics and it can also be used for fabricating relatively large-sized products. However, the disadvantage is its long and multistep process for production.

7.2.1.5 Miscellaneous methods

Unidirectional porous microstructures have been prepared using various miscellaneous methods, such as bubbling, magnetic field, and filament winding methods. In the bubbling methods, Nakahira et al. (2003) prepared unidirectional porous Al_2O_3 ceramics using an aqueous electrophoretic deposition (EPD) process, suspending Al_2O_3 particles in water and depositing the product on a carbon cathode by applying direct voltage. H_2 gas evolved on the cathode by water splitting and the bubbles formed unidirectional pores perpendicular to the deposit. After firing, unidirectional pores ranging from 60 to 170 μm were obtained. Song et al. (2008) used ethanol bubbles to fabricate unidirectional porous hydroxylapatite (HAp) ceramics by a very simple process: warming the suspension, dispersing HAp particles in an ethanol-water mixed solution, and evaporating the ethanol and water as bubbles from the suspension. After drying and firing, unidirectionally aligned pores (120–130 μm) were formed in the matrix of porous HAp ceramics.

In a magnetic field method, unidirectional porous Al_2O_3 ceramics were prepared by Miyagawa and Shinohara (1999). They prepared Al_2O_3 slurry with Ni wires and casted it on a plaster mold. The Ni wires were settled and unidirectionally aligned in the casted product by applying a magnetic field. They were converted to continuous pores after acid-leaching of the Ni wires. An interesting point of this technique is the tunability of the pore intervals by changing the magnetic field strength, but it is difficult to prepare large-sized products.

The filament winding method was originally developed to fabricate continuous fiber reinforced dense ceramic composites. On the other hand, Zhang et al. (2001) fabricated porous Al_2O_3 ceramics with unidirectionally aligned continuous pores by this method using flammable cotton fibers. The reinforced fibers were converted to pores by firing without distinct size change. Although the bending strength (155 MPa) was a quite satisfactory result, the multistep process and small-sized products are disadvantages of this method.

7.2.2 Lotus ceramics prepared by extrusion methods

As described in Section 7.2.1, all the prior techniques for obtaining unidirectional porous microstructures have the disadvantage of being unable to fabricate large-sized products. In ceramic industries, extrusion is a very common process and is widely used to form large-sized products. Therefore, a simple extrusion method is considered applicable for this purpose. In an extrusion process, particles with high aspect ratios are known to orient to reduce the flow resistance and to cause orientation in the green compact. Such anisotropy in the green product is generally unfavorable because it causes inhomogeneous shrinkage in the drying and firing stages. We, however, thought to intentionally use this phenomenon for fabrication of unidirectional porous ceramics using flammable fibers as the pore former. Unidirectional porous ceramics were fabricated by this method for Al_2O_3 (Isobe et al., 2006a,b), cordierite (Okada et al., 2007), and mullite (Okada et al., 2009; Popa et al., 2013b).

7.2.2.1 Lotus-type porous alumina ceramics

Alumina powder (average particle size = 0.7 μm) was mixed with 52 vol% carbon fibers (14 μm in diameter and 600 μm in length), adding water, methyl cellulose, and polycarboxylic acid to form ceramic paste. Nylon fibers and other flammable fibers may be used instead of carbon fibers. The resulting paste was molded into an Auger-type extruder, and formed a green body. The diameters of the extruder barrel and inner aperture used were 50 and 10 mm, respectively. After drying at room temperature, the extruded dried body was degreased at 1100 °C for 48 h and fired at 1600 °C for 2 h. The resulting lotus-type porous alumina ceramics were used for measurements of mechanical strengths and permeabilities.

7.2.2.2 Lotus-type porous mullite ceramics

The starting powders used were Al_2O_3 (4.7 μm), Kentucky kaolin clay (1.0 μm), and Chinese earthen clay (20 μm). The Chinese clay was used to color the products yellow-brown as a consideration for application to building materials. The bulk composition of SiO_2, Al_2O_3, and Fe_2O_3 was 37.1, 57.7, and 5.2 mass%, respectively. These powders were dry mixed with 20 mass% rayon fibers of average fiber diameters 8.1, 9.6, 16.8, 20.0, and 37.6 μm, and 800 μm lengths. Then, the mixtures were kneaded with 40 mass% of water and molded into an extruder. The diameters of the extruder barrel and inner aperture were 95 and 50 mm, respectively. Tubular green bodies with inner and outer diameters of 30–50 mm were extruded using an outlet die. The maximum length of the extruded green bodies was 2100 mm. The extruded green bodies were dried at room temperature for 3 days and at 120 °C for 18 h, then fired at 1500 °C for 4 h. The pore diameters of the ceramics measured by Hg porosimetry were 9.4–30 μm. The resulting pore diameters were close to those of the fibers, with slightly larger pore diameters in the thinner fibers and smaller in the thicker fibers. Their small differences are attributed to the small shrinkage (about 2%) of bodies from the dried stage to fired stage.

7.2.2.3 Microstructures

The microstructures of the cross sections perpendicular and parallel to the extruded directions of lotus-type porous alumina ceramics prepared using carbon fibers and nylon fibers as the pore formers are shown in Figure 7.3. The microstructures of both products showed cylindrical pores unidirectionally aligned parallel to the extruded direction. The degree of ordering of cylindrical pores was higher in the products using carbon fibers (Figure 7.3a and b) than in those using nylon fibers (Figure 7.3c and d). This is caused by the difference of the character of both fibers: hard and brittle in the carbon fibers, while soft and tough in the nylon fibers. Carbon fibers have the disadvantages of brittleness, causing fibers to break apart during kneading, and high fiber cost, which is an important factor for fabrication of large-sized products. For fabrication, any other kinds of flammable fiber can be used, but because of their thermal decomposition, cellulose-based fibers such as rayon fibers are preferable from the standpoint of environmental impact, especially in fabrication of large-sized products.

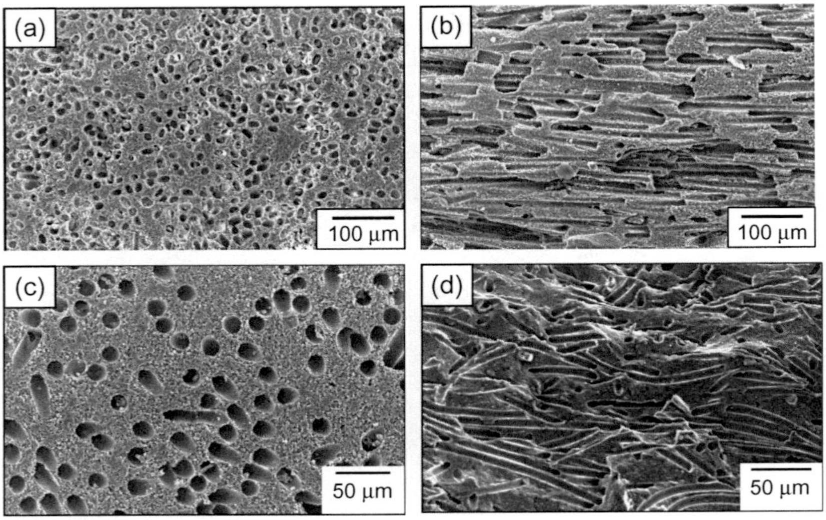

Figure 7.3 Microstructures of lotus-type porous alumina ceramics: (a) perpendicular cross section (carbon fiber), (b) parallel cross section (carbon fiber), (c) perpendicular cross section (nylon fiber), and (d) parallel cross section (nylon fiber).
Isobe et al. (2006a,b).

7.3 Properties of lotus ceramics

7.3.1 Mechanical properties

Changes of bending strengths of lotus-type porous alumina ceramics with unidirectionally aligned pores (lotus-type) and conventional-type porous alumina ceramics with randomly distributed pores prepared by the conventional sintering method using a spherical organic pore former (conventional-type) are shown in Figure 7.4 as a function of porosity (P). The relationships between the bending strength (σ) and P of these ceramics can be fitted by the Knudsen equation (Pekor et al., 2010), represented by the following Equation (7.1),

$$\sigma = \sigma_0 \exp(-bP) \tag{7.1}$$

where σ_0 and b are the fracture strength at $P=0$ and an empirical constant obtained from a slope of semi log plot, respectively. Although the σ_0 value of the conventional-type formed by die pressing was higher than the lotus-type formed by extrusion due to the higher packing of the green body, its decreasing slope ratio (b) was clearly higher in the conventional-type (0.039) than in the lotus-type (0.019). Thus, the bending strength of the lotus-type was almost twice as high as that of the conventional-type at $P=40\%$, owing to the anisotropic porous microstructure of the lotus-type.

Since mechanical strength is an important property for any actual application, the bending strengths of large-sized lotus-type porous mullite ceramics were also investigated.

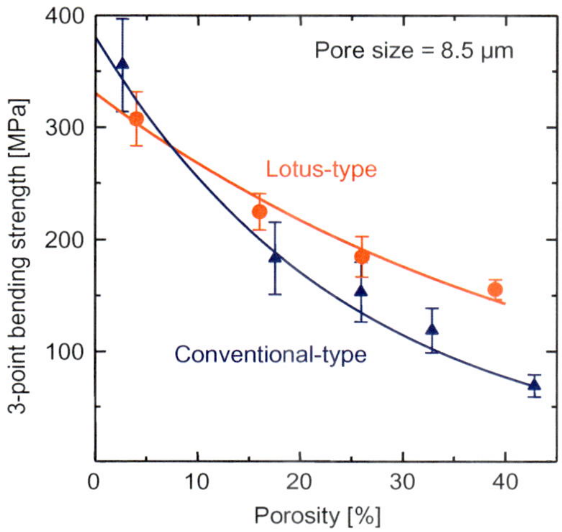

Figure 7.4 Changes of bending strengths of lotus-type and conventional-type porous alumina ceramics as a function of porosity. (Isobe et al., 2006b).

The size of tubular specimens used was 400 mm in length, 30 mm in outer diameter, and 20 mm in inner diameter. The average bending strength obtained was 22.5 MPa at the porosity of 46.7%. The bending strength of conventional-type porous mullite ceramics at a similar porosity of 48% was reported to be 17 MPa (Ma and Tian, 2013). Here, the conventional-type specimens were smaller in pore size (2.7 μm) and smaller in specimen size ($3 \times 4 \times 35$ mm) compared with those of the lotus-type specimens. Since strength of porous ceramics becomes higher with smaller pore size and also smaller specimen size (Davies, 1973) in the measurements, the observed bending strength of the lotus-type porous mullite ceramics is concluded to be significantly higher than the conventional porous mullite ceramics. The well-controlled porous microstructure of the lotus-type specimens is considered to be the reason for the good mechanical strength of lotus-type porous alumina ceramics. The good mechanical strengths of the lotus-type porous mullite ceramics are related to the high resistance to freezing phenomena, passing >100 cycles of temperature changes from -30 to 30 °C in the freezing and thawing test (Japanese Industrial Standards, 1991).

7.3.2 Permeability properties

Permeability is one of the most important properties for porous ceramics. Permeabilities of porous alumina ceramics were measured for lotus-type and conventional-type specimens. The permeability of the lotus-type was measured for thinned specimens perpendicularly cut to the extrusion direction. Changes of permeability values of the lotus-type and conventional-type are shown in Figure 7.5 as a function of porosity. The dashed line in the figure is calculated values for an ideal model (capillary permeability model), having uniform cylindrical pores aligning completely parallel to the gas flow direction. This ideal permeability (μ) is calculated by the following Equation (7.2),

$$\mu = d^2 P / 32 \quad (7.2)$$

where d is the pore size.

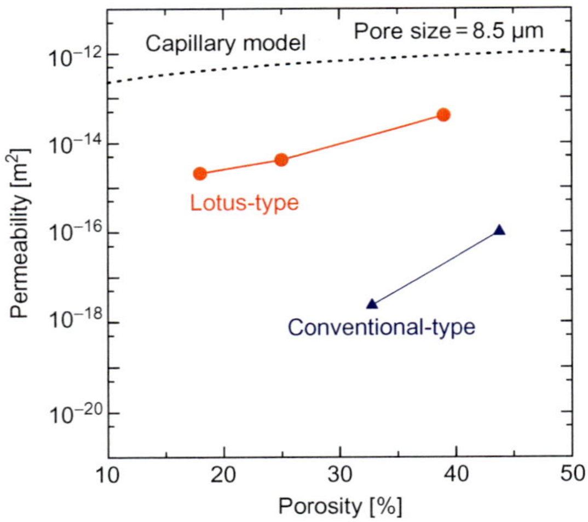

Figure 7.5 Changes of permeabilities of lotus-type and conventional-type porous alumina ceramics as a function of porosity. Isobe et al. (2007).

Clear differences are observed between the permeability values of the lotus-type and conventional-type porous ceramics in Figure 7.5. The first difference is the permeability values; those of the lotus-type are about 10^4 times higher than those of the conventional-type. Thus, it is clear that the lotus-type has excellent permeability compared with the conventional-type. The second difference is the minimum porosity required to obtain a permeable specimen. The permeability was observed only for porosities higher than 30% in the conventional-type, while it was observed at porosities as low as 18% in the lotus-type. This is caused by the difference in the aspect ratios of pore formers and the degree of control of the porous microstructures. This obtained result is reasonable because the calculated minimum porosity for obtaining through holes in the ceramic matrix from percolation theory is 31% for spherical pore formers with random distribution. On the other hand, through holes are formed in much lower porosities of 18% for the lotus-type because the fibers used for the pore former are aligned unidirectionally and contact each other to form through holes by the extrusion process even in low porosity conditions. This is another advantage of the lotus-type compared with the conventional-type. Furthermore, the observed permeability values of the lotus-type are only about one order lower than the calculated permeability values for the ideal pore model. This difference is confirmed to be negligibly small when the corrected pore size value was used for this calculation. A detailed explanation was given in Isobe et al. (2007). Thus, it is concluded that the lotus-type porous alumina ceramics have excellent permeability properties because of the well-controlled porous microstructure.

Permeability properties are also investigated for large-sized lotus-type porous mullite ceramics. Since they are a tubular shape, the permeability was measured from inside to outside of the tube. The observed permeability values were unexpectedly high and only about 20% lower than those of the lotus-type porous alumina ceramics. From these results, the resulting through holes are thought to be aligned mostly unidirectionally but also connected three-dimensionally, showing high permeability from inside to outside of the tube. This type of microstructure is rather suitable for water rise by capillary tension and wetting the surface of porous ceramics. As another application to make use of such high permeability, a ceramic bubbler was developed using the lotus-type porous mullite ceramics and was reported to form microbubbles more easily and effectively than commercial bubblers (Popa et al., 2013a).

7.3.3 Capillary rise properties

Green trees show excellent water rise abilities up to their tops, owing to three major mechanisms: root pressure by osmosis pressure difference in the roots, capillary force in the body and branches, and suction force by water vapor evaporation in the leaves (Johansson and Kifetew, 2010). Thus, lotus ceramics having unidirectionally aligned cylindrical pores are expected to have good water rise ability due to capillary force because they are mimicking the microstructure of xylems and tracheids in the body and branches of trees. Equilibrium capillary rise height (h_{eq}) is represented using the following equation,

$$h_{eq} = 2\gamma \times \cos\theta/\rho \times g \times r \tag{7.3}$$

where γ is the surface tension of water, θ is the contact angle between water and the pore wall, ρ is the density of water, g is the gravitational acceleration, and r is the effective pore radius. The relationship between the capillary rise height and effective pore radius for water can be written as $h_{eq}=(1.49\times10^{-5}\cos\theta/r)\,\mathrm{m}$, assuming $\gamma=73$ mN/m, $\rho=1\times10^3$ kg/m^3, and $g=9.8$ m/s^2. Thus, the capillary rise height increases with the decreasing of contact angle θ and smaller effective pore radius r of the porous ceramics. Considering the applications of this property for counteracting urban heat island effects, the capillary rise rate is also an important factor, and this can be calculated using the following equation (Fries and Dreyer, 2008),

$$\begin{aligned} t = &- \left(8\eta/\rho \times g \times r^2\right)h(t) \\ &- \left(16\eta \times \gamma \times \cos\theta/\rho^2 \times g^2 \times r^3\right)\ln\left(1 - (\rho \times g \times r/2\gamma \times \cos\theta)h(t)\right) \end{aligned} \tag{7.4}$$

where $h(t)$ is the capillary rise height at the time t and η is the viscosity of water. Since we fabricated unidirectionally porous mullite ceramics using the extrusion method (see Section 7.2.2), their capillary rise abilities were investigated (Okada et al., 2009; Popa et al., 2013b).

The capillary rise curve of the lotus-type porous mullite ceramics prepared using 20 μm rayon fibers is shown in Figure 7.6. For the calculation of this curve, it is necessary to simultaneously determine two unknown parameters of θ and r. The h_{eq} value of these ceramics was found to be 1290 mm from the observation of capillary

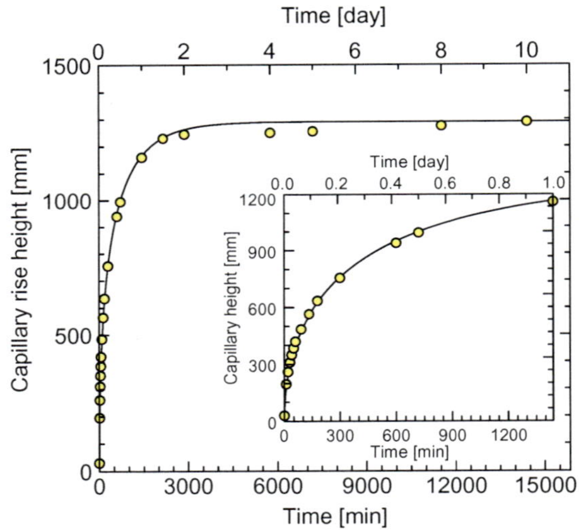

Figure 7.6 Calculated and observed capillary rise height changes for lotus-type porous mullite ceramics using 16.8 μm fibers as a function of time.

rise height after prolonged time. Therefore, Equation (7.3) can be rewritten as $r = (11.6\cos\theta)$ μm for these ceramics. Thus, the parameter is only θ and is easily evaluated by fitting it with the observed data using the trial-and-error method. The best fitting obtained is shown by a solid line in Figure 7.6. The parameters used were $h_{eq} = 1290$ mm, $r = 4$ μm, and $\theta = 69.5°$. The resulting calculated curve showed a very good fitting with the observed values from the beginning to the final stages. Since the maximum capillary rise height value reported for conventional porous ceramics was only about 400 mm (Kubota et al., 2003), the resulting h_{eq} of the lotus-type porous mullite ceramics is much higher than the reported data. Such a distinct enhancement of capillary rise ability is attributed to the well-controlled unidirectional porous microstructure of lotus-type porous mullite ceramics.

Are there any possibilities to further enhance capillary rise ability? Judging from Equation (7.3), two possibilities are considered: lowering the contact angle θ and/or decreasing the effective pore radius r. Figure 7.7 shows examples of those effects for capillary rise curves by fixing one of the two parameters. Compared with the parameters of the present lotus-type porous mullite ceramics, about 50 cm increase of capillary rise height is calculated by lowering θ only $<10°$ at a constant r value (Figure 7.7a), while nearly 70 cm increase of capillary rise height is calculated by decreasing r from 3.9 to 2.5 μm at a constant θ value (Figure 7.7b). Which parameters are easier to change from the present values? Since rayon fibers that are thinner and thicker than the present 20 μm diameter were commercially available, four lotus-type porous mullite ceramics were prepared using four kinds of rayon fibers with 8.1, 9.6, 16.8, and 37.6 μm diameters. Their observed capillary rise height data and calculated capillary rise curves are shown in Figure 7.8 (Popa et al., 2013b). The obtained h_{eq} values were 1780, 1670, 1320, and 950 mm for the corresponding lotus-type porous mullite ceramics, respectively. Thus, the h_{eq} value can be largely enhanced by using

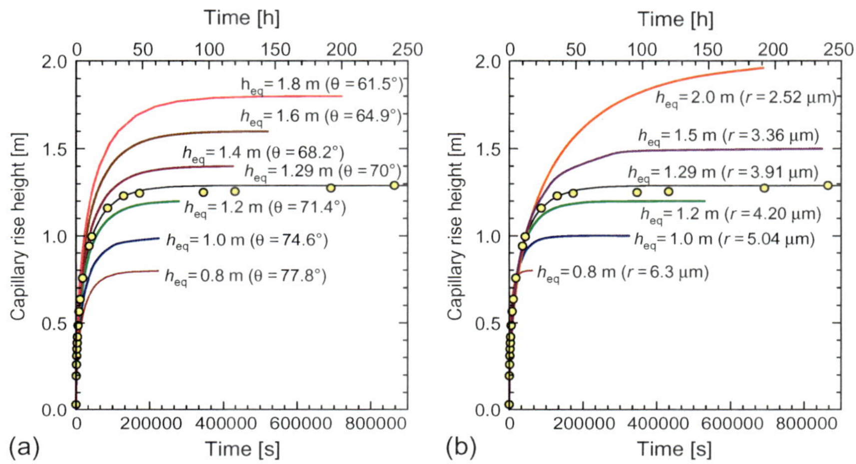

Figure 7.7 Calculated and observed capillary rise height changes for lotus-type porous mullite ceramics using 16.8 μm fibers as a function of time. (a) Constant effective pore radius ($r = 3.91$ μm) and (b) constant contact angle ($\theta = 70°$).

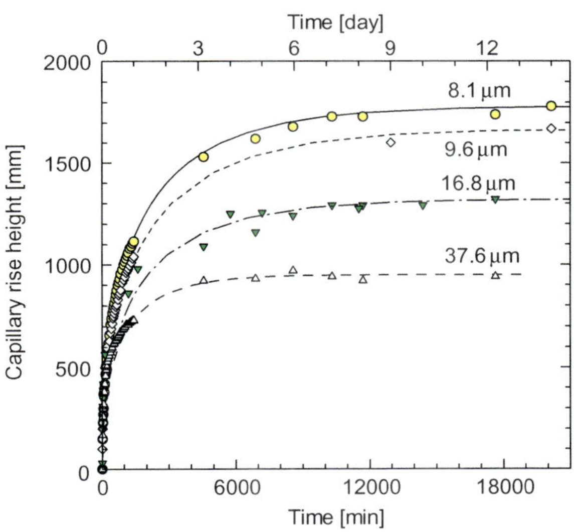

Figure 7.8 Calculated and observed capillary rise height changes for lotus-type porous mullite ceramics using 8.1, 9.6, 16.8, and 37.6 μm fibers as a function of time.
Popa et al. (2013b).

thinner fibers, as expected from the calculation in Figure 7.7b. It is thus confirmed that the resulting lotus-type porous mullite ceramics are very high-potential materials suitable for application to counteracting urban heat island effects by the excellent capillary rise ability of water much higher than 1000 mm, wetting the surface of ceramics, and cooling by latent heat of water vapor evaporation.

7.4 Passive cooling wall using lotus ceramics for counteracting urban heat island effects

The effectiveness of wetting porous ceramics by capillary rise was investigated by comparing the surface temperatures of wetted and non-wetted lotus-type porous cordierite ceramics under sunny conditions (Okada et al., 2007). On one summer day in Tokyo, the resulting surface temperatures of the non-wetted sample varied from 36 to 41 °C by absorbing sun heat, while those of the wetted sample ranged from 27 to 31 °C, suppressing the increase in temperature by about 10 °C. Thus, the cooling effect by wetting the surface of lotus ceramics is recognized to be very effective.

A passive cooling wall (PCW) system using lotus ceramics was investigated by a research group comprised of a university (Tokyo Institute of Technology) and companies in Japan. The companies in the group were a ceramic company fabricating lotus ceramics advised by a ceramics laboratory of the university, a chemical engineering company developing a water supply system, a general construction company setting up the system, and a planning company designing a PCW by referring to the simulation results calculated by an architecture laboratory at the university. The concept of a PCW is considered by expecting cooling effects such as cooling of air through the PCW (cool air); cool radiation spot close to outside dew point temperature (cool radiation); and shielding of direct, global, and reflected solar radiation (sun shadow space). The PCW was arranged by numbers of louvers made of lotus-type porous mullite ceramic tubes.

The mock-ups of the PCW system prepared in 2008 and 2009, along with the related temperature data, are shown in Figure 7.9. As shown in the arrangement chart of PCW in Figure 7.9a, the mock-up prepared in 2008 was arranged by eight lines of ceramic tubes consisting of 11 lotus-type porous mullite ceramic tubes in each line, corresponding to a wall size of 1100 mm width, 1300 mm height, and 550 mm thickness. The capillary rise ability of the lotus-type porous mullite ceramic tubes used was about 1000 mm (Figure 7.9d). This mock-up faced west. Two mock-ups, A and B, prepared in 2009 consisted of 4 × 11 lotus-type porous mullite ceramic tubes, corresponding to a wall size of 1100 mm width, 1800 mm height, and 270 mm thickness. The capillary rise ability of these lotus-type porous mullite ceramic tubes was about 1300 mm. These heights can be confirmed in the thermograph spectrum in Figure 7.9d. Compared with the outside temperature (31.2 °C), the surfaces of the wetted ceramic tubes clearly lowered the temperature, while those of the dried surfaces showed elevated temperature as a result of solar heat. The details of temperature changes, along with the time of the day at various measurement points, are shown in Figure 7.9c. The points from (1) to (4) corresponded to the surface temperatures of ceramic tubes facing south (red line), the temperature in the north (blue line), the temperature in the vicinity of the ceramic tube facing north (green line), and the surface temperature of the tube facing west (ocher color line), respectively. The temperatures of point (1) showed several °C lower than the outside temperatures, even though exposed to solar light. The temperatures of point (2) lowered effectively to the same temperature as the outside dew point temperature. In this direction,

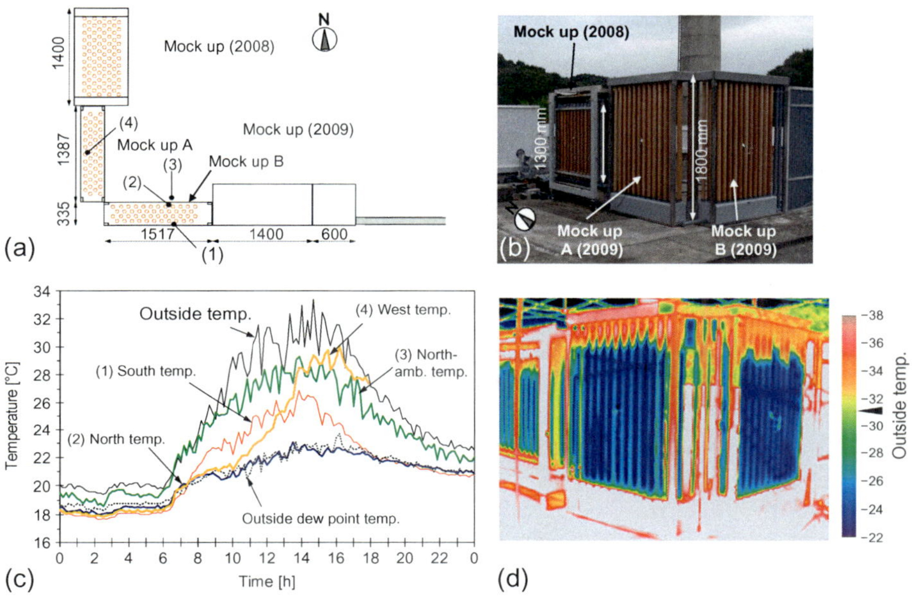

Figure 7.9 Mock-up of passive cooling wall (PCW) and temperature data: (a) arrangement chart, (b) photograph, (c) temperature change, and (d) thermograph.

temperatures of the atmosphere in the vicinity of the ceramic tubes were also lowered 3–4 °C. On the other hand, a lowering of the temperatures at point (4) of nearly 10 °C was achieved in the forenoon, while the temperature was only slightly lowered in the evening after exposure to the afternoon sun. In this way, the PCW system is confirmed to be effective at suppressing temperature increase by wetting ceramic tubes and evaporating water vapor.

The greater Tokyo area, including many satellite cities, is home to more than 30 million people. In this area, many urban redevelopment plans are being considered. Some of the plans place great significance on the environment. For such urban redevelopment plans, the PCW has the possibility of being used as an effective way to suppress urban heat island effects. The PCW system is certain to be introduced in some urban redevelopment plans in the near future.

7.5 Ideas for further enhancing cooling effects using the capillary rise property

The present lotus-type porous mullite ceramics have excellent capillary rise abilities, and their cooling effects can be further enhanced by improving their properties. Considering the two important parameters of contact angle θ and effective pore radius r for higher capillary rise height, the contact angle of the present lotus-type porous mullite ceramics is unexpectedly high: $\theta = 70°$. The following two reasons are considered for this high contact angle.

One reason that the resulting mullite ceramics have a high contact angle is their rough surfaces. In the present ceramics, the addition of 5 mass% of Fe_2O_3 for coloring caused liquid phase sintering, and this resulted in rough surfaces. This can be observed in the microstructure of wall surfaces of cylindrical pores of the specimens, with the poking through of well-developing prismatic mullite crystals here and there due to the crystal growth aided by the liquid phase and the growth space in the pores (Okada et al., 2009). If this is the main reason for the high contact angle of lotus-type porous mullite ceramics, enhancement of capillary rise will be possible by replacing the material with glass, which has a smooth surface and a low contact angle close to 0°. Glass powder can be used for fabrication of lotus-type porous products.

The other possible reason for the high contact angle is the effect of pores. Pores are filled by air, and air behaves completely hydrophobically, i.e., its contact angle is 180°. The surface of porous ceramics is heterogeneous and consists of pores (hydrophobic) and ceramic matrix (considerably hydrophilic). Apparent contact angle (θ^A) for a heterogeneous surface comprised of two such different phases is generally estimated using the Cassie–Baxter equation (Marmur, 2003), which is represented by the following Equation (7.5),

$$\cos\theta^A = R_S f_1 \cos\theta_1 + f_2 \cos\theta_2 = R_S f_1 \cos\theta_1 + f_1 - 1 \tag{7.5}$$

where R_S is the roughness ratio of the wet surface area; f_1 and f_2 are the surface area ratios of ceramic matrix; and θ_1, and θ_2 are the contact angles of the ceramic matrix and pore with water, respectively. In this equation, the apparent contact angle θ^A becomes lower with higher R_S, higher f_1 (i.e., lower porosity), and lower contact angle of ceramic matrix θ_1. In a general sense, the θ^A value tends to become higher with higher porosity, as it is more difficult to lower the value in porous ceramics than dense ceramics. It is therefore necessary to design optimum porosity for this application.

There is a certain possibility to further enhance the capillary rise ability by developing lotus ceramics taking into consideration the aforementioned conditions. This is one of the important points to enhance the effectiveness of the cooling ability of PCW. It is also important to investigate more details on the suitable design of a PCW system that is more effective for cooling through detailed and precise simulation of heat distribution and heat balance. To extend applications of lotus ceramics, development of materials with capillary rise ability for cement-based building materials is also important because most buildings and houses are made with those materials. Recently, geopolymers (Prouis, 2014), alkali aluminosilicate gels, have attracted interest as a new, environmentally friendly material that solidifies at an ambient temperature without firing at high temperatures. Porous geopolymers with capillary rise ability were investigated using an extrusion process by using polylactic acid fibers as the pore formers (Okada et al., 2011). Compared with the present lotus ceramics, the most distinct advantage of these materials is their ability to fabricate without heating at high temperatures, making them an energy-saving material. It is therefore considered that a similar technique can be used for various cement-based materials because large-sized panel-shaped cement-based products were widely manufactured using an extrusion process. Studies on such different viewpoints are very important for the purpose of extending applications of PCW in various urban redevelopments in order to suppress severe urban heat island effects in big cities around the world.

Acknowledgements

Part of the contents of this paper involved work through a research project supported by the Grant for Application of Industrial Technology Innovation of the New Energy and Industrial Technology Development Organization of Japan (NEDO) under Contract No. 0827001. We thank project members Professor Akira Hoyano, Dr Shinji Yamamura, Mr Koshiro Nakajima, Mr Hiroyuki Akagawa, Mr Taisuke Kurata, and Professor Yoshikazu Kameshima. We are also grateful to Mr Shuhei Uchiyama, Mr Yuki Okayasu, Dr Toshihiro Isobe, Dr Chengala D. Madhusoodana, Dr Ken-ichi Katsumata, Dr Catalin Popa, Professor Nobuhiro Matsushita, and Professor Akira Nakajima for their various support.

References

Brandt, M.S., Fuchs, H.D., Stutzmann, M., Weber, J., Cardona, M., 1992. The origin of visible luminescence from "porous silicon": a new interpretation. Solid State Commun. 81, 307–312. http://dx.doi.org/10.1016/0038-1098(92)90815-Q.

Chen, C.Y., Ozasa, K., Katsumata, K., Maeda, M., Okada, K., Matsushita, N., 2012. Bioactive titanium oxide-based nanostructures prepared by one-step hydrothermal anodization. J. Phys. Chem. C116, 8054–8062. http://dx.doi.org/10.1021/jp210783w.

Davies, D.G.S., 1973. The statistical approach to engineering design in ceramics. Proc. Brit. Ceramic Soc. 22, 429–452.

Fries, N., Dreyer, M., 2008. An analytic solution of capillary rise restrained by gravity. J. Colloid Interface Sci. 320, 259–263. http://dx.doi.org/10.1016/j.jcis.2008.01.009.

Fu, Q., Rahaman, M.N., Dogan, F., Bal, B.S., 2008. Freeze casting of porous hydroxyapatite scaffolds. I. Processing and general microstructure. J. Biomed. Mater. Res. B86, 125–135. http://dx.doi.org/10.1002/jbm.b.30997.

Gao, P.Z., Hu, P.F., Wang, W.X., Gong, W.W., 2010. Influence of temperature on the carbonization context, oxidation properties and mechanism of a porous biomorphic carbon template. J. Ceram. Process Res. 11, 297–302.

Gong, D., Grimes, C.A., Varghese, O.K., Hu, W.C., Singh, R.S., Chen, Z., Dickey, E.C., 2001. Titanium oxide nanotube arrays prepared by anodic oxidation. J. Mater. Res. 16, 3331–3334. http://dx.doi.org/10.1557/JMR.2001.0457.

Greil, P., 2001. Biomorphous ceramics from lignocellulosics. J. Eur. Ceram. Soc. 21, 105–118. http://dx.doi.org/10.1016/s0955-2219(00)00179-5.

Guo, S., Wu, Z.B., Zhao, W.R., 2009. TiO_2-based building materials: above and beyond traditional applications. Chin. Sci. Bull. 54, 1137–1142. http://dx.doi.org/10.1007/s11434-009-0063-0.

Hu, L., Wang, C.A., Huang, Y., Sun, C., Lu, S., Hu, Z., 2010. Control of pore channel size during freeze casting of porous YSZ ceramics with unidirectionally aligned channels using different freezing temperatures. J. Eur. Ceram. Soc. 30, 3389–3396. http://dx.doi.org/10.1016/jjeurceramsoc.2010.07.032.

Isobe, T., Tomita, T., Kameshima, Y., Nakajima, A., Okada, K., 2006a. Preparation and properties of porous alumina ceramics with oriented cylindrical pores produced by an extrusion method. J. Eur. Ceram. Soc. 26, 957–960. http://dx.doi.org/10.1016/jjeurceramsoc.2004.11.015.

Isobe, T., Kameshima, Y., Nakajima, A., Okada, K., Hotta, Y., 2006b. Extrusion method using nylon 66 fibers for the preparation of porous alumina ceramics with oriented pores. J. Eur. Ceram. Soc. 26, 2213–2217. http://dx.doi.org/10.1016/jjeurceramsoc.2005.04. 014.

Isobe, T., Kameshima, Y., Nakajima, A., Okada, K., Hotta, Y., 2007. Gas permeability of the porous alumina ceramics with unidirectionally aligned pores by extrusion method. J. Eur. Ceram. Soc. 27, 53–59. http://dx.doi.org/10.1016/jjeurceramsoc.2006.02.030.

Japan Meteorological Agency, 2013. http://www.jma.go.jp.

Japanese Industrial Standards, 1991. Method of test for resistance of exterior materials of buildings to freezing and thawing, JIS A1435.

Johansson, J., Kifetew, G., 2010. CT-scanning and modelling of the capillary water uptake in aspen, oak and pine. Eur. J. Wood Wood Prod. 68, 77–85. http://dx.doi.org/10.1007/s00107-009-0359-4.

Kubota, T., Sugimoto, H., Komiya, H., 2003. Development of multifunctional green wall system using ceramic boards with pump-up ability. Report Obayashi Tech. Inst. 67, 1–6.

Lam, D.C.C., Lange, F.F., Evans, D.G., 1994. Mechanical properties of partially dense alumina produced from powder compacts. J. Am. Ceram. Soc. 77, 2113–2117. http://dx.doi.org/10.1111/j.1151-2916.1994.tb07105.x.

Lee, B.T., Sarkar, S.K., 2009. Fabrication of super-high-strength microchanneled Al_2O_3–ZrO_2 ceramic composites with fibrous microstructure. Scripta Mater. 61, 686–689. http://dx.doi.org/10.1016/j.scriptamat.2009.05.047.

Li, A.P., Müller, F., Birner, A., Nielsch, K., Gösele, U., 1998. Hexagonal pore arrays with a 50–420 nm interpore distance formed by self-organization in anodic alumina. J. Appl. Phys. 84, 6023–6060. http://dx.doi.org/10.1063/1.368911.

Ma, Q.S., Tian, H., 2013. Fabrication and characterization of porous mullite ceramics from pyrolysis of alumina powders filled silicone resin. J. Wuhan Univ. Tech. 28, 1082–1084. http://dx.doi.org/10.1017/s11595-013-0823-1.

Marmur, A., 2003. Wetting of hydrophobic rough surfaces: to be heterogeneous or not to be. Langmuir 19, 8343–8348. http://dx.doi.org/10.1021/la0344682.

Masuda, H., Fukuda, K., 1995. Ordered metal nanohole arrays made by a two-step replication of honeycomb structures of anodic alumina. Science 268, 1466–1468. http://dx.doi.org/10.1126/science.268.5216.1466.

Miyagawa, N., Shinohara, N., 1999. Fabrication of porous alumina ceramics with uni-directionally-arranged continuous pores using a magnetic field. J. Ceram. Soc. Jpn. 107, 673–677.

Nakahira, A., Nishimura, F., Kato, S., Iwata, M., Takeda, S., 2003. Green fabrication of porous ceramics using an aqueous electrophoretic deposition process. J. Am. Ceram. Soc. 86, 1230–1232.

Nakajima, H., 2005. Fabrication of lotus-type porous metals, intermetallic compounds and semiconductors. New. Front. Proc. Eng. Adv. Mater. 502, 367–372.

Okada, K., Shimizu, Y., Kameshima, Y., Nakajima, A., 2005. Preparation and properties of carbon/zeolite composites with corrugate structure. J. Porous. Mater. 12, 289–299. http://dx.doi.org/10.1007/s10934-005-3126-x.

Okada, K., Kameshima, Y., Nakajima, A., Madhusoodana, C.D., 2007. Preparation of lotus-type porous ceramics with high pump-up ability and its cooling effect by water vapor evaporation. J. Heat I. Inst. Int. 2, 1–5.

Okada, K., Uchiyama, S., Isobe, T., Kameshima, Y., Nakajima, A., Kurata, T., 2009. Capillary rise properties of porous mullite ceramics prepared by an extrusion method using organic fibers as the pore former. J. Eur. Ceram. Soc. 29, 2491–2497. http://dx.doi.org/10.1016/jjeurceramsoc.2009.03.012.

Okada, K., Imase, A., Isobe, T., Nakajima, A., 2011. Capillary rise properties of porous geopolymers prepared by an extrusion method using polylactic acid (PLA) fibres as the pore formers. J. Eur. Ceram. Soc. 31, 461–467. http://dx.doi.org/10.1016/jjeurceramsoc.2010.10.035.

Pekor, C., Groth, B., Nettleship, I., 2010. The effect of polyvinyl alcohol on the microstructure and permeability of freeze-cast alumina. J. Am. Ceram. Soc. 93, 115–120. http://dx.doi.org/10.1111/j.1551-2916.2009.03398.x.

Peng, H.X., Fan, Z., Evans, J.R.G., Busfield, J.J.C., 2000. Microstructure of ceramics foams. J. Eur. Ceram. Soc. 20, 807–813. http://dx.doi.org/10.1016/s0955-2219(99)00229-0.

Popa, C., Katsumata, K., Isobe, T., Matsushita, N., Nakajima, A., Kurata, T., Okada, K., 2013a. Preparation and characterization of lotus ceramics with different pore sizes and their implications for generation of microbubbles for CO_2 sequestration applications. Ceram. Int. 39, 1443–1449. http://dx.doi.org/10.1016/j.ceramint.2012.07.087.

Popa, C., Okayasu, Y., Katsumata, K., Isobe, T., Matsushita, N., Nakajima, A., Kurata, T., Okada, K., 2013b. Capillary rise properties of porous mullite ceramics prepared by an extrusion method with various diameters of fiber pore formers. J. Mater. Sci. 48, 941–947. http://dx.doi.org/10.1007/s10853-012-6819-y.

Prouis, J.L., 2014. Geopolymers and other alkali activated materials: why, how and what? Mater. Struct. 47, 11–25. http://dx.doi.org/10.1617/s11527-013-0211-5.

Saggio-Woyansky, J., Scott, C.E., Minnear, W.P., 1992. Processing of porous ceramics. Am. Ceram. Soc. Bull. 71, 1674–1682.

Song, H.Y., Islam, S., Lee, B.T., 2008. A novel method to fabricate unidirectional porous hydroxyapatite body using ethanol bubbles in a viscous slurry. J. Am. Ceram. Soc. 91, 3125–3127. http://dx.doi.org/10.1111/j.1551-2916.2008.02584.x.

Therese, G.H.A., Kamath, P.V., 2000. Electrochemical synthesis of metal oxides and hydroxides. Chem. Mater. 12, 1195–1204. http://dx.doi.org/10.1021/cm990447a.

Ueno, S., Lin, L.M., Nakajima, H., Yasuda, E., 2008. Effects of impurities on formation process during solidification for porous alumina and its compressive strength. J. Ceram. Soc. Jpn. 116, 137–140. http://dx.doi.org/10.2109/jcersj2.116.137.

Ueno, S., Akatsu, T., Nakajima, H., 2009. Fabrication of porous magnesium spinel with directional pores by unidirectional solidification. Ceram. Int. 35, 2469–2473. http://dx.doi.org/10.1016/j.ceramint.2009.02.021.

Yang, T.Y., Ji, H.B., Yoon, S.Y., Kim, B.K., Park, H.C., 2010. Porous mullite composite with controlled pore structure processed using a freeze casting of TBA-based coal fly ash slurries. Resour. Conserv. Recycl. 54, 816–820. http://dx.doi.org/10.1016/j.resconrec.2009.12. 012.

Yoshino, M., 1991. Development of urban climatology and problems today. Energy Build. 15, 1–10.

Zhang, G.J., Yang, J.F., Ohji, T., 2001. Fabrication of porous ceramics with unidirectionally aligned continuous pores. J. Am. Ceram. Soc. 84, 1395–1397.

Innovative evaporative cooling walls

8

A. Carbonari, B. Naticchia, M. D'Orazio
Università Politecnica delle Marche, Ancona, Italy

8.1 Introduction

In this chapter we will present the basics of evaporative cooling walls. Water evaporative cooling systems exploit the latent heat of evaporation of water to reduce sensible heat of air, and therefore its temperature. Based on this principle, both direct and indirect approaches can be developed. Their adoption may dramatically decrease (and sometimes even cancel) cooling loads in warm and hot seasons. So these walls are the best candidates for adoption in mild climates with relatively long summer seasons, where the use of highly insulated walls to limit heating loads may cause internal overheating in hot periods and make the indoors quite uncomfortable and unhealthy. Not only can this technology considerably decrease incoming cooling loads, which is beneficial during warm and hot seasons, but it also keeps high insulation standards against heat dispersion in winter. More specifically, evaporative cooling walls represent an example of a hybrid solution deriving from the combination of ventilated façades and water evaporative cavity walls.

In this chapter we will look into this technology, starting with an overview of the state of the art, then going through the basics of evaporative cooling systems and, finally, focusing on the issue of evaporative cooling walls. In particular, Section 8.4 will be devoted to highlighting some important concepts that are useful for their design, and will describe the experience gained by the authors while working in this field. Likely future trends, suggested sources of information, and conclusions will end the chapter.

8.2 Scientific background

Water evaporative cooling systems are usually classified in direct and indirect cooling, which will be thoroughly explained in Section 8.3. On one hand, using direct approaches, cooling water evaporates directly in the supply air used to ventilate conditioned environments, thus lowering the sensible heat of that air and increasing its humidity content. On the other hand, indirect approaches exploit water evaporation to obtain low temperatures on a separating wall between a secondary duct and either a primary duct or the environment, so that the separating wall acts as a heat exchanger.

Ventilated façades are a particular kind of cavity wall whose cavity is cooled by means of either forced or natural ventilation. As a result, they are able to enhance

Eco-efficient Materials for Mitigating Building Cooling Needs. http://dx.doi.org/10.1016/B978-1-78242-380-5.00008-X

the passive cooling of buildings. Their performance is highly dependent on good design, because geometry and physical properties of ventilated façades must be optimized according to environmental parameters and building typologies (e.g., building use and height). Due to this need, several authors reported how performance varies with geometrical, environmental, and building typology (Ciampi et al., 2003a) (Eicker et al., 2008). Insights into experiments (Saelens and Hens, 2001) enabled authors to learn about the topic and to develop mathematical models and algorithms for simulating a number of different ventilated façade typologies operating in several contexts (Balocco, 2002; Ciampi et al., 2003b). In conclusion, it was found out that the most critical parameters affecting the behavior of ventilated façades are air cavity width, insulating material distribution, solar radiation intensity, outer surface thermal resistance and roughness of the slabs delimiting the air cavity, because they clearly influence the rate of ventilation sparked off, which in turn is responsible for cooling (Ciampi et al., 2003b). These authors validated their model for six different ventilated wall (VW) stratifications and estimated that typical energy savings as high as 40% in summer cooling can be reached, provided that VWs are carefully designed. In addition, ventilated façades are usually provided with rather thick thermal insulation layers, which add to the insulation determined by the air cavity, in fact providing quite good insulation properties in winter, too. As a result, they represent a valid solution to be adopted in mild climates.

Traditionally, latent heat of evaporation was used to cool a building roof because of the quite straightforward way water can be conveyed there and stored on a flat or gently sloped roof surface, typical of hot climates. The most commonly used approaches fall under the category of indirect cooling. One example is given by roof ponds (Sadineni et al., 2011), which are made of a shallow pool of water included in the stratification of a flat roof. Then, a movable top thermal insulation is held close on summer days to shield solar radiation and held open on summer nights to cool down the water pool. The reverse behavior is adopted in the winter. But water shortage requires less water-consuming approaches, such as evapo-reflective roofs, which include a layer of high thermal capacity rock bed in water under a reflective aluminum sheet enclosing an air gap between the water surface and the aluminum reflector (Alvarado et al., 2009). Comparing this system to a regular concrete roof, indoor temperatures were reduced by up to 8 °C (Cheikh and Bouchair, 2004). Still within the category of indirect approaches, the lower surface of a ventilated roof cavity was kept wet so that water could evaporate into air flowing throughout (Manzan and Saro, 2002). The efficiency of this system is conditioned upon the fact that unsaturated outdoor air is available and passes through the roofing ventilated cavity. In fact, it would act as a solar protection.

Similarly, indirect approaches have been hypothesized for application in envelopes, too. Basically, evaporative cooling walls represent an example of a hybrid solution deriving from the combination of ventilated façades and water evaporative cavity walls, which are capable of reducing high cooling loads thanks to water evaporation taking place in their cavity during normal operation in summer. In fact, the case study that will be presented in Section 8.4 integrates natural ventilation with a water evaporative cooling system.

Numerical analyses of a ventilated façade where saturated air is forced to pass along the air space were performed (Cappelli D'Orazio et al., 1999). Technically, air was saturated by direct evaporative cooling before being forced into the air cavity. This feature strongly improves the performances of the envelope, with respect to reference external walls, regardless of the wall's density, thermophysical properties, and geometrical features. However, this study was just numerical and no technology was prototyped to the purpose. Another more mature technology of an indirect cooling prototype was characterized by a set of polycarbonate, hollow panels of 4 mm arranged vertically and equally spaced, which act as a heat exchanger (Gòmez et al., 2012). Outdoor air flows through the cross section of the panels, while exhaust air from a conditioned space flows upwards inside the hollow vertical strips of the panels so as to produce heat transfer from the outdoor air stream to the exhaust air stream in summer conditions. When needed, the exhaust air was subject to adiabatic cooling by spraying water, until its saturation. Of course, the effectiveness of this solution is affected by the flow rate of outdoor air and water content in exhaust air, showing meaningful cooling capacity on the whole.

A different example falling in the category of direct cooling approaches was proposed in the form of a direct water evaporator, which cooled air passing through the main duct ventilating a test room subject to solar radiation (Kant and Mullick, 2003). This solution was found to be able to ensure comfort conditions in the room during summer months.

Alternative solutions where water is used to cool down building components and reduce cooling loads have been proposed, too. The adoption of water evaporative cooling towers was surveyed for the case of high rise buildings (i.e., 100 floors or more) (Wang and Xu, 2008). The towers acted as an indirect evaporative cooling system, where the main cooling tower airflow was put in contact with a secondary water-flow rate, which exploits the use of water generated by evaporation (Costelloe and Finn, 2007).

The first application where the use of water sliding along envelopes was proposed is a greenhouse. In this case, the evaporation of a thin layer of water sliding along a movable canvas was shown to lead to considerable temperature reductions. More generally, provided that an external water film is moving over the external shade cloth during the summer, room temperatures were found to decrease considerably by several degrees, thanks to water evaporation (Ghosal et al., 2003). Also, off-the-shelf solutions in this field have been proposed on the market, where "wet-wall" modules are on sale (GAG, 2014). These products are considered superstructures that cool down the perimeter of a greenhouse and include special solutions to avoid leaks and minimize the number of devices needed in operation mode (e.g., bottom sump tanks are removed). The "wet-wall" solution is available both in aluminum and PVC. Interestingly enough, similar solutions based on the same concept, such as "evaporative cooling pads," are also available. Other authors suggested that this principle could be embedded in buildings, similar to what is usually done with a nuclear reactor. In other words, vertical walls might be cooled by a liquid film falling in a plane channel, along which air mass is allowed to flow down while reducing temperatures and draining heat from the whole wall (Aiello and Ciofalo, 2009). This system is particularly flexible in that its parameters can be optimized according to design considerations and amount of coolant available.

More environmentally friendly solutions ask for the creation of roof lawn gardens, because they favor the spread of natural environment and contribute to energy conservation (Onmura et al., 2001). Thanks to the evaporative cooling effects of vegetation, the surface temperature was found to decrease by about 30 °C during the day, with a concurrent 50% heat flux reduction through the roof. So this approach looks outstanding where maintenance issues are affordable.

Much concern in literature is also devoted to the optimization of some key parameters of those sub-systems and components, whose optimal choice is determined by design conditions, but also restricted by technology constraints. When evaporative passive systems must be put in place, transferring water to air is not an easy task. In this case, porous cooling elements are often placed in the main duct introducing cooling air into buildings. So a comparison of the performances obtained with different degrees of porosity was useful to determine the best material to be adopted for high porosity evaporators and the amount of air mass flow rate that can be fed thanks to the relative heat exchange (Ibrahim et al., 2003). Another passive approach is that of a porous roofing material, whose evaporative cooling effect was exploited to retain moisture and moderate surface temperature (Wanphen and Nagano, 2009). Comparing its performances with the ones of a nonporous material and testing the prototypes when subject to cyclic dynamic testing (dynamic variation of temperature, relative humidity, and solar radiation), it was found that daily average temperatures were substantially decreased. Porous treatments or roughened surfaces are very important because they spark off and enhance heat transfer simply by letting a thin film of liquid flow down their external surface, assuming a fully passive behavior. Experimental campaigns carried out on these configurations showed that the best performances in terms of heat transfer are obtained in case thin rather than thick layers of porous coating are applied on the surface (Leu et al., 2004). In addition, their application for the cooling of rooms should take into account that many parameters must be sized accordingly. The most important parameters were found to be ambient wind speed and thickness of the porous plate. The most influencing ambient conditions were relative humidity and indoor temperature (Chen, 2011). There is no simple linear relationship between these parameters and final performances, so they must be sized according to actual design conditions.

Additional studies surveyed how cooling performances are affected by the type of surface materials, since they are responsible for triggering the evaporation process. Many possible candidates were found to provide good performances for indirect evaporative cooling systems, such as metals, fibers, ceramics, zeolite, and carbon (Zhao et al., 2008). Rather than the material itself, the critical parameter was found to be the porous nature of the coating layer because it may increase the capillary force; even the use of wicked metals (sintered, meshed, grooved, or whiskered) gave good results, which were generally measured in terms of heat transfer coefficient improvement with respect to the case of a flat surface (Kim et al., 2004). A different but efficient approach to increase heat transfer coefficients may be spraying water droplets on microporous coated surfaces; this method was tested in the case of flat and cylindrical heaters and led to increments of heat transfer as high as 400% when compared with that of the uncoated surface cooled by dry air. Hence, the combination of microporous coating and water spraying is a key factor for highly efficient evaporative cooling.

Another enhancement in terms of heat transfer coefficients involved sprinkling water on walls or roofs coated with TiO_2 (super-hydrophilicity of a photocatalyst) in order to maintain a thin film of water on the external surface of buildings (He and Hoyano, 2008a). The way this passive technology can be integrated in building components was then studied further. A passive cooling wall (PCW) prototype proposed by He and Hoyano (2008a) was made up of moist void bricks and was capable of absorbing water and allowing wind penetration, thus reducing the surface temperatures by means of water evaporation. A simulation tool regarding this kind of solution was then developed and validated. Numerical simulations were evaluated on two different scenarios: one where the PCWs were made up of dry brick and another one where the bricks were wet (He and Hoyano, 2008b). Given the successful results obtained with the wet strategy, the above-mentioned technology was then enhanced by a passive cooling strategy, called the passive evaporative cooling wall (PECW), involving porous ceramics capable of passively absorbing water while allowing wind penetration, thus reducing the temperature of the wall (He and Hoyano, 2010).

Another application has been developed lately, where passive cooling similar to those described above is applied at the urban scale. It is expected that water evaporation from moist envelopes can reduce local temperature in urban areas (Saneinejad et al., 2012). The first numerical analyses showed that important reduction of air temperature in street canyons were pursued, thanks to heat and mass exchanges between the canyon walls and outdoor air. Thanks to these successful preliminary results, more insights in the technology were proposed in order to learn more about its feasibility. Again, the ambition of building a PECW led to the construction of pipe-shaped ceramics that possess a capillary force to absorb water up to a level higher than 1.30 m (He, 2011). The total height of the ceramic pipes was 1.80 m. After validating a numerical model with experimental results, several simulations in summer conditions showed that the wall temperature decreased about 10 °C with respect to standard walls. Numerical simulations relative to urban heat island mitigation through several cooling technologies showed that the adoption of evaporative cooling at the urban scale would determine meaningful comfort improvements (Saneinejad et al., 2014). Such benefits consisted of a 3.5 °C reduction of the Universal Thermal Climate Index (UTCI) during a typical meteorological summer day.

These late developments clearly show that the application of evaporative cooling can bring benefits not only to buildings but also to the surrounding urban environments, thanks to consequent reduction of district temperature and overall energy consumption.

8.3 Fundamentals of evaporative cooling

A very clear description of the fundamentals of evaporative cooling was given by Mathews et al. (1994); in this contribution, the authors were also concerned with the way its related mathematical description can be easily integrated within standard dynamic simulation models of buildings. This section will report a summary of that description.

8.3.1 Direct, indirect, and combined evaporative cooling

Direct evaporation of water into an air stream causes its cooling (Figure 8.1a). The addition of water vapor increases the latent heat of the air–vapor mixture. As the process is adiabatic, sensible heat in air is reduced to support water evaporation and dry bulb temperature is reduced as a consequence. The wet bulb temperature of the air remains constant. The lowest possible dry bulb temperature of air leaving a direct cooler is equal to the wet bulb temperature of inlet air, which can be reached provided that inlet air is saturated (Figure 8.1b).

This leads to the definition of the performance factor for direct coolers (PF_D):

$$PF_D = \frac{T_{ent} - T_{out}}{T_{ent} - T_{ent,wb}} \tag{8.1}$$

A performance factor of 1 indicates maximum possible cooling of the air flow.

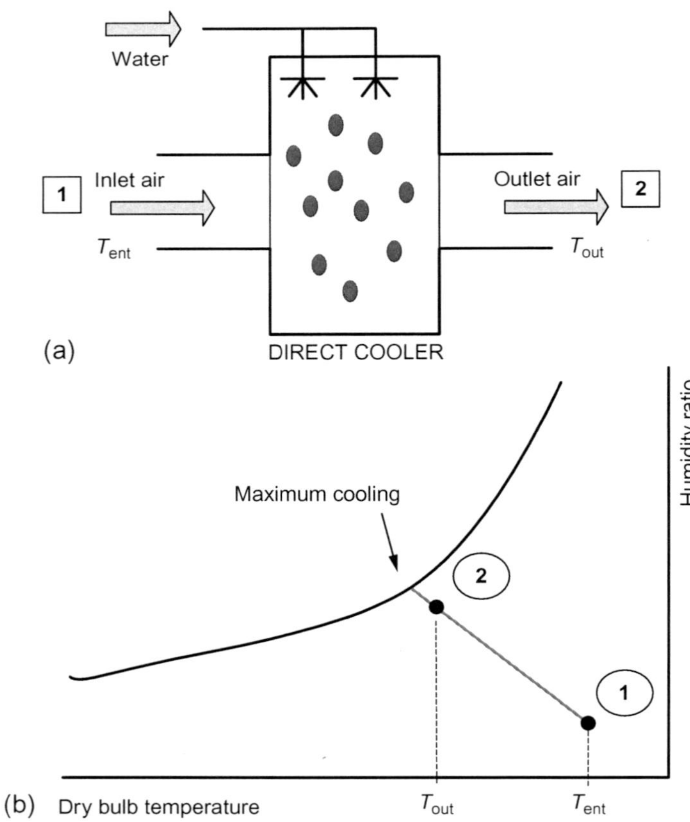

Figure 8.1 Example of direct evaporative cooling (a) and psychrometric diagram relative to that process (b).

In the case of indirect evaporative cooling, the humidity of the air is not increased in the cooling process. Such a cooler uses two airflows, namely primary air that flows into the building and secondary air that cools the primary air (Figure 8.2a). The primary air is cooled sensibly when it flows through one side of a heat exchanger. The other side of the heat exchanger is used for secondary air, which is cooled by direct evaporation of water. The minimum possible outlet temperature of primary air in an indirect cooler is equal to the wet bulb temperature of the secondary airflow (Figure 8.2b). This requires that the secondary air achieves saturated conditions

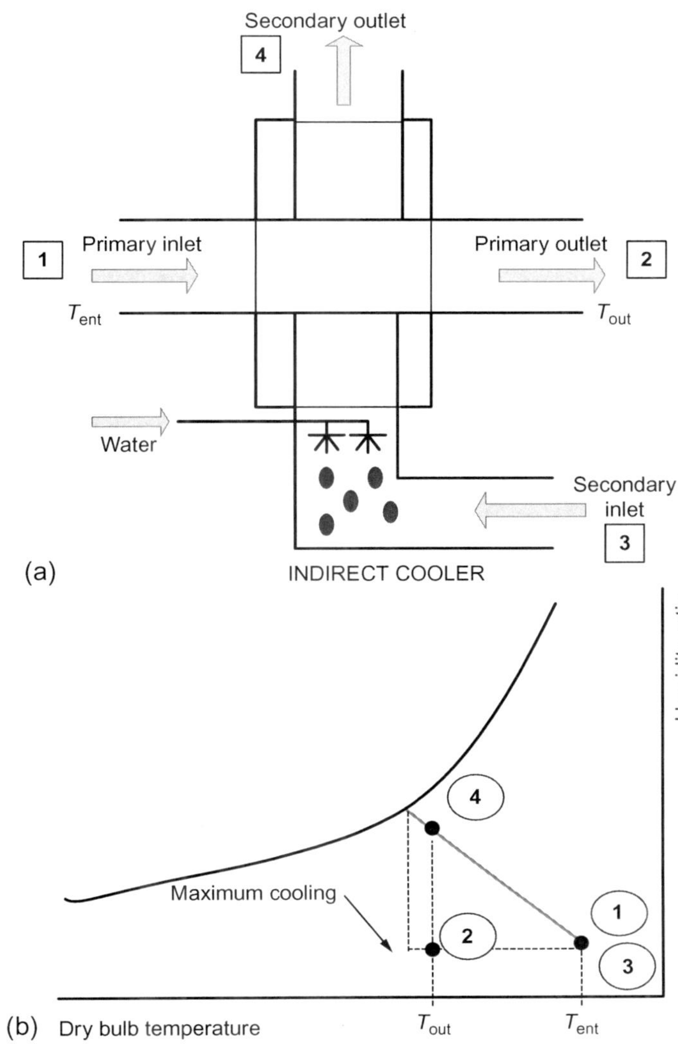

Figure 8.2 Example of indirect evaporative cooling (a) and psychrometric diagram relative to that process (b).

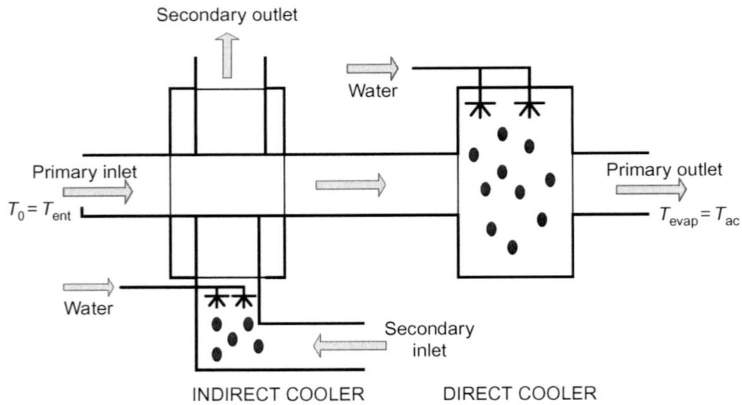

Figure 8.3 Combination of direct and indirect evaporative cooling.

and the heat exchanger is 100% efficient. So the performance factor for indirect evaporative cooling is defined as (PF_I):

$$PF_I = \frac{T_{ent} - T_{out}}{T_{ent} - T_{sec,wb}} \tag{8.2}$$

The combination of direct and indirect evaporative cooling may follow the scheme in Figure 8.3, where air is subject first to indirect cooling (for adjusting its temperature) and then to direct cooling (for adjusting its humidity content).

It was shown that the temperature of outlet air resulting from this process can be computed as (Watt and Brown, 1997):

$$T_{evap} = T_0 - [(PF_D + PF_I - PF_D \times PF_I)T_0 + (PF_D - 1)PF_I \times T_{sec.wb} - PF_D \times T_{1,out,wb}] \tag{8.3}$$

where T_{evap} is dry bulb temperature of the air flowing out of the evaporative cooling system; T_0 is the outdoor air temperature; $T_{sec,wb}$ is the wet bulb temperature of the secondary air used in the secondary side of the cooler; and $T_{1,out,wb}$ is the wet bulb temperature of primary air leaving the indirect first stage. Considering that PF_D and PF_I do not vary with changing weather conditions (Mathews et al., 1994), only constant values need to be specified for a complete hourly simulation. Then the performance factors in Equations (8.1) and (8.2) and the hourly ambient temperatures are substituted in Equation (8.3) to calculate outlet temperature T_{evap}.

8.3.2 Example of integration of the evaporative cooling model within a standard building energy simulator

The mathematical model described in Section 8.3.1 can be easily integrated within a standard energy simulator for buildings. Mathews and Richards (1990) developed and validated the model based on the electrical analogue pictured in Figure 8.4. Thanks to

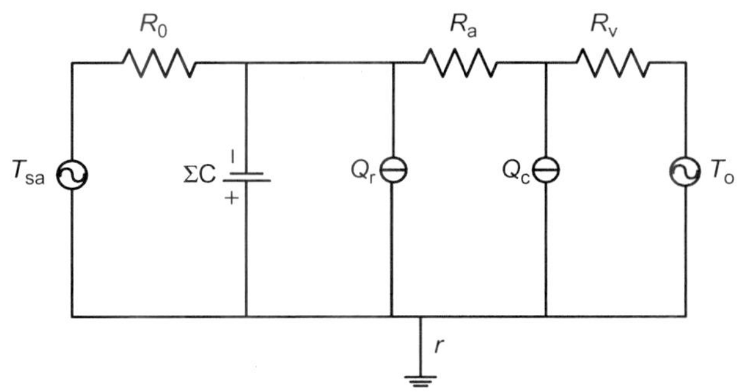

Figure 8.4 Simplified electrical analogue for building zone's transient simulation.

this model, the hourly inside temperatures of any building zones can be estimated and plotted. The model is based on four forcing functions: interior convective heat source ($Q_c(t)$); outdoor air temperature ($T_0(t)$), both acting on the interior air volume i; interior radiative heat source ($Q_r(t)$); and the sol-air temperature ($T_{sa}(t)$), acting on the exterior structure, which accounts for solar radiation on exposed surfaces. The outdoor temperature is the forcing function for natural ventilation and infiltration.

The three thermal resistances in the network account for outer shell of the building (R_0), heat transfer coefficient of inside surfaces (R_a), and outdoor air ventilation and infiltration rate (R_v). The thermal capacitor ($\sum$C) accounts for thermal storage in the building structure. Two concurrent simulation procedures are needed when evaporative cooling is integrated in building simulations. The first procedure provides the hourly temperatures of air leaving the evaporative cooling system, while the second procedure computes the passive hourly indoor air temperature of the building. In order to do that, the scheme in Figure 8.4 must be enhanced to the scheme in Figure 8.5.

This includes heat exchange between indoor (i) and evaporative cooled air (T_{ac}), which is equal to T_{evap} in Equation (8.3), besides c_P (constant pressure specific heat) and m (amount of air flowing into the environment):

$$Q = mc_P(T_i - T_{ac}) \tag{8.4}$$

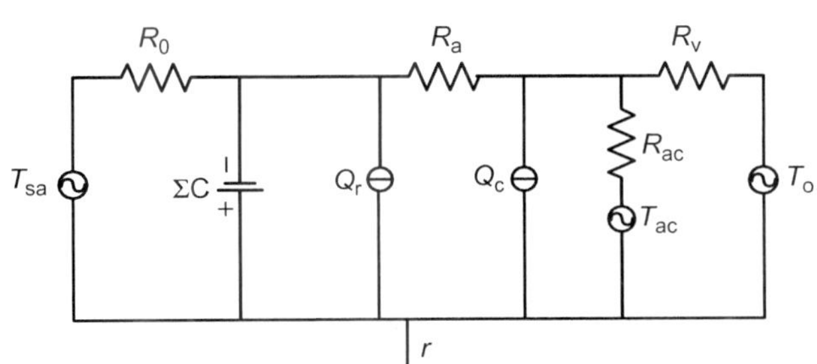

Figure 8.5 Simplified electrical analogue for building zone's transient simulation integrating water evaporative cooling.

R_{ac} in Figure 8.5 is the resistance to heat flowing from the indoor (i) node to the air-conditioned (ac) node and is found by explicating m (ach stands for air change per hour, ρ is air density, and Vol is the room volume):

$$m = \frac{\rho \text{ ach Vol}}{3600} \tag{8.5}$$

Once Equation (8.5) is substituted into Equation (8.4) and R_{ac} is defined as follows:

$$R_{ac} = \frac{3600}{\rho \text{ ach Vol } c_P} \tag{8.6}$$

heat transferred will be equal to:

$$Q = \frac{\Delta T}{R_{ac}} \tag{8.7}$$

In conclusion, the new heat source Q, determined by the water evaporative cooling source, can be integrated in the thermal analogue of Figure 8.4 (Mathews and Richards, 1990) by adding a resistance R_{ac} and a temperature difference as in Equation (8.4), where T_i is already included in the reference electrical network, whereas T_{ac} is computed through the two combined simulations and taking advantage of Equation (8.3) (Mathews et al., 1994). The only inputs needed by the simulator are direct and indirect performance factors and hourly airflow rates. The performance factors allow us to estimate T_{ac} from Equation (8.3), while R_{ac} is computed from specified airflow rates. When only one cooler is simulated, the performance factor of the other cooler must be set to zero. In case both performance factors are null, air changes are introduced at the outdoor air's temperature value.

8.4 Design of an evaporative cooling wall

Basically, an evaporative wall (EW) is a ventilated façade with air-spray devices and an absorbing sheet useful for absorbing liquid water and heat during water evaporation. There are several ways to realize an EW. Figure 8.6a shows a possible functional model of the envelope, coming from the experience gained by the authors in the development of an active evaporative cooling envelope (Naticchia et al., 2010), while Figure 8.6b depicts its typical operational conditions in summer periods. The functional stratification from exterior to interior is made up of an external covering layer with an inner waterproofing finishing, a ventilated cavity equipped with a water spraying system, a porous absorbing insulation layer, and the main bearing structure of the wall. The porous insulation layer is fixed on a regularization layer towards the interior, providing a smooth surface. This is supported by the bearing structure of the wall.

The system integrated in the wall sprays fine water particles on a porous layer, whose latent heat of evaporation will be exploited to absorb incoming heat fluxes.

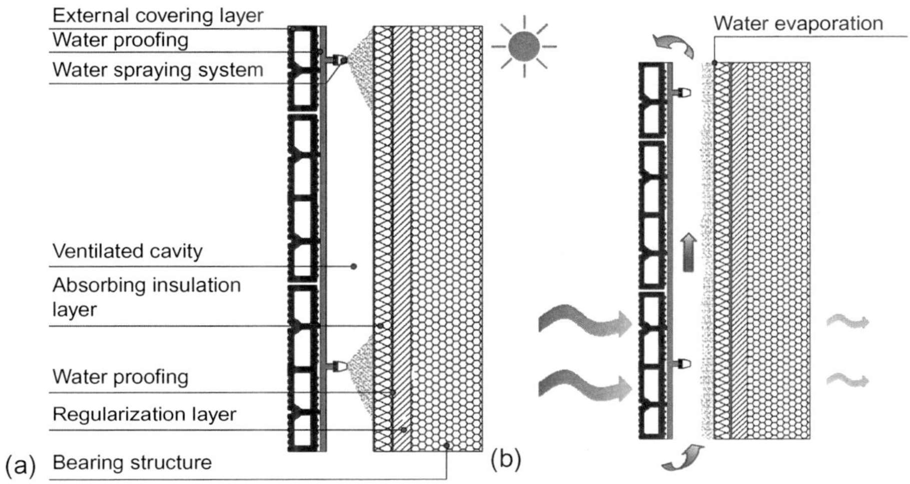

Figure 8.6 Functional (a) and operational (b) model of the active evaporative cooling facade.

The ascending airflow in the cavity facilitates the water's phase change from liquid to gas and its ejection to the exterior. The porous insulating layer absorbs and retains the fine water particles atomized on its external surface by the spraying system, in order to gradually release them into the ascending air flux until its complete drainage. In fact, the spraying system is an active tool for enhancement of the insulation performance of standard ventilated façades in warm seasons. The waterproofing finishing (e.g., polyethylene) applied on the inner side of the exterior layer is used to prevent water absorbance that would otherwise reduce the whole cooling potential and would generate wet spots on the cover. The water spraying system should be scheduled in order to ensure that the porous layer is continuously wet. It should also be noted that this solution can be seamlessly attached to existing façades of buildings during restoration.

8.4.1 Design of the absorbing insulation layer

This layer is required to provide insulation during both summer and winter, to absorb and retain water sprayed by the system on the opposite cavity's side, and to release such water in the form of vapor to the up streaming airflow, thus carrying heat away.

Table 8.1 reports a list of candidates whose overall features comply with the aforementioned requirements. The final choice is affected by the context and by the wall's technology. Insulating materials are classifiable according to their chemical or physical structure. For the purpose of the evaporative cooling wall technology, a material that would keep its insulating and mechanical features, even with high water content, is required. Moreover, the material should also absorb water in a short time (during the few minutes when spraying is on) and then gradually release it into the air cavity. Materials that degrade when exposed to water cannot be adopted. The remaining

Table 8.1 List of candidate materials to be considered for the production of the insulating porous layer

ID	Specimen	Thermal conductivity (W/m K)	Specific heat (J/kg K)	Vapor diffusivity resistance coefficient	Density (kg/m^3)	Thickness (m)
1	Calcium silicate (Lowd.CS)	0.13	1100	3/4	500	0.02
2	Calcium silicate (Highd.CS)	0.17	1100	3/4	875	0.013
3	Cellulose (CL_1)	N.A.	N.A.	4.9	N.A.	0.03
4	Cellulose (CL_2)	N.A.	N.A.	4.9	N.A.	0.015
5	Wood wool (WW)	0.078	1550	5	370	0.035
6	Wood-mineral (WW+MW)	0.09	1550	5	370	0.035
7	Wood-Polystyr. (WW+EPS)	0.09	1550	5	370	0.035
8	Mineral wool (MF)	0.052/0.057	840	1	30	0.015

two properties, that is the absorption capacity and speed of evaporation of several insulation materials, must be measured and assessed on the basis of required specific performances. The insulation materials that currently dominate the European market, belonging to the groups of inorganic fibrous materials (glass wool and stone wool) on one hand and organic foamy materials (expanded and extruded polystyrene, polyurethane foam) on the other, do not comply with some of the aforementioned requirements (Papadopoulos, 2005). In particular, organic foamy materials present a high resistance to vapor transmission and low water absorption (EN 13163, 2001) (EN 13164, 2001); on the contrary, inorganic fibrous materials are affected by wetness. For example, stone wool loses much of its insulating ability when vapor or water penetration takes place (Karamanos et al., 2008). On the contrary, wood wool (WW) is a very promising material, as it owns low resistance to vapor transmission and rather high water absorption features. Similar to the WW specimen are two other materials: the first is made of two external WW boards with mineral wool in between (labeled "WW+MW") and the second is a multilayer insulation board made of two external WW boards with expanded polystyrene in between (labelled "WW+EPS") (EN 13168, 2001).

A good candidate is also calcium silicate, because of its high capillary absorption coefficient (currently used as capillary active interior insulation) (EN 14306, 2009), which can be found with different densities. Table 8.1 lists two representative values (the low density one is labeled Lowd.CS and the high density one is Highd.CS). Interestingly enough, cellulose insulation panels have high moisture storage capacity and water vapor permeability, since their capacity is dependent on their thickness (in the table labeled "CL_1" and "CL_2"). In some cases, absorbing materials must be fire resistant, so mineral wool fiber (labeled "MF") could be used (EN 13162, 2001).

Of course, the choice is not restricted to the options listed in Table 8.1; additional solutions could be suggested. Irrespective of the amount of possible solutions, water absorption and release must be evaluated, because they affect the wall's functioning and its relationship with the spraying system. To this end, several standards for water absorption measurements are available. A rough classification could be provided according to the phenomenon triggering water absorption, which can be obtained either by partial immersion of specimens in water or by capillarity. Examples of standards belonging to the first group are the EN 1609 (1999), EN 12807 (1997), and EN ISO 15148 (2002). Partial immersion is required by EN 1609:1999 and is intended to estimate the behavior of thermal insulating products when exposed to rain. This is the case when PECWs are considered, whereas the functional model in Figure 8.6 exploits capillarity to absorb water when sprayed as fine particles. For that reason, other standards like the European EN 1925 (1999), EN 15801 (2009), and the Italian UNI 10859 (2000) are more suitable. The last two standards impose laying a bedding layer (e.g., filter paper) saturated with water between the water laying in a vessel and the bottom surface of the test specimens. Then, the procedure for estimation of water absorption is similar for all the standards: at specified time intervals from the beginning of the test (10th, 20th, 30th, 60th minute, 2nd, 4th, 6th, 24th hour), the specimens must be taken out of the water, wiped, and weighed. The water absorption is determined by measuring the change in mass at each step; the water absorption coefficient is given by the ratio between the mass increase at a precise step and the square root of the time needed.

Figure 8.7 shows the results from a laboratory trial. Cubic specimens of 0.05 m side were kept at constant induced climate conditions (25 °C and 50% humidity) and in contact with water, separated by a multiple layer of interposed paper filters. Their weight increment per unit surface (measured in g/cm^2) versus the square root of elapsed time (measured in $s^{1/2}$) was computed by:

$$W = \frac{m_t - m_i}{S} \tag{8.8}$$

That is the ratio of the difference between the final and initial weight of a specimen on the size of its contact area on the water. The water absorption coefficient was computed by:

$$A_w = \frac{m_{30} - m_i}{t_{30}^{1/2}} \tag{8.9}$$

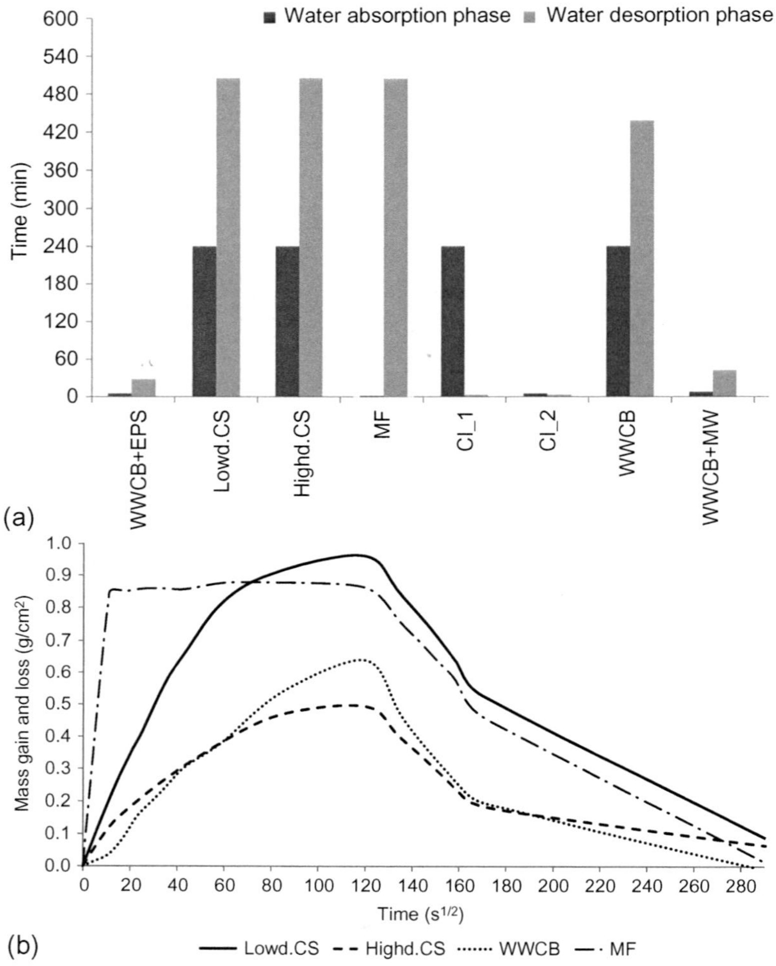

Figure 8.7 Results from the water absorption tests: comparison of absorption and desorption time durations (a) and water intake curves (b).

This is the difference between the weight of a specimen after 30 minutes of contact with water and its initial mass, divided by the square root of the time lag, equal to 30 minutes. Figure 8.7-a shows the absorption time, that is the time elapsed until the weight increase between two consecutive weighs becomes lower than 1%. On the same figure, the desorption time, i.e., the time lag necessary to remove by evaporation the water previously absorbed, is shown. It was determined by letting the same specimen dry under steady state conditions for 20 hours (at 23 °C). It was noticed that each specimen absorbs practically no more water from the 4th to the 24th hours. After this absorption test, the mass decrease versus time in a conditioned environment was recorded, for a drying time period long enough to let the specimens release all the previously absorbed water. Figure 8.7b plots the total mass increase versus the square root of time, where the low density silicate calcium and mineral fiber proved to have the highest water absorption coefficients. However, the latter showed a bad mechanical resistance in wet conditions and therefore it is not durable in evaporative cooling conditions. Looking at the whole absorption–desorption plots of the best performing materials in Figure 8.7b, it can be noticed that the calcium silicate, mineral fiber, and WW cement board had the slowest water desorption phases, comparable to the time needed for phase change. By cross-checking these tests, the best overall performance seems to be offered by calcium silicate insulation: it quickly absorbs a large amount of water and slowly releases it back. In addition, low density calcium silicate presents a good mechanical resistance to the water sprayed on it.

8.4.2 Design of the spraying system

The spraying system installed on the outer boundary surface of the air cavity is required to keep the absorbing layer wet, by sprinkling on it at the optimal rate necessary for its saturation (but no higher than that) in order to avoid water sinking to the ground. In addition, considering that this system must be installed on the opposite side of the air cavity (usually between 0.05 and 0.1 m), it must guarantee a large water distribution, even if sprayed from a short distance.

Its application to the active water evaporative cooling wall can be done by means of a set of atomizer nozzles, which sprinkle a very fine mist of water on the outer surface of the absorbing layer. This technology is already available to irrigation facilities; they all spray on a circular area or a portion of it, when turned upside down, but with different sizes. For example:

- The "360° Micro Sprinkler" forms a whole circular wet area with a diameter of 2 m.
- The "180° Micro Sprinkler" forms a half circular wet area with a diameter of 2.5 m.
- The "90° Micro Sprinkler" forms a quarter circular wet area with a diameter of 0.67 m.
- The "Micro Fog" forms a circular wet area with a diameter of 1 m.

However, their behavior in a specific wall's cavity against a particular absorbing insulation layer must be assessed through experiments. Such an evaluation was performed by measuring the shape and dimensions of the wet area, in function of the distance from the wet surface (Naticchia et al., 2010). Each test was activated at different distances (0.1, 0.2, 0.3, and 0.5 m) from an absorbent 0.84 m $\times$ 0.59 m paper panel, on which the wet footprints obtained could easily be measured. The rate of spraying was also recorded for each nozzle. The one covering the largest surface from a short distance was the "360° Rainjet Micro Sprinkler," which also offered a very fine mist of water, hence minimizing the total number of nozzles needed and avoiding water sinking. With this selected spraying cooling system, calcium silicate is able to absorb 0.428 g/cm^2 in a 5 minute spraying time lag.

8.4.3 Performance assessment for an active evaporative cooling envelope

The experimental campaign described in the rest of this section concerns an experiment carried out by the authors, where one active evaporative cooling wall was compared with the behavior of its ventilated correspondent, which acted as a benchmark, i.e., an envelope made of the same stratification with the exception of the spraying system. The walls were accommodated on the same composite steel and wood bearing frame structure, vertically split by a 0.1 m thick insulation layer. The exterior of the walls was exposed to outdoor conditions (towards southwest), while the inner side was protected from external air using a thick white blanket and conditioned through an oversized split system, working at 24 °C. The basic stratification of the test structure, common to both the walls, was chosen as follows (Figure 8.8a):

- An external covering layer made of extruded-clay hollow flat tiles (0.49 m × 0.24 m × 0.04 m).
- A ventilated air cavity (0.075 m thick).
- An insulating and absorbing layer made of calcium silicate (0.25 m thick).
- A bearing wood board (0.25 m thick).
- A steel frame bearing structure.
- A vapor barrier (polyethylene foil).
- An internal thermal insulation layer made of expanded polystyrene (0.1 m thick).

The reason for choosing calcium silicate is that it gave the best results, as in Section 8.4.1, and optimally coupled with the spraying system mentioned in Section 8.4.2. Calcium silicate was fed by a main pipe connected to a high-pressure water supply and a network of secondary pipes and nozzles positioned in the air cavity at a distance of 0.5 m from each other. In order to obtain only one-dimensional fluxes through the walls, both the perimeter of the bearing structure and the vertical line between the two walls were highly insulated using 0.1 m thick extruded polystyrene boards. A data logger BABUC ABC™ type system was used to record experimental data, which included weather parameters (solar radiation, wind speed and direction, outdoor air temperature, and relative humidity) and temperatures on the several interlayer surfaces. Temperatures were measured with T-type thermocouples (PT-100), which have a temperature range of 65 °C and an error range of 0.5 °C. The thermocouples were placed on the external covering surface, in the middle of the air cavity, on the inner and outer surfaces of the absorbent layer, and on the internal surface of the insulating layer (as shown by the marks in Figure 8.8b). Water sprays were scheduled three times a day. The sprayed water temperature was about 13–15 °C.

Figure 8.9 depicts the experimental results obtained on a representative day, when the spraying system was activated. On that day, three water sprays for a duration of 5 minutes were performed: at 10:45 a.m., at 12:05 a.m., and at 1:35 p.m. In Figure 8.9a it can be noticed that the surface temperatures of the VW are higher than those of the EW, with a difference of several degrees. In Figure 8.9b, from 10:00 a.m. to 5:00 p.m. the air cavity temperature courses of the EW are lower than the standard one by about 5–10 °C. The air ventilation duct temperature in the EW is reduced

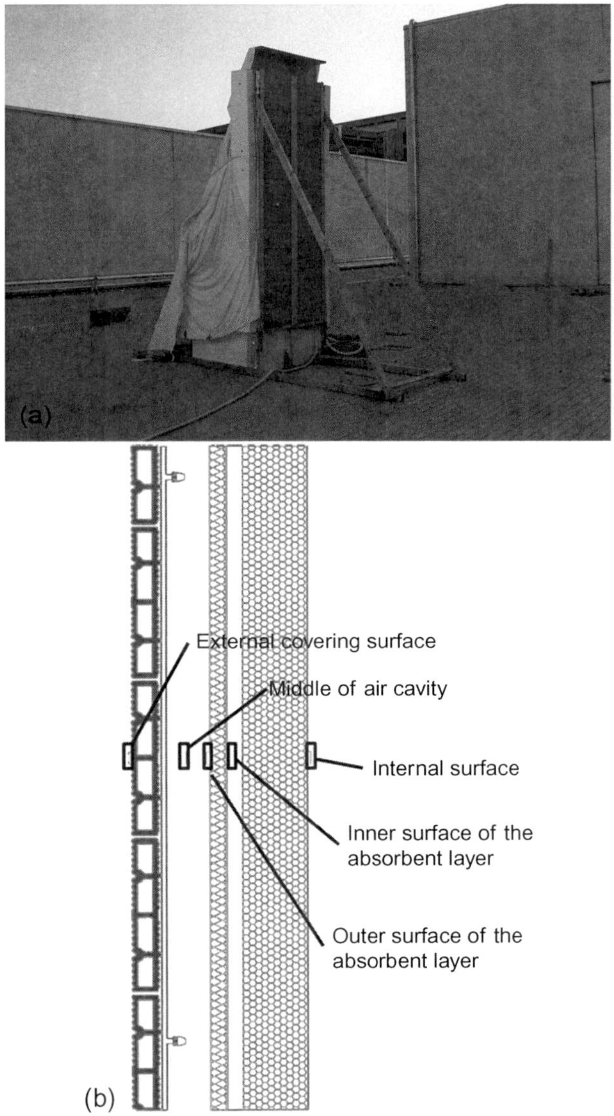

Figure 8.8 Experimental setup for testing the active evaporative cooling wall (a) and position of temperature probes (b).

because of the water emission in the cavity, which is very likely the consequence of the cooling evaporative system. In the VW, the air temperature increased from 8:00 a. m. to 4:00 p.m. when the external temperature was high under solar radiation. The absorbent insulation works as a "cooling pad," which reduces the temperature of

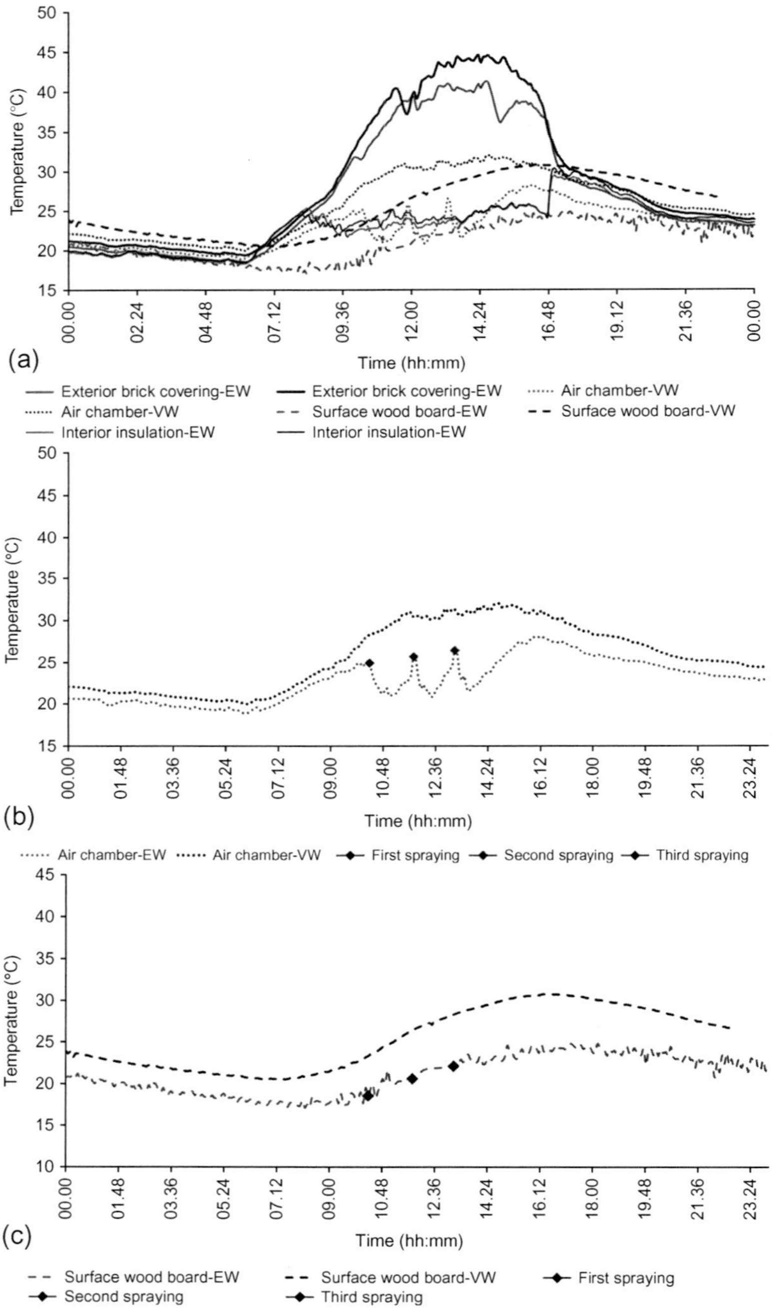

Figure 8.9 Overview of temperature plots during a representative daily test (a) for both walls (active evaporative cooling vs benchmark), air temperature in the wall's cavity (b), temperature plots on the inner side of the absorbing layer (c) and incoming heat fluxes (d).

(Continued)

Figure 8.9 Continued

the entire wall. When the sprayers are activated, the air temperature in the ventilated chamber drops abruptly and is maintained lower than the benchmark VW by 10 °C during the evaporative effect. The reduction in the peak temperature is 5 °C after the spraying. Figure 8.9c compares the temperature on the inner surface of the calcium silicate panel. In the benchmark this ranged between 25 °C and 33 °C, while in the water-EW it ranged between 22 °C and 27 °C, with a decrease in peak temperatures of about 6 °C. Hence the evaporative cooling system coupled with the natural ventilation in the cavity can dramatically decrease the incoming flux, with respect to a standard ventilated façade. Finally, Figure 8.9d estimates incoming heat fluxes for the two walls, exploiting the temperature courses measured on the external surfaces of the absorbing layer and the value of its thermal resistance. It can be noticed that in the water EW not only are the heat fluxes practically canceled, but also that they tend towards negative values, in fact absorbing heat even from the interior. Observing the temperature and heat flux peak during the daytime, it is clear that there is a difference of about 3–4 W/m^2.

Then, the significance of these improvements was verified by means of a statistical analysis. Descriptive statistics were performed on the time series recorded when three water sprayings were performed for a duration of 5 minutes at 10:30 a.m., 12:15 p.m., and 1:45 p.m. The data contain the following temperature courses:

- On the outer finishing ("Outfin").
- In the middle of the ventilated air ("Intair").
- Between the calcium and polystyrene layers ("Calsil").
- On the inner polystyrene layer ("Eps").

Table 8.2 Descriptive statistics comparing the active evaporative cooling wall and its benchmark

	Name of time series			
	Outfin (°C)	**Intair (°C)**	**Calsil (°C)**	**Eps (°C)**
Standard wall				
Mean	26.94	24.36	24.23	22.56
Maximum	42.3	30.63	29.18	29.12
Minimum	18.86	19.56	19.88	18.09
Std. Dev.	7.66	3.50	3.09	2.78
Skewness	0.82	0.33	0.13	0.20
Kurtosis	2.21	1.75	1.55	2.24
Jarque–Bera	40.37	23.93	26.19	8.84
Evaporative wall				
Mean	25.72	21.59	20.33	22.35
Maximum	39.81	28.08	24.14	28.64
Minimum	18.08	18.01	16.74	17.81
Std. Dev.	7.08	2.27	2.45	2.79
Skewness	0.73	0.07	0.01	0.06
Kurtosis	2.02	1.95	1.34	1.94
Jarque–Bera	37.12	13.39	33.00	13.51
	Text (°C)	**Wind (m/s)**	**RH (%)**	**Irr (W/m^2)**
Weather				
Mean	23.17	1.61	83.74	217.80
Maximum	27.42	4.58	90.60	526.00
Minimum	18.40	0.00	67.80	0.00
Std. Dev.	2.90	1.40	5.02	207.37

The same nomenclature was adopted for the water-EW. The descriptive statistics in Table 8.2 suggest the following comments:

- Mean temperatures are significantly higher for the standard VWs than for the water-evaporative VW.
- In the first case their courses decrease from the exterior to the interior, while in the second case they decrease from both extreme surfaces towards the interior as a consequence of the internal cooling effect triggered in the air cavity.
- The temperature ranges for all the variables are higher for the standard wall than for the evaporative one, and in the second case all the values are more stable around their mean, as also confirmed by their standard deviations.
- For the EW, skewness is closer to 0 than for the standard wall time series and temperature distributions are also better balanced around their mean.
- Each distribution is platykurtic, that is to say flatter than a normal distribution, and none is normal shaped (Jarque–Bera test).

Table 8.3 **Descriptive statistics on the differences between the active evaporative cooling wall and its benchmark**

	Name of time series			
	Doutfin (°C)	**Dintair (°C)**	**Dcalsil (°C)**	**Deps (°C)**
Differences (stand evap.)				
Mean	1.21	2.76	3.89	0.2
Maximum	6.70	7.26	5.60	0.76
Minimum	−2.23	1.25	2.28	1.64
Std. Dev.	1.10	1.70	0.81	0.28
Skewness	1.90	1.40	0.19	−2.06
Kurtosis	10.95	3.49	2.09	11.72
Jarque–Bera	937.42	98.22	11.74	1121.21

In Table 8.3, the statistics on the differences between the standard and the EWs are shown:

- Temperature differences on the external layer ("Doutfin").
- Temperature differences in the internal air cavity ("Dintair").
- Temperature differences on the calcium silicate layer ("Dcalsil").
- Temperature differences on the polystyrene ("Deps").

It can be noticed that the highest difference between these temperatures pertains to the interface between the porous-absorbing ("calsil") and the polystyrene layers ("eps"). This value is followed on the list by the "internal air layer temperature difference," which is directly affected by water spraying. The statistical significance of these differences computed on the variable mean values were tested through a one-way analysis of variance (ANOVA), which performs a Fisher test on the null hypothesis that all the time series come from the same population (i.e., there is no statistical difference in the drawn samples) (Wonnacott and Wonnacott, 1990). It showed that "Dcalsil" and "Dintair" are statistically significant at the 99% level, while the differences for the variable "Doutfin" are only significant at the 95% level, as this variable is the least influenced by water spraying. These statistics represent a further confirmation that the insertion of the water spraying system significantly reduces the stratification temperatures and improves thermal performance.

8.5 Future trends

The continuous improvement of living standards is facilitating the massive penetration of air conditioning devices, resulting in constant increase of energy consumption in several countries. For example, although recent EU regulations addressed the reduction of energy consumption of buildings, the growth of energy demand in hot and moderate climates (southern countries) will not stop in the near future without valid alternatives to mechanical air cooling systems.

For that reason, EWs can provide comfort in non-air conditioned buildings while keeping energy consumption low, and can considerably decrease cooling loads of thermostatically controlled buildings.

However, it is clear that further research is necessary in order to optimize the existing EW systems and develop new prototypes. Ranked in order of importance, the critical issues are as follows: (1) water flowing devices and water recovery/distribution systems; (2) durability of absorbing layers; and (3) whole life economic and environmental impact.

As far as water distribution and water recirculation are concerned, the current research outcomes are quite far from satisfactory. The influence on the cooling effectiveness of different spray nozzles and of different water spraying rates should be properly investigated. In addition, it is necessary to understand how moisture adsorption/absorption and desorption processes can interact with the movement of air inside gaps, and how the behavior of the EW can be adjusted during the day in respect to ambient air temperatures and radiative fluxes. Moreover, the approaches for water delivery and spray and their potential of creating even distribution of water across the surface of the layers are themselves issues to be further addressed. In terms of durability of absorbing layers, the current research results are still in their preliminary phase, too. The EW's efficiency is strongly dependent on absorption and desorption capacity, but this property is dependent on the history of the absorption/desorption process.

In terms of economic impact, the life cycle cost of EWs must be analyzed, including production, installation, (energy) costs of buildings with EWs, and the treatment and recovery of the water in their life of service. In fact, economic and environmental impacts relating to EWs still need to be investigated.

Finally, technical and legislative aspects related to the production and market diffusion of these technologies should be analyzed.

8.6 Sources of information and advice

To the author's knowledge, EWs have not been marketed for their application in buildings. They have been considered in an extensive collaborative research effort as a great opportunity for retrofitting existing buildings into low-energy buildings (IEA, 2013). The final report produced as an outcome from this research and more data and information can be retrieved from the official website: http://www.iea.org/etp/buildings/. This publication covers a general overview and some insights into cost-effective solutions to improve the building sector, with particular reference to energy efficiency. Future trends and challenges related to the implementation of such new technologies are also discussed.

On the contrary, other references focus on the theoretical basis of the physical phenomenon itself (Watt and Brown, 1997). More insights on the chemistry related to the evaporative process can be found in books related to the discipline, such as Genskow et al. (2007).

Finally, some authors looked into the technology challenges and potential applications in construction. However, there are no books fully devoted to EWs; rather, they are usually included in books covering general aspects of energy efficiency (Givoni, 1994) and multiphase systems (Faghri and Zhang, 2009).

8.7 Conclusions

In this chapter, the most common cutting-edge technologies currently under development and useful for exploiting the latent heat of evaporation of water were considered. In particular, a strong focus was directed towards the application of water evaporative cooling in building envelopes, which was shown to considerably decrease cooling loads in the warm season. There are both numerical simulations and experimental campaigns showing the great advantages that would be provided by such technology, and marketed applications are already known in the field of greenhouse conditioning. Then, the basics of this approach as split into direct and indirect cooling were provided, even referring to some very common numerical models whose implementation in calculation produced very accurate results. In order to get readers more accustomed to this technology and the related implementation challenges generated, Section 8.4 was dedicated to the recommendations useful for designing evaporative cooling walls and to the performances experimentally measured from a full-scale prototype. Both this experiment and other evidences from literature led us to conclude that some technology aspects (i.e., water spraying, absorbing layers, and whole life impact) must be analyzed further in order to make evaporative cooling walls ready for industrial application. But it's worth investing more in this field, given the high cooling load reduction that this technology is capable of providing, thus contributing to converting the existing stock into very low-energy buildings.

References

Aiello, G., Ciofalo, M., 2009. Natural convection of a hot vertical wall by a falling liquid film. Int. J. Heat Mass Transfer 52, 5954–5961.

Alvarado, J.L., Terrell Jr., W., Johnson, M.D., 2009. Passive cooling systems for cement-based roofs. Build. Environ. 44, 1869–1875.

Balocco, C., 2002. A simple model to study ventilated facades energy performance. Energy Build. 34, 469–475.

Cappelli D'Orazio, M., Cianfrini, C., Corcione, M., 1999. Energy saving by evaporative air-cooling processes in building-envelope ventilated air spaces. Heat Technol. 17, 21–27.

Cheikh, B.H., Bouchair, A., 2004. Passive cooling by evapo-reflective roof for hot dry climates. Renew. Energy 29, 1877–1886.

Chen, W., 2011. Thermal analysis on the cooling performance of a wet porous evaporative plate for buildings. Energy Convers. Manage. 52, 2217–2226.

Ciampi, M., Leccese, F., Tuoni, G., 2003a. Cooling of buildings: energy efficiency improvement through ventilated structures. In: Tiezzi, E. (Ed.), The Sustainable World. WIT Press, Southampton, pp. 199–221.

Ciampi, M., Leccese, F., Tuoni, G., 2003b. Ventilated facades energy performance in summer cooling of buildings. Sol. Energy 75, 491–502.

Costelloe, B., Finn, D., 2007. Thermal effectiveness characteristics of low approach indirect evaporative cooling systems in buildings. Energy Build. 39, 1235–1243.

Eicker, U., Fux, V., Bauer, U., Mei, L., Infield, D., 2008. Facades and summer performance of buildings. Energy Build. 40, 600–611.

EN 12807, 1997. Thermal Insulating Products for Building Applications—Determination of Long Term Water Absorption by Immersion. CEN—European Committee for Standardization.

EN 13162, 2001. Thermal Insulation Products for Buildings. Factory made Mineral Wool (MW) Products. Specification. CEN—European Committee for Standardization.

EN 13163, 2001. Thermal Insulation Products for Buildings. Factory made Products of Expanded Polystyrene (EPS). Specification. CEN—European Committee for Standardization.

EN 13164, 2001. Thermal Insulation Products for Buildings. Factory made Products of Extruded Polystyrene foam (XPS). Specification. CEN—European Committee for Standardization.

EN 13168, 2001. Thermal Insulation Products for Buildings. Factory made Wood Wool (WW) Products. Specification. CEN—European Committee for standardization.

EN 14306, 2009. Thermal Insulation Products for Buildings Equipment and Industrial Installations. Factory made Calcium Silicate (CS) Products. Specification. CEN—European Committee for Standardization.

EN 15801, 2009. Conservation of Cultural Property—Test Methods—Determination of Water Absorption by Capillarity. CEN—European Committee for Standardization, December 2009.

EN 1609, 1999. Thermal Insulating Products for Building Applications—Determination of Short Term Water Absorption by Partial Immersion. CEN—European Committee for Standardization.

EN 1925, 1999. Natural Stone Test Methods—Determination of Water Absorption Coefficient by Capillarity. CEN—European Committee for Standardization.

EN ISO 15148, 2002. Hygrothermal Performance of Building Materials and Products—Determination of Water Absorption Coefficient by Partial Immersion. CEN—European Committee for Standardization.

Faghri, A., Zhang, Y., 2009. Transport Phenomena in Multiphase Systems. Elsevier, Amsterdam.

GAG—Gothic Arc Greenhouses, 2014. Evaporative Cooling, USA. Available from: http://www.gothicarchgreenhouses.com/greenhouse_cooling.htm.

Genskow, L.R., Beimesch, W.E., Hecht, J.P., Kemp, I., Langrish, T., Schwartzbach, C., Smith, F.L., 2007. Perry's Chemical Engineers, eighth ed. McGraw-Hill Professional, New York.

Ghosal, M.K., Tiwari, G.N., Srivastava, N.S.L., 2003. Modeling and experimental validation of a greenhouse with evaporative cooling by moving water film over external shade cloth. Energy Build. 35, 843–850.

Givoni, B., 1994. Passive Low Energy Cooling of Buildings. John Wiley and Sons, US.

Gòmez, E.V., Gonzàlez, A.T., Rey Martìnez, F.J., 2012. Experimental characterization of an indirect evaporative cooling prototype in two operating modes. Appl. Energy 97, 340–346.

He, J., 2011. A design supporting simulation system for predicting and evaluating the cool microclimate creating effect of passive evaporative cooling walls. Build. Environ. 46, 584–596.

He, J., Hoyano, A., 2008a. A numerical simulation method for analyzing the thermal improvement effect of super-hydrophilic photocatalyst-coated building surfaces with water film on the urban/built environment. Energy Build. 40, 968–978.

He, J., Hoyano, A., 2008b. A 3D CAD-based simulation tool for prediction and evaluation of the thermal improvement effect of passive cooling walls in the developed urban locations. Sol. Energy 83, 1064–1075.

He, J., Hoyano, A., 2010. Experimental study of cooling effects of a passive evaporative cooling wall constructed of porous ceramics with high water soaking-up ability. Build. Environ. 45, 461–472.

Ibrahim, E., Shao, L., Riffat, S.B., 2003. Performance of porous ceramic evaporators for building cooling application. Energy Build. 35, 941–949.

IEA, 2013. Transition to Sustainable Buildings—Strategies and Opportunities to 2050. France, International Energy Agency, Paris.

Kant, K., Mullick, S.C., 2003. Thermal comfort in a room with exposed roof using evaporative cooling in Delhi. Build. Environ. 38, 185–193.

Karamanos, A.K., Hadiarakou, S., Papadopoulos, A.M., 2008. The impact of temperature and moisture on the thermal performance of stone wool. Energy Build. 40, 1402–1411.

Kim, J.H., You, S.M., Stephen, U.S., Choi, U.S., 2004. Evaporative spray cooling of plain and microporous coated surface. Heat Mass Transfer 47, 3307–3315.

Leu, J.-S., Jang, J.-Y., Chou, Y., 2004. Heat and mass transfer for liquid film evaporation along a vertical plate covered with a thin porous layer. Heat Mass Transfer 47, 3307–3315.

Manzan, M., Saro, O., 2002. Numerical analysis of heat and mass transfer in a passive building component cooled by water evaporation. Energy Build. 34, 369–375.

Mathews, E.H., Richards, P.G., 1990. A tool for predicting hourly air temperatures and sensible energy loads in buildings at sketch design stage. Energy Build. 14, 61–80.

Mathews, E.H., Kleingeld, M.K., Grobler, L.J., 1994. Integrated simulation of buildings and evaporative cooling systems. Build. Environ. 29, 197–206.

Naticchia, B., D'Orazio, M., Carbonari, A., Persico, I., 2010. Energy performance evaluation of a novel evaporative cooling technique. Energy Build. 42, 1926–1938.

Onmura, S., Matsumoto, M., Hokoi, S., 2001. Study on evaporative cooling effect of roof lawn gardens. Energy Build. 33, 653–666.

Papadopoulos, A.M., 2005. State of the art in thermal insulation materials and aims for future developments. Energy Build. 37, 77–86.

Sadineni, S.B., Madala, S., Boehm, R.F., 2011. Passive building energy savings: a review of building envelope components. Renew. Sustain. Energy Rev. 15, 3617–3631.

Saelens, D., Hens, H., 2001. Experimental evaluation of airflow in naturally ventilated active envelopes. J. Therm. Envelope Build. Sci. 25, 101–127.

Saneinejad, S., Moonen, P., Defraeye, T., Derome, D., Carmeliet, J., 2012. Coupled CFD, radiation and porous media transport model for evaluating evaporative cooling in an urban environment. J. Wind Eng. Ind. Aerodyn. 104–106, 455–463.

Saneinejad, S., Moonen, P., Carmeliet, J., 2014. Comparative assessment of various heat island mitigation measures. Build. Environ. 73, 162–170.

UNI 10859, 2000. Cultural Heritage—Natural and Artificial Stones—Determination of Water Absorption by Capillarity. Ente Italiano di Unificazione—UNI, 31 October 2000.

Wang, S., Xu, X., 2008. Effects of alternative control strategies of waterevaporative cooling systems on energy efficiency and plume control: a case study. Build. Environ. 43, 1973–1989.

Wanphen, S., Nagano, K., 2009. Experimental study of the performance of porous materials to moderate the roof surface temperature by its evaporative cooling effect. Build. Environ. 44, 338–351.

Watt, J.R., Brown, K.B., 1997. Evaporative Air Conditioning Handbook, third ed. Chapmann and Hall, UK.

Wonnacott, T.H., Wonnacott, R.J., 1990. Introductory Statistics, fifth ed. John Wiley and Sons, New York.

Zhao, X., Liu, S., Riffat, S.B., 2008. Comparative study of heat and mass exchanging materials for indirect evaporative cooling systems. Build. Environ. 43, 1902–1911.

Part Three

Roofing materials for reducing building cooling needs

High-albedo roof coatings for reducing building cooling needs

9

A.L. Pisello
CIRIAF – University of Perugia, Perugia, Italy

9.1 Introduction

The main characteristics of high-albedo coatings for application on buildings' roofs consist of suitable superficial optic-energy features that will keep their surface cooler when exposed to solar radiation, compared to other materials with low-albedo finishing. This property is the cause of important benefits in buildings, as well as at urban and global scales. In fact, the high-albedo feature, together with high thermal emittance, characteristic of the large majority of construction materials, can preserve the roofing surface from summer overheating and prevent premature damaging phenomena due to thermal stress. For this reason, this chapter will discuss (i) the physical assessment of such coatings, (ii) the associated thermal behavior of roofs, (iii) the principal procedures to quantify these properties, (iv) the most common and innovative materials and techniques, and (v) their performance during their life cycle.

In particular, their effect at building scale will be dealt with. To this aim, it is important to clarify that high-albedo coatings for roofs are able to reflect the solar radiation by reducing the heat transfer into the building. This effect is also the reason why these coatings are named "cool coatings," because of their high solar reflectance (or albedo) and high thermal emittance. If the building is a free-floating structure, where the air-conditioning systems are not operating, the application of cool coatings is able to decrease the surface temperature of the roof, decreasing the relative indoor thermal gain. Several simulation and field monitoring campaigns were carried out in recent years in order to quantify such effects in summer and winter conditions (Pisello et al., 2014a,b; Pisello and Cotana, 2014; Synnefa et al., 2006, 2007a,b). All these studies agree about their effect in reducing the indoor operative temperature up to 4 °C and then to optimize indoor thermal comfort conditions (Akbari and Konopacki, 2005; Bozonnet et al., 2013; Pisello and Cotana, 2014; Santamouris et al., 2008; Santamouris and Kolokotsa, 2013). If a building is provided with an air-conditioning system, the same high-albedo coating contributes to decreasing the energy consumption for cooling up to 40%, depending on a large variability of boundary conditions, for example, the roof insulation level, weather conditions, indoor occupancy schedules and internal gains, and ventilation-infiltration volumes (Haberl and Cho, 1998; Synnefa et al., 2007a,b). Specific analysis of important works demonstrating this phenomenon is reported in Section 9.5 of this chapter. Extending this consideration into wider energy and environmental perspectives, high-albedo roofs are also able to reduce the peak energy demand of a building sector, which can reduce electricity fees and social

Eco-efficient Materials for Mitigating Building Cooling Needs. http://dx.doi.org/10.1016/B978-1-78242-380-5.00009-1

constraints due to the increasing probability of summer blackout events (Akbari et al., 1997a,b; Levinson et al., 2005). These same coatings, being able to preserve more constant temperature conditions in the roof, are also able to increase the durability of such materials, for example, coatings and membranes, by decreasing fatigue stresses on them (Berdahl et al., 2008; Cool Roofs project, 2010; Gartland, 2008; Levinson et al., 2007).

A key benefit produced by high-albedo coatings and, more in general, high-albedo surfaces exposed to solar radiation in urban areas is represented by their effect on the improvement of the urban microclimate and the mitigation of urban heat island phenomenon. In this field, several simulation studies at micro- and mesoscale levels have been performed in order to show how the increase of urban albedo could produce up to 3 °C of air temperature reduction in the United States, in Greece, and in Canada (Sailor, 1995; Taha, 1997, 2008a,b; Taha et al., 2000). This same mitigation also produces a consequent reduction of electricity demand for cooling buildings, with further benefits at district and urban scales (Synnefa et al., 2008).

By enlarging the analysis perspective up to global climate scale, the terrestrial increase of surface albedo is able to produce a worldwide cooling effect that might mitigate the global warming phenomenon, as documented in several recent studies (Akbari et al., 2009; Cotana et al., 2014; Taha et al., 1999).

Therefore, a threefold beneficial effect could be achieved by implementing high-albedo coatings over roofs. This theme has represented the research focus of physicians, engineers, economists, and climatologists over the course of the last decades. The main research and innovation results made the market ready for implementing such technologies almost at the same price of traditional low-quality black "hot-mop" membranes for roofs (Sproul et al., 2014). This acknowledged technical and economical result represented a powerful trigger for policy makers and decision makers all over the world in adopting these high-albedo solutions in new constructions and major renovations.

In this perspective, this chapter will describe the principal aspects covered in this research area and their main technological issues by clarifying the key technical and theoretical features of high-albedo coatings for roof application.

9.2 Physical characteristics of high-albedo roof coatings

High-albedo coatings, as previously mentioned, are able to limit the increase of surface temperature of a roof, in opposition to a high absorptive finishing. In this view, those coatings are usually applied over flat and low-sloped roofs in general, in order to avoid summer overheating. The two main properties affecting their behavior are (i) solar reflectance or albedo and (ii) thermal emittance. This section concerns the analysis of these two optic-energy properties according to ASTM and the final assessment of the solar reflectance index (SRI), globally describing the cooling potential of a coating for building roofs or urban pavings.

9.2.1 Solar reflectance

The solar reflectance (SR) is the surface capability to reflect solar radiation back to the hemisphere where the solar source is located, integrated over the entire solar spectrum, including specular and diffuse reflection components. In fact, most of the solar energy arriving to the terrestrial level is between 300 and 2500 nm, and its intensity varies with varying wavelengths in the solar spectrum, with a peak around 600 nm. The ultraviolet region holds about 5% of the energy. The visible region contains about 44% of the solar energy, and more than 50% is irradiated by the solar source in the infrared region of the spectrum (Ferrari et al., 2013). The radiation arriving to several areas in our planet depends on atmosphere conditions and perturbation, air humidity, and differential absorption of water, ozone, and carbon dioxide. Important variations occur depending on the air mass, which in turn depends on the period of the year and the terrestrial location (Ferrari et al., 2013). Therefore, solar reflectance is a property that is largely dependent on the reference solar spectrum used for the calculations, as analyzed in Ferrari et al. (2013), and it is a number varying from 0 to 1.

Because a large part of the solar radiation is in the nonvisible regions of the spectrum, a high-albedo coating is not just a light color surface but should be able to reflect the solar radiation in the overall spectrum. A typical "cool" white surface of an elastomeric coating reflects in both the VIS and the NIR regions, and it is characterized by a solar reflectance of about 85%. Titanium dioxide is responsible for the color and for the large visible reflectance capability. Reflectance gradually decreases with increasing wavelength, due to the presence of C—H bonds and the vibration of hydrogen atoms in the coating.

9.2.2 Thermal emittance

Thermal emittance (ε) characterizes the surface capability to reemit the previously absorbed heat away from itself. It is described through a number varying between 0 and 1, representing the comparative/relative emittance with respect to a blackbody operating in similar conditions. In common construction applications, the thermal emittance of a surface is usually higher than 0.8–0.85 except for those layers that are based on metallic components, for example, aluminum shingles. The thermal emittance of a roof or wall component is mainly affected by the characteristics of the layer that is exposed to the solar radiation. Therefore, contrary to other envelope properties, for example, thermal transmittance and thermal capacity, for both SR and ε, the most external layer definitely determines roof cooling potential.

9.2.3 Solar reflectance index

The SRI represents the "coolness" of a coating, or of a surface in general, since it accounts for the two key properties previously described, that is, solar reflectance and thermal emittance. The reference international standard ASTM 1980 (ASTM, 2011) reports the procedure to calculate this index, starting from the evaluation of the steady-state surface temperature under the sunlight (T_s), the apparent sky temperature T_{sky}, the solar absorbance α, the irradiance I, and the thermal emissivity ε, as follows (see Equation (9.1); ASTM, 2011):

$$\alpha I = \varepsilon\sigma\left(T_s^4 - T_{sky}^4\right) + h_c(T_s - T_a) \tag{9.1}$$

High values of the solar reflectance and thermal emittance induce lower surface temperatures and higher SRI values. SRI is a sort of coolness scale where an absorptive standard black surface is characterized by SRI = 0 and the fully reflective standard white material corresponds to SRI = 100. However, SRI values of new and outstanding materials recently available in the market could be lower than 0 and higher than 100. Therefore, for a surface with known solar reflectance and thermal emittance values, and once convection conditions are known/defined, SRI can be calculated by iteration, solving Equation (9.2) for T_s as indicated:

$$\mathrm{SRI} = 100\frac{T_b - T_s}{T_b - T_w} \tag{9.2}$$

In this equation, T_b and T_w are the steady-state surface temperatures of the black and white reference surfaces, respectively. The standard conditions described in the ASTM standard (ASTM, 2011) refer to the numerical values of the following parameters, used in Equation (9.3):

$I = 1000\ \mathrm{W/m^2}$
$T_a = 310\ \mathrm{K}$
$T_{sky} = 300\ \mathrm{K}$
$h_c = 5$, 12, and 30 $\mathrm{W/m^2\,K}$ for low, medium, and high ventosity, respectively.

By taking into account medium wind conditions, the standard black with SRI = 0 is characterized by solar reflectance and thermal emittance values of 0.05 and 0.9, respectively. The standard white has solar reflectance of 0.8 and thermal emittance of 0.9 (Cool roof material database, 2000).

9.3 Thermal-energy assessment of high-albedo roofs

The mechanism that explains cool roofs' behavior has to be explained through the thermal-energy balance of the structure, or that of a generic surface, exposed to the solar radiation, when it is characterized by those physical processes determining its surface temperature (Incropera and DeWitt, 1996; Martin-Dominguez et al., 2011). The main thermal contribution reaching the roof (or any other sun-exposed surface) is produced by solar radiation, which is partially reflected by that surface, according to the physical characteristics described in Section 9.2. In particular, the part of the radiation that is reflected in the visible region of the spectrum determines the surface's visible color and its gloss finishing. Another large part of this radiation is reflected in the near-infrared region. The remaining part of the radiation becomes a solar gain for the surface and for the building in the case of a roof surface, because it is absorbed. Then, part of this radiation is reemitted outward by the roof surface, according to the view factors of the surface (or the roof) and the sky (Siegel and Howell, 2002). This part of the radiation is emitted by the roof with a longer wavelength (around 10,000 nm), therefore in the infrared section of the spectrum. Additionally, the absorbed heat portion enters the roof structure through a conduction process and goes from the warmer to the colder environment. In summer conditions, it contributes to overheating of the indoor spaces of the building and increases the energy demand for cooling (Pomerantz et al., 2000). Finally, the roof exposed to the solar radiation also exchanges energy through a convection process over its surface. In fact, the roof albedo is able to influence the thermal conditions over the roof and to produce non-negligible local effects with varying solar reflectance. For instance, external condensing units of the heat pumps in cooling mode located over the roof could be affected by the same roof albedo. High-albedo surfaces tend to keep the roof environment cooler than low-albedo surfaces by producing a further contribution in energy efficiency improvement of the heat pumps due to the lower thermal gap they have to satisfy, that is, "active cool roof effects" (Pisello et al., 2013b).

9.3.1 Thermal balance of a roof under the sun

As previously mentioned, a cool roof coating is able to reduce its surface temperature with respect to a noncool coating due to its high solar reflectance and high thermal emittance. These two properties then play a crucial role in determining the thermal-energy performance of the roof and the adjacent indoor environment. In fact, high solar reflectance produces lower heat gain through the roof into the building. This phenomenon is simply described by the thermal balance of the roof (Figure 9.1), or by any other surface exposed to solar radiation, and is given by Equation (9.3):

$$(1-R)I = \varepsilon\sigma\left(T_s^4 - T_{sky}^4\right) + h_c(T_s - T_a) - k\frac{dT}{dx} \tag{9.3}$$

where I is the surface insulation measured in W/m^2; R is the solar reflectance of the irradiated surface [0;1]; ε is the thermal emittance of the irradiated surface [0;1]; σ is the Stefan–Boltzmann constant equal to 5.6685×10^{-8} W/m^2 K^4; T_s is the superficial temperature of the roof measured in K; T_{sky} is the conventional sky temperature measured in K; h_c is the convection coefficient measured in W/m^2 K; T_a is the air temperature measured in K; k is the thermal conductivity of the surface exposed to the solar

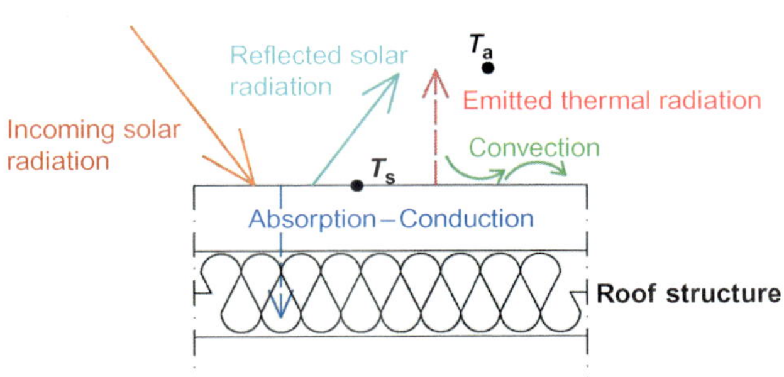

Figure 9.1 Scheme of the energy balance of a roof exposed to solar radiation.

radiation (or the roof), in W/m K; and dT/dx is the temperature gradient in the roof layer.

Equation (9.3) shows that solar reflectance, thermal emittance, and its insulation level are the main factors influencing the roof thermal performance. By considering modern highly insulated roofs, the two superficial properties, R and ε, determine its thermal behavior, during the day and during the night, respectively. Figure 9.2 reports the roof surface thermal profiles of cool brick tiles compared with traditional noncool tiles. These measurements refer to a test field located in central Italy where a cool tile for sloped roof application was prototyped and continuously monitored in summer and winter conditions (Pisello et al., 2014a,b). The figure shows that during daytime, the cool tile is able to decrease the surface temperature up to 15 °C with respect to the noncool tiles. Their solar reflectance values differ by about 56%, from 18% to 74% for the noncool and cool tiles, respectively. On the other hand, since they show

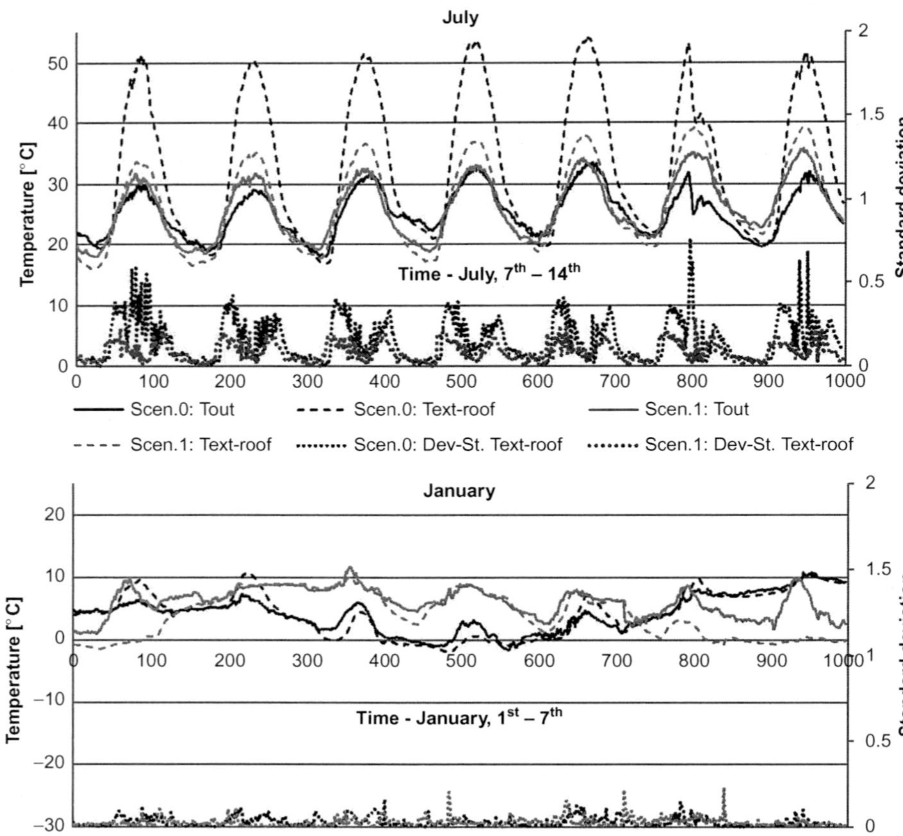

Figure 9.2 Surface temperature profiles of a traditional clay tile roof (scenario 0) and of a cool roof (scenario 1) as monitored during the coldest and the hottest periods in central Italy. Pisello et al. (2014a).

similar emittance values, due to the same clay-based composition of the substrate, their thermal behavior at nighttime is consistent. This shows how thermal emittance is another key property of cool roofs and cool coatings in general. This case study involved the analysis of a noninsulated traditional roof, where also the conduction process plays an important role in the thermal exchange phenomenon. As shown in Figure 9.2, during the winter (January data are shown in the figure), the surface temperature is not significantly affected by the application of the cool tile, since the solar irradiance is weak in that period, and daily overheating was not relevant, particularly during cloudy days.

9.4 How to measure high-albedo properties of roof coatings

In Section 9.3, the main thermal-energy exchange between the roof surface and its surroundings has been described. In particular, the thermal-energy balance of a cool surface characterized by high solar reflectance and high thermal emittance was showed. In this section, codes, standards, and instrument procedures for measuring the main properties of a cool coating are dealt with, taking into account the recent research and institutional developments in this field.

Several experimental in-lab and in-field studies have demonstrated that the implementation of high-albedo roofs in building results in energy saving for cooling by more than 20%, depending on climate conditions, indoor occupancy, and roof thermal characteristics (e.g., insulation level; Zinzi and Fasano, 2009). Important monitoring studies conducted in California and Florida show that the United States could save about US$750 million if an overall albedo increase of 10% is carried out (Akbari et al., 2005b; Levinson and Akbari, 2010; Parker et al., 1998). Additionally, important CO_2 emissions reduction could be achieved through the same strategy, by retrofitting a large amount of urban roofs. For this reason, many governmental authorities and technical organizations have supported the elaboration of new standards for quantifying and supporting the implementation of high-albedo roofs and urban pavings (Akbari et al., 2001), as well as for their intrinsic benefit in reducing the summer peak electricity demand. First of all, ASHRAE standards 90.1-2004 and 90.2-2007 determine incentives (i.e., credit) for constructions with cool roofs. The same standard ASHRAE 90.1-2010, recently updated, strongly recommends the implementation of cool roofs in several warm regions in the United States. The progress achieved in these regions was primarily determined by the work of the Heat Island Group that was created in California, within the Lawrence Berkeley National Laboratory (Akbari et al., 1997b). The same group also prepared the way for the Cool Roof Rating Council (CRRC) in 1998, which links public and private institutions in order to control technical quality and to provide reliable measurement procedures on high-albedo solutions. In fact, California Tile 24 action in the Building Energy Efficiency Program, started in 2005, prescribed minimum values of solar reflectance and thermal emittance to be reached in low-sloped roofs on both residential and nonresidential buildings (Akbari et al., 1997b). These values should be measured and labeled by the CRRC organization, which assumed the responsibility of ensuring the reliability of measured values. In Europe, the same path has been outlined, and the European Cool Roofs Council (EU-CRC; Synnefa et al., 2013), with its technical committee, links private companies and research and government institutions in order to monitor the quality of high-albedo solutions through a specific label and to promote the implementation of "cool" techniques in Europe (Kolokotsa et al., 2013). Finally, even international certification programs such as LEED protocol have taken into account the effect of high-albedo roofs and urban pavements and given distinct credits according to their potential for energy saving and urban heat island mitigation.

The following subsection will describe each parameter measurement procedure and the specific reference standard.

9.4.1 Solar reflectance

The measure of solar reflectance (SR) could be operated through a wide variety of methods and several instrument procedures are recognized by CRRC and EU-CRC. In particular, solar reflectance values can be measured by a (i) spectrophotometer with an integrating sphere, (ii) reflectometer, and (iii) pyranometer.

Spectrophotometer with an integrating sphere (Figure 9.3). The use of this instrument to measure SR is reported in the reference standard ASTM 903 (ASTM, 1996), describing the standard test method for solar absorbance, reflectance, and transmittance of materials covering an area of approximately 0.1 m^2 (ASTM, 1997). The use of the integrating sphere allows analyzing the spectral near-normal hemispherical reflectance in the range 250–2500 nm. Then, the final values are calculated through a weighted average value with respect to the standard solar spectral irradiance.

The spectrophotometer comprises three main parts: (i) a light source, emitting radiation in the range 250–2500 nm; (ii) the diffraction grating and mirrors; and (iii) the integrating sphere that is internally coated with a highly reflective finish. The solar reflectance $\rho(\lambda)$ and the solar transmittance $\tau(\lambda)$ of a planar sample are computed and compared with a standard white sample having known reflectance. The solar absorbance $\alpha(\lambda)$ is calculated as the difference: $1-\rho(\lambda)-\tau(\lambda)$. Because the

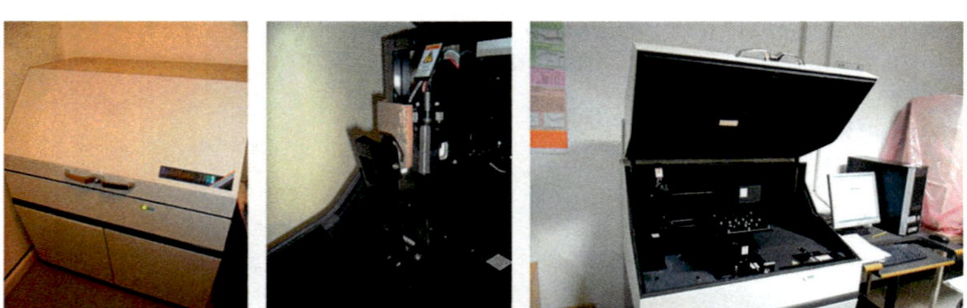

Figure 9.3 Spectrophotometer with an integrating sphere for measuring solar reflectance of coatings.

reflectance value varies with changes of the reference solar spectral irradiance spectrum, the choice of such spectrum gives distinct measured properties for the same sample. With this in mind, Levinson et al. (2010a,b) defined $R_{g,0}$ as the solar reflectance of a homogeneous surface calculated with reference to the global horizontal clear sky air mass one solar spectrum AM1GH, and they showed that this value well rates the reflectance of roofs and other surfaces exposed to solar radiation.

Reflectometer. The measurement procedure determining solar reflectance through reflectometer is reported in ASTM 1549 (ASTM, 2009): standard test method for determination of solar reflectance near ambient temperature using a portable solar reflectometer. Because this is a field measurement, the characterized area of the sample is around 2 cm^2. This instrument is able to measure the reflectance by varying the air mass setup, and it is calibrated with a blackbody cavity, representing null reflectance, and with a surface of known reflectance. The instrument is positioned over the target surface with a 2.5 cm of diameter opening and it is kept fixed until a constant value is registered. The reference specimen with known reflectance has to be previously characterized through solar spectrophotometer according to ASTM E903 (ASTM, 1997), as specified before. This measurement method is particularly suitable for characterizing reflectance properties of opaque and homogeneous materials through fast and relatively simple field measurements. To account for natural roof local differences, which are typical and realistic on heterogeneous roofs, further procedures have been developed to measure reflectance by spectrometer in several random positions (at least 30 points) over the roof surface and therefore to calculate the final value as the average of all determinations (Levinson et al., 2014).

Pyranometer. Reflectance measurement by pyranometer is another in-field procedure described in the ASTM E1918-06 standard (standard test method for measuring solar reflectance of horizontal and low-sloped surface in the field; ASTM, 2006). It is used to measure clear sky global solar reflectance of a horizontal or low-sloped surface (slope $< 2{:}12$) during the central hours of the day, when the sun angle to the normal surface is lower than 45°. Given these peculiar conditions, this method is relatively simple to apply in the summer season, whereas in winter, these conditions could never occur. This measuring technique is also potentially compromised by the presence of shadows, background and instrumental errors, and leveling differences due to the field portable equipment. The measurement requires accurate but affordable instruments, able to detect global solar radiation over the solar spectrum between 300 and 2500 nm. In order to optimize the precision of the albedo measurements, a double-pyranometer apparatus, that is, albedometer, is normally used to reduce the convection differential effects caused by tilting the pyranometer at different angles. For first-class pyranometers, the accuracy could be up to $\pm 0.5\%$ with a resolution of 1 W/m^2. The international reference standard ASTM E1918 describes the albedometer measurement procedure and setup instructions. It should be mounted at 50 cm distance to the surface under evaluation, in order to minimize the errors previously described, and its harm should be 1–1.5 m long (Figure 9.4). Weather conditions, sky clarity, and sun height significantly affect the measured albedo values during the course of the day; therefore, the aforementioned conditions should be respected in order to estimate reliable albedo values. In particular, the stand should be aligned toward the sun,

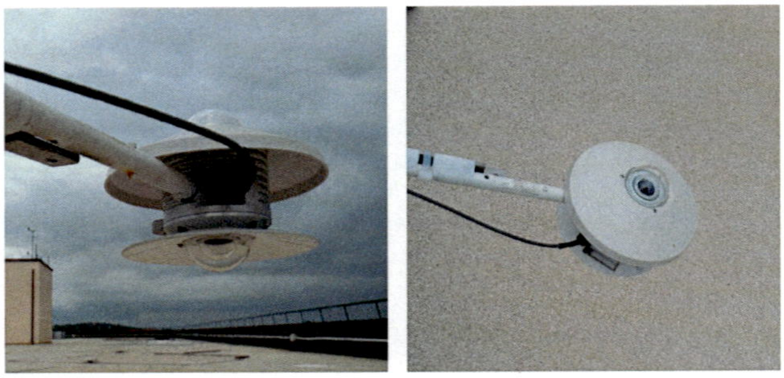

Figure 9.4 Double pyranometer (albedometer) for measuring incoming and reflected radiation.

in order to minimize the shadows created by the instrument itself over the sample, whereas the pyranometer should be placed parallel to the surface plan. When upward-oriented, the pyranometer measures the radiation arriving to the sample surface, while a downward orientation captures the reflected radiation. The measured albedo is in fact the ratio between reflected and incoming global solar radiation values. Several measurements should be carried out during the course of the day and in several weather conditions, in order to estimate the effect of such environmental variations on the albedo quantification. For instance, the presence of clouds near the sun radically affects such measurements, whereas clouds in the horizon line produce lower disturbance. Because ASTM E1918 considers that only measurements carried out in large surface areas are reliable, the modified version of the standard proposed by Akbari et al. (2008) also takes into account those conditions; smaller surface areas of 1 m^2 constitute the target samples to be characterized through the albedometer. In this modified standard (E1918A), the measurement of albedo is determined by considering three measurements of reflected solar radiation. In the first measurement, the targeted small area is covered with a white mask (a solar opaque white surface). In the second configuration, the same area is covered with a black mask (solar opaque black surface). The final configuration uses the uncovered sample. The reflectance of both white mask and black mask should be measured by the spectrophotometer, as described above. The benefits of using the E1918A protocol are (i) the use of small samples, (ii) the simplicity of the technical setup, and (iii) the assumption that most tested samples behave as Lambertian reflective surfaces, such as in the case of using white and black masks. In Akbari et al. (2008), the difference between albedo measurements carried out by the E1918 methodology with respect to E1918A protocol is always lower than 0.02 for single-colored samples and lower than 0.03 for multicolored surfaces.

9.4.2 Thermal emittance

The thermal emittance ε describes the surface capability to emit infrared radiation to the sky and to the surrounding environment in the thermal region of the spectrum. All surfaces at ambient temperature (around 300 K) emit almost completely the radiation between 4000 and 80,000 nm, or the infrared thermal portion. The emittance is also named emissivity and these terms are used interchangeably by most scientists and technicians; its value varies between 0 and 1. Its meaning is expressed by the Stefan–Boltzmann Equation (9.4):

$$q = \varepsilon \sigma T_s^4 \tag{9.4}$$

where q is the radiation emitted by a surface, σ is the Stefan–Boltzmann constant, T_s is the surface absolute temperature in Kelvin, and ε is the surface hemispherical emittance.

The main reference standards used to measure the thermal emittance are (i) the ASTM C1371 (ASTM, 2010) that describes the use of portable emissometers and (ii) the ASTM E408 (ASTM, 2008) that describes the inspection meter techniques. Recent developments in spectrometry made possible the measurement of the thermal emittance as a function of wavelength, through Fourier transform infrared, in the range 2500–10,000 nm. In this case, the spectral emittance is calculated by Equation (9.5):

$$\varepsilon(\lambda) = 1 - \rho(\lambda) - \tau(\lambda). \tag{9.5}$$

where ρ is the spectral reflectance and τ is the spectral transmittance measured as a function of the wavelength λ.

Portable emissometer. Portable differential thermopile emissometer is the instrument usually selected to measure hemispherical emittance of a surface or a coating applied over a surface (Figure 9.5). The instrument is primarily calibrated with high-emittance and low-emittance known samples, and it could be suitable for measurements over high-conductivity samples at ambient temperature. In order to prevent temperature increase, the sample is positioned over a heat sink and it should have a surface larger than 15 cm^2.

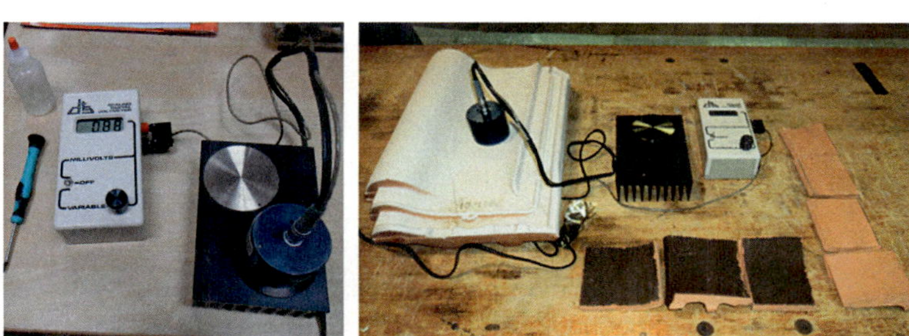

Figure 9.5 Portable emissometer used to measure thermal emittance of clay tiles (calibration procedure and measurement).

Inspection meter techniques. ASTM Standard E408 (2008) provides the description of two test methods (A and B), also detailing the limitations and accuracy of each one. Although these methods are not frequently used, they involve the estimation of emittance through thermal reflectance measurement and results are not so strongly affected by temperature drift. However, they generate erroneous estimations of the thermal emittance according to ASTM C1371 (ASTM, 2010).

9.5 Benefits of high-albedo roof coatings

As previously mentioned, high-albedo coatings applied to building roofs are becoming increasingly important as a passive strategy to save energy for cooling and to improve indoor thermal comfort conditions. This section concerns the analysis of their behavior through the presentation of several case studies, extracted from significant scientific contributions published in the past years. We aim to demonstrate the threefold benefits produced by using high-albedo roofs in the following aspects: energy saving for cooling at building scale, urban heat island mitigation, and global warming mitigation. Because the purpose of this chapter is to focus on building scale, the following subsection will deal with the analysis of high-albedo roof behavior (i) in test-room applications (Pisello et al., 2014a,b), (ii) through computational fluid dynamics tools (Foucquier et al., 2013; Xamán et al., 2010) and/or dynamic validated simulations (Rosado et al., 2014; Wang et al., 2012), and (iii) through continuous monitoring of real buildings. Important contributions to this issue were addressed in the framework of the EU-funded project "the Cool Roofs project" (Synnefa and Santamouris, 2012), which linked all the stakeholders of high-albedo roofs in order to clarify and disseminate technical information about such techniques, to overcome possible policy barriers (Kolokotsa et al., 2013), and to promote research and innovation activities around new materials and innovative solutions for energy saving in buildings and for mitigating urban heat island and global warming phenomena. Therefore, this section is organized by the analyses of several case studies, properly selected in order to provide an exhaustive literature review of the actual state of the art involving high-albedo roofs.

9.5.1 Thermal analysis of roofs as building components

Previous studies focused on both the transient behavior and the stationary behavior of building roofs where the heat transfer through this component is investigated with varying superficial reflectance characteristics.

Concerning the first studies conducted in stationary conditions, Reagan and Acklam (1979) estimated the daily average heat gain through the roof by the total equivalent temperature difference method for a specific roof in Arizona. Then, they evaluated the reduction of the heat gain (up to 50%) with increasing the roof lightness. Assem (2011) developed a series of correlations between the roof heat gain and its reflectance. In particular, he found that the heat gain could be reduced up to 42% with increasing solar reflectance.

Dynamic analyses were performed by taking into account the intrinsic variability of weather conditions and keeping constant the indoor air temperature as typical of controlled conditions. Ahrab and Akbari (2013) performed a numerical study by simulating several roofing configurations in different climate conditions in the United States, and they also studied moisture problems related to roof overcooling in the presence of high-albedo coatings in cold cities. Interesting results were demonstrated in Ahrab and Akbari (2013) where the heat flux through several roof slabs was measured in Mexico. The authors showed that the high-albedo coating was able to decrease the roof surface temperature by 13 °C with a corresponding heat gain from 180 to 80 W/m^2 for the bare slab and the white coated slab, respectively.

9.5.2 Test-room buildings

Important results were also achieved by scientists performing several interesting experimental campaigns in prototype test buildings (also known as test rooms, test cells, test chambers, etc.). Those cells allowed researchers to monitor realistic, dynamic, controlled indoor microclimate and outdoor weather conditions (Figure 9.6, Pisello et al., 2014a,b). Summan and Verma (2003) built experimental cells with different roofs in India. They found that the cool coating was able to decrease the indoor air

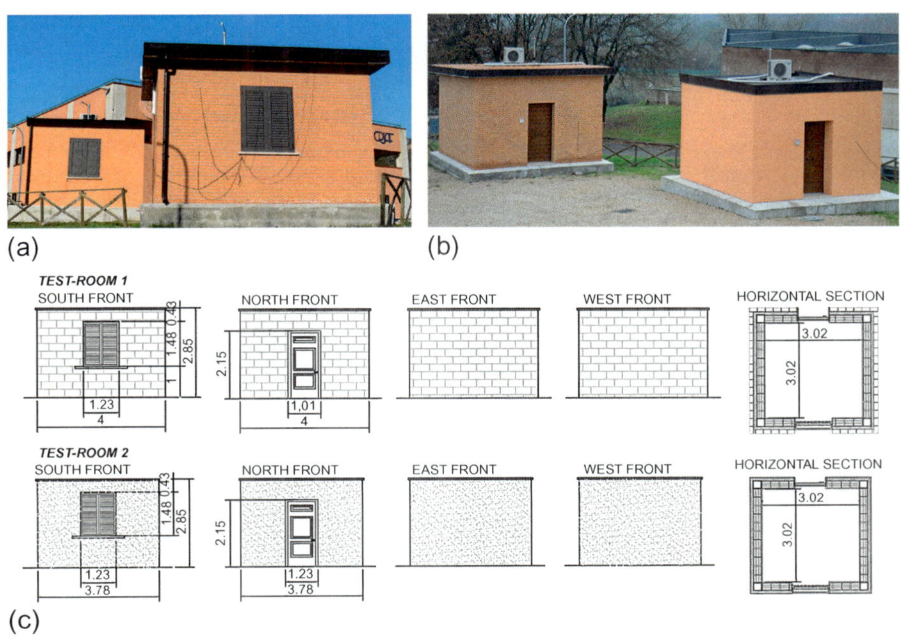

Figure 9.6 Test-room prototype building architecture at the University of Perugia, Italy. Pisello et al. (2014a,b).

temperature by 1.5 °C in the case of an insulated roof and 2.8 °C in the case of a concrete noninsulated roof. Larger differences were found by Guo et al. (2012) who simulated the sun radiation with infrared heat lamps. The tested reflecting coatings for walls reduced the surface temperature by 4.5 °C.

9.5.3 Numerical simulation studies

The analysis of the effect of high-albedo roof coatings was also carried out through numerical dynamic simulations, as a reliable tool to investigate a variety of roof configurations and boundary conditions. In particular, computational fluid dynamics and building dynamic thermal-energy simulation represent the two main instruments used for this purpose.

Santamouris et al. (2008) evaluated the effect of cool roof coatings on the energy requirement for cooling in residential buildings located in different cities around the world in order to estimate possible winter penalties with varying climate conditions. Results of simulations reported that high-albedo roofs are able to reduce cooling energy demand by 18% to 93% and peak cooling demand by 11% to 27% with varying climates. They also showed that the relative winter penalties are largely lower than summer benefits in every climate and that this effect is damped down on high-insulated roof structures. In the framework of the EU-funded project about cool roofs (Synnefa and Santamouris, 2012), Zinzi and Agnoli (2012) also focused on the impact of high-reflectance roofs in residential buildings located in the Mediterranean Basin, showing how cool roofs are able to reduce cooling energy demand up to 40% in Egyptian areas.

Interesting contributions were also produced by researchers implementing computational fluid dynamic models (CFD models), with a relatively theoretical gain in this approach in comparison to previous building simulations. In fact, CFD tools were mostly used to perform heat transfer analysis in cavities with varying external coating reflectance. Nevertheless, Sekar et al. (2012) studied the influence of high-albedo coatings for roofs on the indoor thermal comfort in three-dimensional models, estimating an average decrease of about 7 °C in the indoor air temperature.

9.5.4 Field monitoring studies of real buildings

As expected, several studies were based on the monitoring of real full-scale experiments, using buildings of distinct typologies and uses and where high-albedo roofs were applied in order to compare energy savings for cooling, indoor thermal comfort, and so on. Several studies demonstrated the benefits of such passive techniques, as previously reported, in office, residential, and university buildings, with varying climate conditions and internal gains. Almost all the works agreed with the previously presented results, showing that cool roofs represent useful and relatively simple strategies to save energy for cooling even in temperate climates, such as the London area. Additionally, the more the internal gains increase (e.g., in office buildings), the more the high-albedo roof produces benefits. Several interesting results were also achieved by the development and the industrial production of high-albedo tiles for applications in historic buildings or in traditional architectural style buildings, where, therefore, a modern white reflective coating could not be installed (Libbra et al., 2011). The majority of in-field monitoring campaigns were carried out by installing both indoor and outdoor monitoring setups, consisting of surface temperature probes, air temperature probes, energy meters, and albedometers. In this perspective, Pisello et al. (2014b) monitored an office building in central Italy in order to estimate the passive contribution of a high-albedo roof and its active contribution, consisting of its capability to reduce the air temperature over the roof, where the external units of the heat pumps are located. The same authors also studied the effect of using high-albedo clay tiles in a traditional residential building, where the main contribution was identified as the surface temperature reduction up to 20 °C during the summer, with almost negligible winter penalty, because of the northwest orientation of the sloped cool roof.

Experimentally monitored data were also used to calibrate and validate dynamic simulation models used for real buildings in different operative conditions. In this field, Bozonnet et al. (2013) calibrated a detailed model of a free-floating university building in central France. Aligned with previous studies, they found an important effect produced by high-albedo roof coating in non-insulated roofs and a non-negligible disturbance produced by the variable occupancy schedules that could affect the quality of estimations caused by the passive strategy.

9.6 Materials and techniques

High-albedo roof solutions mainly differ for the (i) location where they are applied, for example, for different construction contexts; (ii) period of construction, driving the development of specific increasingly sophisticated techniques; and (iii) roof configuration, often associated with the building use and architecture. In fact, cool coatings and membranes are often installed in low-sloped to flat roofs (inclination lower than 9°) of classic industrial or commercial large-size buildings. Sloped roofs are in general visible from the ground and they are more common in residential or traditional architectures. In those cases, high-albedo coatings were applied over roof shingles, metal layers, and natural stone or clay tiles. This section will describe the most common high-albedo (cool) coatings and commercial products already available within the market. Then, a final subsection will specifically address innovative products that are under development.

9.6.1 High-albedo roof products

As previously mentioned, the most diffused products concerning high-albedo coatings involve asphalt shingles, metal roof coverings, clay- or cement-based tiles, and optimized bitumen roofing systems (Levinson et al., 2005). Cool coatings could also be characterized as a function of the binder nature, for example, cementitious or elastomeric coatings or the carrier, water- or solvent-based. Cementitious coatings are usually applied over waterproofing membranes, whereas elastomeric coatings might be the actual waterproofing membranes or be directly attached to them. They are

Table 9.1 Thermal-optical properties of the most common high-albedo coatings for roofs

Material name	Solar reflectance (%)	Thermal emittance (%)	SRI (−)
Black coating	4–8	80–90	−10; 3
White coating	70–85	80–90	81–108
Clay tile (red natural)	28–44	85–90	24–48
White engobe clay tile	70–80	85–90	83; 100
White concrete tile	60–75	85–90	68; 92
Gray engobe clay tile	18–33	85–90	12; 34
White polyurethane membrane	65–85	80–90	74; 108
Black bitumen membrane	4–8	80–90	−10; 3
Unpainted metal roof	20–60	10–45	−112; 45
White painted metal roof	6–75	80–90	−7; 92

characterized by a solar reflectance (albedo) higher than 65% at the moment of the installation, and their thermal emissivity is always above 80%.

Coating carriers are also important to classify high-albedo coatings. In fact, they are used to improve the coating workability and the application over roof structures. However, the carrier usually evaporates during the coating drying process.

Table 9.1 reports some reference values of solar reflectance, thermal emittance, and SRI, with reference to previous studies and the measurements carried out by the authors in the laboratory network under the EU-CRC framework.

9.6.1.1 Tiles

Tiles represent the most typical roofing covering in European countries and in traditional architectures in general. They come in single elements and they have different shapes and configurations, depending on the local construction techniques Figure 9.7 reports four tile typologies characterized by different shape and finishing, typically used in classic architectures with sloped roofs. The high-albedo coating could be applied over the tile during the industrial production procedure or simply over the roof, after its installation. The most common materials for tiles are natural clay (mostly in Europe), cement-based materials, and low-cost asphalt shingle

Figure 9.7 (a–d) Four tile typologies characterized by different shape and finishing, typically used in classic architectures with sloped roofs.
Pisello et al. (2013a).

components. Typical clay tiles found in European countries are mostly used around the Mediterranean Basin and often present important constraints due to the necessity to preserve their visual appearance. In this case, since such tile is usually fired during the production process, the high-albedo coating can be applied on the surface of the raw tile as the engobe. It is then fired with the original piece, without modifying the tiles's properties and optical characteristics. This procedure assures a relatively higher stability and higher resistance to the weathering phenomena, meaning large durability of the solution. Light color tiles usually have around 20% to 40% of solar reflectance, while a white tile could have solar reflectance around 60% to 75%. Thermal emissivity values are usually higher than 80%.

9.6.1.2 Asphalt shingles

This solution represents a low-cost roofing typology highly resistant even under severe weather conditions. Tiles are basically made of bitumen, sometimes covered by aggregates. Their typical solar reflectance is less than 15%, given the high absorption potential of asphalt-based materials, but they could be covered by high-albedo coatings with important increases in solar reflectance capability. The principal use of this shingle is carried out in the United States, where important developments were achieved by the Lawrence Berkeley National Laboratory group. In fact, Levinson et al. (2007) prototyped 2-layer and 3-layer systems with highly reflective substrates, exposing titanium dioxide in the external layer and covering with cool coatings (i.e., a coating with high NIR reflectance and similar visual appearance with respect to classic materials of the same color; Levinson et al., 2010c). This way, asphalt shingles could be characterized by solar reflectance values up to 30% to 35% with a green finishing layer.

9.6.1.3 Metal roof layers

Metal roofs are typically covered with metal sheets or tiles made of galvanized steel or aluminum. They usually present a preapplied coating and can be colored in the external surface for final roof covering. They are mainly used in residential and commercial buildings and, most of all, in cold climate conditions, since they can be easily installed in highly sloped roofs. Metal layers covered with high-albedo coatings could be characterized by solar reflectance values up to 60% to 70%. The negative point is, as previously mentioned, the low thermal emittance, usually around 30%. In order to raise this limit, sheets prepainted with cool coatings are actually available in the market, with 80% to 90% thermal emittance values.

9.6.1.4 Bitumen-based membranes

Bitumen membranes are usually applied as waterproof systems for roofs, given their chemical characteristics, elasticity, and their relatively high resistance to thermal deformation and other fatigue effects. If not covered, they are characterized by solar reflectance lower than 20% but, if covered with high-albedo coatings, their reflectance could reach values around 60% to 80% depending on the characteristics of the coating. Thermal emittance is around 80% to 90% for both covered and uncovered membranes.

9.6.2 New materials in high-albedo roof coatings

This section describes the main innovation efforts around high-albedo or "cool" coatings. These research and development activities aim to develop (i) cool materials with similar visual appearance to classic noncool finishing, (ii) thermochromic materials, and (iii) high-thermal capacity systems through the inclusions of phase changing materials.

9.6.2.1 Cool-colored coatings

With the purpose of developing reflective coatings with similar visual appearance to classic coatings, colored (nonwhite) cool materials have been proposed in the last decade and are now available in the market. Cool-colored materials for roofing systems come in general as coatings, shingles, tiles, and membranes, and they consist of complex inorganic colored pigments that are neither light color nor white but are still able to increase their reflectance capability in the nonvisible region of the spectrum, that is, in the NIR range. Therefore, two visually equivalent coatings could perform as cool or noncool coatings depending on their ability to reflect that part of the solar radiation that corresponds to more than 50% of total solar irradiance. Research in this field was prompted by the necessity to preserve the visual appearance of traditional architectures and natural colored materials, for example, the clay red color of tiles and brickworks and the dark brown of wood finishing. Figure 9.8 shows a typical solar reflectance spectrum of cool-colored materials for application as engobes for cool dark tiles, in which the use of specific pigments allows an increase in the reflectance profile in the NIR range, with total increase of the solar reflectance up to 30%. Figure 9.9 shows the results of a prototyping campaign of cool clay tiles for application in historic buildings.

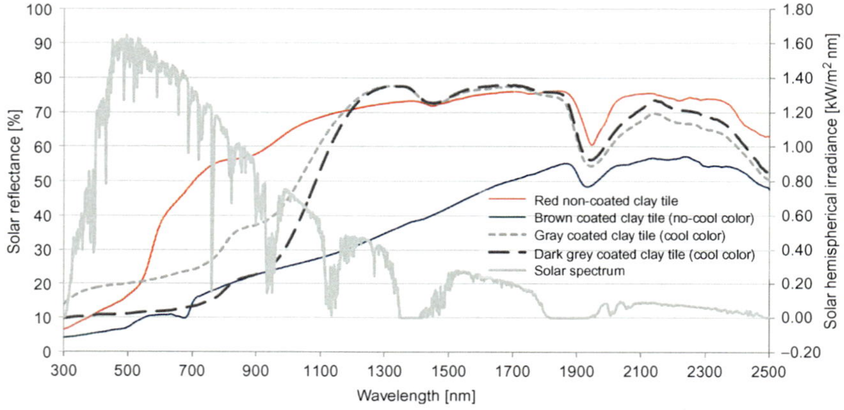

Figure 9.8 Solar reflectance profiles measured through spectrophotometer: comparison between traditional tiles and cool tiles.

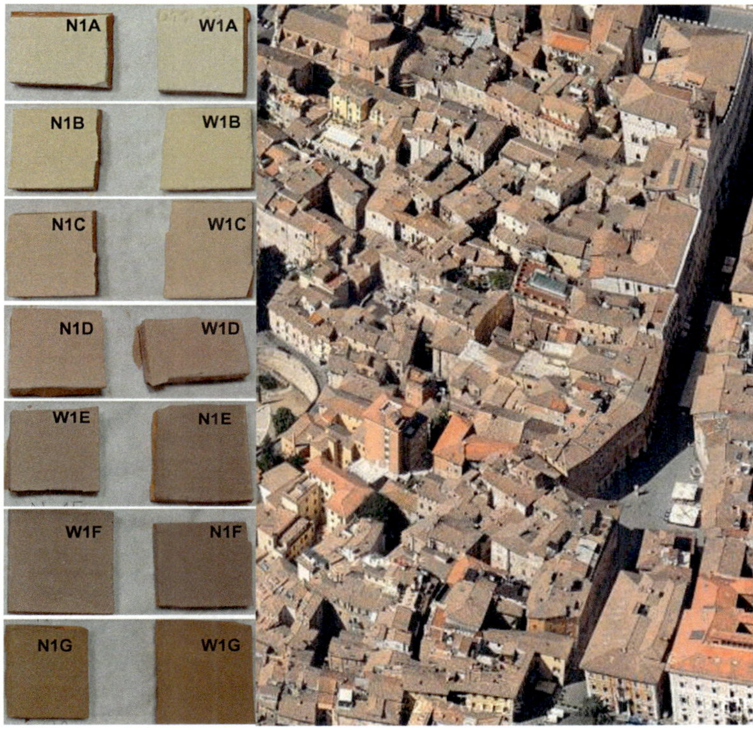

Figure 9.9 Prototyped coated tiles for implementing cool roofs in historic buildings. Pisello et al. (2013a).

9.6.2.2 Thermochromic coatings

Thermochromic coatings are used on building roofs and urban pavings, and are usually produced by incorporating thermochromic pigments into proper binders. All pigments are colored in their cold state, that is, for temperatures lower than the transition temperature, and they become white when the temperature surpasses the transition limit. This temperature is usually set up at about 30 °C. The main advantage of thermochromic paints and coatings is their dynamically variable performance, acting as cool coatings in the summer and as absorptive materials during the winter. More in detail, the thermochromic effect is achieved by a thermally reversible transformation of the internal coating molecular structure, producing consequent modifications in the visible part of the spectral behavior (Ma et al., 2001, 2002). These pigments are usually applied in common white coating mediums, and they are able to affect the thermal balance through the roof structure by increasing their reflectance capability above the transition temperature and reducing the thermal gain through the same roof in warm weather conditions, when they are mostly required.

9.6.2.3 *Coatings with phase change materials (PCMs) inclusion*

Phase change materials (PCMs) are increasingly used in construction in order to store latent heat due to their phase change reaction, producing a sort of extra thermal capacity in a building's envelope (Álvarez et al., 2013). Several studies have been performed to demonstrate and to quantify their effect in improving indoor thermal comfort in both summer and winter conditions. Given their intrinsic dynamic behavior, they are usually tested through monitoring campaigns aimed at comparing the thermal behavior of traditional coatings, cool coatings, and PCM-doped cool coatings under sunlight irradiation, but keeping the same visual appearance (Karlessi et al., 2009). Coatings with PCMs inclusion were demonstrated to lower the surface temperature up to 4 °C and to 2.7 °C for classic and cool coatings, respectively. A key role in PCMs' application is played by the proper choice of the transition temperature and of the concentration of the PCM.

9.7 Aging and weathering of high-albedo roof coatings

Soiling and weathering phenomena gradually modify the surface properties of the investigated coatings (Paroli et al., 1993). Soiling mechanisms consist of the deposition of atmospheric particulate over roof coatings and the presence or growth of microorganisms (Sleiman et al., 2014). Weathering action corresponds to those environmental agents affecting the physical behavior and durability of roof coatings exposed to water, sunlight, and high-temperature fluctuation (Sleiman et al., 2014). Therefore, solar reflectance properties should be specified before and after environmental exposure of such coatings, in order to exhaustively describe their thermal performance as passive cooling solutions during the course of their life cycle. This section gives a brief analysis of several environmental agents affecting high-albedo coatings' durability.

Thermal stress due to daily and seasonal temperature fluctuations represents an important cause of damage for coatings, attributable to differential thermal expansion and to the possible acceleration of chemical reactions responsible for degradation of (when used) polymeric materials.

Solar radiation affects the thermal behavior of the coating and its durability in general. In fact, the UV spectral region of the radiation, in particular, explains much photon energy reaching the material and inducing destructive reactions and color changes on the coating.

Wind pressure can cause structural/mechanical modifications in the coatings and in high-albedo membranes, inducing cracks and other damaging effects.

Rain, snow, and ice are forms in which moisture can touch and penetrate the membranes, inducing both chemical and physical stresses. Firstly, in the case of ice formation, the associated volume expansion is responsible for the generation of internal physical stress on the material. De-icing will cause volume contractions and these cyclic dimensional changes will damage the mechanical strength of the material. Additionally, rain water may bring atmospheric pollutants, which will make contact with the coating and might create corrosive or chemically adverse phenomena.

Atmospheric pollutants are in general responsible for the formation of oxides and hydroxides by combining pollutants with oxygen and water vapor. CO_2, SO_x, and NO_x, mainly produced by combustion processes and traffic, if combined with water, can corrode materials and damage superficial coatings, affecting their solar reflectance capability.

Biological contaminants consist of fungi and other moisture-facilitated organic agents affecting the optical properties of high-albedo coatings. These microorganisms or microplants could deposit over the coating surface, requiring periodic removal and restoration tasks.

9.7.1 Analysis of progressive degradation processes

Several exposure methods were developed to estimate the progressive behavior and the environmental resistance of coatings and membranes, since their thermal performance is radically affected by external agents (Jacques, 2000). In particular, the time performance of such systems is usually tested through (i) natural weathering exposition, (ii) accelerated weathering tests, and (iii) artificial weathering effect production.

Natural weathering exposition consists of simple experimental campaigns where coating or membrane samples are exposed to direct natural sunlight and other natural weathering agents, such as rain and wind. Specific exposition sites (i.e., weathering farms) are selected around the world in order to represent reference environmental conditions. Important environmental and geometric characteristics are controlled and monitored, and the solar reflectance and emittance properties are quantified both before and after the exposure period.

Accelerated weathering tests are in-lab tests that simulate severe outdoor exposure conditions, for example, concentrated sun radiation and high pollutant contents. Artificial weathering tests are used in lab facilities to anticipate and predict the damaging of coatings and membranes, since accelerated degradation conditions are applied. For example, tests are performed to simulate the solar radiation through artificial light sources with similar emission spectrum.

All these procedures are specifically reported in international reference standards, describing the main guidelines for reproducing weathering tests in order to find reliable values of optic-thermal properties of coatings for roof application (ASTM, 2003; ISO, 2004). The international rating systems are progressively implementing the necessity to perform weathering standard tests in order to exhaustively evaluate the life cycle performance of high-albedo commercialized coatings.

9.8 Conclusions

This chapter has explored a wide panorama of physical characteristics and thermal-energy benefits of high-albedo coatings for building roofs aimed at reducing the energy required for cooling a building. The roof as a building component and its thermal balance were presented and briefly discussed in order to explain the behavior of reflective "cool" roof systems. Starting from the analysis of more diffuse, already commercialized, white coatings for roof application, innovative solutions and new findings about weathering and degradation processes were also presented, with the purpose of clarifying the main research contributions around this issue.

The threefold benefit of such technology was also showed, in order to valorize high-albedo roofs and their potential contribution for energy efficiency in buildings, mitigating urban heat island and global warming phenomena.

An exhaustive literature review has been carried out, allowing the reader to focus on specific aspects of high-albedo passive solutions.

Acknowledgments

Part of the information included in the chapter was published in previous works by the author. The author thanks Professor Mattheos Santamouris for being a precious guide in the research into high-albedo coatings and cool roof technologies.

References

Ahrab, M.A.M., Akbari, H., 2013. Hygrothermal behaviour of flat cool and standard roofs on residential and commercial buildings in North America. Build. Environ. 60, 1–11.

Akbari, H., Konopacki, S., 2005. Calculating energy saving potentials from heat island reduction strategies. Energy Policy 33, 751–756.

Akbari, H., Bretz, S., Kurn, D., Hanford, J., 1997a. Peak power and cooling energy savings of high-albedo roofs. Energy Build. 25, 117–126.

Akbari, H., Gabersek, S., Pomerantz, M., Gartland, L., 1997b. Cooling Energy Saving Potentials of Light Colored Roofs for Residential and Commercial Buildings in 11 U.S. Metropolitan Areas. National Technical Information Service, Springfield, VA. http://dx.doi.org/10.2172/508151.

Akbari, H., Pomerantz, M., Taha, H., 2001. Cool surfaces and shade trees to reduce energy use and improve air quality in urban areas. Sol. Energy 70 (3), 295–310.

Akbari, H., Levinson, R., Miller, W., Berdahl, P., 2005a. Cool Colored Roofs to Save Energy and Improve Air Quality. In: Proc. International Conference "Passive and Low Energy Cooling for the Built Environment," Santorini, Greece, May.

Akbari, H., Levinson, R., Rainer, L., 2005b. Monitoring the energy-use effects of cool roofs on California commercial buildings. Energy Build. 37 (10), 1007–1016.

Akbari, H., Levinson, R., Stern, S., 2008. Procedure for measuring the solar reflectance of flat or curved roofing assemblies. Sol. Energy 82 (7), 648–655.

Akbari, H., Menon, S., Rosenfeld, A., 2009. Global cooling: increasing world-wide urban albedos to offset CO_2. Clim. Change 95. http://dx.doi.org/10.1007/s10584-008-9515-9.

Álvarez, S., Cabeza, L.F., Ruiz-Pardo, A., Castell, A., Tenorio, J.A., 2013. Building integration of PCM for natural cooling of buildings. Appl. Energy 109, 14–522.

Assem, E.O., 2011. Correlating thermal transmittance limits of walls and roofs to orientation and solar absorption. Energy Build. 43, 3173–3180.

ASTM, 1996. E 903: Standard Test Method for Solar Absorptance, Reflectance, and Transmittance of Materials Using Integrating Spheres. American Society for Testing and Materials, West Conshohocken, PA.

ASTM, 1997. G7-97: Standard Practice for Atmospheric Environmental Exposure Testing of Nonmetallic Materials. American Society of Testing and Materials, West Conshohocken, PA, USA.

ASTM, 2003. D1669: Standard Practice for Preparation of Test Panels for Accelerated and Outdoor Weathering of Bituminous Coatings. American Society of Testing and Materials, West Conshohocken, PA, USA.

ASTM, 2006. E 1918: Standard Test Method for Measuring Solar Reflectance of Horizontal and Low-Sloped Surfaces in the Field. American Society for Testing and Materials, West Conshohocken, PA.

ASTM, 2008. E408: Thermal Emissivity Measurement Using Inspection Meter Techniques. American Society for Testing and Materials, West Conshohocken, PA.

ASTM, 2009. C1549–09: Standard Test Method for Determination of Solar Reflectance Near Ambient Temperature Using a Portable Solar Reflectometer. American Society for Testing and Materials, West Conshohocken, PA.

ASTM, 2010. C1371: Thermal Emissivity Measurement Using a Portable Emissometer. American Society for Testing and Materials, West Conshohocken, PA.

ASTM, 2011. C1980: Standard Practice for Calculating Solar Reflectance Index of Horizontal and Low-Sloped Opaque Surfaces. American Society for Testing and Materials, West Conshohocken, PA.

Berdahl, P., Akbari, H., Levinson, R., Miller, A.W., 2008. Weathering of roofing materials—an overview. Construct. Build. Mater. 22, 423–433.

Bozonnet, E., Doya, M., Allard, F., 2013. Cool roofs impact on building thermal response: a French case study. Energy Build. 43 (11), 3006–3012.

Cool roof material database. http://energy.lbl.gov/coolroof/ (accessed 12.08.14).

Cool Roofs project (Contract No.: EIE/07/475/SI2.499428): Report on Description of main Cool Roof Products Features, 2010.

Cotana, F., Rossi, F., Filipponi, M., Coccia, V., Pisello, A.L., Bonamente, E., Petrozzi, A., Cavalaglio, G., 2014. Albedo control as an effective strategy to tackle Global Warming: a case study. Appl. Energy 130, 641–647. http://dx.doi.org/10.1016/j.apenergy.2014.02.065.

Ferrari, C., Libbra, A., Muscio, A., Siligardi, C., 2013. Influence of the irradiance spectrum on solar reflectance measurements. Adv. Build. Energy Res. 7 (2), 244–253.

Foucquier, A., Robert, S., Suard, F., Stéphan, L., Jay, A., 2013. State of the art in building modelling and energy performance prediction: a review. Renewable Sustainable Energy Rev. 23, 272–288.

Gartland, L., 2008. Heat Islands: Understanding and Mitigating Heat in Urban Areas. Earth Scan Publications, US, ISBN: 1844072509, 57–83.

Guo, W., Qiao, X., Huang, Y., Fang, M., Han, X., 2012. Study on energy saving effect of heat reflective insulation coating on envelopes in the hot summer and cold winter zone. Energy Build. 50, 196–203.

Haberl, J., Cho, S., 1998. Literature Review of Uncertainty of Analysis Methods (Cool Roofs), Report to the Texas Commission on Environmental Quality. Energy Systems Laboratory, Texas A&M University, College Station, TX.

Incropera, F.P., DeWitt, D.P., 1996. Fundamentals of Heat and Mass Transfer, fourth ed. J. Wiley and Sons, NY.

ISO 11341: 2004. Paints and varnishes—artificial weathering and exposure to artificial radiation—exposure to filtered xenon arc radiation.

Jacques, L., 2000. Accelerated and outdoor/natural exposure testing of coatings. Prog. Polym. Sci. 25, 1337–1362.

Karlessi, T., Santamouris, M., Apostolakis, K., Synnefa, A., Livada, I., 2009. Development and testing of thermochromic coatings for buildings and urban structures. Sol. Energy 83 (4), 538–551.

Kolokotsa, D., Santamouris, M., Akbari, H., 2013. Advances in the Development of Cool Materials for the Built Environment. Bentham Science Publishing, Sharjah, http://dx.doi.org/10.2174/97816080547181130101 eISBN: 978-1-60805-471-8. ISBN: 978-1-60805-597-5.

Levinson, R., Akbari, H., 2010. Potential benefits of cool roofs on commercial buildings: conserving energy, saving money, and reducing emission of greenhouse gases and air pollutants. Energy Effic. 3, 53–109.

Levinson, R., Akbari, H., Konopacki, S., Bretz, S., 2005. Inclusion of cool roofs in nonresidential Title 24 prescriptive requirements. Energy Policy 33 (2), 151–170.

Levinson, R., Berdahl, P., Akbari, H., Miller, W., Joedickec, I., Reillyd, J., Suzukie, Y., Vondran, M., 2007. Methods of creating solar-reflective nonwhite surfaces and their application to residential roofing materials. Sol. Energy Mater. Sol. Cells 91, 304–314.

Levinson, R., Akbari, H., Berdahl, P., 2010a. Measuring solar reflectance—part I: defining a metric that accurately predicts solar heat gain. Sol. Energy 84, 1717–1744.

Levinson, R., Akbari, H., Berdahl, P., 2010b. Measuring solar reflectance—part II: review of practical methods. Sol. Energy 84, 1745–1759.

Levinson, R., Akbari, H., Berdahl, P., Wood, B., Skilton, W., Petersheim, J., 2010c. A novel technique for the production of cool colored concrete tile and asphalt shingle roofing products. Sol. Energy Mater. Sol. Cells 94 (6), 946–954.

Levinson, R., Chen, S., Berdahl, P., Rosado, P., Medina, L.A., 2014. Reflectometer measurement of roofing aggregate albedo. Sol. Energy 100, 159–171.

Libbra, A., Tarozzi, L., Muscio, A., Corticelli, M.A., 2011. Spectral response data for development of cool coloured tile coverings. Opt. Laser Technol. 43 (2), 394–400.

Ma, Y., Zhu, B., Wu, K., 2001. Preparation and solar reflectance spectra of chameleon-type building coatings. J. Sol. Energy 70, 417–422.

Ma, Y., Zhang, X., Zhu, B., Wu, K., 2002. Research on reversible effects and mechanism between the energy-absorbing and energy-reflecting states of chameleon-type building coatings. J. Sol. Energy 72, 511–520.

Martin-Dominguez, I., Lucero-Álvarez, J., Rubin-Zacarías, F., Ledesma-Gallegos, A., Alarcón Herrera, M., 2011. Experimental comparison of heat flow through concrete roofing flagstones with different coatings. In: Proceedings of the Solar World Congress. ISES, Kassel, Germany.

Parker, D.S., Sherwin, J.R., Sonne, J.K., 1998. Measured performance of reflective roofing systems in a Florida commercial buildings. ASHRAE Trans. 104 (1), 963–975, American Society of Heating, Refrigeration, and Air Conditioning Engineers, Atlanta, GA (January).

Paroli, R.M., Dutt, O., Delgado, A.H., Stenman, H.K., 1993. Ranking PVC roofing membranes using thermal analysis. J. Mater. Civil Eng. 5 (1), 83–95.

Pisello, A.L., Cotana, F., 2014. The thermal effect of an innovative cool roof on residential buildings in Italy: results from two years of continuous monitoring. Energy Build. 69, 154–164.

Pisello, A.L., Cotana, F., Nicolini, A., Brinchi, L., 2013a. Development of clay tile coatings for steep-sloped cool roofs. Energies 6 (8), 3637–3653.

Pisello, A.L., Santamouris, M., Cotana, F., 2013b. Active cool roof effect: impact of cool roofs on cooling system efficiency. Adv. Build. Energy Res. 7 (2), 209–221.

Pisello, A.L., Cotana, F., Nicolini, A., Buratti, C., 2014a. Effect of dynamic characteristics of building envelope on thermal-energy performance in winter conditions: in field experiment. Energy Build. 80, 218–230.

Pisello, A.L., Rossi, F., Cotana, F., 2014b. Summer and winter effect of innovative cool roof tiles on the dynamic thermal behavior of buildings. Energies 7 (4), 2343–2361.

Pomerantz, M., Pon, B., Akbari, H., Chang, S.C., 2000. The effect of pavements' temperatures on air temperatures in large cities. LBNL Report-43442.

Reagan, J.A., Acklam, D.M., 1979. Solar reflectivity of common roofing materials and its influence on the roof gain of typical southwestern USA residences. Energy Build. 2, 237–248.

Rosado, P.J., Faulkner, D., Sullivan, D.P., Levison, R., 2014. Measured temperature reductions and energy savings from a cool tile roof on a central California home. Energy Build. 80, 57–71. http://dx.doi.org/10.1016/j.enbuild.2014.04.024.

Sailor, D.J., 1995. Simulated urban climate response to modifications in surface albedo and vegetative cover. J. Appl. Meteorol. 34 (7), 1694–1704.

Santamouris, M., Synnefa, A., Kolokotsa, D., Dimitriou, V., Apostolakis, K., 2008. Passive cooling of the built environment—use of innovative reflective materials to fight heat islands and decrease cooling needs. Int. J. Low-Carbon Technol. 3 (2), 71–82.

Santamouris, M., Kolokotsa, D., 2013. Passive cooling dissipation techniques for buildings and other structures: the state of the art. Energy Build. 57, 74–94.

Sekar, M., Sakthivel, M., Kumar, S.S., Ramesh, C., 2012. Attaining the room comfort by solar reflective paint. Am. J. Environ. Sci. 8, 556–562.

Siegel, R., Howell, J., 2002. Thermal Radiation Heat Transfer, fourth ed. Taylor and Francis, New York.

Sleiman, M., Kirchstetter, T.W., Berdahl, P., Gilbert, H.E., Quelen, S., Marlot, L., Preble, C.V., Chen, S., Montalbano, A., Rosseler, O., Akbari, H., Levinson, R., Destaillats, H., 2014. Soiling of building envelope surfaces and its effect on solar reflectance—part II: development of an accelerated aging method for roofing materials. Sol. Energy Mater. Sol. Cells 122, 271–281.

Sproul, J., Wan, M.P., Benjamin, H.M., Arthur, H.R., 2014. Economic comparison of white, green, and black flat roofs in the United States. Energy Build. 71, 20–27.

Summan, B.M., Verma, V.V., 2003. Measured performance of a reflective thermal coating in experimental rooms. J. Sci. Ind. Res. 62, 1152–1157.

Synnefa, A., Santamouris, M., 2012. Advances on technical, policy and market aspects of cool roof technology in Europe: the cool roofs project. Energy Build. 55, 35–41.

Synnefa, A., Santamouris, M., Livada, I., 2006. A study of the thermal performance of reflective coatings for the urban environment. Sol. Energy J. 80, 968–981.

Synnefa, A., Santamouris, M., Akbari, H., 2007. Estimating the effect of using cool coatings on energy loads and thermal comfort in residential buildings in various climatic conditions. Energy Build. 39 (11), 1167–1174.

Synnefa, A., Dandou, A., Santamouris, M., Tombrou, M., Soulakellis, N., 2008. On the use of cool materials as a heat island mitigation strategy. J. Appl. Meteorol. Climatol. 47, 2846–2856.

Synnefa, A., Pantazaras, A., Santamouris, M., Bozonnet, E., Doya, M., Zinzi, M., Muscio, A., Libbra, A., Ferrari, C., Coccia, V., Rossi, F., Kolokotsa, D., 2013. Interlaboratory comparison of cool roofing material measurement methods. In: 34th AIVC—3rd TightVent-2nd Cool Roofs'—1st venticool Conference. 25–26 September 2013, Athens, Greece.

Taha, H., 1997. Modeling the impacts of large-scale albedo changes on ozone air quality in the South Coast Air Basin. Atmos. Environ. 31, 1667–1676.

Taha, H., 2008a. Meso-urban meteorological and photochemical modeling of heat island mitigation. Atmos. Environ. 42 (38), 8795–8809.

Taha, H., 2008b. Urban surface modification as a potential ozone air quality improvement strategy in California: a mesoscale modeling study. Bound.-Lay. Meteorol. 127 (2), 219–239. http://dx.doi.org/10.1007/s10546-007-9259-5.

Taha, H., Konopacki, S., Gabersek, S., 1999. Impacts of large-scale surface modifications on meteorological conditions and energy use: a 10-region modeling study. Theor. Appl. Climatol. 62 (3–4), 175–185.

Taha, H., Chang, C., Akbari, H., 2000. Meteorological and air quality impacts of heat island mitigation measures in three U.S. cities. Lawrence Berkeley National Laboratory Rep. LBL-44222.

Wang, S., Yan, C., Xiao, F., 2012. Quantitative energy performance assessment methods for existing buildings. Energy Build. 55, 873–888.

Xamán, J., Mejía, G., Álvarez, G., Chávez, Y., 2010. Analysis on the heat transfer in a square cavity with a semitransparent wall: effect of the roof materials. Int. J. Thermal. Sci. 49, 1920–1932.

Zinzi, M., Agnoli, S., 2012. Cool and green roofs, an energy and comfort comparison between passive cooling and mitigation urban heat island techniques for residential buildings in Mediterranean region. Energy Build. 55, 66–76.

Zinzi, M., Fasano, G., 2009. Properties and performance of advanced reflective paints to reduce the cooling loads in buildings and mitigate the heat island effect in urban areas. Int. J. Renew. Sustain. Energy 28, 123–139.

Solar cooling with hydrophilic porous materials for reducing building cooling needs

10

D. Karamanis
University of Patras, Agrinio, Greece

10.1 Introduction

Rapid urbanisation, with buildings and pavements of materials that absorb and store solar radiation, constrained roads and canyons of low wind speeds for convective heat removal, and the released heat of the associated anthropogenic activities have raised temperatures within city islands to higher values than the surrounding rural areas. In Athens, the mean intensity of the island is higher than 10 °C (Santamouris et al., 2001), while it reaches 8.9 °C in London (Kolokotroni and Giridharan, 2008). The urban heat island phenomenon is not limited to large cities, since urban heat islands with intensities higher than 5 °C are observed even in small cities during the summertime (Vardoulakis et al., 2013).

Furthermore, the levels of energy consumed in buildings place the sector among the most significant CO_2 emissions sources in Europe. Concrete is used in huge quantities in buildings, and although it is a low-impact material, its production is responsible for 7–10% of total CO_2 emissions worldwide. During the operation phase, buildings consume more than 40% of the primary energy and emit one-third of global annual greenhouse gases (UNEP, 2009, p. 63). In the summer, electricity consumption is increased due to conventional air conditioning that is used to compensate for the increased indoor temperatures. Depending on the climatic conditions, building characteristics, and operation, air conditioning can be the major contributor to peak electricity demand (e.g., 40% of peak load in Shanghai; Chan et al., 2010). Therefore, the reduction of the use of mechanical (active) HVAC systems and the development of alternative, sustainable, and efficient cooling technologies are vital.

Among the building envelope components, the glazing ratio or wall insulation have a small influence on the ratio of energy demand change due to the urban heat island effect (Bueno et al., 2012). Urban roof is the building component that's most exposed to solar radiation, leading to excessive heat flow indoors. For arid and hot areas, almost 50% of the heat load in the building comes from the roof (Florides et al., 2002; Nahar et al., 1999). Moreover, roofs provide more than 20% of urban free area (Akbari et al., 2003), and heat island mitigation measures can influence directly the cooling loads not only in the last floor but also in the entire building (Spala et al., 2008). Therefore, the application of new and more revolutionary concepts of responsive and

Eco-efficient Materials for Mitigating Building Cooling Needs. http://dx.doi.org/10.1016/B978-1-78242-380-5.00010-8

dynamic technologies, even for the more insulated building roofs with large attenuation in temperature variations, could further improve the energy efficiency of buildings and reduce the cooling loads.

In this direction, some traditional passive cooling methods like green and water spaces and "cool" materials in the urban environment have been revived and are being redeveloped in a more eco-efficient mode. In this context, the most well-developed roof technologies for reducing the heating loads are cool roofs and green roofs (Santamouris, 2014). Both technologies can lower the surface temperatures of roofs and thus decrease the corresponding sensible heat flux to the atmosphere.

However, the practical applicability of these methods depends highly on the local climate and there are important considerations for both technologies. For example, the reflectance of cool coatings reduces even 15% during the first year of the application due to weathering (Bretz and Akbari, 1997). Moreover, coatings of high reflectance are appropriate for the hot Mediterranean climate, but roof reflectivity should be optimized to 0.6–0.7 for colder climates in order to account for winter losses (Kolokotroni et al., 2013). Also, the effectiveness of the high albedo to mitigate high urban temperatures during the night is limited (Saneinejad et al., 2014). Green roofs could produce the same effect as cool roofs if they use continuous water resources for irrigation and they are regularly irrigated and planted with a dense mix of actively transpiring vegetation. Although the cost of green roofs has been highly reduced, the need of water for the irrigation and drainage systems as well as the required intensive maintenance and the dependence on the local climate conditions can limit their worldwide applicability. Therefore, new and more efficient materials and procedures have to be developed (Santamouris, 2014).

Evaporative cooling is the oldest technique of cooling, applied by ancient Egyptians around 2500 B.C. with water-containing porous clay pads over doors and windows for air cooling (Duan et al., 2012). Evaporation, defined as the rate of liquid water change to vapor, is driven by the temperature difference between the dry-bulb and the wet-bulb of the process air and is a function of this difference, solar radiation, pressure, temperature, wind, and physicochemical properties of water.

A few examples of evaporative coolers are humidifiers, air washers, passive down-draft evaporative coolers, and roof surface evaporative coolers (Chan et al., 2010; Leo Samuel et al., 2013). Water films, roof ponds, wetted bag covers, porous roofs, and their combinations with or without reflective surfaces are roof integrated evaporated cooling methods that can be applied in a much higher area and hence provide higher reductions of cooling loads. The earlier studies on the effectiveness of roof water surfaces were performed before the 1940s, where open ponds and spray water were found to be highly effective in reducing heat flux through all types of roofs (Tiwari et al., 1982). Since then, a high number of studies have been performed that have proved the cooling effectiveness of roof evaporation, and some of them have been implemented in practical use (Singh and Tiwari, 1996). A pond of water over the roof with a movable top thermal insulation is among the most studied evaporative systems. During summer, the insulation covers the pond to protect it during daytime and exposes it during the night for nocturnal cooling of the water. In winter, the process happens in reverse. By applying this technique, room temperatures lower than 20 °C have been observed in summer (Sadineni et al., 2011).

Runsheng et al. (2003) proved that gunny bags or cloth floating over a water pond surface with movable insulation is an efficient evaporative-based cooling technique, especially for arid regions. Recently, Spanaki et al. (2011) presented a comprehensive review of roof pond variants and a decision support flowchart of various criteria and parameters to help in the selection of the most appropriate cooling system. Although the cooling results are generally positive for water evaporation systems, specific requirements like the availability of water resources and the humidity increase still need to be addressed. Therefore, the use of materials as supports of water or compounds with phase changes have been proposed. Although there are still issues to be resolved in the latter, their thermal storage property is very promising (Pasupathy and Velraj, 2007).

In the last few years, the use of porous materials for the "free" integrated evaporative cooling of buildings has been started systematically to be studied mainly as a sustainable and alternative way to cool the roof surface of a building or the pavement of outdoor spaces by taking advantage of the properties of porous materials (Liu et al., 1997; Meng, 1996; Meng and Hu, 2005; Okada et al., 2008; Vardoulakis et al., 2011; Wanphen and Nagano, 2009). According to the principle of evaporation cooling of buildings with porous materials, stored water or night-sorbed moisture is evaporated during the hot day and the porous surface temperature is reduced due to the release of the latent heat. Lower surface temperatures contribute to the reduction of air temperature since the intensity of heat transfer through the cold surface is lower while the heat flow inside the building is reduced. Also, there are many indirect benefits, including water retention during heavy rainfall, increase of thermal insulation of the building, and absorption of many polluting elements. Moreover, material degradation due to high summer temperatures is reduced, while relative humidity in winter changes environmental climate to more wet states, resulting in a reduction of the spread of diseases like influenza (Okada et al., 2008).

10.2 Hydrophilic porous materials

10.2.1 Porosity and hydrophilicity

Porous solids are of scientific and technological interest because of their ability to interact with atoms, ions, and molecules not only at their surfaces, but throughout the bulk of the material (Davis, 2002). Because of this, porous materials have been intensively studied in a great variety of applications, ranging from separation and catalysis to photonics, energy conversion and storage, biomedicine, and drug delivery. Their main property to contain closed pores as cavities, channels, or interstices of specific shape and accessibility defines the porosity of the solids. It can be measured as the ratio of the total pore volume V_p to the apparent volume V of the particle. In addition to porosity, porous materials can be characterized by the density of pores, their volume, the pore size or width, and the accessible surface area per unit mass. According to the

International Union of Pure and Applied Chemistry (IUPAC) definition, porous materials are divided into three classes: microporous (pore size < 2 nm), mesoporous (2–50 nm), and macroporous (>50 nm) materials (IUPAC, 1994). Typical materials in the three categories are zeolites, silica gel, and activated carbon (Figure 10.1a). However, the spatial organization of the unit cell (Figure 10.1b) can deliver structures with pore size distribution in two or three regions (Figure 10.1c and d).

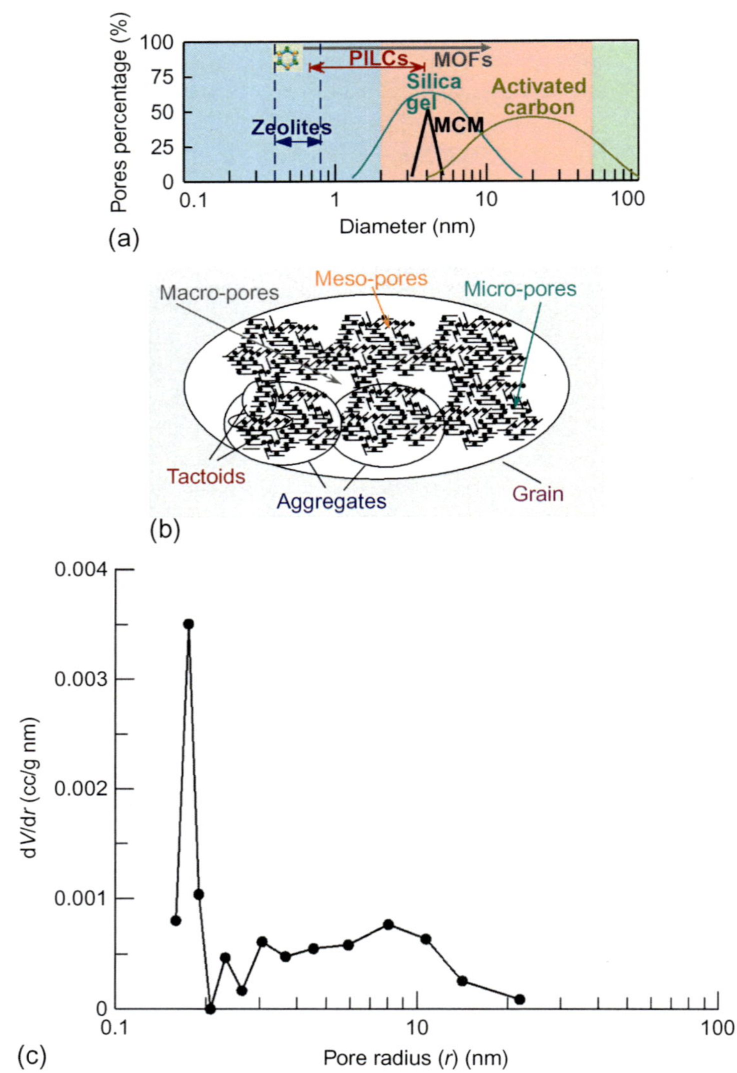

Figure 10.1 (a) Pore size distribution of porous materials. (b) Typical pores in pillared layered clays (PILCs). (c) Pore size distribution of aluminium pillared layered montmorillonite from nitrogen adsorption–desorption isotherms.

(Continued)

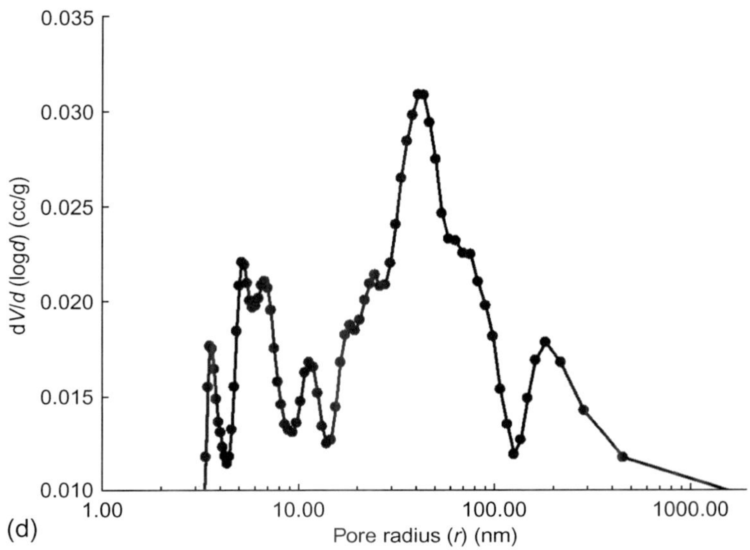

Figure 10.1 Continued. (d) Pore size distribution of modified fly ash by mercury intrusion porosimetry.

For a long time, the majority of porous materials were prepared based on the chemistry of aluminosilicates and aluminophosphates, and zeolites, the best known family of the microporous materials, were produced with a well-defined crystallinity and a narrow and uniform prose size distribution. A large range of zeolite molecular sieves such as zeolites A, X, Y, and ZSM-5 were tailor-made based on the desired application, with excellent properties in catalysis and separation processes. Due to the pure microporous network and the limitations in intracrystalline diffusion, research was later diverted to the development of more open structures with additional porosity and improved diffusional properties (Groen et al., 2003). Materials like pillared layered aluminosilicates and aluminophosphates and ordered mesostructured silica were prepared with the ability to allow larger molecules to enter the pore system or clusters of smaller molecules to be created inside the pores. Recently, the rapid development of the Metal Organic Frameworks was directly related to their ability to be prepared uniformly in a wide range of pore size and porosity without changing their underlying topology (Furukawa et al., 2013). Therefore, new porous hosts to store, manipulate, and react to a wide variety of molecules have emerged.

In addition to porosity, the state of the atoms in the solid that create the space can be important. For example, some materials can adsorb organic pollutants from water and repel water, while atomic isomorphic substitution in the same materials can transform their ability to adsorb water from organic solvents. Therefore, the spatial organization and arrangement of the atoms and functional groups in the solids create their tendency to be hydrophilic (adsorb or be wetted by water) or hydrophobic. The water affinity of a material can be anywhere between the hydrophilic start to the hydrophobic end,

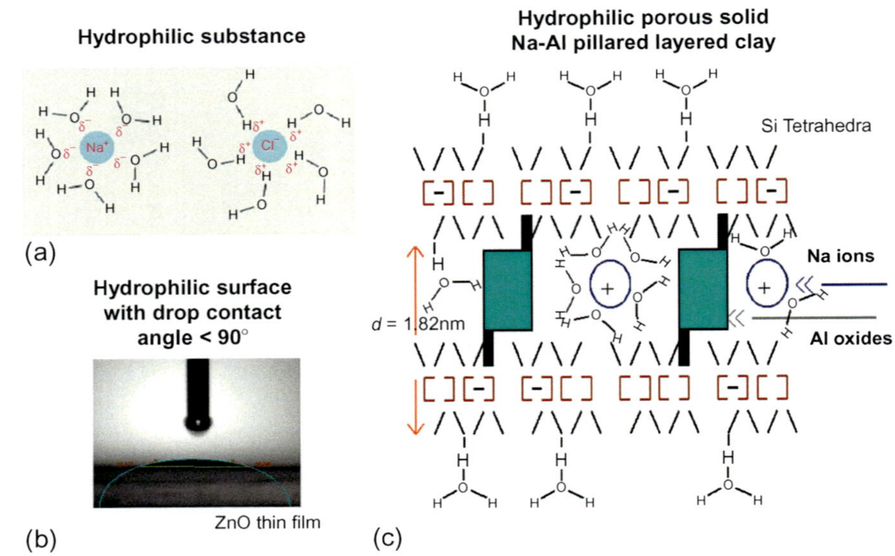

Figure 10.2 (a) Hydration of sodium chlorine ions as an example of a hydrophilic substance. (b) Contact angle measurement with the hydrophilic ZnO thin film. (c) Water vapor adsorption of hydrophilic polar ions and hydrophilic terminated sites in porous pillared layered clays.

while both sites or sites with switchable hydrophilicity to hydrophobicity and vice versa can be present. Historically, the term hydrophilic (from the Greek word "ydrofili") was used to describe the solubility of a substance in water (Figure 10.2a). However, the use of the term has been extended to describe the interaction of the boundary layer of a solid phase with liquid or vapor water. When the boundary is an exposed surface of a thin film, the measure of hydrophilicity is a contact angle defined as the tangent angle between the air, surface, and a water droplet placed on it (Figure 10.2b). The contact angle is related to the three interfacial tensions: solid–vapor, γ_{sv}; solid–liquid, γ_{sl}; and liquid–vapor, γ_{lv} through the equilibrium in Young's equation (Kwok and Neumann, 1999)

$$\gamma_{lv} \cos \Theta_Y = \gamma_{sv} - \gamma_{sl} \tag{10.1}$$

where Θ_Y is the Young contact angle and γ_{sv}, γ_{sl}, and γ_{lv} are the three interfacial tensions.

The contact angle should be measured several times with the sessile-drop technique, and the average angle should be determined together with their standard deviation (Drelich et al., 2011). If the contact angle is less than 90°, the surface is hydrophilic and attracts water. For angles larger than 90°, the surface is hydrophobic and repels water. However, the technique depends on the material structure (film, powder, or bulk specimen with a flat surface), state (dry or receded), and the methodology to deposit the liquid (static or dynamic, rate of liquid movement, or drop volume).

Furthermore, the free energy of hydration ΔG_{sl}, defined as

$$\Delta G_{sl} = \gamma_{sl} - \gamma_{sv} - \gamma_{lv} \tag{10.2}$$

was proposed to be used as the absolute measure of hydrophilicity and hydrophobicity of both molecules and condensed phases (Van Oss, 1994). In this definition, a value of -113 mJ m^{-2} was determined as a cutoff between hydrophilic and hydrophobic materials, which corresponds to a contact angle $\theta \approx 56°$. Therefore, solids with contact angles higher than 56–60° should be considered weakly hydrophobic (Drelich et al., 2011).

In the thermodynamics context of hydrophilicity as an attractive interfacial interaction, hydroscopic solids that adsorb water vapor from their surrounding environment are considered hydrophilic (Figure 10.2c). These solids can be either porous polar substances that attract water molecules through dipole–dipole interaction (hydrogen boning) or nonpolar molecular sieves that attract the water molecules through weak interactions of the van der Waals type. In the first, the polarity is usually due to unreacted or created OH groups like silanols or extra-framework/interstitial cations. Molecular sieves can show high water capacity and, although hydrophobic at low pressures, the initial weak interaction is being followed by water clustering through water cohesive forces and hydrophilic hydrogen bonding.

In view of the above, the hydrophobicity of a porous solid can be also estimated as an index (Ng and Mintova, 2008)

$$\text{Hydrophobicity index} = \frac{\text{Weight loss at } 150\,°\text{C}}{\text{Weight loss at } 400\,°\text{C}} \tag{10.3}$$

Since the loosely bound water will be desorbed at lower temperatures, a hydrophobicity index close to 1 is expected for a highly hydrophobic material. However, this index does not always characterize the water-solid interaction for the different categories of porous materials. In contrast, the hydrophilicity definition through the water vapor adsorption isotherm (Section 10.3) is considered the more accurate method for the affinity of a porous solid to adsorb water.

10.2.2 Micro-, meso-, and macroporous materials

In evaporative cooling of roofs with porous materials, the selection of the appropriate solid is influenced by several parameters:

- High water vapor capacity
- Fast moisture sorption at night
- Stable behavior under thermal and hydrothermal cycles
- Resistance to weathering
- Low price and local availability
- Environmental nontoxicity and easy operation
- Easy preparation into required shape and size for roof application
- Multifunctional ability for CO_2 and pollutants

Porous inorganic solids, which combine many of the above properties, have been extensively studied for water vapor adsorption in similar applications, such as adsorption cooling systems, desiccators, heat storage, and heat pumps. The inorganic porous materials with the prospect of being used in a roof evaporation cooling application can be grouped in three categories. Natural adsorbents and industrial by-products in the first group (e.g., silica sand, volcanic ash, pebbles, siliceous shale), hydrophilic porous materials like silica, zeolites, and clays in the second, and more complex structures like modified clays, mesoporous molecular sieves, porous pillared heterostructures, and metal organic frameworks in the third group. Moving from one category to the other, the preparation complexity and the associated costs increase rapidly. Furthermore, many of the highly hydrophilic materials are not practical yet for large-scale commercial use because of their preparation in very small laboratory quantities or fabrication by almost-impossible-to-upgrade processes. In this context, a brief description on the hydrophilic properties of representative materials in each of the above categories will be presented, with a particular emphasis on the materials tested for building integrated evaporative cooling.

10.2.2.1 Natural materials

Among the most studied natural materials for roof evaporative cooling applications is unconsolidated sand (Dos Santos and Mendes, 2013; Liu et al., 1997; Meng and Hu, 2005; Wanphen and Nagano, 2009). Sand is a low-cost, durable, and locally available material in most of the world. It is mainly composed of silica and secondary phases like calcium carbonate. The sand macroporosity is between 0.25 and 0.5, with a typical value of 0.28, while the specific heat is 800 J/kg K and the moisture isotherm less than 3 (% vol.) up to 99% of relative humidity (Dos Santos and Mendes, 2013). Although nonporous, its packed bed creates macroporosity and when filled with water and heated by the sun, a significant temperature difference can be observed between the bottom and the upper layer of the bed due to water evaporation. In the bed, the driving forces for water flow are capillary and moisture diffusion, which will be analysed in detail in Section 10.4.2. The cooling effect has been found to be better than the effect of a pure water layer due to the higher effective surface area and the siphon potential in the sand layer that enhances the moisture diffusion (Meng and Hu, 2005).

The studies on porous natural materials that exhibit a structural pore size distribution are quite limited. Wanphen and Nagano (2009) investigated the effectiveness of the mesoporous siliceous shale (diatomaceous earth) and the macroporous volcanic ash for roof solar cooling compared to nonporous materials. Siliceous shale with 90% silica had a high porosity (64%), an average pore diameter of 9.4 nm (extended in the entire mesoporous region), and high water vapor adsorption ability (0.242 g/g for relative humidity 60–90%) due to a pore volume of 0.33 cm^3/g. The application of these natural mesoporous materials with added liquid water in either wind tunnel or field experiments was found to lower surface temperature in comparison to mortar concrete. The siliceous shale's ability to adsorb water vapor during the night was mentioned as an additional possible factor contributing to lower surface temperature. Also, the particle size influenced the cooling mechanism, with larger particles working better under solar radiation due to voids ventilation and evaporation enhancement as the siphon potential in the unconsolidated sand.

10.2.2.2 Silica gels

Silica gel is among the most studied materials for water vapor adsorption. It is a partially dehydrated form of polymeric colloidal silicic acid that can be also produced as an aggregate of silica particles (Ng and Mintova, 2008). During the hydrolysis and polymerization of silicon alkoxides, a siloxane network is created with unreacted silanol groups on its surface. The latter contribute to the important hydrophilicity of silica gel. Depending on the preparation conditions, silica gels of different surface area, pore volume, and particle size can be prepared. In all, the structure is amorphous and micro-/meso- porous, with pore apertures from 0.3 to 6 nm. Due to these, silica can adsorb up to 40% of its weight in water. Therefore, silica gels are being commonly applied as desiccants for moisture adsorption. Due to the strong attraction of the water molecules from the silanol groups, temperatures up to 120 °C are needed for water desorption and regeneration.

10.2.2.3 Zeolites

Zeolites are an important class of materials for energy storage, heat-pump application, and solar refrigeration. They are mainly microporous, hydrated aluminosilicates (SiO_4 and AlO_4 tetrahedra) that are found as natural minerals or synthesized in a great number of different forms. The general chemical composition of a zeolite is: $M^{n}{}_{x/n}Si_{1\text{-}x}Al_xO_2 \cdot yH_2O$ with M = e.g. Na^+, K^+, Li^+, Ag^+, NH_4^+, H^+, Ca^{2+}, Ba^{2+}...

Their porosity can be determined from their crystalline structure, because the pores are arranged in a regular way with a limited number (usually one) of district pore sizes of molecular dimensions. Four groups of pores can be categorized (Ng and Mintova, 2008): ultra-large (more than 14-membered ring), large (12-membered ring), medium (10-membered ring), and small (eight-membered ring) pores. In the zeolite structure, the negative framework charge, which is due to aluminium isomorphic substitution, is compensated by the usually exchangeable counterions in the cavities. Therefore, zeolites are highly hydrophilic materials, and in their voids and pores, water molecules can be found either as a hydrated sphere of the exchangeable cations or else they are attracted by OH surface groups. The amount of adsorbed water is usually the measure of zeolites' porosity, and it can be removed by heating. Due to the strong interaction between the charged framework and the water molecules, high temperatures (>200 °C) are needed for complete water removal. These temperatures are suitable for solar heat energy systems but limit the potential zeolite's application in building integrated solar cooling.

10.2.2.4 Clays

Clays are mostly negatively charged phyllomorphous (sheet-like) aluminosilicates, with hydrated interlammelar cations to neutralize the structure and variable amount of water. Due to their rich natural abundance, clays have been used in a large number of industrial and environmental applications. The main clay groups are vermiculite, illite, smectite, and kaolinite, while their structure is mainly composed of Si tetrahedral and Al, Mg, Fe octahedral. Depending on type and size, clay mineral particles may consist of different stacks of lamellae, creating aggregates with pores in all three categories. The clay-water interaction has been extensively studied in different clay groups like smectite and vermiculite, and is evolving through successive stages. At the first stage, the interlammelar cations are hydrated with one layer of water molecules. Upon humidity increase, more hydration layers are formed around the cations that cause the swelling of the structure and increase in the interlammelar space. Further increase to humidity leads to capillary condensation in the pores. On heating, water is driven off and the porosity changes.

In addition to the majority of swelling clays, a nonexpandable clay of particular importance is sepiolite, with pores in the micro-, meso-, and macro- porous region. Sepiolite is a microfibrous, hydrated magnesium silicate with a half-unit cell formula $Si_{12}O_{30}Mg_8(OH,F)_4(OH_2)_4 \cdot 8H_2O$ (Figure 10.3a). It consists of blocks of octahedral magnesium sheets as layers between two tetrahedral silica. SiO_4 tetrahedra are periodically interchanged and determine a regular discontinuity of the sheets, which is the origin of the structural cavities (tunnels) extended along the axis of the microfibers with the silanol groups (Si-OH) (Ruiz-Hitzky et al., 2008).

As a multiporous material, the low-cost and natural phyllosilicate adsorbs the highest amount of water among the clay minerals. Up to 200–250% of its own weight in water can be retained (Karamanis et al., 2011). The ribbons are arranged in staggered rows separated by microporous channels, although larger pores ($d > 0.2$ nm) are also present between fibers. The specific macropore volume (*V*ma) and mesopore volume (*V*me) of the sepiolite have been estimated from mercury intrusion curve as 0.45 cm^3/g and 0.20 cm^3/g, respectively. Nitrogen adsorption-desorption have revealed micropore volume (*V*mi) values of 0.16 cm^3/g and mesopore of 0.37 cm^3/g, with a mean micropore radius of 2 nm and mesopore of 25 nm (Figure 10.3b) (Onal et al., 2008). The total pore volume has been estimated to be around 0.82 cm^3/g from mercury porosimetry and nitrogen adsorption isotherms. In its structure, there are at least three main sites associated with water sorption: the external surfaces, microporous channels, and interfiber micropores and larger pores. In the region of low relative pressures, water is preferentially adsorbed in the highly hydrophilic sites, with the formation of a monolayer followed by multilayer sorption at higher relative pressures and finally by capillary condensation at the zeolitic channels. Detailed analysis of the water vapor adsorption isotherm is presented in the next section.

10.2.2.5 Pillared layered clays

Pillared layered clays are porous materials with a two-dimensional zeolite-like structure, prepared by propping apart the lamellae of expandable layered clays with nano-size pillars of metal polyoxocations (Figure 10.4).

Upon calcination, the hydrated cations are converted to oxide pillars with the creation of micropores. As a result of the dehydroxylation of the polyoxocations,

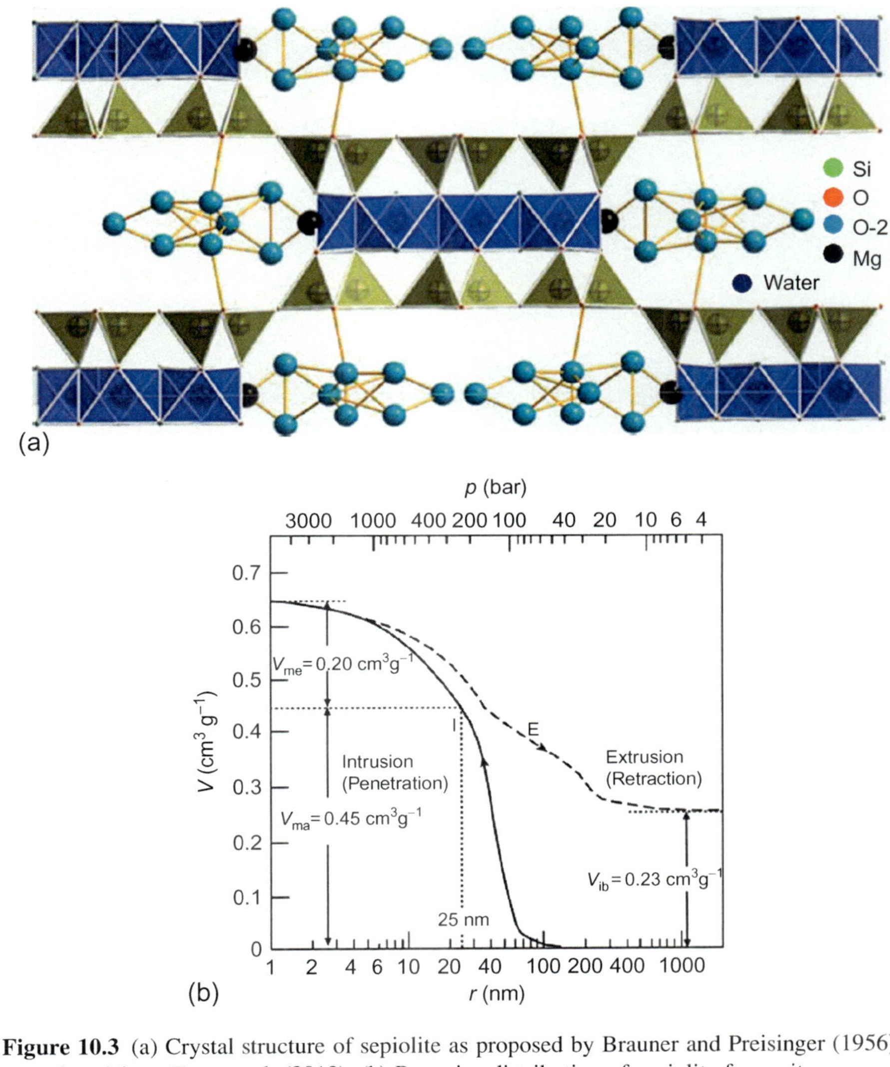

Figure 10.3 (a) Crystal structure of sepiolite as proposed by Brauner and Preisinger (1956) reproduced from Tang et al. (2012). (b) Pore size distribution of sepiolite from nitrogen adsorption–desorption isotherms (r: pore radius; V_{mm}: specific micro-mesopore volume; V_{mi}: specific micropore volume; V_{me}: specific mesopore volume).
(From Onal et al. (2008). Reproduced with kind permission of The Clay Minerals Society, publisher of *Clays and Clay Minerals*).

H_3O^+ ions are liberated in the clay interlayer space and thus the cation exchange capacity of the prepared material is substantially lower than that of the starting clay. Treatment of pillared clay with a base like NaOH or NH_4OH neutralizes the H_3O^+ ions, and this in turn leads to a cation exchange capacity (CEC) restoration of up to 80–90% of the CEC value of the starting clay (Karamanis et al., 1997).

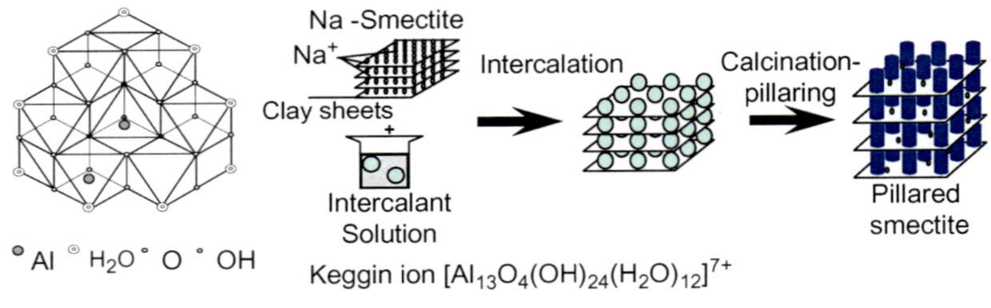

Figure 10.4 Aluminium pillaring of smectite with intercalation of the Keggin ion and calcination.

Therefore, hydrated cations of zeolitic type can be exchanged in the interlayer space in addition to water attracted from structural OH groups (Figure 10.2c). Typical surface area values of 300 m^2/g are found, with micropore volumes of 0.10–0.15 cm^3/g. Depending upon the preparation and drying conditions, extended mesoporosity and macroporosity can also be observed (Figure 10.1). More specifically, voids formed between tactoids as well as between aggregates remain practically intact with freeze-drying of the pillaring precursor (Figure 10.1). Therefore, substantial mesoporosity (up to 75% of the total pore) can be created in addition to microporosity (Karamanis et al., 1997) and can increase the pore volume to more than 0.35 cm^3/g (Bineesh et al., 2012).

The potential use of pillared clays as catalysts and molecular sieves is being extensively investigated, with only a few studies in water adsorption (Sychev et al., 2001; Yamanaka et al., 1990; Zhu et al., 1995). However, pillared layered clays present several advantages as water adsorbents since the preparation is quite simple, from abundant raw sources, and can be scaled-up. The large microporosity can enhance the adsorption at low relative pressures, followed by increased adsorption at the mesopores, exhibiting good thermal and oxidation stability for outdoor use (Zhu et al., 2008; Ng and Mintova, 1995).

10.2.2.6 Mesoporous molecular sieves

Since the main pore sizes of zeolites and pillared clays with well-ordered structures were limited in the microporous region, materials with larger pores were prepared by using self-assembled aggregates of molecules or ions as templates for aluminosilicate gels (Kresge et al., 1992). In this way, ordered mesoporous molecular sieves (MMS) were synthesized, opening up a new direction in material science and technology. By varying the surfactants and the preparation conditions, a great variety of final ordered structures have been produced, with remarkable properties for several applications (Taguchi and Schuth, 2005). MCM-41 is the most common and well-studied MMS and can be prepared from hydrothermal treatment of a silica source (e.g., tetraethylorthosilicate) and a template as cetyltrimethylammonium bromide followed by slow calcination (Figure 10.5; Dimos et al., 2009). The prepared hexagonal arrays are highly ordered, with a diffraction pattern of glass-like silicate walls; uniform

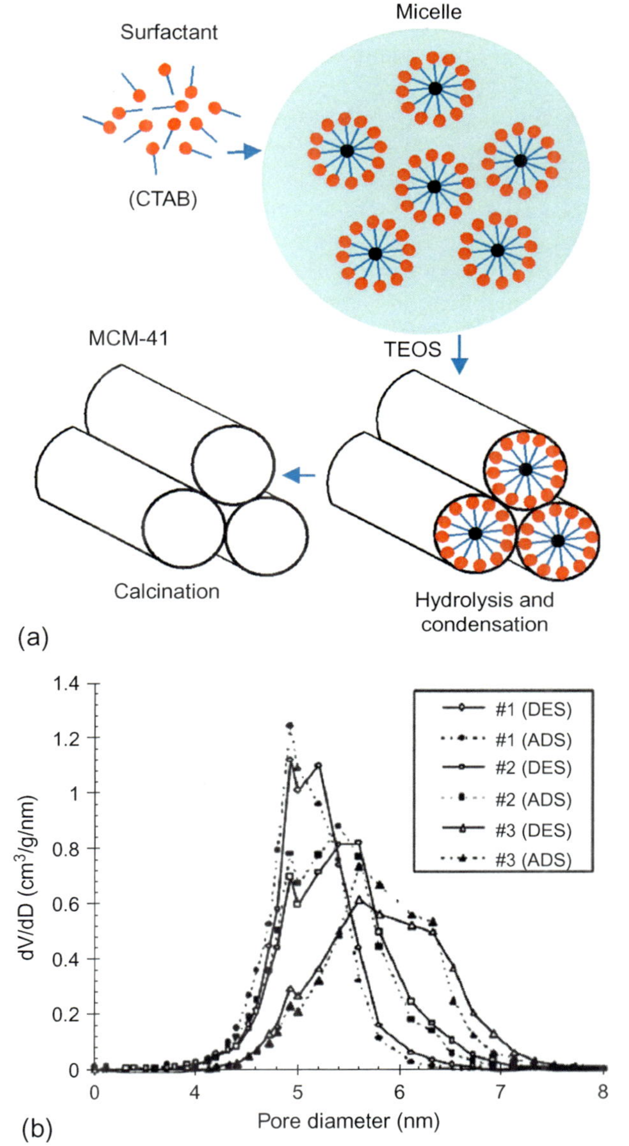

Figure 10.5 (a) Preparation of MCM-41. (b) Pore size distribution of enlarged MCM-41 type materials.
Neimark and Ravikovitch (2001).

cylindrical pores with diameters between 1 and 10 nm; and large surface areas (>500 m^2/g), pore volumes (>0.6 cm^3/g), and number of silanol groups. Due to the extensive mesoporosity and the silanol groups, a large number of water molecules can be adsorbed and confined within the cavities, with interesting and exploitable macroscopic properties.

10.2.2.7 Metal organic frameworks

Due to the need to engineer tailored building blocks with the desired properties for the specific application, the MOFs (metal organic frameworks) materials have emerged through the connection of rigid rod-like organic moieties with inflexible inorganic clusters (Rowsell and Yaghi, 2004). The organic units are ditopic or polytopic organic carboxylates (or other negatively charged molecules), while the secondary building units are metal clusters like Zn_4O tetrahedron. Since the potential combinations of inorganic and organic units are vast, more than 20,000 different MOFs have been reported and studied within the past decade (Furukawa et al., 2013). The structures are highly crystalline and robust, with unique characteristics among crystalline porous materials like the permanent porosity of greater than 50% of the MOF crystal volume. Surface areas range from 1000 to 10,000 m^2/g and volumes can be 10 times higher than zeolites and mesoporous sieves (up to 4.4 cm^3/g for the compound NU-110) (Furukawa et al., 2013). In the MOFs preparation, the solvent itself acts as the main template, while time scale of the hours to days can be used, solvent can be recycled, and the facile quality can be controlled with monitoring by PXRD or microscopy. Their pore aperture is typical up to 2 nm, but with the use of infinite rod-shaped secondary building units with linkers of arbitrary length providing periodicity in the other two dimensions, openings of 9.8 nm have been reached. MOFs are currently being tested for a variety of energy applications like methane uptake and hydrogen storage, gases adsorption and catalytic processes, sensors with optical or magnetic properties, and other applications (Furukawa et al., 2013). Due to these advantages, there are already examples of highly porous MOFs that have been prepared in scaled-up amounts, and several MOFs are already commercially available. However, there are still open matters that need to be resolved, like the comparatively weak thermal and especially hydrothermal stability of MOFs, which is considered a critical issue in water vapor heat pump systems (Henninger et al., 2012).

10.3 Water vapor adsorption on hydrophilic porous materials and their solar interaction

10.3.1 Water vapor adsorption

Water is a fascinating molecule that interacts with other water molecules through dipole–dipole interactions and creates hydrogen bondings. Therefore, it freely exists in nature in three phases (ice, liquid, and vapor). Although it is simple in form, water exhibits some unusual and unique properties: large surface tension, high boiling temperature, and high latent heat of phase changes due to the strong hydrogen bond network. By moving from one phase to the other, its structure is changed from water dimers in the vapor phase, to flickering clusters of molecules with 3–4 hydrogen bonds separated by regions with 2–3 hydrogen bonds in the liquid phase, and eventually up to icosahedral water clusters (Figure 10.6). Further temperature reduction leads to the phase of ice with its many structures, including the two naturally existing hexagonal

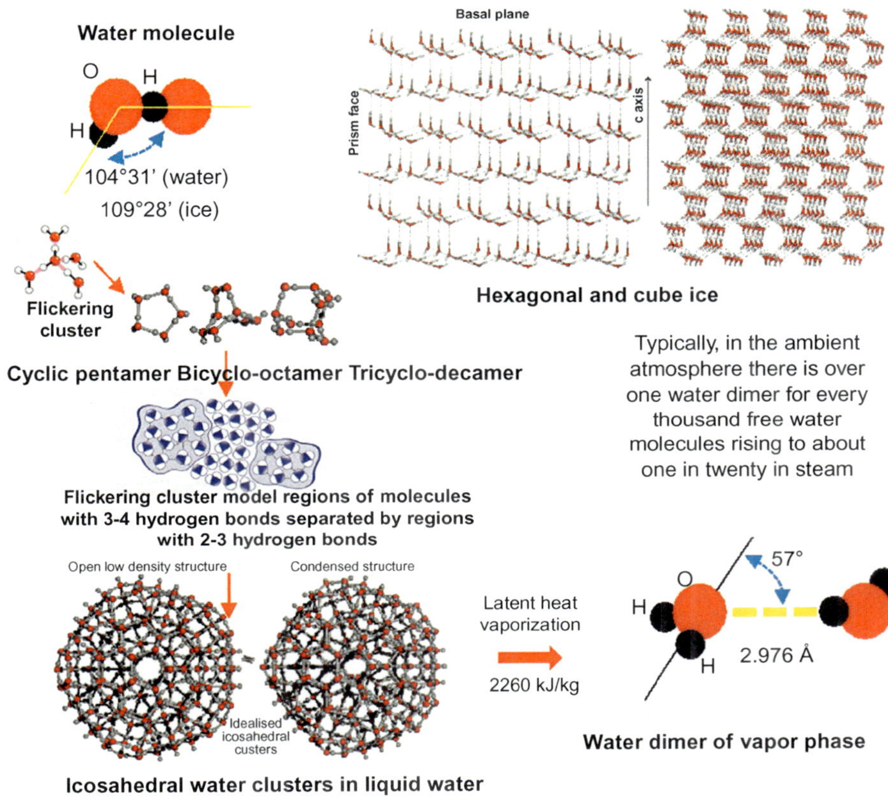

Figure 10.6 Structure of water in the three phases of ice (two representative structures that exist in nature), liquid (model of icosahedral water clusters), and vapor (dimers).
Model pictures from www1.lsbu.ac.uk/water/index2.html after permission from Dr Martin Chaplin.

and cubic structures (Figure 10.6). Phase transitions that take place through environmental heat exchanges shape the Earth's climate and are still under intensive investigation (Makarieva et al., 2013).

When water is adsorbed in a porous material, its properties and phase transitions are strongly affected by the confinement and can be different than those in the free space. This is mainly due to the liquid-pore wall interactions upon water adsorption in the porous matrix and adhesive forces of the water molecule with polar or nonpolar blocks of porous structure, in addition to the liquid–liquid interactions and cohesive forces of free water. Therefore, the isothermal study of the water vapor adsorbed or desorbed on the surfaces of a porous solid as a function of the relative pressure can provide significant information on how the water's heat exchange properties are related to the geometrical constraints and the nature (hydrophilic or hydrophobic) of the porous matrix.

Depending on the driven forces of the water vapor adsorption process, six different isothermic curves are usually observed, categorized from type I to type VI (Figure 10.7). The type I curve is usually observed for materials with high water adsorption capacity and very fast sorption saturation at low partial pressure (P/P_0) due to steric effects, followed by consistent adsorption over a wide range of P/P_0 due to water saturation in the pores. As a rule, the adhesive forces between water molecules and porous walls are much stronger than the water cohesive forces. These adsorbents are classified as very hydrophilic due to high affinity to water even at low P/P_0. The hydrophilic type II isotherms are normally associated with monolayer–multilayer sorption on the nonporous or macroporous surface of a powder. The S-shape is controlled by surface-water interactions and capillary effects in the porous solids. Moreover, hydrophilic materials of different pore size and volume can follow adsorption isotherms of type II or IV, whereby a considerably high sorption capacity of water at low P/P_0 and moderate P/P_0 is measured. A desorption hysteresis is usually observed in type VI as characteristic of mesoporosity and indicating the existence of metastable regions of liquid-like states in the pores. Additionally, the rare type VI step-like isotherm is also characteristic of a hydrophilic sorbent, with stepwise sorption. In contrast, sorption isotherms of types III and V characterize hydrophobic or low hydrophilic materials, with low sorption at low P/P_0, sometimes moderate sorption at the middle P/P_0, and suddenly high water sorption at higher P/P_0. In type V adsorption, which is of interest in building integrated evaporative cooling, the initial low water affinity can be considered from the

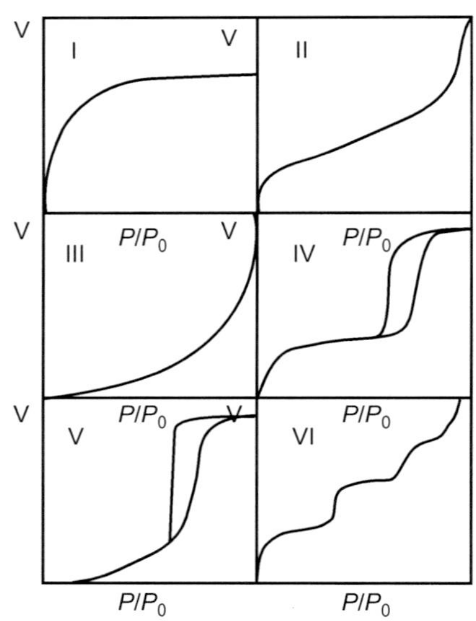

Figure 10.7 Types of water vapor adsorption isotherms. Hydrophilicity is usually associated with types I, II, IV, and VI; hydrophobicity with types II and I; and hydrophobicity/low hydrophilicity with type V.

point of view of a nonpolar or weakly polar framework that resembles a rather hydrophobic porous solid. Due to this hydrophobicity, low thermal temperatures will be needed for water desorption. The step increase in water vapor adsorption for types IV and V is characteristic of water vapor capillary condensation in the pores and is responsible for mesopore and macropore filling. The equilibrium vapor pressure that the condensation is expected to occur can be calculated from the Kelvin equation (Mitropoulos, 2008)

$$\ln \frac{P}{P_0} = -\frac{2\gamma V_\mathrm{L}}{RTr_m} \tag{10.4}$$

where P is the equilibrium vapor pressure in the porous solid, P_0 is the saturated vapor pressure of the liquid, γ is the surface tension and V_L is the molar volume of the bulk liquid adsorbate, R is the gas constant, T is the absolute isothermal temperature, and r_m is the mean radius of curvature of the liquid/gas interface. The latter for a cylindrical pore is given as the ratio of the pore radius to the cosine of the water contact angle (which takes the value of 1 at complete wetting) at the liquid/solid interface. Since the condensation and water molecules clustering are associated with a phase change, heat is released during the water adsorption or is absorbed during desorption. This heat can be utilized in the building integrated evaporative cooling. In addition, the equilibrium vapor pressure in the Kelvin equation is associated with the radius of the water drop inside the porous solid. The curved surface of the drop can be accounted as an added energy in the system (related to the work required to bring the free drop at the meniscus) and due to this, less energy will be needed to convert it to vapor than an identical drop from a plane surface (Johnson, 2002). For weak interactions between water and solid walls, and depending on the temperature and pressure differences, less required energy will result in lower desorption temperatures, which could potentially match those usually observed at buildings' roofs.

In order to calculate the thermodynamic data of the water vapor adsorption from an isothermic curve, the Dubinin–Astakhov equation (Karamanis et al., 2011) can be used as

$$q = q_m e^{\left(\frac{RT\ln\left(\frac{P}{P_0}\right)}{E}\right)^n} \tag{10.5}$$

where q is the sorbed vapor at temperature T and relative pressure P/P_0 and q_m the saturation value, while n and E are the temperature invariant parameters and $RT\ln(P/P_0)$ is the adsorption potential ΔG. For type V isotherms, a logistic function of the form

$$q = \frac{q_m}{1 + e^{\left(k\left(b - \frac{P}{P_0}\right)\right)}} \tag{10.6}$$

or an empirical isotherm model (Kim et al., 2014) can be used to describe the adsorption process as

$$q = \beta K_H \left(\frac{P}{P_0}\right) + (1-\beta)\frac{q_m\left(\frac{K_s P}{P_0}\right)^{1/n}}{1+\left(\frac{K_s P}{P_0}\right)^{1/n}} \tag{10.7}$$

$$K = K_0 e^{\left(-\frac{\Delta H}{RT}\right)} \tag{10.8}$$

$$n = A + \frac{B}{T} \tag{10.9}$$

where q and q_m are the absolute and the maximum adsorbed, respectively, P is the water vapor pressure, P_0 is the saturation pressure of water vapor at the adsorption temperature, K is the Henry constant of both isotherm models, K_0 is the pre-exponential factor of Henry constant, ΔH is the isosteric enthalpy of adsorption at zero loading, A and B are the parameters related with the thermal variation of the heterogeneity coefficient, and n is the heterogeneity parameter. The coefficient β can be obtained empirically, and the proposed functional form for this parameter is given as

$$\beta = e^{\left(-\alpha\frac{P}{P_0}\right)} \tag{10.10}$$

where α is a function of temperature and is defined as

$$\alpha = k_1 + k_2\frac{T}{T_{\text{ref}}} \tag{10.11}$$

with T_{ref} the reference temperature (273 K).

In view of the above and in order to evaluate the potentiality of hydrophilic materials for roof evaporative cooling applications, the kinetics and the adsorbed amount of water vapor can be determined by either dynamic vapor sorption or moisture sorption measurements with desiccators and saturated salt solutions. Among natural materials, siliceous shale and diatomaceous earth exhibit substantial meso-porosity and have been studied as water vapor adsorbents (Maeda et al., 2009; Nakabayashi et al., 2011). However, both siliceous shale and diatomaceous earth are hydrophobic, with low sorption capacities up to 70% of relative humidity and increase at very high pressures due to capillary condensation at meso- and macro-pores. Therefore, their evaporative cooling action will be mainly due to the retention of liquid water in macroscopic pores and their capillary porosity as packing beds.

Water vapor sorption on zeolites has been extensively studied and characterized by several techniques and methods. As shown in Figure 10.8a, the isotherms of NaX, NaY are type I and characterize strong hydrophilic solids with monolayer pore coverage from very low relative pressures. Due to these shapes and the high regeneration temperatures shown in the thermogravimetric graph of Figure 10.8b, typical hydrophilic zeolites are more appropriate for solar driven thermal adsorption storage applications than building integrated evaporative cooling.

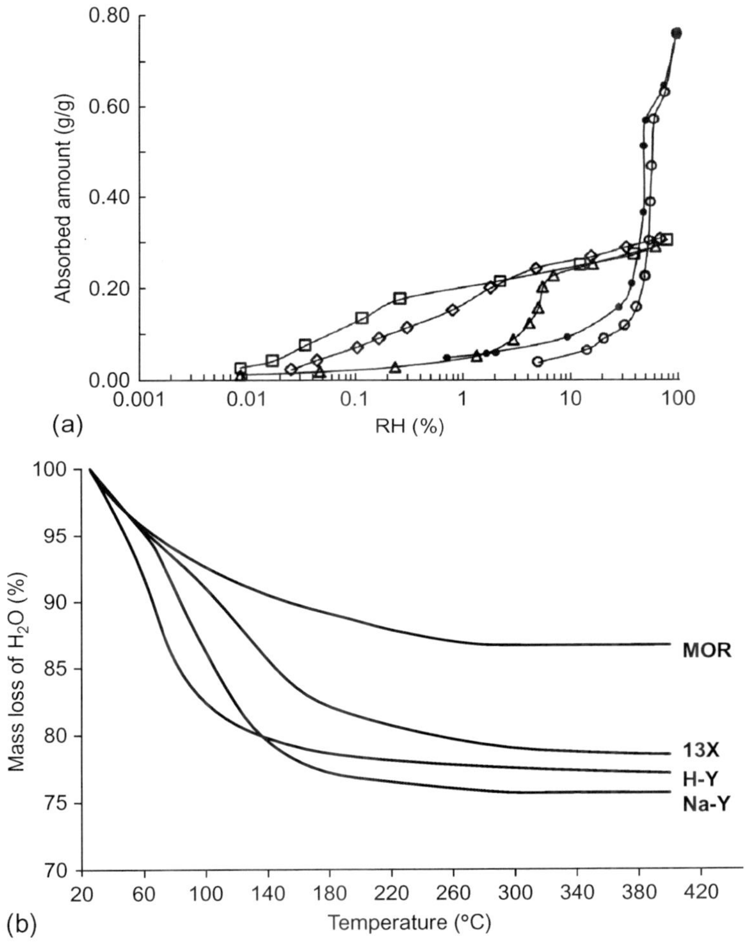

Figure 10.8 (a) Water adsorption isotherms as a function of the relative humidity, RH, for (from left to right): NaX, NaY, SAPO-34, and the silocoaluminophosphate MCM-41. Filled symbols denote desorption (Jänchen and Stach, 2012). (b) TG analysis of the percentage mass loss (due to water) of zeolites NaX, MOR, NaY, and H–Y when undergoing a temperature ramp from 20 to 400 °C at 5 °C min^{-1} (Whiting et al., 2013).

However, the invention of MOFs, with their huge surface area and volume, introduced a new approach of zeolitic pores' affinity for water vapor adsorption. Depending on the nature of the linker and the metal cluster as well the pore size of the MOF structure, water vapor adsorption isotherms of all the types in Figure 10.7 have been observed (Jeremias et al., 2014; Küsgens et al., 2009). Water vapor adsorption on HKUST-1 follows a zeolitic type VI shape, with two initial steps at low pressure due to the two energetically different pores of the solid (big pores first with the presence of chemisorption on free copper sites, and then the small with the more hydrophobic nature). Saturation is reached at P/P_0 around 0.4 and is kept up to the last increase due to condensation at 0.9–1.0 (adsorbed amount around 0.7 g/g) (Küsgens et al., 2009). MIL101 is a mesoporous MOF and exhibits a type V isotherm. The increase step is shown at $P/P_0=0.4$ with the same hydrophilicity for both the cages, with sizes 2.9 and 3.4 nm (Küsgens et al., 2009). Due to the materials' stability, high porosity (SBET > 3000 m^2/g), and water vapor loadings up to 1.2 g/g, MOFs of the MIL101 type could be considered appropriate for heat transformation applications, and further investigation is needed.

In contrast to the recent development of MOFs at lab scale, pillared clays have been studied for more than 25 years,and their preparation at industrial level increases their suitability for large-scale applications. Starting from the raw clay, water adsorption in Na-montmorillonite proceeds through a type II isotherm, with increased uptake for relative pressure over 80% (Figure 10.9a). The sorption of water vapor on clays follows different stages. At low pressure ($P/P_0<0.1$), sorption starts at the external clay surfaces and is followed by the hydration of the exchangeable cations as pressure increases ($0.1<P/P_0<0.5$) (Grim, 1968). At $P/P_0>0.5$, higher hydration layers are formed with the resultant swelling of the clay lamellae for expandable clays like montmorillonite and kaolinite. The ionic radius of the exchangeable cations as well as the charge distribution on the clay octahedra are among the parameters influencing the multistep water vapor adsorption on clays (usually of type VI). The insertion of polyoxocations (i.e., Keggin ion) in the interlammelar space of expandable clay transforms it into a zeolitic hydrophilic material with a type I isotherm (Vardoulakis et al., 2011). After calcination, restoration of the cation exchange capacity, and creation of (meso- and macro-) porosity, hydrophilic structural sites of strong water–solid interactions drive the water adsorption at low P/P_0. Upon monolayer covering of the microporous sites, water will be adsorbed at the mesoporous and macroporous volume for higher P/P_0, following a multilayer covering of type II (Figure 10.9a). Depending on the preparation conditions, water adsorption can be up to 0.18 g/g at the hydrophilic sites covering around $P/P_0=0.5$ and can reach maximum adsorption of 0.3 g/g at higher relative humidity (Malla et al., 1989). In this frame, the pillared clays combine the strong hydrophilic behavior of zeolites at low relative humidity and the increased water vapor adsorption quantities of mesoporous materials at higher pressures. Recent studies also indicate that their pore size can be tailored by varying the pillar density with a subsequent increase in the pore diameter and volume (Herling et al., 2012). Furthermore, pillared clays can be scaled up through the preparation of larger grain sizes with additional capillarity using retained liquid water.

The same type II behavior in water vapor adsorption is also followed by the sepiolite clay (Figure 10.9a). At low P/P_0, water is adsorbed in the highly hydrophilic zeolitic sites and the maximum amount competes well with the corresponding silica gel (type I isotherm in Figure 10.9a). At higher pressures, monolayer forming is followed by multilayer adsorption and eventually by capillary condensation in the zeolitic channels (Karamanis et al., 2011). In addition, the water vapor kinetics of both pillared clays and sepiolite are comparable to silica gel for the time frame of night in order to be fully covered for building integrated solar cooling (Figure 10.9b). These properties of sepiolite and pillared clays, combined with their availability and affordability, make these materials very attractive for solar cooling applications.

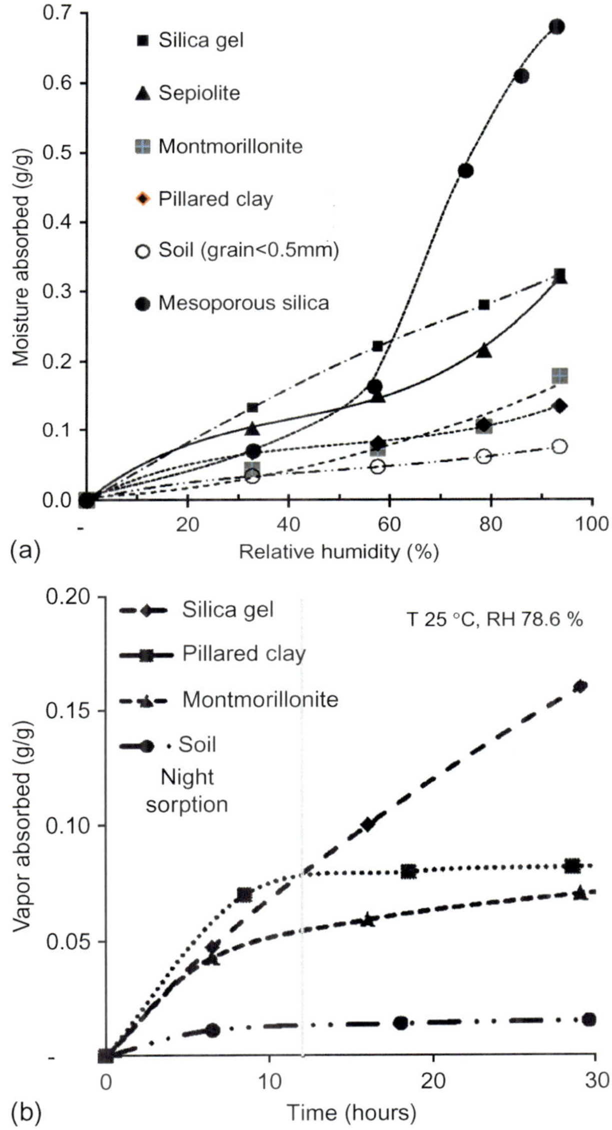

Figure 10.9 (a) Water vapor adsorption isotherms of silica gel, sepiolite, Na-montmorillonite, Al-pillared montmorillonite, mesoporous silica, and soil at 25 °C. (b) Water vapor isothermic adsorption kinetics of silica gel, Na-montmorillonite, and Al-pillared montmorillonite at 25 °C.

Towards the increase of the water vapor capacity of porous solids, mesoporous silica is a porous material candidate with exceptional properties. It is a well-ordered structure with cylindrical arrays but shows hydrophobic or low hydrophilic behavior with type V adsorption isotherm and weak vapor-pore walls interactions (Figure 10.9a). However, there are silanol groups (single, $(SiO)_3Si\text{-}OH$; hydrogen-bonded,

$(SiO)_3Si$-OH-OH-$Si(SiO)_3$; and geminal, $(SiO)_2Si(OH)_2$) on its surface, but with a lower concentration (2.5–3.0 OH/nm^2) in comparison to other silicas (<6 OH/nm^2) (Zhao et al., 1997). Due to the few silanol groups in a well-ordered structure with a very large surface, water is adsorbed through weak interactions at low P/P_0, followed by water clustering and capillary condensation around 0.4–0.6 of the relative pressure. Therefore, the amount of water adsorbed on mesoporous silica increases through hydrogen bonding and reaches values of 0.6–0.8 g/g.

In order to gain further insight into the suitability of hydrophilic porous materials for roof evaporative cooling applications, the removal of water from the porous host as a function of temperature can be evaluated by thermogravimetric measurements. Such results are presented in Figure 10.10 for (a) Al-pillared montmorillonite, (b) sepiolite, and (c) mesoporous silica.

In the thermographic curve of Al-pillared clay, there are two major endothermic peaks of temperature where the physisorbed water is released from the modified air-dried or freeze-dried clays (Figure 10.10a). The first temperature is shown at around 32 °C and corresponds to the surface moisture removal, while the second is around 56.8 °C and corresponds to the desorption of interlammelar water. It is evident that higher temperatures are needed for the complete desorption of the physisorbed water due to the strong hydrophilic sites. In sepiolite, the desorption of surface and zeolitic water is completed at around 100 °C, with an overall mass reduction of 11.6% for samples stored at room atmosphere (Karamanis et al. 2011). According to the DTG deconvolution analysis, there are three weight loss intervals: reduction due to the removal of loosely bound water vapor around 20–30 °C, 4.2% water loss from the external surfaces at 30–60 °C, and 6.5% from water removal from the channel-like voids in the structure at 60–100 °C (Karamanis et al., 2011). A similar thermogravimetric behavior is observed in the well-ordered mesoporous silica, with one major endothermic peak of temperature around 55 °C where the physisorbed and condensed water (after storing the sample at 93% of relative humidity) is released from the pores (Figure 10.10c). In the proposed application of materials integration in building surfaces, the temperature rise of an absorbing building surface with 0.05 reflectance can be about 34–50 °C warmer than the ambient air in full sunlight (Karamanis et al., 2013). Therefore, well-ordered mesoporous structures exhibit higher sorption capacity and lower desorption temperatures than the corresponding properties of pillared clays and sepiolite. However, the low cost of natural and modified clays and the tunable capability of the latter point to their further research optimization towards the direction revealed from the mesoporous solids.

10.3.2 Solar interaction

Due to the need to implement the hydrophilic porous materials outdoors, the interaction of the solids with solar radiation is a parameter that needs to be analysed and optimized. In general, porous solids with water and condensed vapor are penetrated by radiation to some depth, with the infrared part of the solar spectrum to interact within a surface film of thickness depending on the porous material composition. The interaction of the

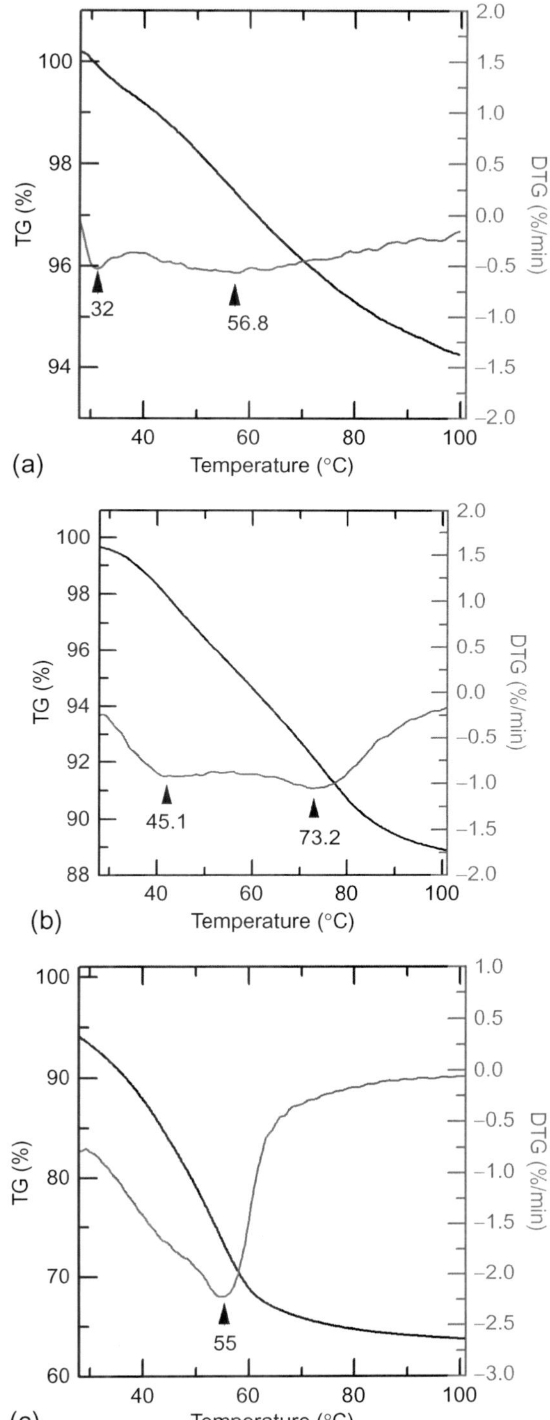

Figure 10.10 Low temperature TG and DTG curves of the (a) Al-pillared montmorillonite stored at room atmosphere, (b) sepiolite stored at room atmosphere, and (c) mesoporous silica after overnight storing at 93% relative humidity.

nonreflected solar radiation may induce changes in the electronic, vibrational, and rotational states of atoms and molecules. Porous materials with moisture in the liquid or vapor phase exhibit changes in the electronic state at wavelengths in the range of 0.2–0.7 μm (ultraviolet and visible rays); changes in the vibrational state that correspond to wavelengths in the range of 0.7–100 μm (IR); and changes in the rotational state that correspond to wavelengths above 100 μm (microwaves). Due to these changes, surface radiation exposure leads to energy absorption, heat generation, and increased material temperature. In the first region of electronic state change, the selection of appropriate colored compounds can increase the scattering of light at the solid particles' surface (reflection) and reduce the increase of the surface temperature.

Moreover, the near infrared region of the electromagnetic spectrum (0.7–2.5 μm) accounts for more than 45% of the surface impinging radiation and strongly influences the heat generation inside the porous material. An appropriate strategy to keep the surfaces "cool" is adding scattering spheres in the form of inorganic pigments (particles of titanium oxide) in order to increase NIR reflectance. This direction has resulted in the development of "cool" materials with the advantages and disadvantages discussed in Section 10.1. Moreover, the NIR region also involves the response of the molecular bonds (stretching and bending) O—H, C—H, C—O, and N—H (Cen and He, 2007). Among these bonds, O—H of water can be in excess within porous hydrophilic materials at will and interact with the surface impinging solar radiation, transforming the radiation to sensible and latent heat for phase changes.

In Figure 10.11a, the absorption spectra of the mesoporous silica samples after the adsorption of water vapor at different humidities are shown. The optical characterization of the samples with predetermined adsorbed water vapor was conducted by a UV–VIS–NIR spectrophotometer (fitted with an integrating sphere that collects both specular and diffuse radiation) over the solar spectrum (200–2500 nm) (Karamanis et al., 2013). The ASTM G173-03 reference spectrum can be used to normalize the data. According to the spectra, four main maxima located at c. 970 nm, c. 1190 nm, c. 1450 nm, and c. 1930 nm are clearly shown in the NIR region of the spectrum for the mesoporous silica sample pre-equilibrated at 75% relative humidity, indicating the water vapor condensation within the mesopores. These maxima correspond to the second overtone of the OH stretching band ($3\nu_{1,3}$), the combination of the first overtone of the O—H stretching and the OH-bending band ($2\nu_{1,3}+\nu_2$), first overtone of the OH-stretching band ($2\nu_{1,3}$), and combination of the OH- stretching band and the O—H bending band ($2\nu_{1,3}+\nu_2$), respectively (Büning-Pfaue, 2003). In the mesoporous silica sample pre-equilibrated at 33% of RH, the intensity of the maxima is much lower due to the absence of condensed water and the reduction of adsorbed vapor. By normalizing the absorption data with the solar spectral irradiance ASTM G-173, an increase of the NIR to the total absorption ratio from 31% to 42% was calculated for the two samples at 33% and 93% RH, respectively. The same behavior is also observed with the sepiolite material (Figure 10.11b), although higher temperatures are needed to remove all the water from the pores.

In addition, the creation of titanium oxide on the surface of the mesoporous silica increases the solar absorbance in the UV region due to the semiconducting properties

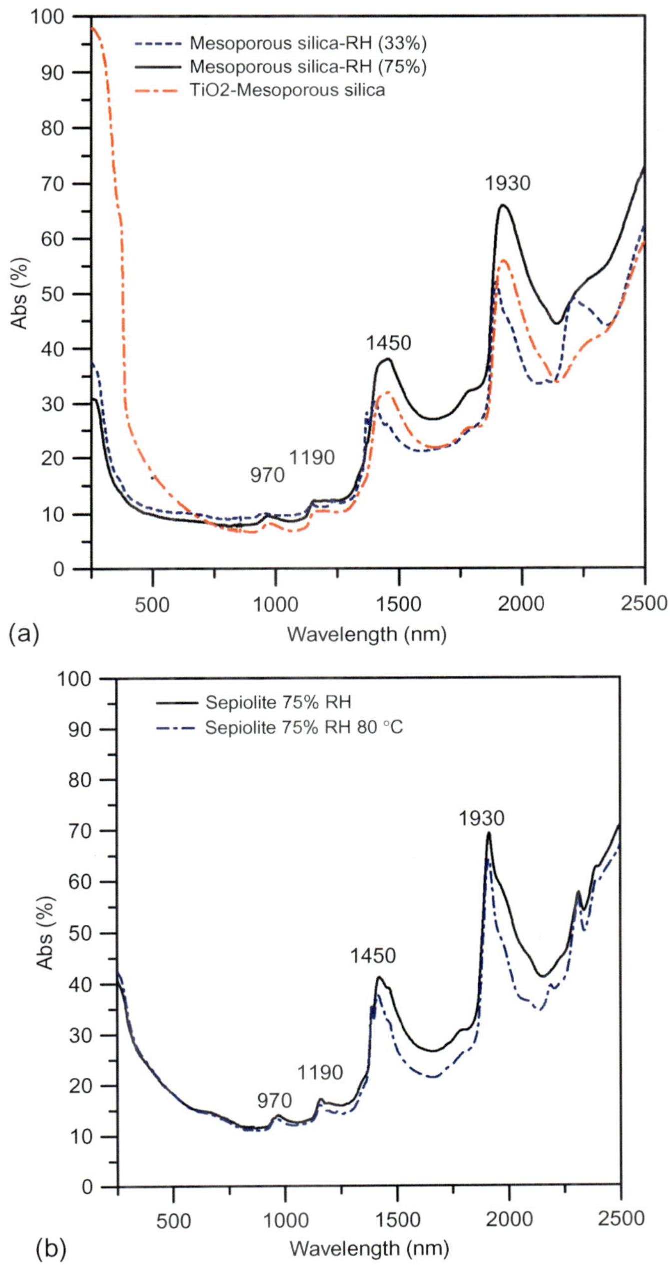

Figure 10.11 (a) Comparison of absorbance spectra of a mesoporous silica pre-equilibrated at 33% and 75% RH. The spectra of a composite with low concentration (7%) TiO_2 buildup on the mesoporous silica surface and marble dust with similar reflectance are also shown. (b) Absorbance spectra of sepiolite pre-equilibrated at 75% RH and heated at 80 °C for 2 min.

of TiO_2, but the effect is limited in the NIR due to the intense absorbance of the porous matrix. Nevertheless, the integration of titanium oxide on the composite is evident and should be viewed as an additional photoresponsive centre in the porous host for outdoor environmental applications.

10.4 Solar evaporative cooling

10.4.1 The concept

The evaporative cooling effect with solids of enhanced porosity and water vapor adsorption capacity can be maximised by optimising the porosity characteristics in accordance with the outdoor application conditions of solar radiation, temperature, and relative humidity. A porous medium can have extensive porosity in the macroscale, with an underlying hydrophilic pore structure in the microscale. While the first are macroporous, the latter can be micropores and/or mesopores (Figure 10.12). As shown in Section 10.3.2, micropore filling is the dominant mechanism for water

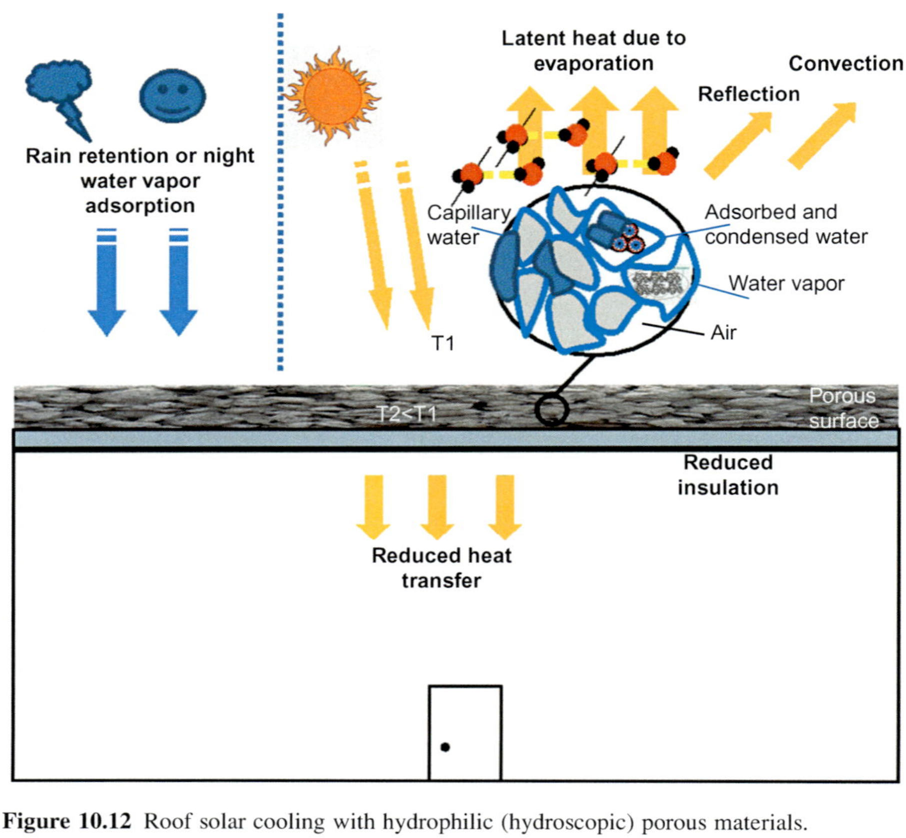

Figure 10.12 Roof solar cooling with hydrophilic (hydroscopic) porous materials.

vapor adsorption in materials like zeolites, aluminophosphates, and metal organic frameworks. In this case, water vapor is strongly attached on the solid wall and higher temperatures for water desorption are required. In contrast, the capillary condensed water in the mesopores is located in a dense liquid-like state with chemical potential lower than its bulk saturating value (Horikawa et al., 2011). In hydrophilic solids, condensation occurs at unsaturated bulk pressure $P < P_0$, while oversaturated pressures are required to push a liquid in hydrophobic pores (Brovchenko and Oleinikova, 2008). Due to the confinement, the liquid–vapor transition occurs at lower temperatures in comparison to the bulk liquid–vapor transition. Therefore, the condensed water can be removed by absorption of the IR part of the solar radiation, which elevates surface temperatures up to 70 °C during hot summers (Karamanis et al., 2013). The mechanism of water removal from the multiporous matrix resembles the natural evapotranspiration process. In leaves, the transpirational flux of water exits the veins as liquid and travels towards the stomata in both the vapor and liquid phases before exiting the leaf as vapor (Rockwell et al., 2014). In a similar manner, moisture from rainwater retention or condensed water in the multipores is escaping a porous structure through diffusion and convection flow in the vapor phase and through capillary action, gravity, and the result of external pressure gradients in the liquid phase. Transport can be accomplished by only one phase or simultaneously by both the liquid and vapor phases. Water starts to vaporise from the surface of the moist material when the absorbed energy has increased the temperature enough for the water vapor pressure to exceed the partial pressure in the surrounding air. Depending on the porous composition, heat can be generated inside very porous materials as well and thermal transfer in the solid is facilitated. The diffusion in the pores can be a rapid or a slow process depending on the nature of the material being irradiated and on its moisture content at any time. It may thus be the limiting rate in the evaporative cooling operation, or if moisture diffusion is rapid, the rate of heat absorption on the surface or the rate of vaporization may be the controlling factor (Lof, 1962).

The environmental parameters of relative humidity, temperature, solar radiation, rainfall, and wind speed that define the boundary conditions at the material-atmosphere interface are time variable and influence the desorption process. Thus, from a fundamental point of view, there are three basic mechanisms that drive the moisture transport in and out of porous materials: adsorption/desorption; condensation/evaporation; and capillary/diffusion action, all of which can be optimized for the evaporative cooling application as a function of the particular environmental conditions.

10.4.2 Experimental

In order to assess the efficiency of non-natural hydrophilic porous materials prepared in lab quantities, lab experiments with simulated environmental conditions can be performed. In this context, a wind tunnel of controllable environmental conditions can be developed and used (Vardoulakis et al., 2011; Wanphen and Nagano, 2009). Typically, the wind tunnel consists of the setting entrance, the contraction zone, the diffuser, the test section, and the fan housing (Figure 10.13). The setting entrance

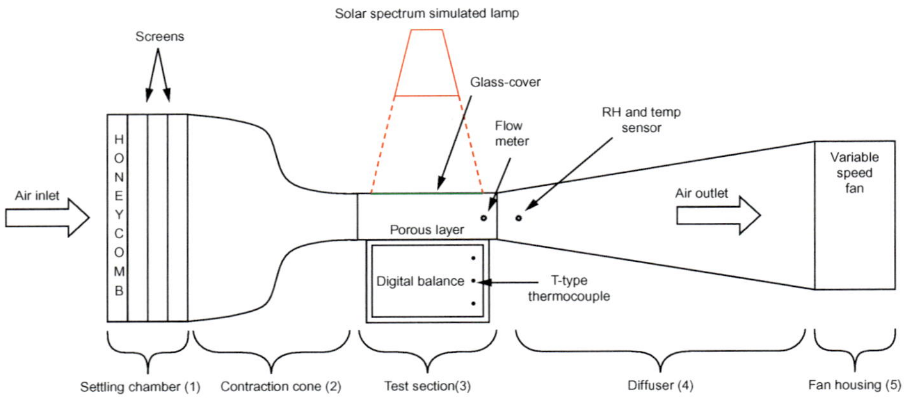

Figure 10.13 Wind tunnel of adjustable and controllable environmental conditions for the assessment of the evaporating cooling properties of porous materials.
Vardoulakis et al. (2011).

section moderates the airflow, while the contraction cone increases the mean velocities at the test section. The diffuser part minimises any air turbulence in the wind flow created from the axial fan with the wind speed regulator and flow meter. In the test section, porous materials are put in an insulated holder where thermocouples measure the surface and bulk temperatures under irradiation from a solar spectrum simulated lamp with the simultaneous recording of their mass variation. Humidity and temperature are controlled and data-logged by sensors while they are adjusted by air-conditioning and humidification.

In an environmental chamber with wind tunnel apparatus and irradiation from a metal halide lamp (500 W/m^2), Wanphen and Nagano (2009) investigated the performance of various natural porous materials for evaporative cooling applications with the addition of liquid water. Their experiments showed that porous materials like natural siliceous shale, with high moisture adsorption ability over 70% of relative humidity, could lower surface temperature when compared to general roofing like mortar concrete. They concluded that, in addition to color selection, latent heat flux played an important role in lowering surface temperature.

In order to extend the principle of evaporative cooling to inorganic hydrophilic materials of low cost and better hydroscopic properties, we tested the surface temperature reduction of modified clays with the addition of liquid water in the wind tunnel and low simulated irradiation (with a metal halide lamp of 100 W/m^2) (Vardoulakis et al., 2011). By varying the preparation conditions, different modified clay materials were prepared, characterized, and tested for solar cooling. The tested materials possessed pores in the microporous and mesoporous region. In dark wind flow conditions of low speed to simulate summer conditions, the sample with the extended mesoporosity exhibited the higher evaporation rate, with a 24-hour integrated value of 6.67 kg/m^2 and two linear rates of moisture evaporation: a rate of 0.493 kg/m^2/h for the first 6 h and a slower rate of 0.208 kg/m^2/h for the rest of the 24 h. The two periods, one with constant drying rate and the other with a reducing rate, are usually observed in drying experiments with constant environmental conditions (Hall et al., 1984). In stage I, the drying rate can be higher than the evaporation rate of a free water volume under the same conditions due to the increase of the effective surface both on or near the layers of the irradiated surface. The material's pores provide a larger effective evaporating surface area while vapor can evaporate and diffuse to the surface through the material voids. In stage II, the drying rate is limited by the unsaturated flow within the porous solid and towards the drying surface. Pillared clays followed the two stages with enhanced evaporation.The water evaporated from the surface that was relatively wet and the rate was constant for the first 6 h. When the surface was getting dryer, the evaporation rate reduced due to water transportation to the free front from the undersurface layer in a rate similar to the reduced evaporation rate in order to keep it constant for the rest of the studied period. In this case, the transport rate was similar to the second evaporation rate from undersurface levels and subsequent surface evaporation. The total latent heat of vaporization was 15.1 MJ/m^2 and the latent heat flux was 174 W/m^2 (with the assumption of a constant evaporation rate). The convective drying of added water in the aluminium modified samples revealed that the surface temperature can be reduced up to 6.3 °C due to the evaporation cooling effect.

The effect was further observed in cyclic experiments with simulated solar radiation, where the maximum surface temperature of the aluminium modified clay was 3.8 °C lower than the value of a reference soil sample. The temperature reduction was even higher within the material, with a maximum middle temperature reduction of 6.5 °C (Figure 10.14). Due to the different but low measured values of reflectance in the samples, further experimental and theoretical analysis revealed that the reflection of the incident low intensity radiation was not the primary responsible mechanism for attaining the lower temperatures with the modified clay samples.

Similar behavior was also observed in a modified by-product from the hydrothermal transformation of coal fly ash to zeolitic form (Karamanis and Vardoulakis, 2012). Water vapor adsorption isotherms revealed that the hydrophobic fly ash (type II) was transformed to type IV for the modified hydrophilic material. By adding liquid water, the maximum difference of temperature increase under simulated solar irradiation was observed between the treated fly ash and the concrete with values of 5.0, 5.4, and 7.5 °C for the surface, middle, and bottom position, respectively (Karamanis and Vardoulakis, 2012). All the above experiments were performed with liquid water, with water vapor adsorption a contributing component in the evaporative cooling of added liquid water rather than the driving mechanism.

In order to replace the precipitation and retention by night humidity adsorption for nontropical climates, we have recently shown that the surface evaporative cooling by vapor sorption is possible on the hydrophilic natural clay mineral of sepiolite (Karamanis et al., 2012). The test was performed in the wind tunnel with low incoming radiation. Two 24-hour cycles were used, with two 12-hour periods with the lamp on and the raising of the relative humidity from 52% to 70% during the two 12-hour periods with the lamp off. The weight variation curve of sepiolite during the cycles is shown in Figure 10.15a. The weight of sepiolite followed an almost linear reduction

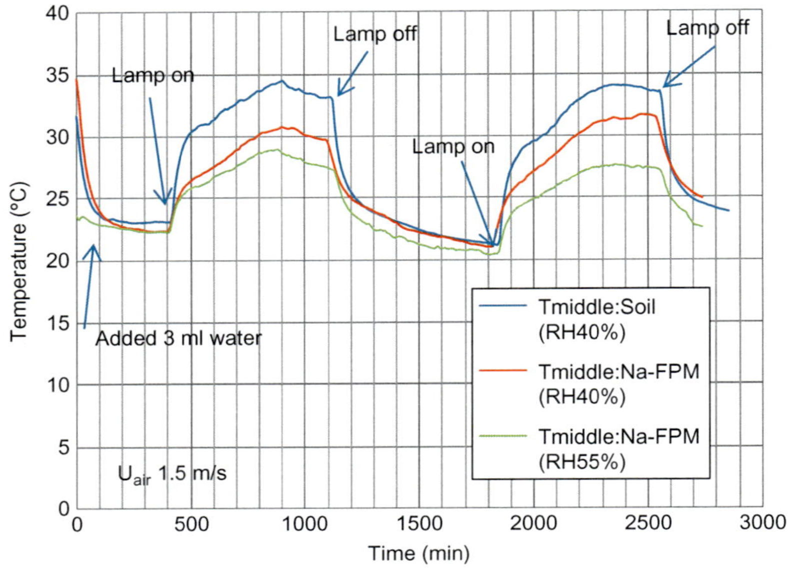

Figure 10.14 The temperature variation in the middle position of a pillared clay sample (Na-FMP) and typical soil with cyclic experiments of two continuous cycles with simulated solar radiation of low intensity.
Vardoulakis et al. (2011).

of the weight of the material with the lamp on and an exponential increase with the lamp off. The reduction was due to the evaporation and desorption of water vapor upon material irradiation, while mass increase was due to water vapor adsorption upon relative humidity increase. Moreover, the temperature increase in sepiolite compared to concrete was measured in cyclic experiments with simulated solar radiation of two continuous cycles (Figure 10.15b). In this experiment, the maximum difference of temperature increase under simulated solar irradiation between the sepiolite sample and concrete was around 5 °C for the bottom position (Karamanis et al., 2012).

Since the principle of water vapor condensation and evaporative cooling was validated with low intensity irradiation, a mesoporous silica with high water vapor capacity was recently tested under higher solar simulated intensity from two 80 W xenon lamps (300 W/m^2) (Karamanis et al., 2013). The material selection was based on the criteria of porosity and hydrophilocity described in the previous sections. In addition, the comparison was performed with a white natural material with very high IR reflectance values at the same conditions as sepiolite experiments (Karamanis et al., 2013). It was observed that the difference in temperature increase under simulated solar irradiation between mesoporous silica and marble dust with comparable reflectance was almost 5 °C in the first irradiation hours and reduced to 2 °C at the end of irradiation (Figure 10.16). With a latent heat of 2300 kJ/kg for water vaporization at the attained material temperatures, the silica mass reduction

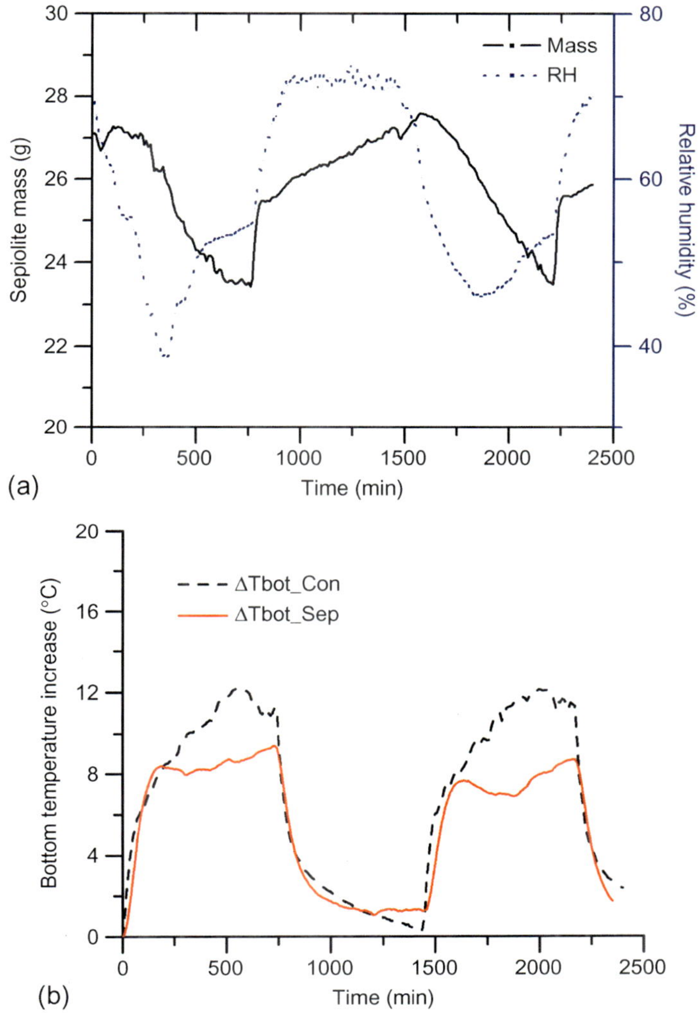

Figure 10.15 (a) Increase of the sepiolite mass due to relative humidity increase in the tunnel and decrease due to low simulated solar irradiation (50 W/m^2). (1st cycle: 0–720 min-lamp on-RH 52 ± 5%, 720–1440 min-lamp off-RH 70 ± 3%; 2nd cycle: 1440–2160 min-lamp on, 2160–2400 min-lamp off) (b) Increase of the temperature at the bottom position of the Sep and Concrete materials cell (3 cm) due to low simulated solar irradiation (Karamanis et al., 2012).

corresponded to half of the absorbed incoming radiation for water evaporation. Therefore, the low temperature solar heat transformation in hydroscopic mesoporous materials with the associated evaporation cooling can be considered as an additional mechanism towards the reduction of roof heat fluxes in buildings. Furthermore, the in situ buildup of TiO_2 or ZnO nanoparticles on open porous hosts does not affect

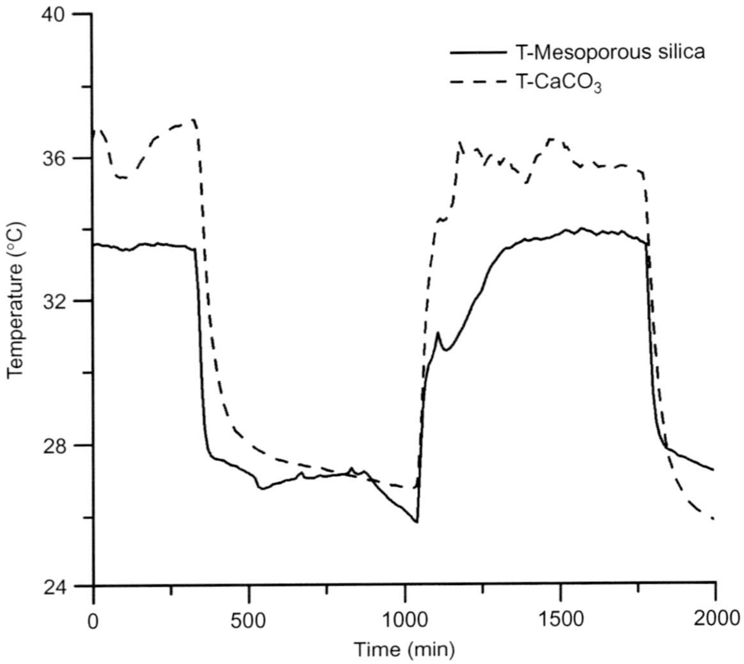

Figure 10.16 Temperature increase inside the mesoporous silica and marble dust samples due to the simulated solar irradiation (average 300 W/m^2 and wind tunnel flow rate of 430 ± 35 m^3/h). Karamanis et al. (2013), reproduced with kind permission of Taylor & Francis Ltd, publisher of *Advances in Building Energy Research* (www.tandfonline.com).

drastically the water vapor adsorption properties of the composites (Karamanis et al., 2012). The VIS reflectance of the composite are even increased due to the in situ growth of the white semiconducting nanoparticles. Furthermore, the UV part of the solar spectrum can be utilised in semiconductor absorption for pollutants' photodegradation. However, the site competition or synergy between water vapor and air pollutants needs further combined research and development in addition to technological implementation and building integration of low-cost porous materials and their combinations.

10.5 Future trends

The adverse effects of (micro) climate change have called for urgent environmental mitigating actions beyond the typical considerations. Reducing the building cooling load is a major prerequisite for more integrated and adaptation measures. In this direction, active systems require long development to reduce the size, cost, maintenance, and ease of use (e.g., heat pumps have been developed over the last 40 years in Europe). In buildings design, any system such as solar cooling will be much more applicable when incorporated at the building housing even from the design phase. In addition, modern building requires a reduction of the energy and environmental costs of embedded materials. In a reassessing building design frame, it should be considered that the water vapor evaporation–condensation mechanism is already an important natural process that drives Earth's life. In the macroscopic model, evaporation consumes about 25% of solar energy input, while condensation drives the hydrologic cycle through annual precipitation. Therefore, the development of new eco-efficient energy materials as low-cost thermoresponsive materials (especially in the IR region) in order to be exploited in evaporative cooling applications in the city environment is of primary importance for the reduction of the heat island effect due to direct solar irradiation and the anthropogenic urban longwave radiative heat.

Acknowledgements

Thanks are due to our colleagues, Prof. Mihalakakou, Prof. N. Ökte, Dr E. Vardoulakis, Dr S. Krimbalis, Dr E. Chalkia, Mrs E. Kyritsi, Mr G. Gorgolis, Mr V. Kapsalis, and Mr N. Skandalos, with whom we have worked in this review; Universities of Ioannina, Western Greece, and Patras; and our funding agencies Hellenic General Secretariat of Research & Technology (GSRT) programs of ARISTEIA I, Hrakleitos II, Bilateral Program Greece—Turkey ("Operational Program Education and Lifelong Learning" co-funded by the European Social Fund (ESF) and National Resources), and Latsis and Empirikion Foundations.

References

Akbari, H., Rose, S.L., Taha, H., 2003. Analyzing the land cover of an urban environment using high-resolution orthophotos. Landsc. Urban Plan. 63, 1–14.

Bineesh, K.V., Kim, M., Lee, G., Selvaraj, M., Hyun, K., Park, D., 2012. Production of elemental sulfur and ammonium thiosulfate by the oxidation of H_2S containing water vapour and ammonia over V/Zr-PILC catalysts. J. Ind. Eng. Chem. 18, 1845–1850.

Brauner, K., Preisinger, A., 1956. Struktur und entstehung des sepioliths. Tscher. Mineral. Petrogr. Mitteil. 6, 120–140.

Bretz, S.E., Akbari, H., 1997. Long-term performance of high-albedo roof coatings. Energy Build. 25, 159–167.

Brovchenko, I., Oleinikova, A., 2008. Interfacial and Confined Water. Elsevier, Amsterdam, p. 115.

Bueno, B., Norford, L., Pigeon, G., Brittera, R., 2012. A resistance–capacitance network model for the analysis of the interactions between the energy performance of buildings and the urban climate. Build. Environ. 54, 116–125.

Büning-Pfaue, H., 2003. Analysis of water in food by near infrared spectroscopy. Food Chem. 82, 107–115.

Cen, H., He, Y., 2007. Theory and application of near infrared reflectance spectroscopy in determination of food quality. Trends Food Sci. Technol. 18, 72–83.

Chan, H.Y., Riffat, S.B., Zhu, J., 2010. Review of passive solar heating and cooling technologies. Renew. Sust. Energ. Rev. 14, 781–789.

Davis, M.E., 2002. Ordered porous materials for emerging applications. Nature 417, 813–821.

Dimos, K., Stathi, P., Karakassides, M., Deligiannakis, Y., 2009. Synthesis and characterization of hybrid MCM-41 materials for heavy metal adsorption. Microporous Mesoporous Mater. 126, 65–71.

Dos Santos, G.H., Mendes, N., 2013. Numerical analysis of passive cooling using a porous sandy roof. Appl. Therm. Eng. 51, 25–31.

Drelich, J., Chibowski, E., Desheng Meng, D., Terpilowski, K., 2011. Hydrophilic and superhydrophilic surfaces and materials. Soft Matter 7, 9804–9828.

Duan, Z., Zhan, C., Zhang, X., Mustafa, M., Zhao, X., Alimohammadisagvand, B., Hasan, A., 2012. Indirect evaporative cooling: past, present and future potentials. Renew. Sust. Energ. Rev. 16, 6823–6850.

Florides, G.A., Tassou, S.A., Kalogirou, S.A., Wrobel, L.C., 2002. Measures used to lower building energy consumption and their cost effectiveness. Appl. Energy 73, 299–328.

Furukawa, H., Cordova, K.E., O'Keeffe, M., Yaghi, O.M., 2013. The chemistry and applications of metal-organic frameworks. Science 341, 1230444-1–1230444-12.

Grim E.r., R., 1968. Clay Mineralogy. McGraw-Hill, New York, pp. 384.

Groen, J.C., Peffer, L.A.A., Perez-Ramirez, J., 2003. Pore size determination in modified micro- and mesoporous materials. Pitfalls and limitations in gas adsorption data analysis. Microporous Mesoporous Mater. 60, 1–17.

Hall, C., Hoff, W.D., Nixon, M.R., 1984. Water movement in porous building materials–VI. Evaporation and drying in brick and block materials. Build. Environ. 19, 13–20.

Henninger, S.K., Jeremias, F., Kummer, H., Janiak, C., 2012. MOFs for use in adsorption heat pump processes. Eur. J. Inorg. Chem. 2012, 2625–2634.

Herling, M.M., Kalo, H., Seibt, S., Schobert, R., Breu, J., 2012. Tailoring the pore sizes of microporous pillared interlayered clays through layer charge reduction. Langmuir 28, 14713–14719.

Horikawa, T., Do, D.D., Nicholson, D., 2011. Capillary condensation of adsorbates in porous materials. Adv. Colloid Interface Sci. 169, 40–58.

IUPAC, 1994. Recommendations for the characterization of porous solids. Pure Appl. Chem. 66, 1739–1758.

Jänchen, J., Stach, H., 2012. Adsorption properties of porous materials for solar thermal energy storage and heat pump applications. Energy Procedia 30, 289–293.

Jeremias, F., Frohlich, D., Janiak, C., Henninger, S.K., 2014. Water and methanol adsorption on MOFs for cycling heat transformation processes. New J. Chem. 38, 1846–1852.

Johnson, A.E., 2002. Analysis of the Young equation and use of the Kelvin equation in calculating the Gibbs excess. Colloids Surf. A Physicochem. Eng. Asp. 202, 33–39.

Karamanis, D., Vardoulakis, E., 2012. Application of zeolitic materials prepared from fly ash to water vapor adsorption for solar cooling. Appl. Energy 97, 334–339.

Karamanis, D., Aslanoglou, X., Assimakopoulos, P.A., Gangas, N.H., Pakou, A.A., Papayanakos, N.G., 1997. An aluminium pillared montmorillonite with fast uptake of strontium and cesium from aqueous solutions. Clays Clay Miner. 45, 709–717.

Karamanis, D., Ökte, A.N., Vardoulakis, E., Vaimakis, T., 2011. Water vapor adsorption and photocatalytic pollutant degradation with TiO_2 sepiolite nanocomposites. Appl. Clay Sci. 53, 181–187.

Karamanis, D., Vardoulakis, E., Kyritsi, E., Ökte, N., 2012. Surface solar cooling through water vapor desorption from photo-responsive sepiolite nanocomposites. Energy Convers. Manage. 63, 118–122.

Karamanis, D., Kyritsi, E., Krimpalis, S., Vardoulakis, E., Gorgolis, G., Kapsalis, V., Mihalakakou, G., Ökte, N., 2013. Cooling roofs through low temperature solar-heat transformations in hydrophilic porous materials. Adv. Build. Energy Res. 7, 235–243.

Kim, Y., Thu, K., Choon Ng, K., 2014. Adsorption characteristics of water vapour on ferroaluminophosphate for desalination cycle. Desalination 344, 350–356.

Kolokotroni, M., Giridharan, R., 2008. Urban heat island intensity in London: an investigation of the impact of physical characteristics on changes in outdoor air temperature during summer. Sol. Energy 82, 986–998.

Kolokotroni, M., Gowreesunker, B.L., Giridharan, R., 2013. Cool roof technology in London: an experimental and modelling study. Energy Build. 67, 658–667.

Kresge, C.T., Leonowicz, M.E., Roth, W.J., Vartuli, J.C., Beck, J.S., 1992. Ordered mesoporous molecular sieves synthesized by a liquid-crystal template mechanism. Nature 359, 710–712.

Küsgens, P., Rose, M., Senkovska, I., Fröde, H., Henschel, A., Siegle, S., Kaskel, S., 2009. Characterization of metal-organic frameworks by water adsorption. Microporous Mesoporous Mater. 120, 325–330.

Kwok, D.Y., Neumann, A.W., 1999. Contact angle measurement and contact angle interpretation. Adv. Colloid Interface Sci. 81, 167–249.

Leo Samuel, D.G., Shiva Nagendra, S.M., Maiya, M.P., 2013. Passive alternatives to mechanical air conditioning of building: a review. Build. Environ. 66, 54–64.

Liu, W., Peng, S.W., Mizukami, K., 1997. Moisture evaporation and migration in thin porous packed bed influenced by ambient and operating conditions. Int. J. Energy Res. 21, 41–53.

Lof, G.O.G., 1962. Solar energy for the drying of solids. Sol. Energy 6, 122–128.

Maeda, H., Kato, S., Ishida, E.H., 2009. Preparation of hydrothermally solidified mesoporous materials from diatomaceous earth for moisture control application. Int. J. Appl. Ceram. Technol. 6, 431–436.

Makarieva, A.M., Gorshkov, V.G., Sheil, D., Nobre, A.D., Li, B.-L., 2013. Where do winds come from? A new theory on how water vapour condensation influences atmospheric pressure and dynamics. Atmos. Chem. Phys. 13, 1039–1056.

Malla, P., Yamanaka, S., Komarneni, S., 1989. Unusual water vapour adsorption behaviour of montmorillonite pillared with ceramic oxides. Solid State Ionics 32 (33), 354–362.

Meng, Q., 1996. Study on cooling of building roof by passive water evaporation. Acta Ener. Sol. Sinica 17, 256–262.

Meng, Q., Hu, W., 2005. Roof cooling effect with humid porous medium. Energy Build. 37, 1–9.

Mitropoulos, A.Ch., 2008. The Kelvin equation. J. Colloid Interface Sci. 317, 643–648.

Nahar, N.M., Sharma, P., Purohit, M.M., 1999. Studies on solar passive cooling techniques for arid areas. Energy Convers. Manage. 40, 89–95.

Nakabayashi, S., Nagano, K., Nakamura, M., Togawa, J., Kurokawa, A., 2011. Improvement of water vapour adsorption ability of natural mesoporous material by impregnating with chloride salts for development of a new desiccant filter. Adsorption 17, 675–686.

Neimark, V., Ravikovitch, P.I., 2001. Capillary condensation in MMS and pore structure characterization. Microporous Mesoporous Mater. 44–45, 697–707.

Ng, E., Mintova, S., 2008. Nanoporous materials with enhanced hydrophilicity and high water sorption capacity. Microporous Mesoporous Mater. 114, 1–26.

Okada, K., Matsui, S., Isobe, T., Kameshima, Y., Nakajima, A., 2008. Water-retention properties of porous ceramics prepared from mixtures of allophone and vermiculite for materials to counteract heat island effects. Ceram. Int. 34, 345–350.

Onal, M., Yilmaz, H., Sarikaya, Y., 2008. Some physicochemical properties of the white sepiolite known as pipestone from Eskisehir, Turkey. Clays Clay Miner. 56, 511–519.

Pasupathy, A., Velraj, P.R., 2007. Effect of double layer phase change material in building roof for round thermal management. Energy Build. 40, 193–203.

Rockwell, F.E., Holbrook, N.M., Stroock, A.D., 2014. The competition between liquid and vapour transport in transpiring leaves. Plant Physiol. 164, 1741–1758.

Rowsell, J.L.C., Yaghi, O.M., 2004. Metal–organic frameworks: a new class of porous materials. Microporous Mesoporous Mater. 73, 3–14.

Ruiz-Hitzky, E., Darder, M., Aranda, P., 2008. An introduction to bio-nanohybrid materials. In: Ruiz-hitzky, E., Agira, K., Lvov, Y. (Eds.), Bio-inorganic Hybrid Nanomaterials: Strategies, Synthesis, Characterization and Applications. Wiley VCH Verlag Gmbh, Weinheim, pp. 1–32.

Runsheng, T., Etzion, Y., Erell, E., 2003. Experimental studies on a novel roof pond configuration for the cooling of buildings. Renew. Energy 28, 1513–1522.

Sadineni, S.B., Madala, S., Boehm, R.F., 2011. Passive building energy savings: a review of building envelope components. Renew. Sust. Energ. Rev. 15, 3617–3631.

Saneinejad, S., Moonen, P., Carmeliet, J., 2014. Comparative assessment of various heat island mitigation measures. Build. Environ. 73, 162–170.

Santamouris, M., 2014. Cooling the cities – a review of reflective and green roof mitigation technologies to fight heat island and improve comfort in urban environments. Sol. Energy 103, 682–703.

Santamouris, M., Papanikolaou, N., Livada, I., Koronakis, I., Georgakis, C., Argiriou, A., Assimakopoulos, D.N., 2001. On the impact of urban climate on the energy consumption of buildings. Sol. Energy 70, 201–216.

Singh, A.K., Tiwari, G.N., 1996. Energy conservation in a cinema hall under hot and dry condition. Energy Convers. Manage. 37, 531–539.

Spala, A., Bagiorgas, H.S., Assimakopoulos, M.N., Kalavrouziotis, J., Matthopoulos, D., Mihalakakou, G., 2008. On the green roof system. Selection, state of the art and energy potential investigation of a system installed in an office building in Athens, Greece. Renew. Energy 33, 173–177.

Spanaki, A., Tsoutsos, T., Kolokotsa, D., 2011. On the selection and design of the proper roof pond variant for passive cooling purposes. Renew. Sust. Energ. Rev. 15, 3523–3533.

Sychev, M., Prihod'ko, R., Stepanenko, A., Rozwadowski, M., de Beer, V.H.J., van Santen, R. A., 2001. Characterization of the microporosity of chromia- and titania-pillared montmorillonites differing in pillar density II. Adsorption of benzene and water. Microporous Mesoporous Mater. 47, 311–321.

Taguchi, A., Schuth, F., 2005. Ordered mesoporous materials in catalysis. Microporous Mesoporous Mater. 77, 1–45.

Tang, Q., Wang, F., Tang, M., Liang, J., Ren, C., 2012. Study on pore distribution and formation rule of sepiolite mineral nanomaterials. J. Nanomater. 382603, 1–6.

Tiwari, G.W., Kumar, A., Sodha, M.S., 1982. A review—cooling by water evaporation over roof. Energy Convers. Manage. 22, 143–153.

UNEP, 2009. Buildings and Climate Change. Sustainable Buildings & Climate Initiative, France, pp. 63.

Van Oss, C.J., 1994. Interfacial Forces in Aqueous Media. Marcel Dekker, Inc, New York.

Vardoulakis, E., Karamanis, D., Assimakopoulos, M.N., Mihalakakou, G., 2011. Solar cooling with aluminium pillared clays. Sol. Energy Mater. Sol. Cells 95, 2363–2370.

Vardoulakis, E., Karamanis, D., Fotiadi, A., Mihalakakou, G., 2013. The heat island effect in a small Mediterranean city of elevated summertime temperatures. Sol. Energy 94, 128–144.

Wanphen, S., Nagano, K., 2009. Experimental study of the performance of porous materials to moderate the roof surface temperature by its evaporative cooling effect. Build. Environ. 44, 338–351.

Whiting, G., Grondin, D., Bennici, S., Auroux, A., 2013. Heats of water sorption studies on zeolite–$MgSO_4$ composites as potential thermochemical heat storage materials. Sol. Energy Mater. Sol. Cells 112, 112–119.

Yamanaka, S., Malla, P.R., Komarneni, S., 1990. Water adsorption properties of alumina pillared clay. J. Colloid Interface Sci. 134, 31–58.

Zhao, X.S., Lu, G.Q., Whittaker, A.K., Millar, G.J., Zhu, H.Y., 1997. Comprehensive study of surface chemistry of MCM-41 using 29Si CP/MAS NMR, FTIR, pyridine-TPD, and TGA. J. Phys. Chem. B 101, 6525–6531.

Zhu, H.Y., Gao, W.H., Vansant, E.F., 1995. The porosity and water adsorption of alumina pillared montmorillonite. J. Colloid Interface Sci. 171, 377–385.

Cool green roofs for reducing building cooling needs

11

B.-S. Lin, Y.-J. Lin
National Taiwan University, Taipei, Taiwan

11.1 Introduction

The dense development of cities in recent years has caused numerous environmental problems. In urban areas with a large amount of stereoscopic space, the green roof has become one of the main strategies used to increase the urban green space and to improve the quality of the urban environment.

Green roofs provide substantial efficacy for ecosystem services, including the regulation of the microclimate (D'Orazio et al., 2012; Kumar and Kaushik, 2005; Lin et al., 2013; Wong et al., 2003, 2007), management of storm waters (Lee et al., 2013; Volder and Dvorak, 2014), noise absorption (Renterghem and Botteldooren, 2009; Yang et al., 2012), air purification (Baik et al., 2012; Yang et al., 2008), and biodiversity upgrade in cities (Francis and Lorimer, 2011). The exacerbation of the urban heat island effect and global climate anomalies has caused green roofs and their ability to regulate microclimate to become the focus of considerable research. Green roofs have been studied in diverse terrestrial ecoregions in Europe, the Americas, Asia, Oceania, and Africa, revealing that the ecosystem services provided by green roofs are currently recognized and valued by numerous countries worldwide (Blank et al., 2013).

Green roof is the outermost layer of the rooftop and can be viewed as another floor of the structure (Sailor et al., 2008). Compared with other structural layers, the green roof layer is a "live" structure covered by vegetation and growing mediums. Evapotranspiration and shading are the main mechanisms by which green roofs achieve temperature moderation, resulting in highly stable temperatures inside the building and, thus, energy conservation. Figure 11.1 shows that an extensive green roof with a 10 cm thick growing medium layer can block approximately 90% of the heat flux entering a building, thus reducing the ceiling surface temperature (Lin, 2012). On the same green roof, the cooling effect of different plant species with varied characteristics differed considerably (Lin, 2010). According to Cox, thermal insulation values vary for different plant types (Berardi et al., 2014). Regarding growing mediums, Sailor and Hagos (2011) indicated that thermal properties vary substantially as a function of growing medium design. The density, thickness, and moisture content of the growing medium layer determine the cooling potential of a green roof (Barrio,

Eco-efficient Materials for Mitigating Building Cooling Needs. http://dx.doi.org/10.1016/B978-1-78242-380-5.00011-X

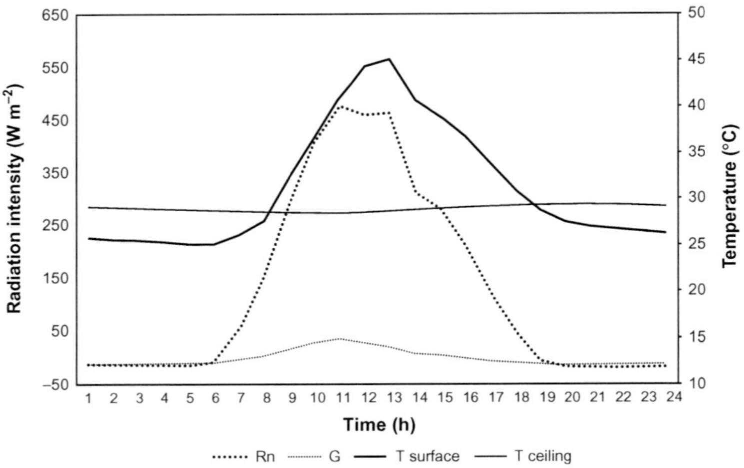

Figure 11.1 Variances in heat flux, bared rooftop surface temperature, and the surface temperature of the ceiling under an extensive green roof in summer.
Source: This figure is extracted from our research report in 2012.

1998). Therefore, the choice of plant species and the formula and thickness of the growing mediums can determine not only the success of the green roof construction but also its cooling efficiency (Lin, 2010; Lin et al., 2011; Sailor et al., 2008; Wong et al., 2007). In addition, community-scale green roof installation can extend its cooling benefits to the ground to improve the comfort of urban street environment (Lilliana et al., 2013).

Energy-saving effectiveness of green roofs is observed not only in air-conditioning in summer but also in heating in winter. However, the energy-saving effect provided by green roofs varies with the regional climate. In Athens, under Mediterranean climatic conditions, a green roof on a nursery school building could markedly decrease electricity consumption in air-conditioning in summer by 6–19% for the entire building and by 12–87% for the top floor alone. However, the green roof did not decrease the electricity consumption from heating in winter (Sfakianaki et al., 2009). Regarding the temperature reduction effect, the thermal effectiveness of an extensive green roof and the regional climate are closely related; the daytime cooling effectiveness was higher in the tropical island climate than in the subtropical island climate in summer (Lin et al., 2013).

In addition to the regional climate, building materials available in a particular region affect the cooling ability of green roofs. The energy-saving effectiveness of green roofs is approximately 40% for commercial buildings in Athens (Spala et al., 2008); in Canada, however, this effectiveness in a high-occupancy LEED Gold standard building is 5.4% in July (Feng and Hewage, 2014). Different levels of insulation of the roof slab resulted in different cooling effectiveness of the green roof. With a decrease in the amount of insulation, the energy-saving effectiveness of green roofs increases (Niachou et al., 2001; Theodosiou, 2003). In other words, old buildings with poor insulation have benefitted most from a green roof.

Studies investigating the effectiveness of green roofs by measuring the change in temperature reduction have also shown that the cooling effectiveness varied with the regional climate and building materials. In Singapore, the maximum temperature reduction achieved with green roof installation is approximately 18 °C (Wong et al., 2007). On hot and humid Réunion island, India, the maximum temperature reduction achieved is approximately 38 °C (Morau et al., 2012).

In this chapter, we summarize the types of green roof system, materials and properties of each layer, design and construction process, and propose design principles for green roofs with the primary function of cooling that can serve as a reference for future planners and designers.

11.2 Green roof types

Green roofs can be classified into three categories according to the maintenance requirement: extensive, semi-intensive, and intensive green roofs (see Figures 11.2–11.4). This categorization of green roofs is the most widely accepted and commonly used system.

The three types of green roofs differ in structural load, thickness of the growing medium layer, the type of vegetation they support, cost, required maintenance, applicable roof type, and accessibility. The three categories and their characteristics are summarized in Table 11.1.

Figure 11.2 Extensive green roof.
Source: Photographed by authors.

Figure 11.3 Semi-intensive green roof.
Source: Photographed by authors.

Figure 11.4 Intensive green roof.
Source: Photographed by authors.

Table 11.1 Green roof categories and characteristics

	Extensive green roof	Semi-intensive green roof	Intensive green roof
Layer	Vegetation layer, growing medium layer, filter membrane layer, drainage (and retention) layer, root barrier layer, and waterproof layer		
Load	50–100 kg m^{-2}	80–150 kg m^{-2}	250–1000 kg m^{-2}
Growing medium depth	Below 15 cm	Below 15 cm	15–100 cm
Vegetation	Limited variety	Limited variety	Unrestricted variety
Cost	Low	Average	High
Maintenance required	Low	Average	High
Applicable roof type	Flat and sloped	Flat	Flat
Accessibility	Typically not accessible except for maintenance	Permit low-density usage	Open access; mainly used for leisure and recreational purposes

11.3 Materials and properties

The structure of a green roof, from top to bottom, includes the following layers: vegetation, growing mediums, filter membrane, drainage (and retention), root barrier, and waterproof layer (Luckett, 2009).

11.3.1 Vegetation layer

The vegetation layer is the most crucial layer of a green roof. The choice of plant species is limited by the climate of the region, the thickness of the growing medium layer, the purpose of the green roof, and the amount of maintenance required. The types of plants chosen and their characteristics can be roughly classified by their suitability for extensive or intensive green roofs.

11.3.1.1 Types of vegetation and their characteristics

Plant choice is relatively limited for extensive green roofs. Suitable plants must exhibit resistance to heat, drought, wind, cold, barren soil, and pests and disease (Dunnett and Kingsbury, 2004; Getter and Rowe, 2006). Morphologically, they are low creeping plants with shallow roots. Plants suitable for extensive green roofs must self-propagate and self-seed and have high coverage (White and Snodgrass, 2003). Therefore, commonly chosen plants include sedum, succulents, ground cover plants, mosses, and low shrubs (see Figure 11.5). Any native species with shallow roots can also be considered (Getter and Rowe, 2006).

Intensive green roofs can support almost all plant species. Generally, any commonly used species can be chosen, and these plants often have a high ornamental value.

11.3.1.2 Thermal transfer characteristics of the vegetation layer

The vegetation layer is the uppermost layer of the green roof structure and is therefore the outermost layer of a building that is exposed to solar radiation. When solar radiation reaches the plant canopy, energy is transferred through absorption, reflection, or transmission. The absorption, reflection, and transmission rates differ depending on plant species (Jones, 1992; Miller, 1977).

A high reflection rate results in considerable energy being reflected back to the atmosphere; a low transmission rate results in the transmission of an insubstantial amount of energy into the building. Therefore, a vegetation layer with high reflection and low transmission rates can reduce the amount of energy that is transferred into the building interior and thereby serve its purpose of cooling the building and conserving energy.

Generally, plants with high reflection rates are light in color and have waxy leaves (Jones, 1992; Lin and Lin, 2010). Plants with low transmission rates have dense foliage and high plant height. Plant height and foliage density are essential characteristics of vegetation that influence the heat transfer of green roofs (Alexandri and Jones, 2007; Sailor, 2008).

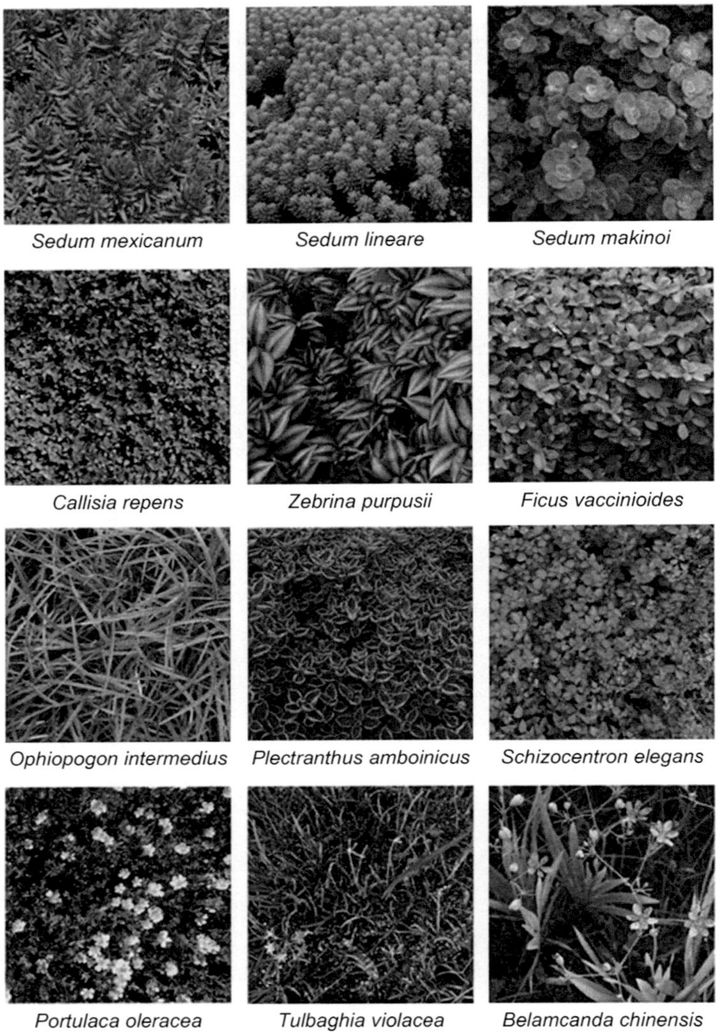

Figure 11.5 Commonly used plant species in extensive green roofs.
Source: Photographed and edited by authors.

Furthermore, stomatal resistance and photosynthesis-type plants also influence heat transfer within the vegetation layer (Jim, 2014a; Lin, 2010; Sailor, 2008). Plants with low stomatal resistance tend to exhibit more transpiration, leading to high latent heat. Crassulacean acid metabolism (CAM) plants engage in CAM photosynthesis and close their stomata during the day to reduce water loss and reduce evapotranspiration. Hence, evapotranspiration of CAM plants is low, resulting in low latent heat in the day. CAM plants have a lower temperature reduction effect than well-watering

C3/C4 plants do. Sedum and succulents constitute a large fraction of the CAM plants and are commonly used in extensive green roofs.

11.3.2 Growing medium layer

The growing medium layer is another vital layer in the structure of green roofs. The type and composition of growing mediums affect the choice of vegetation and their growth and the structural load on the roof.

11.3.2.1 Types and characteristics of growing mediums

To satisfy the needs of plant growth and reduce the roof load, the growing mediums used on green roofs are primarily lightweight and soilless, which can be categorized as organic or inorganic mediums (see Figure 11.6).

Commonly used inorganic growing mediums include perlite, pumice, river sand, expanded clay, and ceramsite; commonly used organic growing mediums include peat, coconut coir, tree bark, sawdust, grain hulls, synthetic fiber soil, and livestock manure. Both organic and inorganic growing mediums must be lightweight, but they also must exhibit high stability, water retention, and aeration.

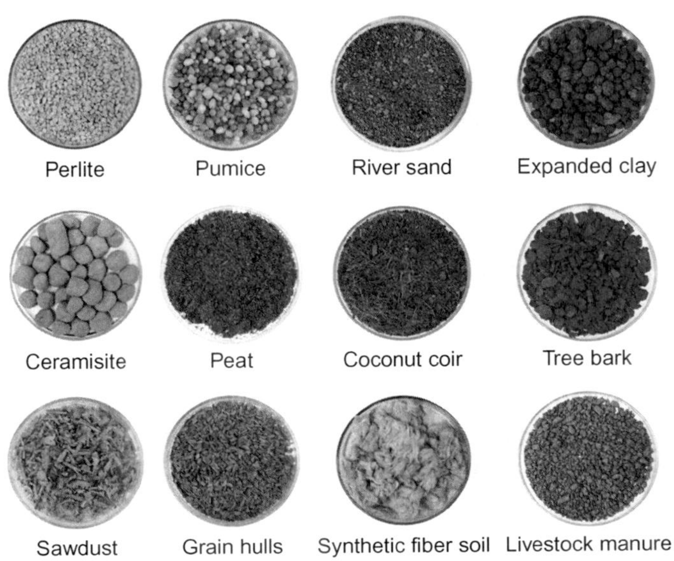

Figure 11.6 Commonly used growing mediums in green roofs.
Source: Photographed and edited by authors.

11.3.2.2 Growing medium formulas and thermal transfer characteristics

The growing medium layer is composed of two or more components in specific ratios. These ratios and compositions differ according to local climate conditions, the availability and price of the components, the green roof type, and plant species. The mixed medium blend likely exhibits excellent water retention and aeration characteristics.

The thermal conductivity (k, W/m K) and specific heat capacity (C_p, J/kg K) of the medium blend differ from those that characterize a single medium. These thermal characteristics depend on the moisture content of the mixed mediums. A high moisture level results in a large k and C_p of the medium (Sailor et al., 2008).

11.3.3 Filter membrane layer

Underneath the growing medium layer lies the filter membrane layer, which is often composed of nonwoven fabric. This prevents the loss of the growing medium through runoff and the blockage of rooftop drainage pipes caused by this runoff.

11.3.4 Drainage (and retention) layer

The purpose of the drainage (and retention) layer is to eliminate excess water from the growing medium layer to prevent the plant roots from rotting. It also retains water to reduce the need for irrigation or watering.

Materials used in the drainage (and retention) layer include aggregates, drainage boards, and composite drainage and water retention boards. Aggregates are expensive and cause high roof loads; therefore, their use is limited. Drainage boards as well as composite drainage and water retention boards have recently become more common. They are lightweight, can bear a high load, are easy to install, and are available in a variety of configurations.

11.3.5 Root barrier layer

The root barrier layer prevents the roots from penetrating and damaging the roof structure of a building. A chemical is added to the middle layer of the root barrier sheet to inhibit root growth. This inhibitor causes the cutinization of the root system and prevents downward growth. Commonly used materials include fabrics and thermal plastic.

Fabrics are lightweight and do not increase the roof load and are suitable for use in extensive green roofs. Thermal plastics comprise composite plastics. They are more costly than fabrics are but exhibit more favorable root inhibition characteristics and can prevent the penetration of large roots.

11.3.6 Waterproof layer

The waterproof layer prevents moisture from penetrating and damaging the roof structure of a building. Commonly used materials include leakproof and waterproof coatings and polyvinyl chloride (PVC) waterproofing membranes.

Leakproof and waterproof coatings are highly adhesive and can penetrate concrete structures to form an effective moisture barrier. They are easy to apply and exhibit high ductility. They can be easily spread into cracks and corners.

PVC waterproofing membranes are also easy to install. They are fire-resistant, wear-resistant, and durable.

11.4 Design principles for reducing cooling needs

The cooling effect of green roofs is undeniable. In green roof design and construction process, all specifications, from the characteristics of the site to the spatial configuration of the green roof and the structural design of the green roof, can affect the cooling efficiency.

In the following sections, we first depict each step of the design and construction process and then the design principles for green roofs with the primary function of cooling.

11.4.1 Design and construction process

The process of green roof design and construction involves determining the purpose of the structure, analyzing the work site, deciding on the type of green roof, designing and estimating the project costs, constructing the green roof, and maintaining the roof (see Figure 11.7).

11.4.1.1 Determine the purpose of the green roof

Before beginning the construction of a green roof, the designer must first ascertain the functions of the green roof. Possible functions include

- a space for recreational or social activities,
- the provision of environmental benefits (e.g., energy conservation, stormwater management, and increased urban biodiversity),
- a vegetable garden,
- a recreational or healing garden,
- beautification of the rooftop,
- compliance with laws or regulations.

11.4.1.2 Site analysis

Site analysis primarily involves environmental analysis and demand analysis. The investigation and analysis of these two dimensions lead to the assessment of a suitable roof style and become the basis for the subsequent spatial design of the green roof.

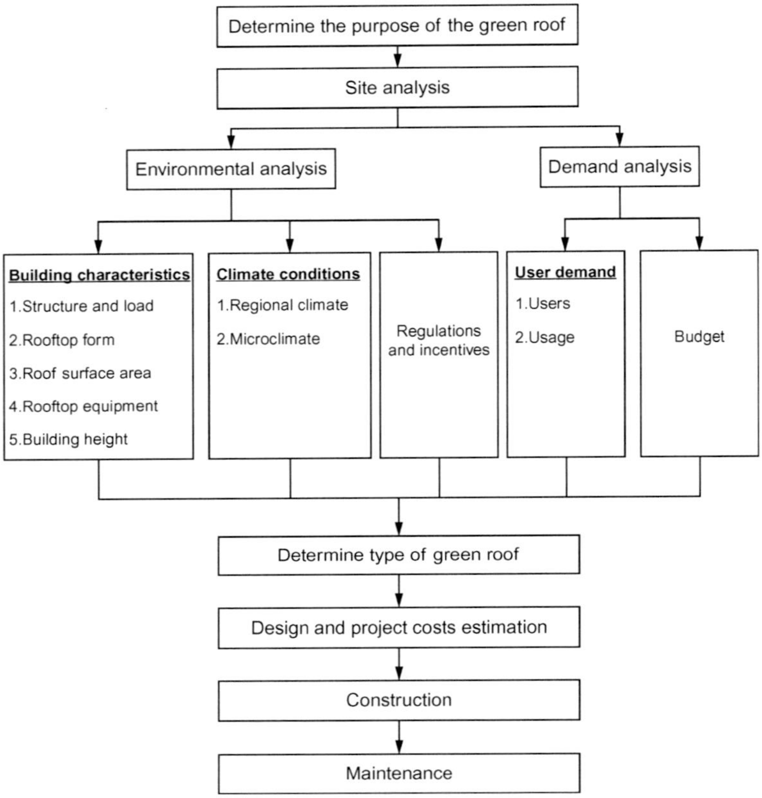

Figure 11.7 Green roof design and construction procedure.
Source: Drawn by authors.

Environmental analysis

A. Building characteristics

The main assessment categories of building characteristics include building structure and load, rooftop form, roof surface area, rooftop equipment, and building height.

(1) Structure and load

The structure of a building determines the load that the roof can bear. A high load capacity leads to considerable flexibility in the selection of green roof types.

Common building structures include those that are composed of wood, brick, steel, concrete, and reinforced concrete.

(2) Rooftop form

The rooftop form is typically either flat or sloped.

A flat roof has a drainage slope of less than 10%. The rooftop can be used as a venue for various activities. Therefore, the choice of green roof type is highly flexible.

A sloped roof cannot be a venue for activities. Only extensive green roofs can be placed on sloped roofs. Care must be taken during construction to reduce the loss of growing medium.

(3) Roof surface area

The surface area of the roof determines the amount of area available for the green roof.

(4) Rooftop equipment

Rooftops often hold equipment rooms, ventilation equipment, exterior cooling units, and water towers. The type, placement, and area occupied by the equipment will affect the design of the green roof. The affected dimensions include the following:

- Spatial design of the green roof: The amount and distribution of equipment on the roof often results in the division of the rooftop space. This division of space and the lack of aesthetic appeal of the equipment often increase the complexity of green roof design.
- Microclimate conditions: The heat created by ventilation equipment and exterior air-conditioning units results in harsh microclimate conditions on rooftops. This limits the choice of rooftop plant.

(5) Building height

Considerations related to building height include the height of the building itself as well as that of surrounding buildings. Building height influences the delivery of green roof materials, the cost of maintenance, wind flow fields on the roof, and the visual value of the scenery.

B. Climate conditions

(1) Regional climate

Regional climate conditions include rainfall, snowfall, temperature, typhoons, and rainstorms. These weather conditions directly affect the type of vegetation chosen, the growing medium formula, and the use of drainage and retention boards in the green roof. The amount of rainfall and snowfall will increase the live load of the building and will affect the type and design of the green roof chosen based on the load.

(2) Microclimate

The assessment of microclimate factors includes wind, irradiation, and shadows.

- Wind: The wind velocity and direction will affect the choice and configuration of plant. Wind velocity increases as the building height increases and the distance to the edge of the roof decreases; therefore, tall plants should not be placed near the edge of the roof.
- Irradiation and shadows: Shadows on a rooftop are primarily cast by taller buildings nearby, although structures such as water towers or equipment rooms that are taller than the building may also cast shadows. In addition, the reflection of light from nearby buildings increases the thermal energy, rendering the microclimate conditions on rooftops harsh.

C. Regulations and incentives

The assessment should include a compilation of all regulations related to rooftop construction and incentives related to green roofs.

Demand analysis

A. User demand

While considering the primary purposes of the roof, interviews should be conducted to determine who will use the green roof and their demands regarding the use of the roof.

(1) Users: Includes the number and demographics of present and future users.

(2) Usage: Includes direct and indirect uses of the roof. Direct uses include rest and relaxation as well as gardening; indirect uses include intercepting rainwater runoff and reducing cooling needs.

B. Budget

The budget is one of the major considerations for the construction of a green roof. Designers can develop an appropriate design according to the client's budget.

Determine type of green roof

The choice of green roof type must integrate the site analysis, particularly building structure and load, with rooftop form, user demand, and budget.

Design and project costs estimation

(1) Design
The designer is responsible for designing the rooftop space according to the demand of the client. The designer's work involves three stages:
  A. Schematic design
  At the initial design stage, the client is provided with rough diagrams of the design, which can include floor plans, sectional elevations, and perspective drawings. This provides the client with a clear understanding of the design concept and elements and allows the client to confirm the design plan.
  B. Design development
  Detailed designs are created according to finalized design proposals. Floor plans, sectional elevations, and perspective drawings are produced.
  C. Construction document
  The construction documents, including drawings and written specifications, are submitted.

(2) Project cost estimation
Based on the designs, project costs are estimated, including the amount of construction undertaken, the unit of construction, the cost per unit, and the total construction cost. Subsequently, construction budget documents are produced.

Construction

During construction, the designer must supervise work progress and quality to ensure that construction procedures and material specifications meet the requirements of the construction documents.

Maintenance

The maintenance plan is devised, including maintenance duties, methods, and frequency.

11.4.2 Design principles for reducing cooling needs

Designing green roofs for cooling purposes begins with the selection of the green roof type. Spatial configuration and the design of the green roof structure are crucial structural attributes in the design process. The three design principles for lowering the temperature can be categorized as follows:

11.4.2.1 Type of green roof

The primary concerns in choosing a green roof are building structures, load, and rooftop form. After these concerns are addressed, other requirements can be considered when choosing the green roof type.

(1) If the primary purposes of the green roof are cooling and energy conservation, an extensive green roof should be chosen.

The cooling effect of extensive green roofs is substantial. Extensive green roofs have the characteristics of large planting areas, low maintenance, and low implementation costs. Therefore, if the cost of maintenance is a consideration, extensive green roof is the optimal choice.

(2) If the primary purposes of the roof are cooling and energy conservation, but roof access is to be provided, a semi-intensive green roof should be chosen.

Semi-intensive green roofs and extensive green roofs share similar characteristics. However, semi-intensive roofs allow people to access the roof and use the space. Therefore, when the usage intensity is low, semi-intensive roofs can be chosen.

(3) If a green roof is to provide cooling and energy conservation, but usage intensity is high, an intensive green roof should be chosen.

Various types of green roofs exert distinct cooling effects (Susca et al., 2011; Teemusk and Mander, 2010). Overall, the cooling effect of an intensive green roof is more beneficial than that of an extensive green roof because the soil layer of the intensive green roof is thicker, which provides a more effective heat barrier. In addition, intensive green roofs can support trees, which can provide large areas of shade.

11.4.2.2 Spatial configuration

When designing a green roof, the space can be partitioned according to function. This partitioning can produce a planting area, a scenic area, and a rooftop equipment area, among others.

These functional partitions exhibit distinct thermal characteristics because of the differences in material properties. Therefore, the design, size, and layout of these partitions will affect the cooling effects of the green roof.

(1) Extensive green roof
- Install the largest possible planting areas to reduce exposed roof area.
- The periphery of rooftop equipment should undergo simple beautification, or a buffer zone should be established to reduce its influence on plant growth.

(2) Semi-intensive green roof
- Install the largest possible planting areas to reduce exposed roof area. Plant bushes in the sunniest spots to block sunlight and provide shade.
- Simple landscaping should primarily include walkways and benches. The green roof is the optimal location for these landscaping elements; if they must be placed on the exposed roof, then the space they occupy should be minimized and they should not be placed in areas with direct sunlight.
- Rooftop equipment should undergo simple beautification and can be integrated into the design of the landscape structures.

(3) Intensive green roof
- Maximize planting areas to reduce the amount of exposed roof. Plant shrubs or trees should be placed in areas of direct sunlight to block sunlight and provide shade.
- Landscape elements should include cool materials such as water features or planter boxes. Landscaping should include walkways and benches. The green roof is the optimal location for these landscaping elements; if they must be placed on the exposed roof, then the space they occupy should be minimized and they should not be placed in areas with direct sunlight.
- The layout of the landscaping elements should consider the distribution and size of rooftop equipment. Equipment should be integrated into the design as much as possible to maximize the green roof area and partition the rooftop equipment.

11.4.2.3 Green roof structure

Among the layers of the green roof, the vegetation and growing medium layers require the most decisions and are the most crucial for achieving thermal conduction. The design principles of these two layers with cooling as the main function are detailed as follows:

(1) Choose light-colored multilayered plants that provide excellent coverage.
Research has indicated that multilayered plants that provide excellent coverage produce highly beneficial cooling effects. The brightness of plants that is light in color results in high reflectivity, leading to the reflection of solar radiation and achieving the goal of efficient cooling (Lin and Lin, 2010; Lin et al., 2011).

(2) Choose growing mediums with high water retention and low thermal conduction coefficients.
The growing medium layer in green roofs should comprise low-density medium blends that have high water retention and low thermal conduction coefficients (Castleton et al., 2010; Jim, 2014b; Sailor and Hagos, 2011; Sun et al., 2013). This leads to an increased number of air pockets in the medium, an increased evapotranspiration in the medium and plants, and a decrease in the thermal conductivity of the growing medium layer. In addition, research has revealed that the thickness of the growing medium layer can affect cooling. Thick layers result in effective cooling, and a 10 cm growing medium layer can prevent most heat from penetrating the building (Jim and Tsang, 2011).

11.5 Future trends

The microclimate modification effect of green roofs has been demonstrated worldwide across various ecoregions. The cooling effect of green roofs is particularly pronounced and promoted by researchers and local governments in hot and humid regions. Because of the aesthetical and other environmental benefits of green roofs, construction companies can attract customers to green-roofed houses even when the houses are expensive. However, green roof installation in all cities is still at a nascent stage. For citizens, the major concerns associated with green roofs are installation and maintenance costs, roof leaking risks, damage to building structures, shortage of technical support, a lack of promotion and incentives from governments, and a lack of scientific data available for evaluating applicability to local conditions (Fernandez-Cañero et al., 2013; Nicholas et al., 2010; Wong and Lau, 2013; Wu and Lin, 2014; Zhang et al., 2012). Hence, addressing these concerns will be paramount for industries, researchers, and governments. At the end of this chapter, we suggest three aspects for future development:

11.5.1 Establishing standards and guidelines

A standardized environment guarantees the quality of green roofs and further increases the installation possibility. The wherewithal to install green roofs widespread and the guidelines for the planning, construction, and maintenance of green roofs are required.

Germany is recognized as the origin of modern green roofs. The FLL green guideline is the most crucial reference for roof greening. Although this guideline exists, successful green roof installation is closely related to regional climate, local building properties, and the availability of roof greening materials. Thus, regulating a specific guideline for a particular location is necessary.

In addition, investigating and developing materials are crucial. Standardized materials decrease construction and maintenance costs and increase the environmental benefits of green roofs. Properties of roof greening materials include thermal property, appearance property, fireproof property, structural load property, and duration property. The thermal properties of roof greening materials are particularly critical for the cooling effect and are mainly determined using the single-material test. However, many compound materials are widely used in roof greening. Thus, thermal property tests are required for these compound materials, particularly for vegetation and growing medium layers. Material standardization should also be coordinated with building requirements and regulations; that is, local or global building construction trends should be considered.

Different plant arrangements and growing medium blends can exert considerably different temperature reduction effects. Developing optimal formulas for vegetation and growing medium layers may be another relevant research topic in the future.

11.5.2 Information provision and technical support

Making information and technical support accessible to the public would increase the installation rate. Information provision and technical support should follow regional guidelines and contain the design and construction procedures, material lists, regular maintenance, and lists of certified roof greening companies. The local governments and greening associations should assume the responsibility of information provision and technical support. Constructing a consulting platform or agency would benefit both developers and building owners.

11.5.3 Education, legislation, and incentives

Previous studies have indicated that the public is not well aware of green roofs (Rogers, 2003; Wong and Lau, 2013; Wu and Lin, 2014). However, the public has positive attitudes toward green roofs (Jungels et al., 2013). Under this circumstance, the green roof-promoting strategy should first shift its focus to raising awareness about green roofs. Integrating information on green roofs and their benefits into environmental education programs would be one of the most effective methods for increasing the installation rate. Introducing legislation and providing incentives would encourage developers and building owners to install green roofs.

Furthermore, a green roof visible at street level has high promotion potential (Jungels et al., 2013). Governments or greening associations could promote green roofs by installing them in public places.

References

Alexandri, A., Jones, P., 2007. Developing a one-dimensional heat and mass transfer algorithm for describing the effect of green roofs on the built environment: comparison with experimental results. Build. Environ. 42, 2835–2849.

Baik, J.J., Kwak, K.H., Park, S.B., Ryu, Y.H., 2012. Effects of building roof greening on air quality in street canyons. Atmos. Environ. 61, 48–55.

Barrio, E.P.D., 1998. Analysis of the green roofs cooling potential in buildings. Energy Build. 27, 179–193.

Berardi, U., GhaffarianHoseini, A., Ali GhaffarianHoseini, A., 2014. State-of-the-art analysis of the environmental benefits of green roofs. Appl. Energy 115, 411–428.

Blank, L., Vasl, A., Levy, S., Grant, G., Kadas, G., Dafni, A., Blaustein, L., 2013. Directions in green roof research: a bibliometric study. Build. Environ. 66, 23–28.

Castleton, H.F., Stovin, V., Beck, S.B.M., Davison, J.B., 2010. Green roofs: building energy savings and the potential for retrofit. Energy Build. 42, 1582–1591.

D'Orazio, M., Di Perna, C., Di Giuseppe, E., 2012. Green roof yearly performance: a case study in a highly insulated building under temperate climate. Energy Build. 55, 439–451.

Dunnett, N., Kingsbury, N., 2004. Planting Green Roofs and Living Walls. Timber Press, Inc, Portland, OR.

Feng, H., Hewage, K., 2014. Energy saving performance of green vegetation on LEED certified buildings. Energy Build. 75, 281–289.

Fernandez-Cañero, R., Emilsson, T., Fernandez-Barba, C., Herrera Machuca, M.Á., 2013. Green roof systems: a study of public attitudes and preferences in southern Spain. J. Environ. Manage. 128, 106–115.

Francis, R.A., Lorimer, J., 2011. Urban reconciliation ecology: the potential of living roofs and walls. J. Environ. Manage. 92, 1429–1437.

Getter, K.L., Rowe, D.B., 2006. The role of extensive green roofs in sustainable development. HortScience 41, 1276–1285.

Jim, C.Y., 2014a. Heat-sink effect and indoor warming imposed by tropical extensive green roof. Ecol. Eng. 62, 1–12.

Jim, C.Y., 2014b. Building thermal-insulation effect on ambient and indoor thermal performance of green roofs. Ecol. Eng. 69, 265–275.

Jim, C.Y., Tsang, S.W., 2011. Biophysical properties and thermal performance of an intensive green roof. Build. Environ. 46, 1263–1274.

Jones, H.G., 1992. Plants and Microclimate: A Quantitative Approach to Environmental Plant Physiology, second ed. Cambridge University Press, New York.

Jungels, J., Rakow, D.A., Allred, S.B., Skelly, S.M., 2013. Attitudes and aesthetic reactions toward green roofs in the Northeastern United. Landsc. Urban Plan. 117, 13–21.

Kumar, R., Kaushik, S.C., 2005. Performance evaluation of green roof and shading for thermal protection of buildings. Build. Environ. 40, 1505–1511.

Lee, J.Y., Moon, H.J., Kim, T.I., Kim, H.W., Han, M.Y., 2013. Quantitative analysis on the urban flood mitigation effect by the extensive green roof system. Environ. Pollut. 181, 257–261.

Lilliana, L., Peng, H., Jim, C.Y., 2013. Green-roof effects on neighborhood microclimate and human thermal sensation. Energies 6, 598–618.

Lin, B.S., 2010. The cooling effects of plant and green land. Unpublished Doctoral Dissertation, Graduate Institute of Horticulture, National Taiwan University, Taipei, Taiwan.

Lin, B.S., 2012. Report on cooling effect of extensive green roof and its simulation. Taipei, Taiwan.

Lin, B.S., Lin, Y.J., 2010. Cooling effect of shade trees with different characteristics in a subtropical urban park. HortScience 45, 83–86.

Lin, B.S., Lin, Y.J., Yu, C.C., 2011. Thermal benefits of extensive green roof. In: Proceedings of 9th Landscape Academic Conference. Taipei, pp. 633–645.

Lin, B.S., Yu, C.C., Su, A.T., Lin, Y.J., 2013. Impact of climatic conditions on the thermal effectiveness of an extensive green roof. Build. Environ. 67, 26–33.

Luckett, K., 2009. Green Roof Construction and Maintenance. McGraw-Hill, USA.

Miller, D.H., 1977. Water at the Surface of the Earth: An Introduction to Ecosystem Hydrodynamics. Academic Press, New York.

Morau, D., Libelle, T., Garde, F., 2012. Performance evaluation of green roof for thermal protection of buildings in Reunion Island. Energy Procedia 14, 1008–1016.

Niachou, A., Papakonstantinou, K., Santamouris, M., Tsangrassoulis, A., Mihalakakou, G., 2001. Analysis of the green roof thermal properties and investigation of its energy performance. Energy Build. 33, 719–729.

Nicholas, S.G.W., Rayner, J.P., Raynor, K.J., 2010. Green roofs for a wide brown land: opportunities and barriers for rooftop greening in Australia. Urban For. Urban Green. 9, 245–251.

Renterghem, T.V., Botteldooren, D., 2009. Reducing the acoustical facade load from road traffic with green roofs. Build. Environ. 44, 1081–1087.

Rogers, E.M., 2003. Diffusion of Innovations, fifth ed. The Free Press, Simon & Schuster, Inc, New York, USA.

Sailor, D.J., 2008. A green roof model for building energy simulation programs. Energy Build. 40, 1466–1478.

Sailor, D.J., Hagos, M., 2011. An updated and expanded set of thermal property data for green roof growing media. Energy Build. 43, 2298–2303.

Sailor, D.J., Hutchinson, D., Bokovoy, L., 2008. Thermal property measurements for ecoroof soils common in the western U.S. Energy Build. 40, 1246–1251.

Sfakianaki, A., Pagalou, E., Pavlou, K., Santamouris, M., Assimakopoulos, M.N., 2009. Theoretical and experimental analysis of the thermal behaviour of a green roof system installed in two residential buildings in Athens, Greece. Int. J. Energy Res. 33, 1059–1069.

Spala, A., Bagiorgas, H.S., Assimakopoulos, M., Kalavrouziotis, J., Matthopoulos, D., Mihalakakou, G., 2008. On the green roof system. Selection, state of the art and energy potential investigation of a system installed in an office building in Athens, Greece. Renew. Energy 33, 173–177.

Sun, T., Bou-Zeid, E., Wang, Z.H., Zerba, E., Ni, G.H., 2013. Hydrometeorological determinants of green roof performance via a vertically-resolved model for heat and water transport. Build. Environ. 60, 211–224.

Susca, T., Gaffin, S.R., Dell'Osso, G.R., 2011. Positive effects of vegetation: urban heat island and green roofs. Environ. Pollut. 159, 2119–2126.

Teemusk, A., Mander, Ü., 2010. Temperature regime of planted roofs compared with conventional roofing systems. Ecol. Eng. 36, 91–95.

Theodosiou, T.G., 2003. Summer period analysis of the performance of a planted roof as a passive cooling technique. Energy Build. 35, 909–917.

Volder, A., Dvorak, B., 2014. Event size, substrate water content and vegetation affect storm water retention efficiency of an un-irrigated extensive green roof system in Central Texas. Sustain. Cities Soc. 10, 59–64.

White, J.W., Snodgrass, E., 2003. Extensive green roof plant selection and characteristics. In: Greening Rooftops for Sustainable Communities, (Ed.), Proceeding of 1st North American Green Roof Infrastructure Conference, Awards and Trade Show. The Cardinal Group, Toronto, pp. 166–176.

Wong, J.K.W., Lau, L.S.K., 2013. From the 'urban heat island' to the 'green island'? A preliminary investigation into the potential of retrofitting green roofs in Mongkok district of Hong Kong. Habitat Int. 39, 25–35.

Wong, N.H., Tay, S.F., Wong, R., On, C.L., Sia, A., 2003. Life cycle cost analysis of rooftop gardens in Singapore. Build. Environ. 38, 499–509.

Wong, N.H., Tan, P.Y., Chen, Y., 2007. Study of thermal performance of extensive rooftop greenery systems in the tropical climate. Build. Environ. 42, 25–54.

Wu, C.J., Lin, B.S., 2014. The Study of public construction impediments to green roof in Taiwan'. In: Proceedings of 12th Landscape Academic Conference. Taipei, C1–5, p1–15.

Yang, J., Yu, Q., Gong, P., 2008. Quantifying air pollution removal by green roofs in Chicago. Atmos. Environ. 42, 7266–7273.

Yang, H.S., Kang, J., Choi, M.S., 2012. Acoustic effects of green roof systems on a low-profiled structure at street level. Build. Environ. 50, 44–55.

Zhang, X., Shen, L., Tam, W.Y., Lee, W.Y., 2012. Barriers to implement extensive green roof systems: a Hong Kong study. Renew. Sustain. Energy Rev. 16, 314–319.

Influence of vegetation damage on urban cooling effects

12

A. Speak
University of Manchester, Manchester, United Kingdom

12.1 Introduction

Increasing the proportion of urban green space is used as a mechanism for counteracting urban heat islands (UHIs) and adapting cities to climate change pressures such as higher summer temperatures and intense precipitation. Urban planners, working at the neighborhood level, and architects, working at the building level, are recognizing the need to consider the advice of urban climatologists when designing sustainable, climate-sensitive urban forms. While the recognition of UHIs and other anthropogenically caused urban weather systems is not necessarily unprecedented (Hebbert and Jankovic, 2013), there is a definite recent surge of interest in the long-term benefits of vegetation above and beyond mere aesthetic considerations. As the noted climatologist Helmut Landsberg put it:

> *The knowledge we have acquired about urban climates should not remain an academic exercise on an interesting aspect of the atmospheric boundary layer. It should be applied to the design of new towns or the reconstruction of old ones. The purpose is, of course, to mitigate or eliminate the undesirable climate modifications brought about by urbanization.*
>
> *Landsberg (1981, 255)*

This chapter will briefly describe how vegetation can impact urban microclimates and improve the thermal comfort of cities, but more importantly will discuss how these benefits cannot manifest when the quantity and/or quality of the vegetation is impaired by damage. The benefits of natural resources such as plants to humanity are termed ecosystem services (Defra, 2007), and when these benefits are hindered, the perception of the vegetation can switch to one of an ecosystem disservice (Lyytimaki and Sipila, 2009).

The urban system—both the built infrastructure and the activities that occur within the cities—can present an inhospitable environment for all but the hardiest of plant species. These processes will be examined along with a consideration of the pressures of a warmer future climate. Finally, some techniques for preventing vegetation damage will be discussed so that the benefits of this essential component of urban design can be promoted.

Eco-efficient Materials for Mitigating Building Cooling Needs. http://dx.doi.org/10.1016/B978-1-78242-380-5.00012-1

12.2 The urban system

Before detailing the consequences of vegetation damage in cities, it is first necessary to characterize the urban setting in which these processes are occurring. Cities share a number of features that serve to produce the microclimatic effects of interest. These include the widespread use of building materials with a high thermal capacity such as brick and concrete, the arrangement of building-lined streets forming urban canyons, and a high degree of surface sealing that forces rainwater into surface gullies and subterranean drains for rapid removal of water from the roads and walkways. One might say that these represent the "gray" component of cities, while "green" is used to describe a land use that incorporates vegetative elements.

12.2.1 The urban green-gray balance

Green spaces in cities are rarely vestiges of pristine natural land around which the city has been built. They are more usually artificially created and highly maintained pockets of vegetation incorporated into the urban form, such as street trees, gardens, parks, and ornamental planters. The amount of green space in any given city is governed by a number of factors including land value, building density, climate, and history. Cultural factors are important, from the tree-lined boulevards of European cities to the dense urban centers of Asian megacities. This results in large differences in the proportion of surface area given over to parks and gardens between cities, ranging from 47% in Singapore and 46% in Sydney to 2.6% in Shanghai and 2.5% in Mumbai (BOP, 2013). The distribution of this green space within cities also varies greatly in nature—from highly florally diverse, interconnected green corridors to isolated parks surrounded by densely built-up areas. A classic example of the latter is New York City's Central Park.

With regard to lowering the high air temperatures synonymous with the UHI effect, urban vegetation achieves its effect by

- reduction of solar heat gains in urban surfaces through leaf shading (Hardin and Jensen, 2007),
- increase of solar reflectance due to the higher albedo of leaf surfaces compared to most building materials (Brown and Gillespie, 1995),
- increase of latent cooling by adding moisture to the air through evaporation from soils and leaf surfaces and transpiration from plants (Solecki et al., 2005).

The higher the proportion of the urban surface area which is given over to vegetation, the greater the cooling effect that can be expected. This leads to a less intense UHI effect and thus a lower amount of cooling degree days, which reflects the amount of energy used to cool homes and business structures with air-conditioning (Akbari and Konopacki, 2005). Tree planting in Manchester, the United Kingdom, can be used to climate-proof high-density residential areas (Hall et al., 2012) with reductions of maximum building surface temperature of between 0.5 °C and 2.3 °C being possible. However, additional planting could only increase tree cover by between 2.8% and 5.3% due to limitations on available space (Hall et al., 2012).

The cooling effect of green roofs reduces heat transfer through the roof, thus reducing the heating and cooling energy consumption in a building (Martens et al., 2008); that is, they provide an insulation function. The indoor air temperature reductions can provide direct energy savings, as every decrease in internal temperature of 0.5 °C may reduce electricity use for air-conditioning by up to 8% (Dunnett and Kingsbury, 2004). Because buildings represent 65% of total urban electricity consumption, significant energy savings could be achieved from the wide-scale greening of roofs (Kula, 2005). Urban electricity use in summer could also be reduced by the effect that large-scale roof greening may have on the UHI, with a 25% reduction in electricity use suggested by Akbari and Konopacki (2005) from the indirect effects on the UHI of cool roof installation and increasing the amount of urban vegetation.

Gill et al. (2007) characterized the urban environment of Manchester into distinct urban morphology types (UMTs) and used these to approximate the amount of evapotranspiring surfaces. This information was used in an energy exchange model, which was run for the baseline 1961–1990 climate and low- and high-emission scenarios for the 2020s, 2050s, and 2080s. Adding or removing 10% green cover had a dramatic effect on surface temperature (Figure 12.1), with addition resulting in surface temperatures being kept at or below the 1961–1990 baseline temperatures up to, but not including, the 2080s high-emission scenario (Gill et al., 2007). 10% removal of green cover from the town center UMT resulted in a surface temperature increase of 8.2 °C by the 2080s high-emission scenario compared to a rise of only 4.4 °C if no change was made to surface cover. An assumption made in this study, however, was that there is no drought stress on the vegetation.

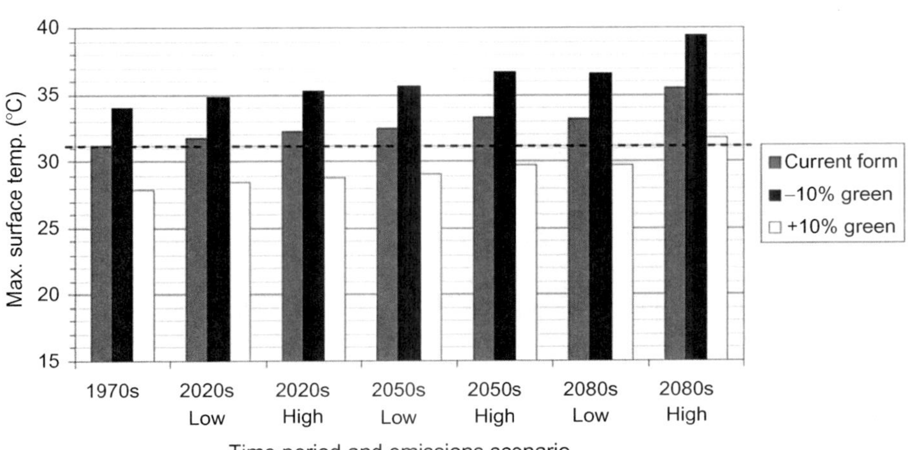

Figure 12.1 Maximum surface temperature for the 98th percentile summer day in town centers, with current form and when 10% green cover is added or removed. Dashed line shows the temperature for the 1961–1990 current form case.
Gill et al. (2007).

12.2.2 Local climate considerations

Generalizations about the impact of vegetation damage will be made in this chapter and a lot of examples come from the temperate climate city of Manchester, the United Kingdom; however, it is important to recognize that these impacts are highly dependent on the specific geographic setting of the city under consideration. The following factors are the two most highly influential in determining the quantity and quality of urban green space:

- *Aridity*. Cities in hot, arid climates, such as Phoenix in the United States and Fes in Morocco, are limited in the amount and type of vegetation that is suitable. Sometimes, in these cities, it is actually preferable to utilize techniques to reduce UHIs that are not so reliant on scarce water resources (Stone, 2012) such as light-colored building materials and cool roofs (Akbari and Matthews, 2012).
- *Latitude*. Cities in mid to high latitudes have pronounced seasonal variations in vegetation amount depending on the proportion of deciduous versus evergreen plant species. Shading from deciduous vegetation is reduced in the winter months, which can be a benefit in these colder months by allowing solar gain. Tropical cities tend to have evergreen.

In addition, altitude, local topography, and proximity to large open water bodies can all influence the climate characteristics of an area with resultant effects on vegetation distribution.

12.3 Causes of vegetation damage

The causes of damage to urban vegetation will be discussed by separating them into ground level and rooftop vegetation and then considering the effects of climate change.

12.3.1 Street trees and parks

Physical damage is perhaps the most common cause of stunted growth, sparse foliage, and mortality in urban vegetation. Cities are the sites of much human pedestrian activity and traffic flow, and vegetation can be trampled and knocked. Parks receive high numbers of visitors, especially when the weather is fine, and this can damage the grass. Vandalism is common and can affect up to 30% of freshly planted street trees in some UK towns, but generally below 5% in central European towns (Pauleit et al., 2002). Utility trenching can negatively affect street trees by damaging the roots. Their urban setting can often increase the vulnerability of plants to synoptic weather phenomena. Persistent strong winds in urban canyons can affect the growth of sensitive plants, and hurricanes can easily uproot solitary trees that aren't protected by neighbors as they would be in a forest (Burley et al., 2008).

Water is a major factor limiting plant distribution and productivity around the globe (Taiz and Zeiger, 1998), and cities present their own unique water-related problems. Street trees are often planted into pavements with minimal opportunities for water to infiltrate to the highly compacted soil. Not only is water supply reduced, but also demand is increased due to higher temperatures and lower humidities, leading to higher water loss rates from evapotranspiration. Water deficits are frequently assumed to be at least partly responsible for the early mortality and poor growth of street trees (Whitlow et al., 1992).

Urban areas are more prone to air, water, and soil pollution than rural areas due to the high concentration of industry and vehicle use. The soil of many gardens, allotments, and parks can be contaminated with heavy metals and polycyclic aromatic hydrocarbons (Culbard et al., 1988; Hursthouse et al., 2004). When these pollutants are phytotoxic, significant losses of plant productivity can occur. Ground-level ozone and photochemical smog are common in cities with large volumes of traffic and high sunlight levels (Hill, 2010), and these are very damaging to leaf tissues (Fuhrer et al., 1997). In a controlled fumigation experiment, diesel fumes were found to cause premature leaf senescence, changes in gas exchange, and growth inhibition, with evidence suggesting that NO_x was the key phytotoxic component of exhaust fumes (Bell et al., 2011). Road salting in the winter can cause roadside soils to become saline (Ramakrishna and Viraraghavan, 2005). Even cigarettes can be harmful when the tree pits are used as ashtrays by the office workers who congregate under the trees. Table 12.1 summarizes the main challenges faced by urban trees in European countries and Figures 12.2–12.5 illustrate some of the tree damages discussed.

Table 12.1 Main challenges for urban trees according to survey respondents (Pauleit et al., 2002)

	Soil	De-icing salt	Air pollution	Vandalism	Drought	Elevated summer temps	Frost damage
Greece	X			X	X	X	
Italy							
Spain	X		X	X			
France	X	X					
Croatia			X	X			
Slovenia	X	X		X	X		
Slovakia	X		X			X	
Austria	X	X		X	X		
Germany	X	X				X	X
The Netherlands	X	X		X			
Belgium	X	X	X				
UK	X	X		X			
Ireland			X	X			
Denmark	X	X	X				
Finland	X	X		X			
Norway	X	X					
Iceland		X	X				X

Figure 12.2 City center tree in the United Kingdom damaged by vandalism.

Estimates of mortality rates for trees in US cities vary from 7 years (Moll, 1989) to 19 to 28 years (Roman and Scatena, 2011), and nearly 39% of newly planted trees in Liverpool died within five years of planting (Gilbertson and Bradshaw, 1990). Compare this to the many decades that trees can live for in rural areas.

12.3.2 Green roofs

Green roofs are broadly classified as intensive or extensive. Intensive green roofs, often called roof gardens, generally have a substrate depth greater than 150–200 mm (Oberndorfer et al., 2007) and are characterized by an ability to support a wide variety of plants, but with the need for high levels of maintenance and possibly artificial irrigation. Extensive green roofs have a thinner substrate, require less maintenance, and are characterized by ground coverage plants of low height such as grasses and sedum. *Sedum* is a succulent genus chosen for its tolerance to drought and its shallow substrate adaptability (Monterusso et al., 2005; VanWoert et al., 2005).

Figure 12.3 Tree protection cage used as a litter bin.

Figure 12.4 The tree on the left is dead and will provide no shade in the summer. Cause of death is unknown.

Figure 12.5 Limited root space in the planter will put limits on the growth of this tree, and irrigation will be required in dry summer months.

As for trees, green roof vegetation is subject to the damaging effects of air pollution, but green roofs are generally inaccessible to the public; therefore, damage from trampling and vandalism is minimal. The biggest threat to green roof health is water shortage. Rooftops present a harsh environment for vegetation due to high wind velocities, full sun exposure, and large diurnal temperature extremes (Getter and Rowe, 2006). Green roof soil moisture content can rapidly decrease to the permanent wilting point, which is the soil moisture level at which a plant wilts and cannot recover its turgidity when placed in a saturated atmosphere for 12 h (Taiz and Zeiger, 1998). This soon causes mortality in non-succulent species such as grasses and ornamental plants, and in prolonged drought conditions, even xerophytic plants such as sedums can suffer. *Sedum acre* and *S. reflexum* were able to sustain photosynthesis, but with a reduced biomass, after four months without water on a 75 mm deep substrate (Durhman et al., 2006).

Figure 12.6 An intensive green roof before (left) and after (right) it was damaged by drought. The vegetation before the damage was akin to that of a meadow, and after the damage, bare substrate and mosses dominate.

Mismanagement can also play a part as was seen in Manchester when grasses and shrubs on an intensive green roof were cut back during a drought period causing large-scale mortality (Speak et al., 2013). Large areas of bare substrate were subsequently exposed (Figure 12.6), which then became colonized by invasive weeds. The green roof industry is relatively young, and some green roofs have been known to fail shortly after construction due to poor choice of plant species or inadequate design leading to drainage issues. Maintenance throughout the green roof's life, such as weeding and feeding, may also be necessary (FLL, 2008). When this is not provided, it can lead to early mortality.

12.3.3 Climate change pressures

While vegetation is touted as a form of climate change adaptation, the reverse can also be true; that is, a warmer climate and changing rainfall patterns can put increased pressure on urban vegetation. Heat waves will become more frequent, and cities such as Paris may see maximum summer temperature rise by as much as 5 °C by the end of the twenty-first century (Lemonsu et al., 2013). Problems with drought are likely to worsen in some parts of the world as a consequence of higher temperatures and lower precipitation in summer (IPCC, 2007). A modeling study investigated water stress to amenity grass in Manchester and found that in a predicted climate for the 2080s, some areas of the conurbation displayed reduced evapotranspiration for 3–5 months leading to increases in surface temperature of up to 15 °C (Gill et al., 2013). Grasses in drier areas of the United Kingdom, such as Cambridge, already suffer from browning in the summer months, which increases with soil temperature (Trudgill et al., 2010). Increases in soil temperature can reduce the hydraulic conductivities and hence water uptake of soil and plants by increasing water viscosity and membrane permeability (Kramer and Boyer, 1995).

The effect of climate change on urban vegetation is complex, with a predicted increase in biomass due to higher CO_2 levels, longer growing seasons, and a positive relationship between temperature and growth in northern forests (Bisgrove and Hadley, 2002; D'Arrigo et al., 2008) but an increase in drought stress. Stressed plants are more prone to attack from pests and disease, and a warmer climate will see an increase in the virulence and geographic spread of some of these pests (Bisgrove and Hadley, 2002). The range of many plant species will shift as climates warm with a general shift northward (Chen et al., 2011), and plant diversity will fall (Thuiller et al., 2005).

12.4 Consequences of vegetation damage

12.4.1 Measurement techniques

Before discussing the consequences, it is worth noting a number of metrics useful for quantifying vegetation damage. The main parameter to consider is the leaf area index (LAI), which is a dimensionless character of vegetation canopies and is defined as the total one-sided area of leaf tissue per unit ground surface area (Bréda, 2003). This provides a measure of the interface between canopy and atmosphere but can be difficult to quantify due to spatial (horizontal and vertical) and temporal variability. There are direct methods, such as leaf harvesting, allometry, and litter collection, and indirect methods which infer LAI from measurements of radiation transmission through the canopy (Bréda, 2003). Damage can be assessed by comparing measured LAIs with those obtained from the same species grown in an optimum or stress-free environment. Other methods of assessing plant health include physiological techniques such as water potential (Taiz and Zeiger, 1998), leaf succulence (ratio of water content of leaf to leaf surface area; Jones, 2011), and foliar nutrient levels (Rahman et al., 2011). On a green roof, vegetation cover is a useful indicator of green roof health and can be estimated via digital analysis of photographic images (Bousselot et al., 2010).

12.4.2 Building level

The main impact at the building level will result from the reduction of shade caused by vegetation damage. When shade trees suffer from reduced growth or premature leaf senescence, LAI becomes smaller allowing greater amounts of solar radiation to reach building and street surfaces. Additional lowered evapotranspirational cooling will also have a net warming effect, and this will be dealt with in the next section. Shading of building surfaces by trees can be highly beneficial for lowering cooling energy demands. It has been observed that three trees adjacent to a house in California can reduce the cooling load by between 17% and 57%. Shading accounts for 10–35% of this effect with the rest predicted to come from evapotranspiration (Akbari et al., 1997). It follows that any reduction in the LAI will have a proportional reduction in this cooling from shading.

Green roofs are being utilized increasingly in climate change adaptation strategies and as a means to reduce the costs associated with mechanical air-conditioning (Castleton et al., 2010). Green roofs may be increasingly used for their insulation properties, which reduce the heat flux through the roof (Del Barrio, 1998) with resultant energy savings well documented. Onmura et al. (2001) demonstrated a 50% reduction in heat flux into a room underneath a roof lawn garden compared to a concrete slab roof in Osaka in summer. Air temperature in a room below a well-irrigated intensive green roof was 2 °C cooler than under a control roof (Harazono et al., 1990/1991). These studies, by revealing relative cooling differences between green and conventional roofs, give an indication of the potential increase in downward heat flux on green roofs when the vegetation is damaged and areas of bare soil substrate are revealed. The effect of foliage height and density on thermal flux was shown in a study by Theodosiou (2003). Model simulations where the foliage height and LAI provided the least shading of the underlying soil resulted in lower cooling. Foliage height and LAI are both intrinsically related to the health of the vegetation. In an experiment comparing building cooling potential of green roofs with and without vegetation, it was noted that a well-watered soil substrate with no plants lacked the ability to lower indoor temperatures, showing that the shade provided by the plants is more effective than cooling by evaporation of water (Schweitzer and Erell, 2014).

The bare substrate on green roofs is often low-albedo, dark soil, which can rapidly absorb thermal energy. Data from Wong et al. (2007) showed surface temperature increases of up to 10 °C and overlying air temperature increases of around 2 °C after a new extensive green roof was constructed in Singapore. The major reason for these increases in temperature relative to the conventional roof was the incomplete foliage cover at the start of the life of the green roof, before the vegetation established. Surface warming was also observed on a green roof during summer months at higher latitudes in Tartu, Estonia (Teemusk and Mander, 2010). Bare substrate patches were given as the reason for this warming. Dark soil substrates can be responsible for elevating thermal absorption over conventional roof surfaces and thus reversing the beneficial impacts that the green roofs were initially constructed for.

Speak et al. (2013) monitored an intensive green roof after it was severely damaged by a combination of drought and mismanagement. The grass on the roof was cut very short in a prolonged dry period and the grass subsequently died, leaving large areas of soil exposed. The effect of vegetation damage on surface temperatures can be clearly seen in Figure 12.7, which shows that surface temperatures on the damaged grass in 2011 were on average 13.3 °C (70%) lower than those on the paved path. The grass was 13.1 °C (67%) cooler than the path in 2012, which was not as warm as the path in the 2011 image. The selected area over the individual plant in 2011 shows average temperatures 18.5 °C (58%) cooler than the paving, with a similar surface temperature of just over 25 °C to the grass in 2012. While not specifically measured in this study, it is expected that heat flux into the building below would have been higher in 2011. The lack of shade provision and large areas of exposed, bare soil would raise the surface temperatures, as seen in Figure 12.7 where the surface temperature of the damaged grass is 5.2 °C warmer, on average, than the healthy plant individual. Increased heat flux into the building below could be modeled with a building energy model that takes into account the U values (heat transfer coefficient) of the different roof components.

Planting trees for shade and installing green roofs can offer significant energy-saving benefits as a result of reducing the heat gain in building surfaces. It is clear, however, that these benefits rely on high LAIs, which are negatively impacted by

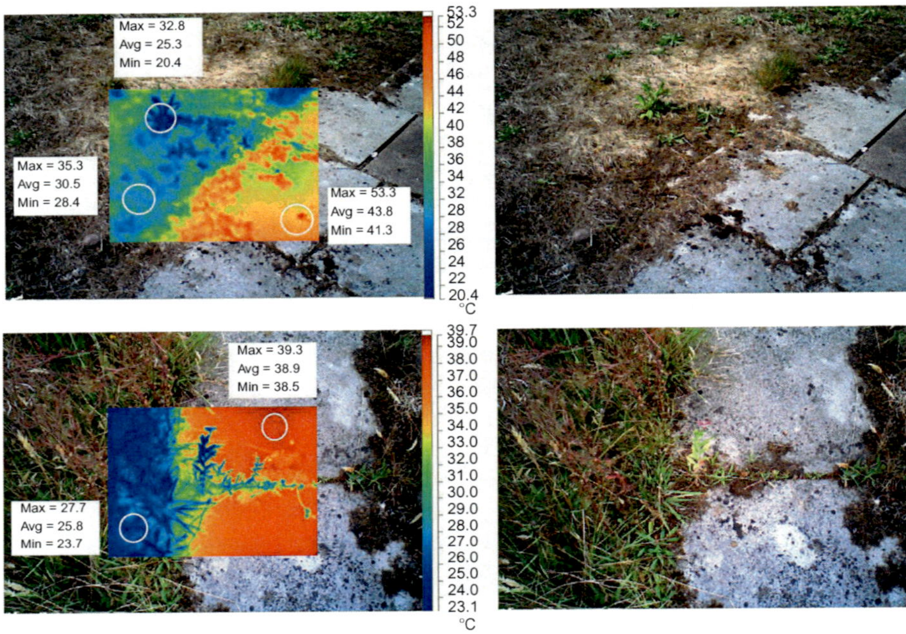

Figure 12.7 Thermal images of a damaged green roof in summer 2011 (top) and when recovered in summer 2012 (bottom; Speak et al., 2013). Insets show minimum, maximum, and average temperatures (°C) in the selected areas.

vegetation damage. LAI was shown to be the main explanatory factor for park cooling island effects in a modeling study by Vidrih and Medved (2013).

12.4.3 Air temperature and thermal comfort

The direct cooling effect of planting shade trees to alter the energy balance of a building is fairly easy to include in energy simulation models. More difficult is to quantify the indirect effects of altering the energy balance of a city (Akbari et al., 2001). Subsequently, it is also difficult to quantify the effect of vegetation damage and reduction in evapotranspirative cooling on the ambient temperature of a street or neighborhood. Akbari et al. (2001) suggested that 5–10% of urban energy demand in the United States is used to cool buildings just to compensate for the 0.5–3 °C increase in temperature attributable to the UHI effect. This pervasive elevation in ambient temperatures in cities also has important implications beyond energy demands for air-conditioning because it is the air temperature that determines the thermal comfort of humans in the city (Gulyas et al., 2006).

Some studies in Manchester have attempted to link reduced tree growth to reduced ambient cooling. Rahman et al. (2011) looked at the growth and leaf physiology of *Pyrus calleryana* trees, which had been growing in three contrasting settings: in pits

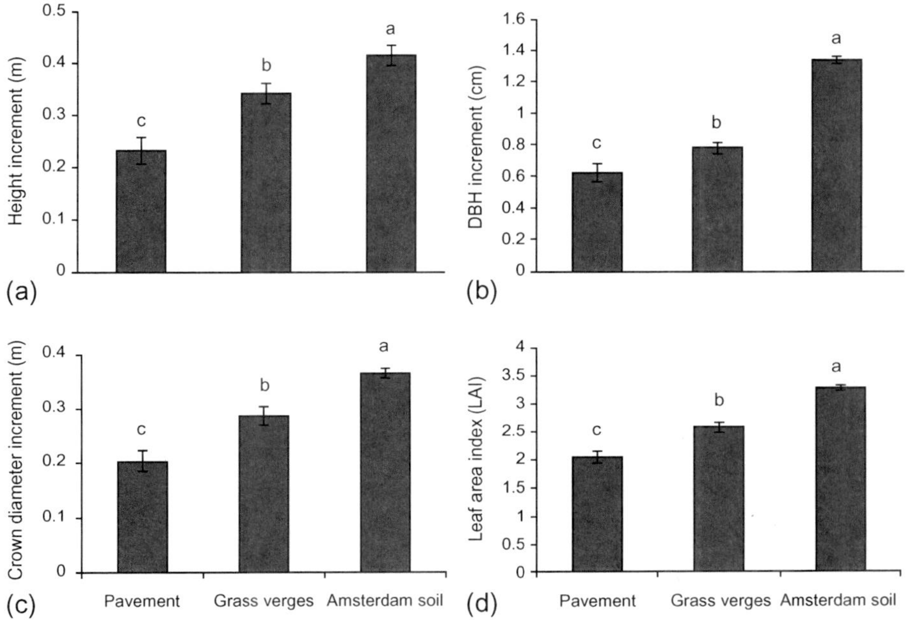

Figure 12.8 Differences in growth and morphology of *P. calleryana* grown on three different planting regimes. Annual growth increments (2004–2010) in (a) height, (b) DBH, (c) crown diameter, and (d) LAI of the crown in May 2010. Graphs show means ± standard error ($n = 15$ for paved streets, 21 for grass verges, and 13 for Amsterdam soil).
Rahman et al. (2011).

within pavements, in grass verges, and in Amsterdam soil. Amsterdam soil is a sand-based (70–80%) mixture that both supports the surrounding area by resisting compaction and supports the needs of the tree by allowing air and water to reach the roots (Couenberg, 1994). Trees grown in this soil had better performance, such as stomatal conductance and leaf water potential, than those that underwent the other two treatments (Rahman et al., 2011). The reduced growth (Figure 12.8) and physiological performance of trees grown in pavements resulted in an estimated five times decrease in peak evapotranspirational cooling (1 kW), compared with those of the Amsterdam soil trees. A further study on *P. calleryana* looked at the effect of growing the trees in pits covered in permeable paving. Young trees in covered pits exhibited growth rates almost half that of trees grown in open pits (Rahman et al., 2013) and provided only 350–650 W of cooling compared with 1 kW for the open pit trees. Lack of soil oxygen hindering deeper rooting, rather than restricted soil moisture content, was the reason for the poor performance (Rahman et al., 2013). The reduced canopy density and canopy spread observed in these trees would also have implications for cooling by shading.

Regarding grass in parks, in periods of summer drought when the grass turns brown or is trampled to bare soil, then evapotranspirative cooling will be reduced. Modeling work has shown that surface temperatures of UMTs with large proportions of grass, such as school playing fields, can be increased by 14–15 °C under short periods of severe drought. The grass can lose over 50% of its evaporative cooling effect and will behave like bare soil (Gill, 2006; Gill et al., 2007).

Few studies have looked at the ambient cooling impacts of green roofs, concentrating more on the surface temperature benefits. In Speak et al. (2013), air temperature at 300 mm above the surface was recorded for a period of 17 months over the green roof and over an adjacent paved roof. The monthly median air temperature over the green roof when healthy was up to 1.06 °C cooler than the conventional roof. This cooling effect was strongest at night with an average difference of 1.58 °C and a maximum difference of 4.2 °C seen. Vegetation resulted in an impairment of the green roof cooling effect with a lower maximum average relative cooling of 0.78 °C observed, and daytime temperatures were higher than over the bare roof for a larger proportion of the day than over the undamaged green roof section. The green roof in this study was observed to have a complex diurnal pattern to its thermal microclimate impacts, which traditional theories (Di Giovanni et al., 2013) based on evapotranspirative cooling, were not able to fully explain. Differences in the thermal capacity of the roof materials and vertical mixing characteristics could have been influencing the patterns seen. Vegetated surfaces can be quite complex in their thermal behavior due to lower wind speeds caused by friction from the plant biomass, evapotranspiration from the water component of the plant system, and plant-canopy interactions with both incoming solar radiation and outgoing long-wave radiation (Erell et al., 2011).

It is clear that evapotranspirative cooling will be reduced when vegetation is damaged and/or under water stress. Both of these conditions may be common in urban areas, so for maximal ambient cooling from vegetation, adequate water supplies and protection for the plants are required.

12.5 Damage prevention techniques

12.5.1 Species choices

Vegetation damage frequently arises from a misinformed choice of plant species for the urban environment. Factors influencing the choice of species are often related to aesthetic properties, cost, and availability when resilience, durability, and potential for ecosystem service provision are more important. Additionally, it is increasingly necessary to consider future climate pressures for a given area and make choices of species based on long-term goals for climate change adaptation (Carter, 2011). There is a wealth of information on the climate preferences and biogeography of common plants (Tivy, 1993), and these can be used to identify species that may be suitable for a location in the future. For instance, the UK Climate Impacts Programme highlight several tree species that should benefit from extra warmth in Britain, such as Silver Maple (*Acer saccharinum*) and Italian Cypress (*Cupressus sempervirens*; Bisgrove and Hadley, 2002).

There is a good deal of research looking at species choices for green roofs (Butler et al., 2012; Dvorak and Volder, 2010; Lundholm et al., 2010; MacIvor et al., 2011), which aim to preserve the ecosystem services they provide while also protecting local biodiversity (Francis and Lorimer, 2011). For example, in the Midwestern United States, there is interest in recreating natural prairies, common to the area, by using local native species (Durhman et al., 2006). These native plants are drought-tolerant but unfortunately rely on deep extensive root systems to obtain water, which are not possible on a green roof with limited substrate depth. Xerophytic plants, such as succulents, are a sensible choice for green roofs because they are physiologically and morphologically adapted to the harsh, dry environment. Plants that utilize crassulacean acid metabolism (CAM) are particularly useful because gas exchange occurs only at night, meaning that leaf stomata can close during the day, thus reducing transpirational water losses (Sayed, 2001). In South Australia, the succulent species, *Carpobrotus rossii*, showed 100% survival on a green roof and exhibited maximum horizontal growth rate, leaf succulence, shoot biomass, and water use efficiency compared with three other native Australian species (Razzaghmanesh et al., 2014). Schweitzer and Erell (2014) showed another succulent, *Aptenia cordifolia*, provided the highest cooling benefit per unit irrigation water in the Eastern Mediterranean. However, they suggested that in such a hot, arid climate, the overall cooling efficiency was low and should not be considered on the basis of building energy conservation alone but on the other benefits that make green roofs attractive, such as air pollution reduction (Speak et al., 2012).

Gill et al. (2013) recommended using trees instead of grass in urban parks because their rooting systems can reach water at greater depths, and thus, evapotranspirative cooling is still provided in droughts. Trees have the additional benefit over grass of providing shade. Armson et al. (2012) showed that tree shade can reduce globe temperatures by up to 5–7 °C, while surface type, such as grass versus concrete, had little effect.

There are sometimes historical reasons for the dominance of a few species in cities. In London, the London plane tree (*Platanus × hispanica*) was widely planted due to its resistance to air pollution thanks to its technique of shedding bark. The planting of only a few species can give cause for concern since species diversity is an important factor in boosting the resilience of urban tree populations to abiotic and biotic stresses (Pauleit et al., 2002). Pauleit et al. (2002) suggested the adoption of standards and regulations to improve the quality of tree selection with a wide dissemination of good practice. Sjöman et al. (2012) identified 27 tree species suitable for warm and periodically dry habitats from a dendroecological study, of which only four are currently used in Northern European cities.

The uptake of new, drought-resistant, choices of plants for climate proofing cities will potentially require a paradigm shift. Quite often, the plants that are best adapted to the urban environment are considered to be "invasive" or "weeds." The buddleia (*Buddleja davidii*) is a classic example of these advantageous, ruderal plant species that are quick to colonize the corners of cities that receive little maintenance. It is a ubiquitous urban plant that can seemingly grow out of thin air in the cracks and mortar of aging buildings (Figure 12.9) and can provide ecosystem services such as

Figure 12.9 A buddleia plant growing out of a brick structure.

shading of building surfaces. It is also very good for biodiversity as its flowers are much loved by pollinating insects. Urban spontaneous vegetation was found to be better for biodiversity than lawn habitats while being similar in terms of climatic regulation services (Robinson and Lundholm, 2012). There may need to be a cultural shift away from viewing such vegetation as an eyesore or nuisance.

12.5.2 Artificial irrigation

An obvious technique to prevent drought damage to vegetation is to water the plants manually. This ensures that evapotranspirative cooling is maintained, even in hot, dry periods and the substrates are prevented from reaching their permanent wilting point. The main problem with this technique is sourcing the water. In many parts of the world, access to fresh water is a major health and political issue, and using municipal water supplies for watering street trees will be a low priority. In the twentieth century, the world's population tripled, while the use of renewable water sources grew sixfold. Future population growth will result in an increasing demand for water and will have serious consequences on the environment (World Water Council, 2010). Demand for water for irrigation is increasing just when it is becoming more scarce.

The solution therefore is to look at alternative sources of water and improve the efficiency of irrigation techniques. Alternative water sources include harvesting rainwater in wet periods for use in dry periods, potentially in storage tanks built into the building structure in the utility spaces. Moritani et al. (2013) suggested that intermittent irrigation of green roofs with municipal wastewater is acceptable. The higher salinity of this water (6–18 dS m^{-1}) can reduce soil water potentials and hence evapotranspiration, but CAM plants in their study responded by increasing water use efficiency.

The efficiency of water use by turf grasses can be improved with irrigation programs that consider evapotranspiration losses, soil moisture retention, rooting depth, infiltration rate, and type of turf as opposed to one that simply advocates occasional watering in the summer months. Plots of Bermuda grass (*Cynodon dactylon*) in a subtropical climate were irrigated in response to data from soil moisture sensors. Irrigation water savings of 42–95% over conventional techniques were obtained (Augustin and Snyder, 1984). Technology-based water management could also consider rainfall, temperature, and wind speed data to calculate evaporative losses, which then trigger computerized irrigation. Also, by promoting deep rooting through thatch control, aeration, moderate fertilization, and infrequent irrigation, significant quantities of water can be saved. Further savings can be achieved by planting drought-resistant grass varieties or using subsurface drip lines (Taylor et al., 2006). Returning grass clippings to a lawn can improve the resilience to drought by retaining moisture and lowering N leaching (Trudgill et al., 2010).

Finally, it is suggested to determine the water loss crop coefficient (K_c) for a wide range of landscape plants (Lowry et al., 2011). K_c is the proportion of the reference evapotranspiration (hypothetical water loss rate for a climatic zone used as a standard measure for comparing individual plants), which is required by a plant species to thrive productively. Establishing a database of K_c for common urban and suburban plants will facilitate the assessment of urban water needs by landscape managers (Pittenger and Shaw, 2007).

12.5.3 Technology

Technology can improve the efficiency of irrigation as mentioned above, and it can also aid in improving the living conditions of urban vegetation in other ways. Research on the preparation of soils or the design of planting pits can feed into the promotion and implementation of state-of-the-art practices (Pauleit et al., 2002). Amsterdam soil has already been mentioned as a simple soil composition blueprint, which can improve subsurface drainage and gas exchange. Pervious concretes are a technique that omits the fine aggregates from normal concrete so that pores are formed within the material. Viswanathan et al. (2011) found that soil under pervious concrete had higher oxygen levels and 2% more volumetric water content than that under impervious concrete. There were no differences in root growth, but this was potentially due to the high clay content of the soil reducing CO_2 exchange.

There are a number of novel green roof substrates under investigation, such as expanded clay, which aim to keep the weight load of green roofs as low as possible while enhancing drainage. To aid proper moisture management, the substrate media must efficiently absorb and retain water, offer good drainage, and have a high porosity (Miller, 2003). Moisture retaining elements include mulching and addition of moisture-retaining materials to the substrate (Dunnett and Kingsbury, 2004).

12.6 Conclusion and future trends

There is currently a distinct lack of research looking specifically at the effects of vegetation damage on urban cooling. Therefore, by necessity, conclusions must be made from comparing green case studies with nongreen ones. The assumption is then that the less green space in an urban environment, the more the thermal microclimate is dominated by the absorption and storage of short-wave solar radiation during the day and reradiation of long-wave radiation at night. Replacement of vegetation with asphalt in the urban climate model ENVI-met (Figure 12.10) increased daytime air temperatures by 2.4 °C on average (Skelhorn et al., 2014). Damage impairs the direct shading effects and the indirect cooling from evapotranspiration and shifts the energy balance of the local area to one dominated by materials such as asphalt in the example.

It is important to quantify the impacts that drought stress has on the cooling ability of vegetation because climate change predictions for many areas indicate a stressful future for urban vegetation. One of the findings of the modeling work by Gill et al. (2007) for the United Kingdom is that, even by the 2020s under a low-emission scenario, drought becomes apparent because of the combined effect of low precipitation and high temperatures. Extra care must be taken by landscape managers, and urban planners to ensure the green spaces of cities are not adversely affected in the future. This may incur considerable costs from tree acquisition, protection, site preparation and maintenance, etc. As can be seen in Table 12.2, average costs per tree in Euros ranges from 13 in Spain to 5000 in Norway (Pauleit et al., 2002). This must be balanced against the long-term energy savings along with other benefits of urban vegetation such as air pollution reduction, rainwater retention, and aesthetics.

The built environment generally has a lifetime of 20–100 years, so climate change adaptation can soon be incorporated into future building stock (Graves and Phillipson, 2000), but the time to start is now.

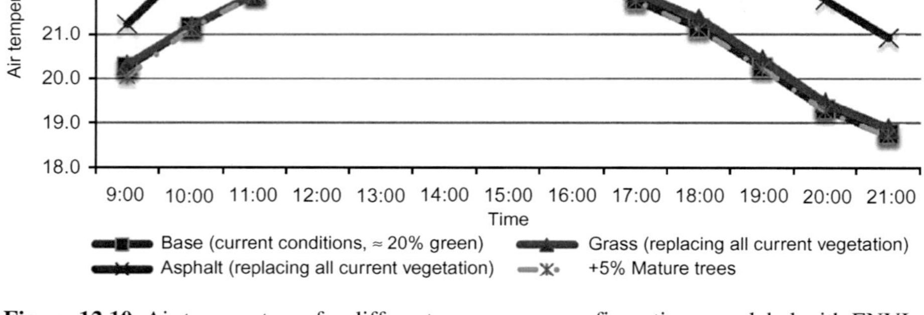

Figure 12.10 Air temperatures for different green space configurations, modeled with ENVI-met in Wythenshawe, Manchester.
Skelhorn et al. (2014).

Table 12.2 Average costs for new tree plantings in Europe (Pauleit et al., 2002)

	Average costs for new street tree plantings (Euro per tree)[a]	Range of costs (Euro per tree)[a]	Average costs by categories
Spain	13	13	
Slovenia	<100	<100	≤250 Euro
UK	120	50–190	
Italy	350	284–413	
Austria	600	470–727	>250–1000
Belgium	770	40–1500	
France	1250	200–2300	
Denmark	1375	750–2000	>1000–2000
Germany	1450	500–2400	
Finland	1670	1667–1681	
Norway	3750	2500–5000	>2000

[a]Overall costs including acquisition of trees and site preparation.

12.7 Sources of further information

Several aspects of this chapter were necessarily summarized, and the reader is directed to the following sources for more in-depth information.

12.7.1 Urban microclimates

Erell, E., Pearlmutter, D., Williamson, T., 2011. Urban Microclimate: Designing the Spaces Between Buildings. Earthscan Ltd, London.

12.7.2 Green roof construction

FLL, 2008a. Forschungsgesellschaft Landschaftsentwicklung Landschaftsbau e. V. (FLL) Guidelines for the Planning, Construction and Maintenance of Green Roofing. English language edition, FLL, Bonn (available at www.fll.de ed.).

References

Akbari, H., Konopacki, S., 2005. Calculating energy-saving potentials of heat-island reduction strategies. Energy Policy 33 (6), 721–756.

Akbari, H., Kurn, D.M., Bretz, S.E., Hanford, J.W., 1997. Peak power and cooling energy savings of shade trees. Energy Build. 25, 139–148.

Akbari, H., Matthews, H.D., 2012. Global cooling updates: reflective roofs and pavements. Energy Build. 55, 2–6.

Akbari, H., Pomerantz, M., Taha, H., 2001. Cool surfaces and shade trees to reduce energy use and improve air quality in urban areas. Sol. Energy 70 (3), 295–310.

Armson, D., Stringer, P., Ennos, A.R., 2012. The effect of tree shade and grass on surface and globe temperatures in an urban area. Urban For. Urban Greening 11 (3), 245–255.

Augustin, B.J., Snyder, G.H., 1984. Moisture sensor-controlled irrigation for maintaining bermudagrass turf. Agron. J. 76 (5), 848–850.

Bell, J.N.B., Honour, S.L., Power, S.A., 2011. Effects of vehicle exhaust emissions on urban wild plant species. Environ. Pollut. 159, 1984–1990.

Bisgrove, R., Hadley, P., 2002. Gardening in the Global Greenhouse: The Impacts of Climate Change on Gardens in the UK. Technical Report, UKCIP, Oxford.

BOP, 2013. World Cities Culture Report 2013 BOP Consulting, London. Available online at http://www.worldcitiescultureforum.com/world-cities-culture-report-2013-high-resolution.

Bousselot, J.M., Klett, J.E., Koski, R.D., 2010. Extensive green roof species evaluations using digital image analysis. HortScience 45 (8), 1288–1292.

Bréda, N.J.J., 2003. Ground-based measurements of leaf area index: a review of methods, instruments and current controversies. J. Exp. Biol. 45 (392), 2403–2417.

Brown, R.D., Gillespie, T.J., 1995. Microclimate Landscape Design: Creating Thermal Comfort and Energy Efficiency. John Wiley & Sons, Chichester.

Burley, S., Robinson, S.L., Lundholm, J.T., 2008. Post-hurricane vegetation recovery in an urban forest. Landsc. Urban Plan. 85, 111–122.

Butler, C., Butler, E., Orians, C.M., 2012. Native plant enthusiasm reaches new heights: perceptions, evidence, and the future of green roofs. Urban For. Urban Greening 11, 1–10.

Carter, J.G., 2011. Climate change adaptation in European cities. Curr. Opin. Environ. Sustain. 3, 193–198.

Castleton, H.F., Stovin, V., Beck, S.B.M., Davison, J.B., 2010. Green roofs; building energy savings and the potential for retrofit. Energy Build. 42, 1582–1591.

Chen, I.-C., Hill, J.K., Ohlemueller, R., Roy, D.B., Thomas, C.D., 2011. Rapid range shifts of species associated with high levels of climate warming. Science 333 (6045), 1024–1026.

Couenberg, E.A.M., 1994. Amsterdam tree soil. In: Watson, G.W., Neely, D. (Eds.), The Landscape Below Ground: Proceedings of an International Workshop on Tree Root Development in Urban Soils. International Society of Arboriculture, Champaign, IL, pp. 24–33.

Culbard, E.B., Thornton, I., Watt, J., Wheatley, M., Thompson, M., 1988. Metal contamination in British urban dusts and soils. J. Environ. Qual. 17, 226–234.

D'Arrigo, R., Wilson, R., Liepert, B., Cherubini, P., 2008. On the 'Divergence Problem' in Northern forests: a review of the tree-ring evidence and possible causes. Global Planet. Change 60, 289–305.

Defra, 2007. An introductory guide to valuing ecosystem services. Report published by the Department for Environment, Food and Rural Affairs Accessed online 17/04/2014 https://www.gov.uk/government/uploads/system/uploads/attachment_data/file/181882/pb12852-eco-valuing.pdf.

Del Barrio, E.P., 1998. Analysis of the green roofs cooling potential in buildings. Energy Build. 27, 179–193.

Di Giovanni, K., Montalto, F., Gaffin, S., Rosenzweig, C., 2013. Applicability of classical predictive equations for the estimation of evapotranspiration from urban green spaces: green roof results. J. Hydrol. Eng. 18 (1), 99–107.

Dunnett, N., Kingsbury, N., 2004. Planting Green Roofs and Living Walls. Timber Press, Portland.

Durhman, A.K., Rowe, B.D., Rugh, C.L., 2006. Effect of watering regimen on chlorophyll fluorescence and growth of selected green roof plant taxa. HortScience 41 (7), 1623–1628.

Dvorak, B., Volder, A., 2010. Green roof vegetation for North American ecoregions: a literature review. Landsc. Urban Plan. 96, 197–213.

FLL, 2008b. Forschungsgesellschaft Landschaftsentwicklung Landschaftsbau e. V. (FLL) Guidelines for the Planning, Construction and Maintenance of Green Roofing. English language edition, available at www.fll.de ed. FLL, Bonn.

Francis, R.A., Lorimer, J., 2011. Urban reconciliation ecology: the potential of living roofs and walls. J. Environ. Manage. 92 (6), 1429–1437.

Fuhrer, J., Skärby, L., Ashmore, M.R., 1997. Critical levels for ozone effects on vegetation in Europe. Environ. Pollut. 97 (1–2), 91–106.

Getter, K.L., Rowe, D.B., 2006. The role of extensive green roofs in sustainable development. HortScience 41 (5), 1276–1285.

Gilbertson, P., Bradshaw, A.D., 1990. The survival of newly planted trees in inner cities. Arboric. J. 14, 287–309.

Gill, S., 2006. Climate change and urban greenspace. Ph.D. Thesis, University of Manchester.

Gill, S.E., Handley, J.F., Ennos, A.R., Pauleit, S., 2007. Adapting cities for climate change: the role of the green infrastructure. Build. Environ. 33 (1), 115–133.

Gill, S.E., Rahman, M.A., Handley, J.F., Ennos, A.R., 2013. Modelling water stress to urban amenity grass in Manchester UK under climate change and its potential impacts in reducing urban cooling. Urban For. Urban Greening 12, 350–358.

Graves, H.M., Phillipson, M.C., 2000. Potential Implications of Climate Change in the Built Environment. FBE Report 2. BRE, Centre for Environmental Engineering, East Kilbride.

Gulyas, A., Unger, J., Matzarakis, A., 2006. Assessment of the microclimatic and human comfort conditions in a complex urban environment: modelling and measurements. Build. Environ. 41, 1713–1722.

Hall, J.M., Handley, J.F., Ennos, A.R., 2012. The potential of tree planting to climate-proof high density residential areas in Manchester, UK. Landsc. Urban Plan. 104 (3–4), 410–417.

Harazono, Y., Teraoka, S., Nakase, I., Ikeda, H., 1990/1. Effects of rooftop vegetation using artificial substrates on the urban climate and the thermal load of buildings. Energy Build. 15–16, 435–444.

Hardin, P.J., Jensen, R.R., 2007. The effect of urban leaf area on summertime urban surface kinetic temperatures: a Terre Haute case study. Urban For. Urban Greening 6 (2), 63–72.

Hebbert, M.J., Jankovic, V., 2013. Cities and climate change: the precedents and why they matter. Urban Stud. 50 (7), 1332–1347.

Hill, M.K., 2010. Understanding Environmental Pollution, third ed. Cambridge University Press, Cambridge.

Hursthouse, A., Tognarelli, D., Tucker, P., Marsan, F.A., Martini, C., Madrid, L., Madrid, F., Diaz-Barrientos, E., 2004. Metal content of surface soils in parks and allotments from three European cities: initial pilot study results. Land Contam. Reclam. 12 (3), 189–196.

IPCC, 2007. Climate Change 2007: The Physical Science Bases. Contribution of Working Group I to the Fourth Assessment Report of the Intergovernmental Panel on Climate Change.

Jones, L.A., 2011. Anatomical adaptations of four *Crassula* species to water availability. Biosci. Horiz. 4, 13–22.

Kramer, P.J., Boyer, J.S., 1995. Water Relations of Plants and Soil. Academic Press, San Diego.

Kula, R., 2005. Green roofs and the LEED green building rating system. In: Greening Rooftops for Sustainable Communities Conference Proceedings. Sustainable Solutions, Inc, Washington, DC, USA, May 4–6, 2005.

Landsberg, H., 1981. The Urban Climate. International Geophysics Series, vol. 28 Academic Press, New York.

Lemonsu, A., Kounkou-Arnaud, R., Desplat, J., Salagnac, J.L., Masson, V., 2013. Evolution of the Parisian urban climate under a global changing climate. Clim. Change 116 (3–4), 679–692.

Lowry Jr., J.H., Ramsey, R.D., Kjelgren, R.K., 2011. Predicting urban forest growth and its impact on residential landscape water demand in a semiarid urban environment. Urban For. Urban Greening 10, 193–204.

Lundholm, J., MacIvor, J.S., MacDougall, Z., Ranalli, M., 2010. Plant species and functional group combinations affect green roof ecosystem functions. PLoS One 5 (3), 11.

Lyytimaki, J., Sipila, M., 2009. Hopping on one leg—the challenge of ecosystem disservices for urban green management. Urban For. Urban Greening 8 (4), 309–315.

MacIvor, J.S., Ranalli, M.A., Lundholm, J.T., 2011. Performance of dryland and wetland plant species on extensive green roofs. Ann. Bot. 107, 671–679.

Martens, R., Bass, B., Alcazar, S.S., 2008. Roof-envelope ratio impact on green roof energy performance. Urban Ecosyst. 11 (4), 399–408.

Miller, C.P.E., 2003. Moisture management in green roofs. Greening Rooftops for Sustainable Communities Conference, Chicago.

Moll, C.W., 1989. The state of our urban forest. Am. For. 95, 61–64.

Monterusso, M.A., Rowe, D.B., Rugh, C.L., 2005. Establishment and persistence of Sedum spp. and native taxa for green roof applications. HortScience 40 (2), 391–396.

Moritani, S., Yamamoto, T., Andry, H., Inoue, M., Kato, K., Saito, H., 2013. Effect of combined water and salinity stress factors on evapotranspiration *of Sedum kamtschaticum* Fischer in relation to green roof irrigation. Urban For. Urban Greening 12, 338–343.

Oberndorfer, E., Lundholm, J., Bass, B., Coffman, R.R., Doshi, H., Dunnett, N., Gaffin, S., Kohler, M., Liu, K.K.Y., Rowe, B., 2007. Green roofs as urban ecosystems: ecological structures, functions, and services. Bioscience 57 (10), 823–833.

Onmura, S., Matsumoto, M., Hokoi, S., 2001. Study on evaporative cooling effect of roof lawn gardens. Energy Build. 33, 653–666.

Pauleit, S., Jones, N., Garcia-Marin, G., Garcia-Valdecantos, J.-L., Riviere, L.M., Vidal-Beaudet, L., Bodson, M., Randrup, T.B., 2002. Tree establishment practice in towns and cities—results from a European survey. Urban For. Urban Greening 1, 83–96.

Pittenger, D.R., Shaw, D.A., 2007. Review of research on water needs of landscape plants. In: 2007 Symposium on Efficient Water Use in the Urban Landscape. New Mexico State University, Las Cruces, NM.

Rahman, M.A., Smith, J.G., Stringer, P., Ennos, A.R., 2011. Effect of rooting conditions on the growth and cooling ability of *Pyrus calleryana*. Urban For. Urban Greening 10, 185–192.

Rahman, M.A., Stringer, P., Ennos, A.R., 2013. Effect of pit design and soil composition on performance of *Pyrus calleryana* street trees in the establishment period. Arboric. Urban For. 39 (6), 256–266.

Ramakrishna, D., Viraraghavan, T., 2005. Environmental impact of chemical deicers—a review. Water Air Soil Pollut. 166, 49–63.

Razzaghmanesh, M., Beecham, S., Kazemi, F., 2014. The growth and survival of plants in urban green roofs in a dry climate. Sci. Total Environ. 476–477, 288–297.

Robinson, S.L., Lundholm, J.T., 2012. Ecosystem services provided by urban spontaneous vegetation. Urban Ecosyst. 15 (3), 545–557.

Roman, L.A., Scatena, F.N., 2011. Street tree survival rates: meta-analysis of previous studies and application to a field survey in Philadelphia, PA, USA. Urban For. Urban Greening 10 (4), 269–274.

Sayed, O.H., 2001. Crassulacean acid metabolism 1975–2000, a checklist. Photosynthetica 39, 339–359.

Schweitzer, O., Erell, E., 2014. Evaluation of the energy performance and irrigation requirements of extensive green roofs in a water-scarce Mediterranean climate. Energy Build. 68, 25–32.

Sjöman, H., Gunnarsson, A., Pauleit, S., Bothmer, R., 2012. Selection approach of urban trees for inner-city environments: learning from nature. Arboric. Urban For. 38 (5), 195–204.

Skelhorn, C., Lindley, S., Levermore, G., 2014. The impact of vegetation types on air and surface temperatures in a temperate city: a fine scale assessment in Manchester, UK. Landsc. Urban Plan. 121, 129–140.

Solecki, W.D., Rosenzweig, C., Parshall, L., Pope, G., Clark, M., Cox, J., Wiencke, M., 2005. Mitigation of the heat island effect in urban New Jersey. Global Environ. Change B: Environ. Hazards 6 (1), 39–49.

Speak, A.F., Rothwell, J.J., Lindley, S.J., Smith, C.L., 2012. Urban particulate pollution by four species of green roof vegetation in a UK city. Atmos. Environ. 61, 283–293.

Speak, A.F., Rothwell, J.J., Lindley, S.J., Smith, C.L., 2013. Reduction of the urban cooling effects of an intensive green roof due to vegetation damage. Urban Climate 3, 40–55.

Stone, B., 2012. The City and the Coming Climate. Cambridge University Press, Cambridge.

Taiz, L., Zeiger, E., 1998. Plant Physiology, second ed. Sinauer Associates Inc, Sunderland, MA.

Taylor, K.S., Anda, M., Sturman, J., Mathew, G.H., 2006. Subsurface dripline tubing—an experimental design for assessing the effectiveness of using dripline to apply treated wastewater for turf irrigation in Western Australia. Desalination 1–3 (5), 375–385.

Teemusk, A., Mander, U., 2010. Temperature regime of planted roofs compared with conventional roofing systems. Ecol. Eng. 36, 91–95.

Theodosiou, T., 2003. Summer period analysis of the performance of a planted roof as a passive cooling technique. Energy Build. 35, 909–917.

Thuiller, W., Lavorel, S., Araujo, M.B., Sykes, M.T., Prentice, C., 2005. Climate change threats to plant diversity in Europe. Proc. Natl. Acad. Sci. U. S. A. 102 (23), 8245–8250.

Tivy, J., 1993. Biogeography: A Study of Plants in the Ecosphere, third ed. Longman Sci & Tech, Michigan.

Trudgill, S., Jeffery, A., Parker, J., 2010. Climate change and the resilience of the domestic lawn. Appl. Geogr. 30, 177–190.

VanWoert, N.D., Rowe, D.B., Andresen, J.A., Rugh, C.L., Xiao, L., 2005. Watering regime and green roof substrate design affect Sedum plant growth. HortScience 40 (3), 659–664.

Vidrih, B., Medved, S., 2013. Multiparametric model of urban park cooling island. Urban For. Urban Greening 12, 220–229.

Viswanathan, B., Volder, A., Watson, W.T., Aitkenhead-Peterson, J.A., 2011. Impervious and pervious pavements increase soil CO_2 concentrations and reduce root production of American sweetgum (*Liquidambar styraciflua*). Urban For. Urban Greening 10, 133–139.

Whitlow, T.H., Bassuk, N.L., Reichert, D.L., 1992. A 3-year study of water relations of urban street trees. J. Appl. Ecol. 29, 436–450.

Wong, N.H., Puay yok, T., Yu, C., 2007. Study of thermal performance of extensive rooftop greenery systems in the tropical climate. Build. Environ. 42, 25–54.

World Water Council, 2010. Water Crisis. Accessed online 15.04.14 at http://www.worldwatercouncil.org/library/archives/water-crisis/.

Technical and economic analysis of green roofs to reduce building cooling needs

13

F. Ascione[1], N. Bianco[1], R.F. De Masi[1], F. de Rossi[2], G.P. Vanoli[2]
[1]University of Naples Federico II, Napoli, Italy; [2]University of Sannio, Benevento, Italy

13.1 Introduction: international framework in matters of energy efficiency in buildings

Recently, the European Community (EC), consistent with a worldwide attention towards a low-carbon future, enacted wide legislation for improving energy performances of new and existing buildings. Starting from the EC Commission Directive 2002/91/EC (2003), all Member States have been required to adapt future building activity and refurbishing objectives according to a common strategy. Some years later, the EC Commission Directive 2010/31/EC (2010) amended the previous indications, and imposed new ambitious targets by establishing the goals of nearly zero-energy buildings. "*The nearly zero or very low amount of energy required should be covered to a very significant extent by energy from renewable sources, including energy from renewable sources produced on-site or nearby.*" The EU Member States have to ensure nearly zero-energy new buildings starting from 31 December 2018 for public buildings and from 31 December 2020 for all others. Recently, the EC Commission Directive 2012/27/EC (2012) enacted some other relevant measures, among them the mandatory energy refurbishment of 3% yearly of the total floor area of buildings owned/occupied by central governments, as well as the request of a long-term strategy for the renovation of the building stock of each Member State.

New attention concerns energy measures for reducing the energy demands for space cooling at the national level. For instance, the Italian DPR 59/2009 (2009) establishes mandatory adoption of energy efficiency actions for reducing the cooling demand of buildings, by imposing solar screens, thermal mass, low periodic thermal transmittance of envelope, low decrement factor, and high time-lag effect. Innovative strategies can be adopted, such as green roofs (explicitly cited) or phase change materials. A complete description of heat transfer phenomena and the energy benefits achievable by adoption of green roofs are discussed in this chapter and Chapter 12. Conversely, this chapter proposes a focused technical and economic investigation. A brief description of thermal phenomena and a summarized state of the art of the scientific literature about the economic investigations of green roofs' profitability are initially proposed.

Eco-efficient Materials for Mitigating Building Cooling Needs. http://dx.doi.org/10.1016/B978-1-78242-380-5.00013-3

13.2 Behaviors of green roofs: heat transfer phenomena and literature state of the art

A green roof identifies a technological roofing solution in which several layers—with a vegetative external coating—are integrated in the building envelope. Examples of green roofs are very ancient, starting from the hanging gardens of Babylon built in 600 BCE. Also the modern styles during the first half of 1900 (e.g., Le Corbusier, Gropius, and F. L. Wright) show greenery integrated in the building roofs.

Energy benefits of green roofs in mitigating heat transfer had already been identified in the 1960s, with a relevant diffusion in Germany and in northern Europe. Green roofs, when accurately designed, provide positive effects for both the indoor microclimate (by reducing the operative temperatures) and with reference to the energy demand for active cooling. Favorable effects can also be achieved during the heating season; thus, green roofs can be adopted in cold and warm climates.

Several kinds of green roofs exist. Usually, these are multilayered structures, with a vegetative layer, a cultural soil, a drainage system, and layers designed for storing a quantity of rainwater, in order to limit the irrigation need. Single-layer solutions are rare today, with some examples of detached houses in the Faroe Islands.

Commonly, two macro-typologies are identified: intensive or extensive green roofs. The second ones require lower maintenance and can also be installed in existing buildings after a static analysis, since the added weight is marginal. Conversely, intensive green roofs can be designed only for new buildings appositely designed, due to the high weight.

In wintertime, the roughness of the canopy and the quasi-stable air layer within the foliage reduce the heat convection on the external side of the roof by lowering the heat losses. Moreover, the added crop layer (especially if dry) increases thermal resistance.

In summer, beneficial effects are related to the thermal mass of the soil (characterized by significant thickness), wetness, and high thermal capacity. Finally, attenuation and time lag of heat transfer are obtained. Moreover, the evapotranspiration phenomenon, concerning both vegetation and soil, allows a favorable evaporative cooling. The evapotranspiration is directly related to solar radiation, moisture content, stomatal resistance of vegetation, resistance of the boundary limit to the vapor diffusion, and amount of foliage. 'Stomatal resistance' and 'leaf area index' are the main parameters:

- Stomatal resistance influences the transpiration of the water through a plant, and, more in detail, across the foliage. This, expressed in s/m, is related to the rate of opening and closing of the plant's stoma, connected to kind of vegetation, phase of life, and health conditions.
- Leaf area index (LAI) is the area of foliage projected on a unitary roof surface (m^2/m^2). The LAI depends on the density of the vegetation leafs. High values favor higher transpiration of water vapor, improving the evaporative cooling.

The combining of the aforementioned phenomena during the cooling season reduces the inner surface temperature of ceiling, with lower operative temperatures and positive effects on indoor comfort and on energy demands for air-conditioning (Ascione et al., 2013). Conversely, the mere economic benefits, induced by installation of green roofs, can be poor if compared to the maintenance costs (e.g., the artificial watering). Therefore, the main target should be the improving of indoor comfort, also according to adaptive criteria, in order to avoid the installation and/or use of active cooling systems.

Many papers highlighted the green roofs' capability in improving the urban environment in terms of reduction of energy demands of existing buildings, mitigation of urban heat island effèct, improvement of air pollution, waste water management, and reduction of noise (Castleton et al., 2010; Saadatian et al., 2013). Through the life cycle assessment (LCA) methodology, Rincòn et al. (2013) have experimentally evaluated that, in Mediterranean continental climate, an extensive green roof with recycled rubber has a significantly lower environmental impact compared to noninsulated conventional roof (7% reduction). The economic benefits of green roofs have been often discussed through the life cycle cost (LCC) analysis, as in Kosareo and Ries (2007). By simulations, Wong et al. (2003) showed that in Singapore, an extensive green roof could achieve a lower life cycle cost of 14.6% compared to a traditional roof. Peri et al. (2012) applied the LCC analysis to green roofs: the need to add a certain amount of substrate—for keeping the thickness constant—has a significant impact on the overall maintenance cost (55%). Using a 50-year LCC analysis supported with economic evaluations, Sproul et al. (2014) have found that white roofs have better performances compared to green roofs in various U.S. regions. However, green roofs are sometimes preferable, having a good impact on storm water management.

The main economic barriers to the green roofs' installation concern the high investment and operating costs, especially if adjustments are needed for application in existing buildings (Berardi et al., 2014; Carter and Keeler, 2008). Claus and Rousseau (2012) have evidenced the advisability of financial incentives. Also, Bianchini and Hewage (2014) indicate that green roofs could be economically feasible only if operating costs are opportunely estimated. These arguments will be explored in the following sections. Through a validated simulation model, for several climatic and economic scenarios, Chan and Chow (2013) have evaluated that a green roof, designed in Hong Kong, could reduce the yearly energy demand from 2.4% to 10%, with favorable payback periods.

Surely, the economic feasibility of green roofs is influenced by technology, vegetation, integration in specific climates, and characteristics of buildings. In hot and dry climates, the economic feasibility of green roofs is marginal (Ascione et al., 2013). In the Eastern Mediterranean regions, only appropriate plants can survive. By experiments, Schweitzer and Erell (2014) have found that the *Aptenia cordifolia* was the most economical of the tested plants, requiring little water.

Sun et al. (2014) proposed a correlation among thermal performances of green roofs, soil depth, and moisture content (guaranteed by proper irrigation). In Beijing, the irrigation costs are balanced by lower air-conditioning requirements. The irrigation allows benefits in terms of biodiversity and evapotranspiration (Rowe et al., 2014). However, irrigation methods and efficiency should be optimised. Green roofs are not always appropriate in all conditions and climates. Therefore, specific studies are required for determining the most suitable substrate parameters and plants, and for quantifying the actual benefits.

13.3 Criteria for suitable feasibility studies

In the next sections, the adopted numerical code for the green roof analysis is EnergyPlus 7.2.0 (DOE, 2012), a widely validated whole building energy simulation program developed by the United States Department of Energy. Crawley et al. (2008) highlighted its potentialities. Olsen and Chen (2003), about the capability of EnergyPlus in predicting energy performances, say: "*both small and full-scale validation studies have demonstrated that EnergyPlus results are well within the accuracy needed for building design.*"

At the EnergyPlus website (http://apps1.eere.energy.gov/buildings/energyplus/), many "Testing and Validation" documents are proposed. According to analytical, comparative, and global heat balance tests, a number of validation studies are proposed, concerning international procedures (ASHRAE RP 865, 2002; ASHRAE RP 1052, 2001a; ANSI/ASHRAE, 2011) and test methodologies of the "International Energy Agency Best Test programs." With reference to both building envelopes and active energy systems, the validations revealed favorable responses.

Presently, EnergyPlus offers various alternative algorithms for solving the heat transfer through the building envelope: (a) Conduction transfer functions (CTF), (b) Conduction finite differences (ConFD), and (c) Combined heat and moisture finite elements (HAMT). These methods require highly variable computational times and sources. Thus, the selection should be based on the specific necessities.

The CTFs are very powerful. These allow the evaluation of the transient heat transfer through the building envelope by directly relating, through the state-space representation, the heat flux at one face of building element to the present and previous temperatures of the inner and outer faces and to the previous values of heat flows. The thermal field inside the component is not completely determined. If nodal temperatures inside the envelope structures are required (e.g., adoption and/or optimisation of phase change materials), other algorithms must be used, such as the ConFD. Conversely, studies requiring a careful evaluation of vapor condensation inside a building component should adopt the "Combined heat and moisture finite element algorithm." Regarding the simulation timestep (i.e., the time interval between two consecutive energy balances), this can vary between 1 and 60 timesteps per hour. In case of green roof modeling, a minimum of 12 timesteps is suggested.

The green roof model implemented in EnergyPlus, and thus in this study, is based on the studies of Sailor (2008), which slightly modifies the FASST methodology of Frankenstein and Koenig (2004). The algorithm takes into account the solar radiation on the external roof surface, as well as the radiation at high wavelength between structure and surrounding environment. The model was tested above all with reference to the evapotranspiration phenomenon, by taking into account the effects of soil and canopy on convective heat transfer. The required simulation inputs are related to the soil peculiarities (e.g., thickness, thermal conductivity, density, specific heat, moisture content at the rated conditions) and to the vegetation (e.g., height, leaf area index, stomatal resistances, spectral characteristics). The physical formulation is based on two main equations, concerning the energy balance influenced by the vegetation (Equation (13.1)) and the heat transfer through the soil (Equation (13.2)).

$$F_{\mathrm{f}}=\sigma_{\mathrm{f}}\left[I_{\mathrm{s}}^{\downarrow}(1-\alpha_{\mathrm{f}})+\varepsilon_{\mathrm{f}}I_{\mathrm{ir}}^{\downarrow}-\varepsilon_{\mathrm{f}}\sigma T_{\mathrm{f}}^{4}\right]+\frac{\sigma_{\mathrm{f}}\varepsilon_{\mathrm{f}}\varepsilon_{\mathrm{g}}\sigma}{\varepsilon_{\mathrm{g}}+\varepsilon_{\mathrm{f}}-\varepsilon_{\mathrm{f}}\varepsilon_{\mathrm{g}}}\left(T_{\mathrm{g}}^{4}-T_{\mathrm{f}}^{4}\right)+H_{\mathrm{f}}+L_{\mathrm{f}} \quad (13.1)$$

$$F_{\mathrm{g}}=(1-\sigma_{\mathrm{f}})\left[I_{\mathrm{s}}^{\downarrow}(1-\alpha_{\mathrm{g}})+\varepsilon_{\mathrm{g}}I_{\mathrm{ir}}^{\downarrow}-\varepsilon_{\mathrm{g}}\sigma T_{\mathrm{g}}^{4}\right]-\frac{\sigma_{\mathrm{f}}\varepsilon_{\mathrm{f}}\varepsilon_{\mathrm{g}}\sigma}{\varepsilon_{\mathrm{g}}+\varepsilon_{\mathrm{f}}-\varepsilon_{\mathrm{f}}\varepsilon_{\mathrm{g}}}\left(T_{\mathrm{g}}^{4}-T_{\mathrm{f}}^{4}\right)+H_{\mathrm{g}}+L_{\mathrm{g}}+K\frac{\partial T_{\mathrm{g}}}{\partial z} \quad (13.2)$$

In Equations (13.1) and (13.2), the following terms are: F_{f}, sum of energy terms at the atmosphere/foliage interface [W m^{-2}]; F_{g}, sum of energy terms at the ground surface [W m^{-2}]; H_{f}, sensible heat flow at the atmosphere/foliage interface [W m^{-2}]; H_{g}, sensible heat flow at the atmosphere/soil interface [W m^{-2}]; L_{f}, latent heat flow at the atmosphere/foliage interface [W m^{-2}]; L_{g}, latent heat flow at the atmosphere/soil interface [W m^{-2}]; σ_{f}, fractional vegetation coverage; $I_{\mathrm{s}}^{\downarrow}$, total incoming solar radiation [W m^{-2}]; α_{f}, foliage absorption coefficient; ε_{f}, foliage infrared emissivity; ε_{g}, soil infrared emissivity; $I_{\mathrm{ir}}^{\downarrow}$, total incoming infrared radiation [W m^{-2}]; σ, constant of Stefan–Boltzman [W m^{-2} K^{-4}]; T_{f} temperature of the foliage [K]; T_{g} temperature of the soil [K]; z height or depth [m].

Sensible (H_{f} and H_{g}) and latent (L_{f} and L_{g}) heat flows are calculated for both foliage ($_{\mathrm{f}}$) and soil ($_{\mathrm{f}}$), according to Frankenstein and Koenig (2004), as in Equations (13.3)–(13.6).

$$H_{\mathrm{f}}=\left(1.1\cdot LAI\rho_{\mathrm{af}}c_{\mathrm{p\text{-}air}}C_{\mathrm{f}}w_{\mathrm{af}}\right)(T_{\mathrm{af}}-T_{\mathrm{f}}) \quad (13.3)$$

$$L_{\mathrm{f}}=l_{\mathrm{f}}\cdot LAI\cdot\rho_{\mathrm{af}}C_{\mathrm{f}}w_{\mathrm{af}}r''(q_{\mathrm{af}}-q_{\mathrm{f,sat}}) \quad (13.4)$$

$$H_{\mathrm{g}}=\left(\rho_{\mathrm{ag}}c_{\mathrm{p\text{-}air}}C_{\mathrm{hg}}w_{\mathrm{af}}\left(T_{\mathrm{af}}-T_{\mathrm{g}}\right)\right) \quad (13.5)$$

$$L_{\mathrm{g}}=C_{\mathrm{eg}}l_{\mathrm{g}}w_{\mathrm{af}}\rho_{\mathrm{ag}}\left(q_{\mathrm{af}}-q_{\mathrm{g}}\right) \quad (13.6)$$

where LAI, leaf area index [m^{2} m^{-2}]; ρ_{af}, air density at the temperature of the foliage [kg m^{-3}]; ρ_{ag}, air density at the temperature of the soil surface [kg m^{-3}]; $c_{\mathrm{p\text{-}air}}$, specific heat of air, constant pressure [W m^{-2}]; C_{f}, dimensionless bulk heat transfer coefficient; C_{hg}, sensible dimensionless bulk heat transfer coefficient at the ground surface; $C_{\mathrm{e,g}}$, latent dimensionless bulk heat transfer coefficient at the ground surface; w_{af}, air velocity (wind) in the canopy [m s^{-1}]; T_{af}, air temperature in the canopy; l_{f}, latent heat of vaporization at the temperature T_{f} [J kg^{-1}]; l_{g}, latent heat of vaporization at the temperature T_{g} [J kg^{-1}]; r'', dimensionless wetness factor of the surface; q_{af}, mixing ratio of the air within the canopy; $q_{\mathrm{f,sat}}$, mixing ratio, at saturation conditions, at the foliage temperature.

In Equation (13.3), the value 1.1 is a dimensionless constant, proposed by Deardorff (1978), for taking into account of '*stalks, stems, twigs, and limbs which exchange heat but do not transpire.*'

The complete model is available in the technical documentation of EnergyPlus. Sailor fully describes the calculations of T_g and T_f, transferred into the CTF resolution scheme of EnergyPlus. The elegant formulation also considers the water absorbed by the soil due to irrigation or rain, and the water runoff. The model confidence was evaluated by means of comparisons with observations in Florida.

In this chapter, hourly energy simulations have been performed in order to accurately estimate the suitability of green roofs in terms of energy savings, operating costs, and reduction of polluting emissions. Technical-economical evaluations are proposed.

The EPBD recast introduces the cost-optimal method for the design of new buildings or energy retrofit. In detail, Member States have to ensure that minimum energy performance requirements for buildings are set, with a view towards achieving cost-optimal levels. This new approach is designed to define energy requirements, in terms of primary energy, which lead to the lowest cost during the estimated economic lifecycle. The comparative methodology framework is detailed in the Commission Delegate Regulation (EU) n. 244/2012 of 16 January 2012 (Commission Delegated, 2012). This methodology includes six main steps for comparing energy efficiency measures, based on the energy performances and investment costs.

The first step is the definition of a reference building (real or virtual building), typical in geometry, shape, dimensions, surface to volume ratio, energy performances of building envelope, and systems.

The second step is the definition of energy efficiency measures for all parameters that have a direct or indirect impact on the energy performance. The third step is the calculation of the primary energy demand resulting from the application of these measures and packages of measures to a reference building in accordance with the common framework provided in Annex I of Directive 2010/31/EU. The fourth step concerns the calculation of the global costs, evaluated as the sum of the present value of the initial investment costs, sum of operational costs and replacement costs (referred to the starting year), and disposal costs if applicable.

For the calculation at macroeconomic level, according to Equation (13.7), an additional cost concerning the greenhouse gas emissions is introduced (i.e., the monetary value of environmental damage). The annual cost ($C_{a,i}$) is the sum of running costs (annual maintenance costs, operating and energy costs) and periodic or replacement costs.

$$C_g(\tau) = C_I + \sum_j \left[\sum_{i=1}^{\tau} (C_{a,i}(j) \times R_d(i) + C_{c,i}(j)) - V_{f,\tau}(j) \right] \tag{13.7}$$

In Equation (13.7), the following terms are defined; τ is the calculation period; $C_g(\tau)$ global cost (referred to starting year τ_0) over the calculation; C_I initial investment costs for measure or set of measures "j"; $C_{a,i}(j)$, annual cost during year "I" for measure or set of measures "j"; $R_d(i)$, discount factor for year "i" to convert a cash flow occurring at a given point in time to its equivalent value at the starting point; $C_{c,i}(j)$ carbon cost for measure or set of measures "j" during year "i"; $V_{f,\tau}(j)$, residual value of measure or set of measures "j" at the end of the calculation period (discounted to the starting year τ_0).

Then (fifth step), a sensitivity analysis has to be done by considering the variation of energy prices and discount rates. Finally, for each reference building, the cost-optimal level is achieved (sixth step) by comparing the global cost calculated for the considered energy efficiency measures and for packages/variants of those measures.

In this study, the philosophy of the cost-optimal methodology has been tailored for evaluating technical and economic feasibility of green roofs, by taking into account all parameters affecting energy performances, savings, and costs. Compared to a base case without roof vegetation, the primary energy saving index (ΔEP) has been calculated in terms of kWh/m^{2}a, as reported in Equation (13.8), where EP^{BS} is the energy request of the reference building and EP^{GR} is the demand of the building refurbished with the green roof.

$$\Delta EP = EP^{BS} - EP^{GR} \tag{13.8}$$

The reductions of operating costs (ΔOC) (€/year) and polluting emissions (ΔPE) (kg $CO_{2\text{-equivalent}}$/year) are also evaluated (Equations (13.9) and (13.10)). The superscripts "BS" and "GR" respectively refer to the base building and the one refurbished with green roof.

$$\Delta OC = OC^{BS} - OC^{GR} \tag{13.9}$$

$$\Delta PE = PE^{BS} - PE^{GR} \tag{13.10}$$

Operating costs and polluting emissions are determined by taking into account the following parameters (some of these are variable depending on the specific country): (a) efficiency of national power generation system; (b) LCA emission factor for natural gas and electricity; (c) electricity tariff; (d) natural gas cost; and (e) water price. Heating and cooling seasons are both considered.

By considering the investment cost of the energy efficiency measures (EEM in the following lines), simple pay-back (SPB) and discounted pay-back period (DSB) have also been calculated, according to Equations (13.11) and (13.12). The time-dependent money value has been taken into account.

$$SPB = \frac{C_I}{\Delta OC} \tag{13.11}$$

$$DPB = N : \sum_{i=1}^{N} \frac{F_i}{(1 + R_d)^i} = C_I \quad \text{with } 0 < N < LF \tag{13.12}$$

The new terms are: F_i = annual cash flow (economic saving), with reference to the i-th year; C_I = initial investment cost of the EEM; N = number of years; R_d = discounting rate; LF = number of years of the cut-off period (i.e., lifetime of the EEM).

Finally, for having a last indication about the profitability of the investment, the net present value (NPV) has been evaluated (Equation (13.13)).

$$\mathrm{NPV} = \sum_{i=1}^{\mathrm{LF}} \frac{F_i}{(1+R_{\mathrm{d}})^i} - C_{\mathrm{I}} \quad (13.13)$$

A discount rate equal to 3% has been considered (fixed value, and thus deterministic approach), in order to take into account the value of money over the time (−5%/year) and the increase of energy prices (+2%/year). Obviously, an energy efficiency measure can be considered only if the NPV will be >0. According to the European Standard EN 15459 (CEN, 2007a), the technical and economical investigations take into account the lifetime of the energy efficiency measures. The BPIE (2013), in a large study concerning the cost-optimality methodology, has proposed lifetimes of some technologies of building envelope and HVAC systems. In this chapter, various scenarios will be considered with reference to funding programs and cost of watering.

13.4 Presentation of the case studies

Three typologies of extensive green roofs have been studied. Three European climatic conditions are considered, as well as three typologies of building, according to various levels of efficiency of the integrated system building/HVAC plant, depending on the construction period (i.e., typical 1980s buildings, typical 1990s buildings, typical 2000s buildings). The building (Figure 13.1) is representative of the tertiary sector, having retail and commercial uses at the ground floor and offices at the first floor. The overall conditioned floor area is 2585 m^2. The gross roof area is 1293 m^2, and the overall height is 7.0 m. Main dimensions and geometrical peculiarities are in Table 13.1. Regarding the thermal envelope, the main components are shown in Figure 13.2. The vertical envelope has opaque walls in hollow blocks, with extruded polystyrene as insulation, with a thickness variable between 4 and 12 cm. Similarly,

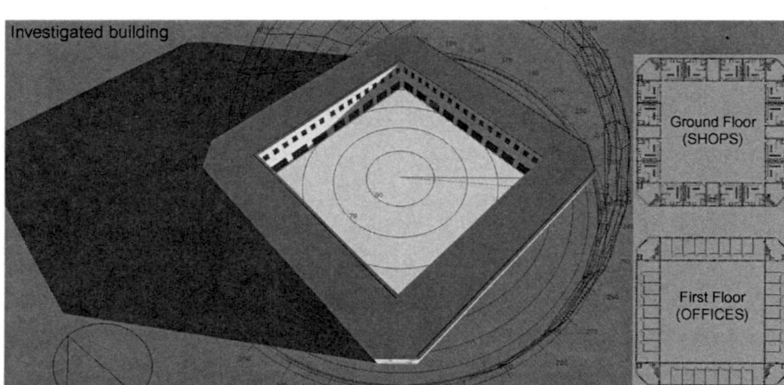

Figure 13.1 Geometrical information of the case studies' buildings.

Table 13.1 **Geometrical characteristics of buildings and thermophysical properties and peculiarities of HVAC systems according to the construction periods**

Main building dimensions and geometry							
Total conditioned area [m^2]		2585.2				Maximum length [m]	51.7
Gross volume [m^3]		9048.1					
Height [m] (2 floors)	7.0	Surface to volume ratio [m^{-1}]			0.54	Maximum width [m]	51.7
Heat transfer area of external walls, roof, and fenestration							
	Total		North	East		South	West
Gross area of vertical walls [m^2]	2349		591.5	582.5		630.9	504.1
Window opening area [m^2]	602.6		156.6	154.4		141.84	149.7
Window-to-wall ratio [%]	26.65		26.5	26.5		22.5	27.5
Gross roof area [m^2] 1293							
1980s building and HVAC peculiarities							
$U_{BASEMENT}$ [W/m^2 K] 0.52 U_{WALLS} [W/m^2 K]	0.45	U_{ROOF} [W/m^2 K]	0.53	$U_{WINDOWS}$ [W/m^2 K]	3.76	$SHGC_{WINDOWS}$	0.76
Nominal efficiency of boiler η [−] (standard hot water boiler fuelled by natural gas)						0.80	
Nominal energy efficiency ratio (EER) of chiller [W_{TH}/W_{EL}]						2.60	
(electric air-cooled chiller, old generation)							
Mechanical ventilation equal to Comfort Category I according to the EN 15251						10 l/s/p + 1 l/s/m^2	
Infiltration rate ACH [h^{-1}]			0.9 ACH (*1.35 ACH during the cooling period*)				

(Continued)

Table 13.1 Continued

1990s building and HVAC peculiarities									
$U_{BASEMENT}$ [W/m² K]	0.33	U_{WALLS} [W/m² K]	0.30	U_{ROOF} [W/m² K]	0.33	$U_{WINDOWS}$ [W/m² K]	2.78	$SHGC_{WINDOWS}$	0.70
Nominal efficiency of boiler η [−] (two stars hot water boiler fuelled by natural gas)								0.90	
Nominal energy efficiency ratio (EER) of chiller [W_{TH}/W_{EL}] (electric air-cooled chiller, medium efficiency)								2.80	
Mechanical ventilation equal to Comfort Category I according to the EN 15251								10 l/s/p + 1 l/s/m²	
Infiltration rate ACH [h^{-1}]					0.6 ACH (*0.9 ACH during the cooling period*)				
2000s building and HVAC peculiarities									
$U_{BASEMENT}$ [W/m² K]	0.24	U_{WALLS} [W/m² K]	0.22	U_{ROOF} [W/m² K]	0.24	$U_{WINDOWS}$ [W/m² K]	2.51	$SHGC_{WINDOWS}$	0.70
Nominal efficiency of boiler η [−] (condensing hot water boiler fuelled by natural gas)								0.98	
Nominal energy efficiency ratio (EER) of chiller [W_{TH}/W_{EL}] (electric air-cooled chiller, new generation)								3.11	
Mechanical ventilation equal to Comfort Category I according to the EN 15251								10 l/s/p + 1 l/s/m²	
Infiltration rate ACH [h^{-1}]					0.3 ACH (*0.45 ACH during the cooling period*)				

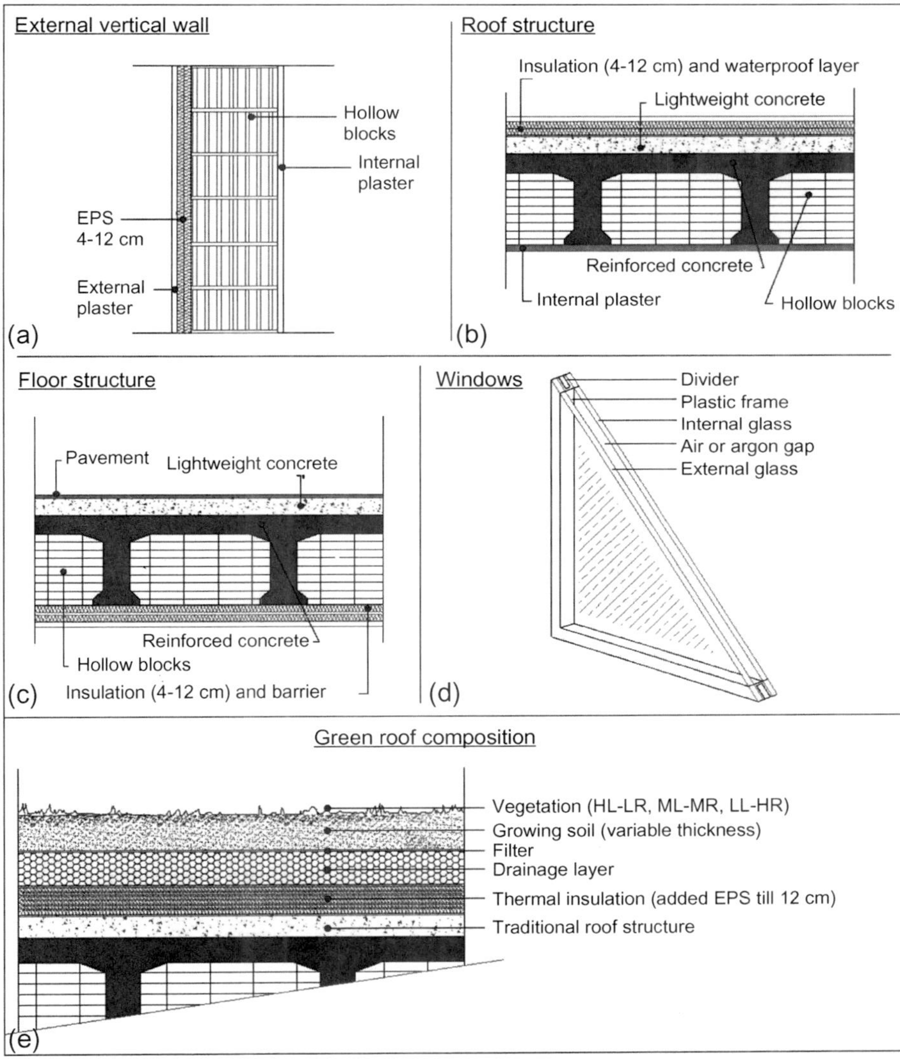

Figure 13.2 Structures of the envelope of the considered buildings.

ceiling, floor on the ground, and roof have a structural frame of beams and joists in reinforced concrete and interposed hollow bricks, according to a common European building technique. Also, for horizontal structures, the thermal insulation has a thickness variable from 4 cm (1980s building) to 12 cm (2000s building). Windows are double glazed, with thicknesses equal to 3/6/3 (inefficient building), 3/12/3 (medium efficient building), and 6/12/6 (efficient building). The cavity is air- or argon-filled.

The building is equipped with centralized systems for the indoor microclimatic control. The HVAC system combines mechanical ventilation and control of sensible loads for the space heating and cooling (Table 13.1). Fan coils operate by mixing outdoor air (OA) and recirculation air. The OA flow rate satisfies the standard EN 15251 (CEN, 2007b) Category I. Indeed, by taking into account the typical occupancy rate for offices, shops, and common spaces, 11 l/s/person are assigned, with an additional amount of 1 l/s/m^2. Moreover, the infiltration rate is variable, with air change rates related to the building air tightness. The infiltration rate has been increased 50% in summer, taking into account the natural ventilation because of more frequent opening of doors and windows.

As cited, three European common climates are considered. The cities—representative of the Southern, Central, and Northern European climates—are respectively Naples (Italy), Stuttgart (Germany), and Copenhagen (Denmark). The hourly weather data are IWEC (ASHRAE, 2001b), while precipitation data are the most recent available (WMO, 2013). Energy and water tariffs are diversified for the considered countries (Eurostat, 2013; GWI, 2013), whose main parameters are in Table 13.2 (weather conditions), Table 13.3 (efficiency of energy systems, tariffs, and emissions factors), and Figure 13.3 (precipitation). The efficiency of thermoelectric conversion has been assumed from the International Energy Agency (IEA, 2008), while emissions factors are those proposed by the Covenant of Mayors (2013, last access).

In order to have various levels of evapotranspiration and thermal mass, three configurations of green roofs have been considered, different in plants, leaf area index, stomatal resistance, and vegetation height (Table 13.4):

1. HL-LR means High LAI and Low Stomatal Resistance.
2. ML-MR means Mean LAI and Mean Stomatal Resistance.
3. LL-HR means Low LAI and High Stomatal Resistance.

Table 13.2 Main peculiarities of the considered climates

	Northern European climate (Reference: Copenhagen)	Central European climate (Reference: Stuttgart)	Southern European climate (Reference: Naples)
Heating degree-days [°Cd]	1364	3490	3653
Heating period	from 01/11 to 31/03	from 16/10 to 30/04	from 01/10 to 30/05
Cooling degree-days [°Cd]	756	106	45
Cooling period	from 16/05 to 15/10	from 01/06 to 30/09	from 01/06 to 30/09

Annual heating and cooling degree-days have been calculated with respect to an 18 °C baseline.

Office and retails are heated for 10 h per day (office from 8.00 to 18.00, shops from 10.00 to 20.00) and cooled for 5 h per day.

Table 13.3 Efficiency of the thermoelectric national system, energy (Eurostat, 2013), and water (GWI, 2013) costs for the considered European countries

	Dimension	Denmark	Germany	Italy
Thermoelectric conversion efficiency	[$kWh_{electric}$/ $kWh_{thermal}$]	0.43	0.39	0.45
Natural gas cost	[€/kWh_{gas}]	0.113	0.292	0.083
Electricity cost	[€/$kWh_{electric}$]	0.300	0.066	0.229
Water price	[€/m^3]	6.445	3.912	1.321

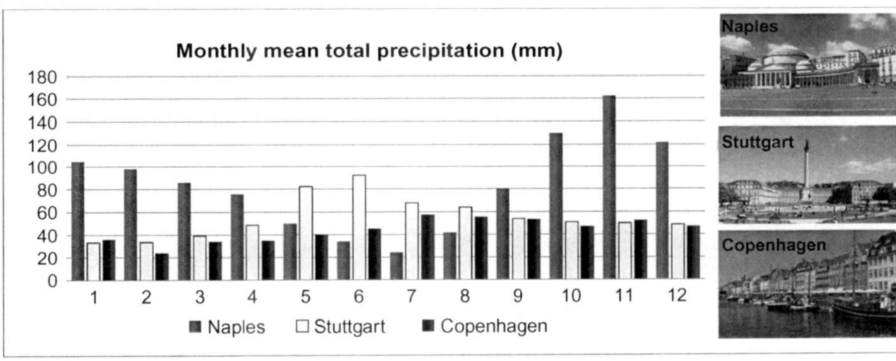

Figure 13.3 Statistical precipitation for the three considered climates (Southern, Northern and Central Europe).

Table 13.4 Main characteristics of the vegetation modeled on the roofs

	Dimension	HL-LR	HL-LR	HL-LR
Height of plants [m]	[m]	0.1	0.25	0.40
Leaf Area Index (LAI) [m^2/m^2]	[m^2/m^2]	0.8	2.0	3.5
Leaf reflectivity [−]	[−]	0.3	0.4	0.5
Leaf emissivity [−]	[−]	0.95	0.95	0.95
Density of dry soil [kg/m^3]	[m^2/m^2]	700	1 100	1 000
Thickness of the soil [m]	[m]	0.10	0.14	0.18
Minimum stomatal resistance	(s/m)	300	200	120
Saturation volumetric moisture content of the soil	(−)	0.40	0.40	0.40
Residual volumetric moisture content of the soil	(−)	0.01	0.01	0.01
Initial volumetric moisture content of the soil	(−)	0.20	0.20	0.20

Table 13.5 **Installation cost per unit of roof area for various typologies of green roofs**

Extensive green roof typology	Unitary cost (€/m^2)
Green roof high LAI and low stomatal resistance (no thermal insulation)	83.25
Green roof high LAI and low stomatal resistance (4 cm thermal insulation)	105.75
Green roof medium LAI and medium stomatal resistance (8 cm thermal insulation)	121.87
Green roof medium LAI and medium stomatal resistance (no thermal insulation)	78.25
Green roof medium LAI and medium stomatal resistance (4 cm thermal insulation)	100.75
Green roof high LAI and low stomatal resistance (8 cm thermal insulation)	116.87
Green roof low LAI and high stomatal resistance (no thermal insulation)	73.25
Green roof low LAI and high stomatal resistance (4 cm thermal insulation)	95.75
Green roof low LAI and high stomatal resistance (8 cm thermal insulation)	111.87

In order to meet the maximum values of thermal transmittance established by the legislations when a building component is refurbished, the green roof typologies have been combined with additional insulation layers characterized by various thicknesses. Table 13.5 shows the unitary installation costs.

13.5 Results and discussion

The feasibility study of the technical-economical convenience of green roofs is proposed in the next subsections. Green roofs are equipped with a layer of thermal insulation, below the drainage layer, in order to always achieve a thickness of 12 cm of polystyrene. This means that, for the 2000s building, the cost of insulation has not been considered because the building was already insulated. Conversely, for the 1980s and 1990s buildings, respectively 8.0 and 4.0 cm of additional insulation are provided. Indeed, during the refurbishing of a roof structure, it makes no sense to install roof vegetation without proper insulation. The cost of artificial watering has been calculated for the period between 1st May and 30th September. The feasibility studies take into account four scenarios:

- DPB^1 = No funding of green roof and full cost of irrigation, according to the water tariff of the country (→ 'No incentives' scenario)
- DPB^2 = Incentive of green roof, in terms of 50% of investment cost, subdivided in five equal annual rates and full cost of irrigation (→ '50% funding' scenario)
- DPB^3 = No funding of green roof and cost-free irrigation, using local well water, source water, or by rainwater recovery (→ 'No incentives, no water cost' scenario)
- DPB^4 = Incentive of green roof, in terms of 50% of the investment cost and cost-free irrigation (→ '50% funding, no water cost' scenario)

13.5.1 Copenhagen (Denmark): northern European climate

(a) Refurbishment of 1980s building. The building has poor thermal insulation and standard systems for heating and cooling. The energy demand for heating (329 kWh/m^{2}a) is much higher compared to the cooling request (6 kWh/m^{2}a). Both energy requests are expressed in terms of primary energy. The installation of a green roof, equipped with an additional thermal insulation layer of 8 cm in polystyrene, allows a reduction of the summer energy requests between 6.9% (HL-LR solution) and 3.2% (LL-HR green roof). Considering the low energy demand for cooling, this translates into scarce economic savings, lower than €130 per year. Conversely, the water cost for irrigation can be higher than €3800 for the HL-LR solution, and about €740 for the LL-HR green roof. Thanks to the addition of thermal insulation, the green roof installation allows a cost reduction of 3.2–3.3% in winter for all considered typologies of green roof. It means an economic savings of around €3300 per year. Globally (Table 13.6), green roofs can be profitable only if water costs are not considered. Thus, rainwater recovery systems are recommended. However, also by considering cost-free irrigation and a 50% incentive, the NPV value, calculated according to a lifetime equal to 50 years, is low, between €2000 and 8000. Conversely, if irrigation cost is considered, the investment cannot be repaid. The best solution is the LL-HR green roof. Indeed, the annual energy savings are slightly lower compared to HL-LR green roof, but both installation and maintenance costs are lower. A global summary is reported in Figure 13.4.

(b) Refurbishment of 1990s building. The roof already has 8 cm of external insulation, so only 4 cm of additional insulating layer are provided. The base building has a primary energy demand for heating around 234 kWh/m^{2}a, and the cooling request is 6.0 kWh/m^{2}a. The green roof installation implies a reduction of energy demands, with primary energy savings in the range of 4.3/0% for cooling, and around 1.6% for heating. The building energy performance is not significantly improved by green roofs, with cost saving never higher than €1200, while the water cost for irrigation (water price around €6.445/m^3) can be higher than €3800 per year (HL-LR solution). Finally, a positive annual cash flow, in terms of cost saving, can be obtained only for green roofs that do not require irrigation costs. The investment cannot be repaid in a reasonable period.

(c) Refurbishment of 2000s building. Green roofs are installed without additional thermal insulation. Thus, the green roof effects concern only the phenomena related to vegetation and soil. The base building has an energy demand for heating of 146 kWh/m^{2}a, while 7 kWh/m^{2}a are requested for the space cooling. Green roofs induce benefits in both seasons, due to the additional thermal resistance related to the soil thickness and to the lower convection in the canopy (winter), while the summer evapotranspiration implies a favorable evaporative cooling. Globally, by neglecting the artificial watering costs, an annual economic savings between €180 and 350 can be achieved. The annual cash flow is poor, and thus the investment, also without considering maintenance cost, cannot be repaid. The better configuration of green roofs, in terms of primary energy saving (−0.8%) and reduction of polluting emissions (−1.0%), is the HL-LR solution. Conversely, if water costs are considered, the preferable solution is the one that minimizes the irrigation (i.e., LL-HR green roof). Figure 13.5 proposes a comparison between the base case and the HL-LR solution.

Table 13.6 shows all simulation outcomes for Copenhagen.

13.5.2 Stuttgart (Germany): central European climate

(a) Refurbishment of 1980s building. The building is poorly insulated and has high-energy demand for heating (296 kWh/m^{2}a), while the cooling primary energy request is lower than

Table 13.6 **Copenhagen: summarized feasibility analysis of considered EEMs**

EEM	ΔEP		ΔOC		WC	$\Delta CO_{2\text{-equiv}}$		C_I	DPB[1]	DPB[2]	DPB[3]	DPB[4]
	(kWh/m²y)	(%)	(€/year)	(%)	(€/year)	t $CO_{2\text{-equiv}}$/year	(%)	(€)	(years)			
1980s building												
HL-LR	29,073	−3.4	3399	−3.3	3825	6.0	−3.5	157,529	*Never*	*Never*	≫50	≈48
ML-MR	28,786	−3.4	3384	−3.3	2563	5.9	−3.4	151,066	≫50	≫50	≫50	≈44
LL-HR	28,326	−3.3	3363	−3.3	743	3.3	−3.3	144,603	≫50	≫50	≫50	≈41
1990s building												
HL-LR	10,565	−1.7	1219	−1.7	3826	2.3	−1.8	136,692	*Never*	*Never*	≫50	≫50
ML-MR	10,435	−1.7	1219	−1.7	2563	2.2	−1.7	130,230	*Never*	*Never*	≫50	≫50
LL-HR	9764	−1.6	1159	−1.6	743	1.9	−1.5	123,766	≫50	≫50	≫50	≫50
2000s building												
HL-LR	3151	−0.8	367	−0.8	3826	0.9	−1.0	107,609	*Never*	*Never*	≫50	≫50
ML-MR	2375	−0.6	270	−0.6	2563	0.6	−0.7	101,146	*Never*	*Never*	≫50	≫50
LL-HR	1574	−0.4	180	−0.4	745	0.3	−0.3	94,683	*Never*	*Never*	≫50	≫50

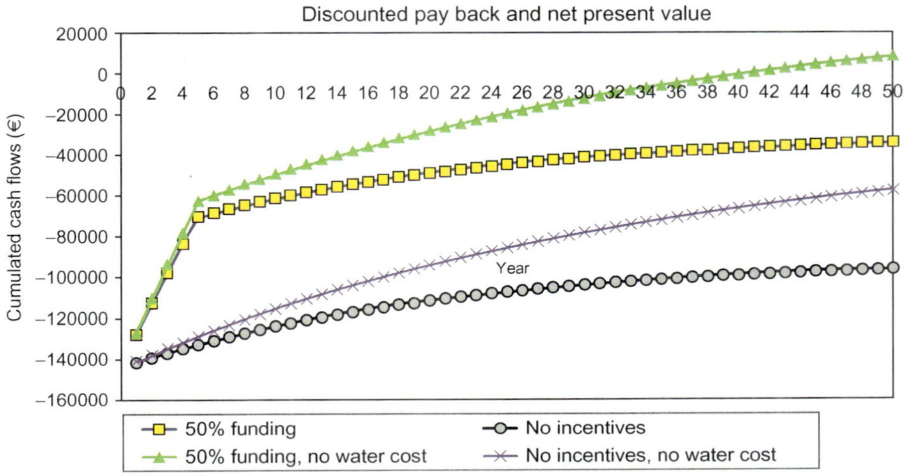

Figure 13.4 Copenhagen: HL-LR green roof applied to 1980s building. Feasibility studies.

13 kWh/m^{2}a. The installation of the green roofs requires a further thickness of 8 cm of polystyrene. The investment costs are reported in Table 13.5. The more profitable green roof solution, under the energy saving point of view, is the HL-LR. The achievable savings are, with reference to the cooling period, between 6.5% and 2.7%. Winter benefits are also achieved, because of the added insulation, with a percentage reduction of the heating primary energy of around 3.5% for all considered typologies of roof vegetation. The irrigation costs are high because of the high tariffs in Germany (around €3.912/m^3). In particular, the HL-LR green roof requires more than €2200 per year for the artificial irrigation, and this means around €175 per year for a roof surface of 100 m^2. The watering costs are higher than the energy savings for the HL-LR and ML-MR solutions. The irrigation expenditure of the LL-HR solution is lower (around €582 per year). Finally, green roofs are effective if the irrigation is free of costs, even if the investment is high compared to the achievable energy savings. Conversely, if irrigation is provided at the price of civil use of water, energy refurbishments by means of green roofs never can be repaid. Finally, to be profitable, green roofs require financial incentives.

(b) Refurbishment of 1990s building. A modern building requires, obviously, a lower energy demand for heating (211 kWh/m^{2}a) compared to the previous one. Conversely, the energy request for cooling is slightly higher. The primary energy demand for cooling is around 13 kWh/m^{2}a. The installation of the green roofs includes an additional 4 cm of polystyrene, in order to reach 12 cm. The summer irrigation costs are around €2260, 1800, and 580 respectively for the HL-LR, ML-MR, and LL-HR solutions. The highest cooling energy savings are obtained by HL-HR green roof, with a percentage decrement equal to 5.4%. The LL-HR green roof allows only a 1.1% energy savings for the space cooling. Some benefits can also be achieved in winter: the primary energy demand for the microclimatic control is reduced around 2.0% for the HL-LR roof vegetation, and around 1.8% for the LL-HR solution. If the installation costs are considered, the investment is never profitable.

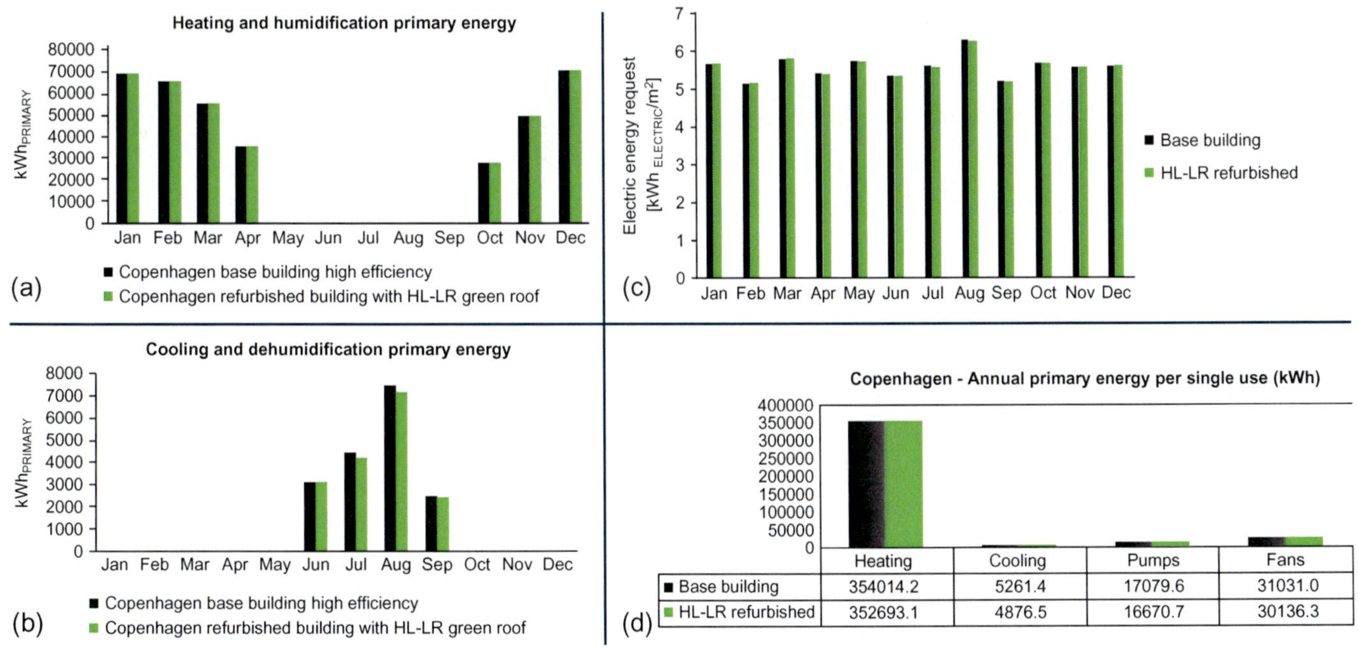

Figure 13.5 2000s buildings in Copenhagen: comparisons of energy demands of the base building and the HL-LR refurbished building.

(c) Refurbishment of 2000s building. This building is already insulated and the HVAC system is efficient. Therefore, the effects of green roofs are evaluated as "standing alone," without the cumulative effect of additional insulation. The energy demand for heating, with reference to the base case building, is 131 kWh/m^{2}a, while the energy request for cooling is 16 kWh/m^{2}a. The more efficient green roof in reducing the cooling demand is the HL-LR one. The base case building and the HL-LR refurbished are compared in Figure 13.6. Of course, high LAI and low stomatal resistance increase the evapotranspiration, and thus the energy demand for cooling is lowered, in term of primary energy, around 5.2%. Slightly better winter performance can be achieved too, with a decrease of around 0.6% due to the lower convection. Minor performances can be achieved if the LAI is reduced and the stomatal resistance is increased. Indeed, the LL-HR vegetation induces only a minimum energy saving for cooling. Conversely, in winter, a lower evapotranspiration allows better performances because the solar radiation increases the external temperature of the roof. Regarding feasibility, the economic indexes show negative trends for all typologies of green roofs and all scenarios about incentives and/or free watering. Finally, the annual savings cannot repay the investment.

All outcomes of the feasibility study for Stuttgart are reported in Table 13.7.

13.5.3 Naples (Italy): southern European climate

(a) Refurbishment of 1980s building. The climate conditions are Mediterranean; the cooling period is May–October. The base building, in order to keep indoor comfort, requires 85 kWh/m^{2}a (heating) and 58 kWh/m^{2}a (cooling) in terms of primary energy. These values are typical of the Italian building stock. In order to meet the Italian prescriptions concerning the refurbishment of buildings (i.e., the compliance to maximum U_{value}), the green roof is insulated with an additional 8 cm of polystyrene. Therefore, the effect of the added thermal insulation is combined with the effects of roof vegetation. The best green roof performances are achieved by HL-LR solution, with annual decrement of energy demand equal to 4.7%. In summer, the cooling demand is lowered 5.1% (minor performances can be achieved with ML-MR and LL-HR solutions, with reduction in primary energy for the space cooling, inclusive of the auxiliaries, respectively around 4.2% and 2.1%). Of course, the HL-LR solution is the one requiring the highest irrigation cost, because of the higher evapotranspiration. The water cost in Italy is moderate, around €1.3/m^3. The watering costs, ranging from €1332 per year (HL-LR) to €229 per year (LL-HR), highly reduce the economic advantages (−€1645 per year for the HL-LR, −€1549 per year for the ML-MR, −€1293 per year for the LL-HR) related to the energy savings. The more convenient solution is the LL-HR, because of the low water need. Conversely, if the irrigation is free of cost, the HL-LR green roof is the best solution. In terms of a mere profitability study, the green roofs are not convenient.

(b) Refurbishment of 1990s building. The building has a thermal insulation of 8 cm and the efficiencies of the heating and cooling systems are typical of the construction period. It requires 57 kWh/m^{2}a of primary energy for winter heating and 55 kWh/m^{2}a for space cooling. The application of a green roof, characterized by an additional 4 cm of insulation, induces a generalized decrement of energy demand for both heating and cooling. The best winter effects are obtained by the LL-HR green roof (−3.1% the annual energy demand) because the scarce amount of evapotranspiration allows that part of the solar radiation to increase the temperature of roof. Conversely, in summer, the HL-LR green roof, with a

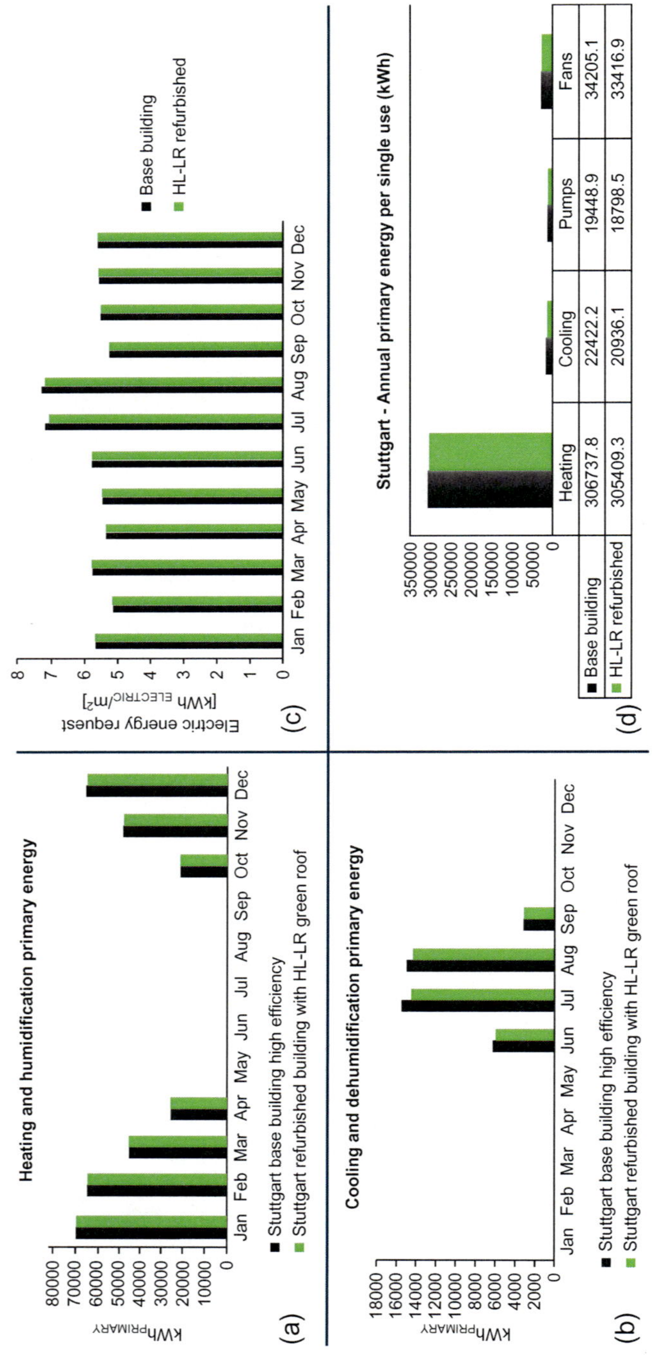

Figure 13.6 2000s buildings in Stuttgart: comparisons of energy demands of the base building and the HL-LR refurbished building.

Table 13.7 **Stuttgart: summarized feasibility analysis of considered EEMs**

EEM	ΔEP		ΔOC		WC	$\Delta CO_{2\text{-equiv}}$		C_I	DPB[1]	DPB[2]	DPB[3]	DPB[4]
	(kWh/m^{2}y)	(%)	(€/year)	(%)	(€/year)	t $CO_{2\text{-equiv}}$/year	(%)	(€)	(years)			
1980s building												
HL-LR	28,497	−3.6	2069	−3.7	2256	6.2	−3.7	157,529	*Never*	*Never*	≫50	>50
ML-MR	28,502	−3.6	2039	−3.6	1798	6.2	−3.7	151,066	≫50	≫50	≫50	>50
LL-HR	27,922	−3.5	1936	−3.4	582	5.8	−3.5	144,603	≫50	≫50	≫50	>50
1990s building												
HL-LR	11,391	−2.0	895	−2.1	2260	2.6	−2.1	136,692	*Never*	*Never*	≫50	≫50
ML-MR	11,325	−2.0	866	−2.1	1799	2.5	−2.1	130,230	*Never*	*Never*	≫50	≫50
LL-HR	10,598	−1.8	734	−1.8	582	2.2	−1.8	123,766	≫50	≫50	≫50	≫50
2000s building												
HL-LR	4186	−1.1	421	−1.5	2257	1.1	−1.4	107,609	*Never*	*Never*	≫50	≫50
ML-MR	3946	−1.0	372	−1.3	1799	1.0	−1.2	101,146	*Never*	*Never*	≫50	≫50
LL-HR	2516	−0.7	174	−0.6	582	0.5	−0.6	94,683	*Never*	*Never*	≫50	≫50

decrement of the cooling energy demand around 4.0%, gives the highest savings. Annually, the best performances are achieved by the HL-LR green roof (−3.1%) even if high irrigation is required, and thus the energy savings (−€896) are nullified by the cost of water (more than €1300). Conversely, the LL-HR green roof drastically reduces the irrigation (€229 per year), but the improvement of energy performance is negligible. For the HL-LR and ML-MR solutions, the watering costs are higher than the money savings obtained because of the lower energy demand.

(c) Refurbishment of 2000s building. Both building envelope and HVAC system are efficient. Therefore, the installation of green roof is cheaper because the thermal insulation not necessary. The base building requires a primary energy demand for heating of 33 kWh/m^{2}a. The energy request for cooling, also in this case inclusive of the auxiliaries and expressed in terms of primary energy, is 59 kWh/m^{2}a. Compared to the cooling requests of the low efficient and medium efficient buildings, this building requires higher energy because of a typical effect of hyper-insulation. The HL-LR solution of green roof allows a decrement of only 0.9% of energy demand for the heating (a full comparison is proposed in Figure 13.7), while slightly better performances are achieved by the LL-HR solution (−1.6%). Conversely, the HL-LR green roof shows the best cooling energy performances, with an energy demand that decreases about 3.2%. By considering the entire year, all solutions are beneficial, even if the costs of artificial irrigation completely nullify the energy savings (and thus the annual cash flows are disadvantageous). Finally, paybacks are not possible.

For Naples, the overall feasibility study is reported in Table 13.8.

The above reported outcomes reveal a good impact of green roofs on the energy demands for cooling and heating. Thus, beyond a lower cooling load—due to evaporative cooling effects, nighttime radiation, and higher thermal capacity—the energy demand for heating can also be reduced by installing greenery on the roof. Indeed, for inefficient buildings, the green roof installation implies an additional thermal insulation. Moreover, for insulated roofs, the vegetation reduces the thermal transmittance because of a lower convection on the external surface of the structure as well as the additional soil layer that, mainly when dry, contributes as added thermal resistance.

Conversely, the economic profitability is debatable. In this regard, Berardi et al. (2014), at the end of their study, give a comparative framework of the economic barriers (i.e., installation costs, cost of watering) and economic benefits (e.g., lower energy demand and costs for the indoor microclimatic control) of green roofs. The same authors summarize exemplificative local policies aimed at promoting the installation of green roofs in various countries.

In the same way, Feng and Hewage (2014)—in a simulative study concerning various typologies of green roofs—underline that these solutions are often not cost-effective in the winter season and/or in cold regions. However, the same authors show that the other achievable benefits (social, environmental, reduction of noise) can contribute to making this building technology profitable. Different criteria for taking into account the social benefits of green roofs are also considered necessary by Clark et al. (2008). These authors talk about the necessity of new research aimed at transferring the social positive effects to the building owners by changing the policies in matters of green roofs. Some considerations will be proposed in the next section.

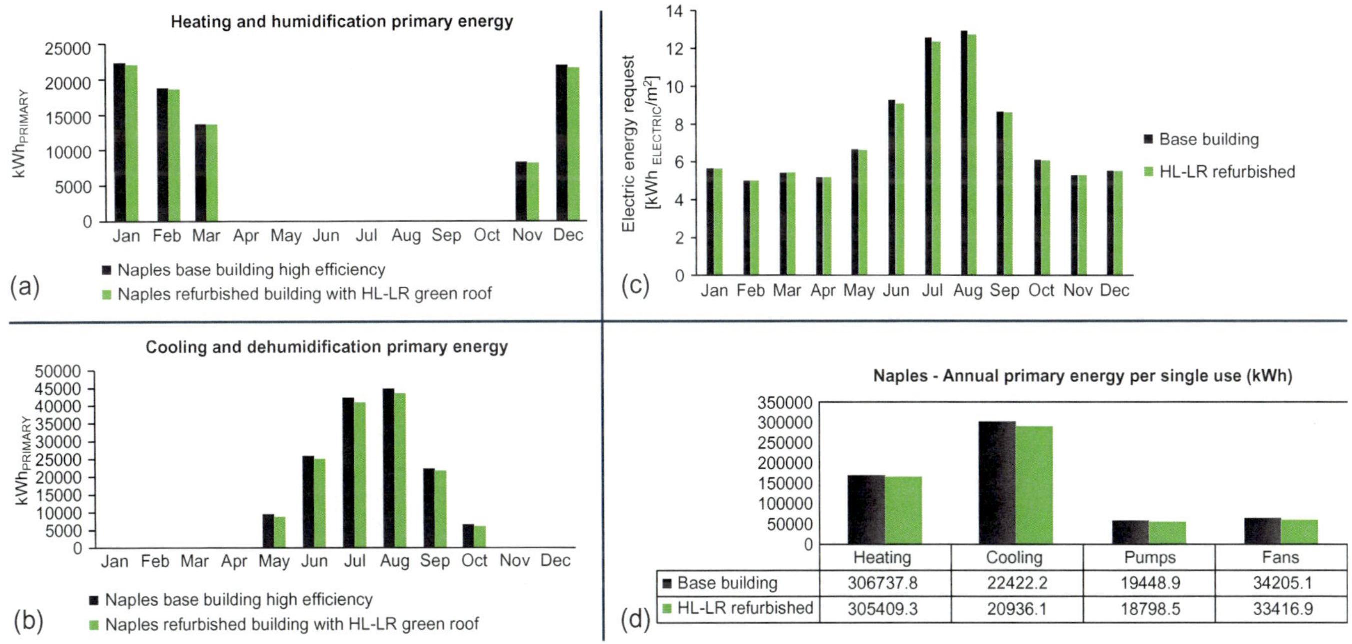

	Heating	Cooling	Pumps	Fans
Base building	306737.8	22422.2	19448.9	34205.1
HL-LR refurbished	305409.3	20936.1	18798.5	33416.9

Figure 13.7 2000s buildings in Naples: comparisons of energy demands of the base building and the HL-LR refurbished building.

Table 13.8 **Naples: summarized feasibility analysis of considered EEMs**

EEM	ΔEP		ΔOC		WC	$\Delta CO_{2\text{-equiv}}$		C_I	DPB[1]	DPB[2]	DPB[3]	DPB[4]
	(kWh/m²y)	(%)	(€/year)	(%)	(€/year)	t $CO_{2\text{-equiv}}$/year	(%)	(€)	(years)			
1980s building												
HL-LR	17,477	−4.7	1646	−4.8	1322	4.9	4.8	157,529	≫50	≫50	≫50	≫50
ML-MR	16,671	−4.5	1549	−4.5	1055	4.6	4.5	151,066	≫50	≫50	≫50	≫50
LL-HR	14,482	−3.9	1293	−3.7	229	3.7	3.7	144,603	≫50	≫50	≫50	≫50
1990s building												
HL-LR	9113	−3.1	896	−3.2	1322	2.7	3.3	136,692	*Never*	*Never*	≫50	≫50
ML-MR	7649	−2.6	732	−2.7	1055	2.2	2.7	130,230	*Never*	*Never*	≫50	≫50
LL-HR	5062	−1.7	438	−1.3	229	1.6	1.5	123,766	*Never*	*Never*	≫50	≫50
2000s building												
HL-LR	5560	−2.3	577	−2.5	1322	1.8	−2.5	107,609	*Never*	*Never*	≫50	≫50
ML-MR	3901	−1.6	396	−1.7	1055	1.2	−1.7	101,146	*Never*	*Never*	≫50	≫50
LL-HR	495	−0.2	23	−0.1	229	0.1	−0.1	94,683	*Never*	*Never*	≫50	≫50

13.6 Rainwater harvesting systems for improving the economics of green roofs

Irrigation costs strongly reduce the economic profitability of green roofs. At its core, the matter involves the ethical use and the waste of water. In the case of green roofs, irrigation is necessary for the life of vegetation and thus for activating evapotranspiration. It is not a waste of a resource, because it implies other benefits beyond the mere reduction of building cooling needs. Water is a precious resource that should not be used, particularly if drinkable, when not necessary. A global approach to water use and strategies for its recovery are required. Presently, European countries require enormous amounts of drinkable water, even higher than 150 L per person per day. This demand can be drastically reduced, as drinkable water is only necessary for drinking, kitchen use, and personal hygiene. Conversely, irrigation, cleaning of surfaces, and so on can reuse rainwater by adopting various systems.

The simpler system for single apartments is the direct connection of a cisterna to the central gutter drainpipe, equipped with proper filters. Conversely, advanced systems provide the collection of rainwater from large surfaces and thus the building roofs and surrounding spaces if characterized by a low risk of pollution. In Table 13.9, three typologies of surfaces collecting the rainwater are identified. A careful evaluation must be done according to the specific situation.

Only some typologies of roofs allow for water recovery, and this depends on the risk of contamination. Copper, zinc, and lead roofing layers are not suitable. If the place is characterized by an absence of polluting factors, properly designed draining pavements for the external environment can also allow the collection of rainwater in local wells connected to centralized tanks. The reservoirs are provided with a mechanism of 'overfill' for discharging excessive water into the sewer. The harvested water can be used for

Table 13.9 Rainwater harvesting systems. Classification of collecting surfaces on the basis of pollution risks

Not polluted surfaces	Partly polluted surfaces	Polluted and highly polluted surfaces
Sports facilities	Transit ways in mixed areas	Roads with more than 500 vehicles/day
Bike paths	Industrial roofs	Tunnels longer than 300 m
Pedestrian walkways	Parking areas at low frequency of use	Roof surfaces with polluting coatings
Courtyards	Roads with traffic up to 500 vehicles/day	Parking areas at medium-high frequency
Private gardens	Market squares	Areas for loading and unloading of materials
Green roofs	Waterproofing surfaces of patios	Areas for maintenance of vehicles
Domestic roofs	Seasonal parking areas and spaces for market	Zones of racking and storage of scrap

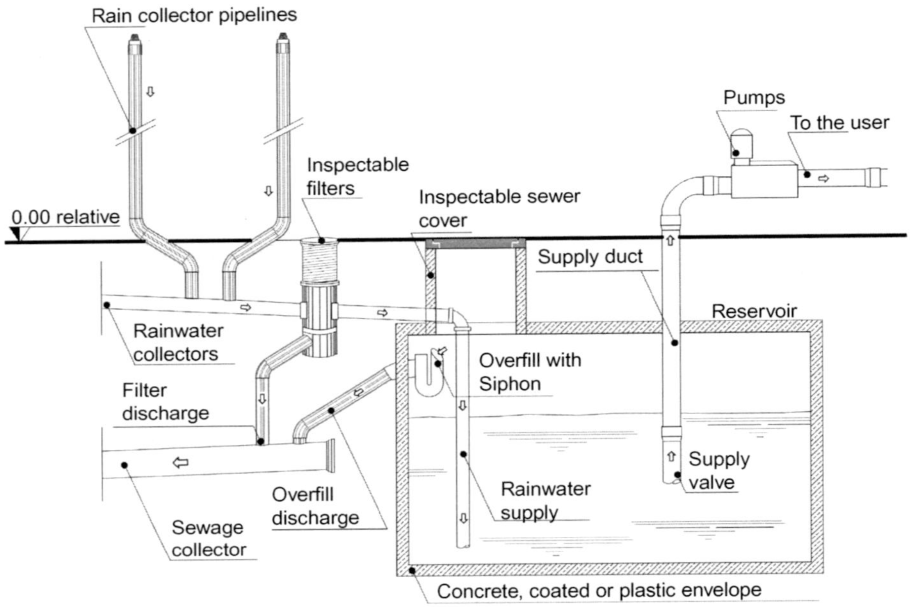

Figure 13.8 Main components of a rainwater harvesting system.

irrigation by smartly managing the control systems for activating taps, regulators, and so on. Figure 13.8 proposes an example of a rainwater harvesting system. In addition to the 'capturing' surface, many other components are necessary, including a smoothing filter on the rainwater supply pipeline in order to purify the water supplied to the tank. The reservoir can be sized according to the Standard DIN 1989 (2001) part 1. Other necessary devices are floating filters, suction hose, pressure pump, pressure hose, automatic switch, ball valve, overflow traps, and controllers. A strong recommendation of the DIN 1989 establishes that risks of possible contamination of drinking water must be absolutely avoided. Thus, the lines of harvested water and the drinkable water must be separated and well-identified with appropriate and evident labels. Other prescriptions concern the risk of frost and consequent damage (if the tanks are buried at depths higher than 1 m this risk can be limited), the sealing of the building envelope, and the protection from noise and fire. Regarding the collection areas, the DIN 1989:2001 cites that surfaces be as little polluted as possible. The same standard provides indication about treatments for improving the quality of harvested water by means of biological, chemical, and physical processes. Typology of filters, maintenance frequency, and operations are related to rainwater use. Various filters can be selected, and these must be placed in the supply line of the reservoir. In addition to rainwater storage, the tank is used also for sedimentation of fine solids.

Filters (both systems with retention and with a separate discharge) must be chosen based on efficiency and purification capability and, if connected to the sewage line, systems to prevent possible backup are required. Additional processes and filtration procedures can be required in particular conditions of collection areas (e.g., flocculation, flotation, biological processes, diaphragm processes, disinfection, and particular filtrations). In addition to frost, light and other radiative/thermal effects must also be avoided. Suitable materials are plastic, concrete, and steel (with proper coatings). Moreover, single tanks may be connected. The standard specifies characteristics in terms of resistance to loads and with reference to the opening of the tanks and storage volume. Position, characteristics, and number of pumps have to be accurately designed. A supplemental feed system, controlled in order to be switched on below a minimum filling grade of the tank, is required. However, a general control unit must supervise the entire system. Other prescriptions concern the pipelines and ventilation ducts (i.e., infiltration of water, leaves, garbage, and entering of small creatures must be avoided) and the risk of refluxes. Also, indications about systems for avoiding the entrance of sewer gases are proposed. Finally, the entire paragraph 16 of the DIN 1989 (2001) concerns the sizing of the reservoir volume (minimum and nominal water storage volumes), taking into account various parameters (i.e., local precipitation and distribution, characteristics of the collection area, rate of drainage, amount of required process water). Three methods are proposed for the sizing. Information about initial operation, operating conditions, inspection, and maintenance are also detailed.

All told, these systems are standardized, safe, ethical, and economical. They also confer benefits connected to the urban green, such as the improvement of livableness and the reduction of pollution. Therefore, these systems can be used to implement more convenient and feasible green roofs.

13.7 Conclusions and future trends

The investigations are designed to evidence relevant aspects of energy performance achievable by the installation of green roofs. Green roofs are efficient in reducing both heating and cooling demands, especially when they are installed to improve the thermal properties of old building envelopes. Indeed, as a refurbishing strategy, the combined effects (i.e., higher insulation, lower convection, thermal mass and capacity, evapotranspiration and thus evaporative cooling) produce significant energy advantages. Otherwise, for buildings already insulated and characterized by satisfactory energy performance, the additional benefits achievable by adoption of roof vegetation are marginal. With reference to the mere building cooling needs, a proper evapotranspiration of plants and soil can be achieved, and it implies a decrement of temperature difference between external roof surface and indoor air, with consequent lower cooling loads. On the other hand, the evapotranspiration requires an abundance of water and in warm climates during the summer season, the rainwater intercepted by the soil is not enough. Finally, in order to make convenient green roofs, rainwater harvesting systems are necessary.

In accordance with the aims of the chapter, technical and economical investigations have been proposed. However, these cannot take into account the larger questions connected to the future development of our countries and our urban areas: What is the value of a better life? How is it possible to quantify higher indoor comfort and thus a better livability of indoor spaces? Indeed, future studies, here highlighted as brief 'food for thought,' will also concern the possibility of improving indoor thermal comfort by means of installation of green roofs, for instance by admitting adaptive criteria, and thus trying to avoid completely the installation of active cooling systems.

Evapotranspiration, time lag effect, and lower decrement factor (caused by the presence of vegetation and soil on the roof) attenuate and reduce heat transmission from the outside to the internal environment in summer. This means lower and more stable (during the 24 h) temperatures of the inner surfaces of roof structures, which is particularly relevant in buildings characterized by a prevalent diurnal use.

Finally, if an adaptive model for thermal comfort is assumed, green roofs are surely relevant solutions. Other developments could analyse the urban effects and thus better air quality in cities, filtration of polluting dusts, benefic effects induced by the drainage of rainwater, and reduction of urban heat island effects and noise. Pacheco-Torgal (2013) and Kolokotsa et al. (2013) have recently reviewed these topics by means of wide research. In addition, there are indubitable values related to the relaxed and esthetic effects of these technologies. Again, what is the economic value of these invaluable arguments?

References

ANSI/ASHRAE, 2011. Standard 140:2011 'Standard Method of Test for the Evaluation of Building Energy Analysis Computer Programs'. ASHRAE, Atlanta, GA.

Ascione, F., Bianco, N., de Rossi, F., Turni, G., Vanoli, G.P., 2013. Green roofs in European climates. Are effective solutions for the energy savings in air-conditioning? Appl. Energy 104, 845–859.

ASHRAE, 2001a. Research project 1052, report titled Development of an Analytical Verification Test Suite for Whole Building Energy Simulation Programs – Building Fabric'.

ASHRAE, 2001b. IWEC – International Weather Data for Energy Calculations. ASHRAE, Atlanta, GA, USA.

ASHRAE, 2002. Research project 865, report titled 'Development of Accuracy Tests for Mechanical System Simulation'.

Berardi, U., GhaffarianHoseini, A.H., GhaffarianHoseini, A., 2014. State-of-the-art analysis of the environmental benefits of green roofs. Appl. Energy 115, 411–428.

Bianchini, F., Hewage, K., 2012. Probabilistic social cost-benefit analysis for green roofs: a life-cycle approach. Build. Environ. 58, 152–162.

BPIE, 2013. Implementing the Cost-Optimal Methodology in EU Countries, Published in March 2013 by the Buildings Performance Institute Europe (BPIE). Available from: http://bpie.eu/cost_optimal_methodology.html#.UxS0s_l5OCk (accessed 15.10.13.).

Carter, T., Keeler, A., 2008. Life-cycle cost—benefit analysis of extensive vegetated roof systems. J. Environ. Manage. 87, 350–363.

Castleton, H.F., Stovin, V., Beck, S.B.M., Davison, J.B., 2010. Green roofs; building energy savings and the potential for retrofit. Energy Build. 42, 1582–1591.

CEN—European Committee for Standardization, 2007a. Standard EN 15459:2008 'economic evaluation procedure for energy systems in buildings'.

CEN—European Committee for Standardization, 2007b. Standard EN 15251:2007 'indoor environmental input parameters for design and assessment of energy performance of buildings addressing indoor air quality, thermal environment, lighting and acoustics'.

Chan, A.L.S., Chow, T.T., 2013. Energy and economic performance of green roof system under future climatic conditions in Hong Kong. Energy Build. 64, 182–198.

Clark, C., Adriaens, P., Talbot, F.B., 2008. Green roof valuation: a probabilistic economic analysis of environmental benefits. Environ. Sci. Technol. 42, 2155–2161.

Claus, K., Rousseau, S., 2012. Public versus private incentives to invest in green roofs: a cost benefit analysis for Flanders. Urban For. Urban Green. 11, 417–425.

Commission Delegated Regulation No 244/2012 of 16 January 2012, 2012. 'Supplementing Directive 2010/31/EU of the European parliament and of the council on the energy performance of buildings by establishing a comparative methodology framework for calculating cost-optimal levels of minimum energy performance'. Off. J. Eur. Union Legis. 81, 18–36.

Commission and Council Directive 2002/91/EC of 16 December 2002 on the energy performance of buildings, 2003. Off. J. Eur. Union Legis. 1, 65–71.

Commission and Council Directive 2010/31/EU of 19 May 2010 on the energy performance of buildings (Recast), 2010. Off. J. Eur. Union Legis. 153, 15–35.

Commission and Council Directive 2012/27/EU of 25 October 2012 on energy efficiency, amending Directives 2009/125/EC and 2010/30/EU and repealing Directives 2004/8/EC and 2006/32/EC, 2012. Off. J. Eur. Union Legis. 315, 1–56.

Covenant of Mayors—Committed to Local Sustainable Energy 'Technical Annex to the SEAP Template Instructions Document: the Emission Factors'. Available from: http://www.eumayors.eu (accessed 20.12.13.).

Crawley, D.B., Hand, J.W., Kummert, M., Griffith, B.T., 2008. Contrasting the capabilities of building energy performance simulation programs. Build. Environ. 43 (4), 661–673.

Deardorff, J.W., 1978. Efficient prediction of ground surface temperature and moisture, with inclusion of a layer vegetation. J. Geophys. Res. 83, 1889–1903.

DIN—Deutsches Institut für Normung, 2001. DIN 1989:2001 'Rainwater harvesting systems—Part I. Planning, installation, operation and maintenance'.

DOE, 2012. U.S. Department of Energy. 2012. Energy Plus simulation software. Version 7.2.0.

Eurostat, 2013, Energy Price Statistics, Available from: http://epp.eurostat.ec.europa.eu/statistics_explained/index.php/Energy_price_statistics (accessed 13.12.13.).

Feng, H., Hewage, K., 2014. Energy saving performance of green vegetation on LEED certified buildings. Energy Build. 75, 281–289.

Frankenstein, S., Koenig, G., 2004. 'FASST Vegetation Models', U.S. Army Corps of Engineers—Engineer Research and Development Center (ERDC/CRREL), Technical Report TR, 04–25.

GWI—Global Water Intelligence, 2013, Combined Tariffs. Available from: http://www.globalwaterintel.com/archive/12/9/market-profile/global-water-tariffs-continue-upward-trend.html (accessed 20.01.14.).

IEA—International Energy Agency, 2008. Energy Efficiency Indicators for Public Electricity Production from Fossil Fuels, OECD /IEA, July 2008, In Support of G8 Plan of Action. Available from: http://www.iea.org/publications/freepublications/publication/En_Efficiency_Indicators.pdf (accessed 16.03.14.).

Italian Republic Presidential Decree n. 59/2009 of 2 April 2009 (in Italian), 'Attuazione dell'articolo 4, comma 1, lettere a) e b), del decreto legislativo 19 agosto 2005, n. 192, e successive modificazioni, concernente attuazione della direttiva 2002/91/EC sul rendimento energetico in edilizia', 2009. Official Bulletin n. 132.

Kolokotsa, D., Santamouris, M., Zerefos, S.C., 2013. Green and cool roofs' urban heat island mitigation potential in European climates for office buildings under free floating conditions. Sol. Energy 95, 118–130.

Kosareo, L., Ries, R., 2007. Comparative environmental life cycle assessment of green roofs. Build. Environ. 42, 2606–2613.

Olsen, E.L., Chen, Q., 2003. Energy consumption and comfort analysis for different low-energy cooling systems in a mild climate. Energy Build. 35 (6), 560–571.

Pacheco-Torgal, F., 2013. Eco-efficient construction and building materials research under the EU Framework Programme Horizon 2020. Constr. Build. Mater. 51, 151–162.

Peri, G., Traverso, M., Finkbeiner, M., Rizzo, G., 2012. The cost of green roofs disposal in a life cycle perspective: covering the gap. Energy 48, 406–414.

Rincón, L., Coma, J., Pérez, G., Castell, A., Boer, D., Cabeza, L.F., 2014. Environmental performance of recycled rubber as drainage layer in extensive green roofs. A comparative life cycle assessment. Build. Environ. 74, 22–30.

Rowe, D.B., Kolp, M.R., Greer, S.E., Getter, K.L., 2014. Comparison of irrigation efficiency and plant health of overhead, drip, and sub-irrigation for extensive green roofs. Ecol. Eng. 64, 306–313.

Saadatian, O., Sopian, K., Salleh, E., Lim, C.H., Riffat, S., Saadatian, E., Toudeshki, A., Sulaiman, M.Y., 2013. A review of energy aspect of green roofs. Renew. Sustain. Energy Rev. 23, 155–178.

Sailor, D.J., 2008. A green roof model for building energy simulation programs. Energy Build. 40 (8), 1466–1478.

Schweitzer, O., Erell, E., 2014. Evaluation of the energy performance and irrigation requirements of extensive green roofs in a water-scarce Mediterranean climate. Energy Build. 68, 25–32.

Sproul, J., Wan, M.P., Mandel, B.H., Rosenfeld, A.H., 2014. Economic comparison of white, green, and black flat roofs in the United States. Energy Build. 71, 20–27.

Sun, T., Bou-Zeid, E., Ni, G., 2014. To irrigate or not to irrigate: analysis of green roof performance via a vertically-resolved hygrothermal model. Build. Environ. 73, 127–137.

WMO, 2013. World Weather Information Service, Official Observations. Official Forecasts. Available from: http://worldweather.wmo.int (accessed 12.01.14.).

Wong, N.H., Tay, S.F., Wong, R., Ong, C.L., Sia, A., 2003. Life cycle cost analysis of rooftop gardens in Singapore. Build. Environ. 38, 499–509.

Part Four

Phase-change materials (PCMs) and chromogenic smart materials for reducing building cooling needs

Phase-change materials for reducing building cooling needs

14

L.F. Cabeza[1], L. Navarro[1], C. Barreneche[1,2], A. de Gracia[1], A.I. Fernández[2]
[1]University of Lleida, Lleida, Spain; [2]University of Barcelona, Barcelona, Spain

14.1 Introduction

Eco-efficiency has been proposed as one of the main tools to promote a transformation from unsustainable development to one of sustainable development (Yadong, 2013). It is based on the concept of creating more goods and services while using fewer resources and creating less waste and pollution. According to the WBCSD (2000) and Lovins (2008), the critical aspects of eco-efficiency are reduced material intensity of goods or services, reduced energy intensity of goods or services, reduced dispersion of toxic materials, improved recyclability, maximum use of renewable resources, greater durability of products, and increased service intensity of goods and services.

Therefore, eco-efficient materials are those materials that ensure the use of less material in a given application, ensure the use of less energy both in the production or transformation of the material (low embodied energy materials) and in the operation time within the application (low operational energy), are nontoxic and recyclable, and last more time within the application.

Similarly, eco-efficient systems are those that ensure a reduction in the material intensity of goods or services, a reduction in the energy intensity of goods or services, reduced dispersion of toxic materials, improved recyclability, maximum use of renewable resources, greater durability of products, and increased service intensity of goods and services.

Finally, one can define materials that help in achieving more eco-efficient systems as those materials that help systems in meeting the requirements listed above.

Within this context, eco-efficient PCM are not only those PCM that fulfill the requirements listed but also any material that help a system in becoming eco-efficient.

Cabeza et al. (2013) reviewed low-carbon and low embodied energy materials in buildings. This review stated that embodied energy in building materials has increased its importance in the life cycle of a building compared to operating energy, due to the better energy performance of the buildings. There is also a general agreement that embodied energy is difficult to quantify and that, since there is no generally accepted methodology for its measurement or calculation, today's data disagree between authors and studies.

PCMs have never been looked at as eco-efficient materials. The aim of this chapter is to evaluate the eco-efficiency of PCM when used in buildings to mitigate cooling demand.

Eco-efficient Materials for Mitigating Building Cooling Needs. http://dx.doi.org/10.1016/B978-1-78242-380-5.00014-5

14.2 Phase-change materials

Materials to be used for phase-change thermal energy storage (TES) must have a large latent heat and high thermal conductivity. They should have a melting temperature lying in the practical range of operation, melt congruently with minimum subcooling, and be chemically stable, low cost, nontoxic, and noncorrosive. Materials that have been studied during the last 40 years are salt hydrates, paraffin waxes, fatty acids, and eutectics of organic and nonorganic compounds (Cabeza et al., 2011). A classification of materials used as PCM as organic and inorganic is presented in Figure 14.1 and in Table 14.1.

According to Kenisarin and Mahkamov (2007), the following PCM properties to be used for latent heat storage were highlighted as desirable:

1. A high value of the heat of fusion and specific heat per unit volume and weight
2. A melting point that matches the application
3. A low vapor pressure (<1 bar) at the operational temperature
4. A chemical stability and noncorrosiveness
5. Not be hazardous, highly inflammable, or poisonous
6. A reproducible crystallization without degradation
7. A small subcooling degree and high rate of crystal growth
8. A small volume variation during solidification
9. A high thermal conductivity
10. Availability and abundance

According to Kenisarin and Mahkamov (2007), the PCM to be used in the design of thermal storage system should have desirable thermophysical, kinetic, and chemical properties:

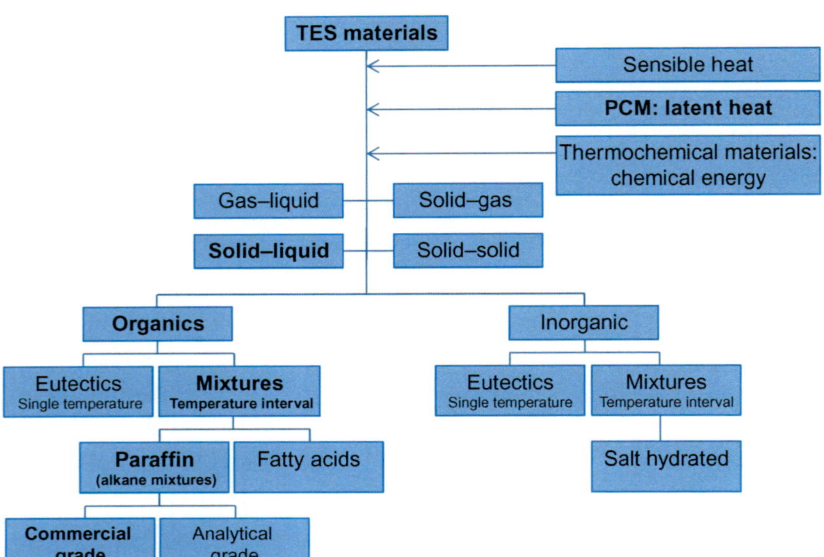

Figure 14.1 Classification of energy storage materials (Kenisarin and Mahkamov, 2007).

Table 14.1 Classification of phase-change materials

	Organic	Inorganic
Materials	High aliphatic hydrocarbon Acid/esters or salts Alcohols Aromatic hydrocarbons Aromatic ketone Freons Polymers	Crystalline hydrates Molten salts Metals Alloys
Advantages	Noncorrosives Low or no subcooling Chemically and thermally stable Adjustable transition zone	Higher phase-change enthalpy Higher thermal conductivity Nonflammable Inexpensive
Disadvantages	Lower phase-change enthalpy Low thermal conductivity Flammability Low density Highly volatile Volume change	Subcooling Corrosion Phase separation Phase segregation, lack of thermal stability

Adapted from Cabeza et al. (2011).

- Thermophysical properties
  - Melting temperature in the desired operating temperature range to assure storage and extraction of heat in an application with a fixed temperature range
  - High latent heat of fusion per unit volume to achieve high storage density compared to sensible storage
  - High specific heat to provide additional significant sensible heat storage
  - High thermal conductivity of both solid and liquid phases to assist the charging and discharging energy of the storage system
  - Small volume change on phase transformation and small vapor pressure at operating temperature to reduce the containment problem
  - Congruent melting of the PCM for a constant storage capacity of the material with each freezing/melting cycle
  - Reproducible phase change to use the storage material many times (also called cycling stability)
- Nucleation and crystal growth
  - High nucleation rate to avoid supercooling of the liquid phase and to assure that melting and solidification proceed at the same temperature
  - High rate of crystal growth, so that the system can meet demand of heat recovery from the storage system
- Chemical properties
  - Complete reversible freeze/melt cycle
  - No degradation after a large number of freeze/melt cycle
  - No corrosiveness to the construction materials
  - Nontoxic, nonflammable, and nonexplosive material for safety: for environmental and safety reasons

- Economics
  - Abundant
  - Available
  - Cost-effective: to be competitive with other options for heat and cold storage

Many substances have been studied as potential PCMs, but only a few of them are commercialized as so (Barreneche et al., 2014; Cabeza et al., 2011; Farid et al., 2004; Kenisarin and Mahkamov, 2007; Mehling and Cabeza, 2008). It is not the aim of this chapter to list all potential materials studied as PCMs, but as an example, Table 14.2 shows a very short list of PCMs and their thermophysical properties, classified as shown in Table 14.1.

Barreneche et al. (2014) developed a new database to classify PCM and to help in their selection for a given application. Figure 14.2 shows all PCMs that can be used for active and passive systems in buildings (see Section 4). Melting temperature versus latent heat of fusion of all these substances is plotted and the different colors present different groups or natures of PCM: eutectic, organic, fatty acid, paraffin, and inorganic. Depending on the requirement of the system to implement the PCM, the selection of the substance with the highest latent heat of fusion for a given melting point may be performed easily in the graph. Different bubble colors mean different building temperature ranges depending on the application. It can be observed that a wide range of values of latent heat is reported for a given application like cold storage, which includes ice as one of the materials with the highest latent heat.

PCMs are reported for cooling applications in buildings with melting temperatures of up to 21 °C. All substances available for this application (commercial and not commercial) are plotted in Figure 14.3. Different bubble colors mean different natures of PCMs and most of them have a latent heat within the range 100–250 kJ/kg. It is interesting to note that the shape and size of the bubbles are quite diverse. This may be due to several reasons: few or even single data reported for a given substance or, the opposite, a broad set of values for melting temperature and/or latent heat for a given substance.

On the other hand, PCMs reported to control the comfort temperature in indoor buildings are plotted in Figure 14.4. The temperature range selected was between 20 °C and 30 °C and these PCMs are commonly used in passive systems. This window can be larger or shorter depending on the climate conditions where the PCMs system have to be implemented. Again, some of the PCMs show narrow ranges of melting temperature and/or latent heat (Climsel24 or paraffin C17) meanwhile other PCMs like 1-dodecanol or paraffin C18 show broader bubbles.

14.3 Eco-efficient phase-change materials

However, for a PCM to be able to be considered as an eco-friendly material, it needs to fulfill the cited properties in Section 14.2.

Table 14.2 Selection of materials for comfort applications in buildings and their thermophysical properties

Material	Type	Melting temperature (°C)	Heat of fusion (kJ/kg)	Thermal conductivity (W/m K)	Density (kg/m^3)
n-Heptadecane	Organic	19	240	0.21	760 (Liquid)
61.5 mol% capric acid +38.5 mol% lauric acid	Fatty acid mixture	19.1	132	–	–
Paraffin C_{16}–C_{18}	Organic	20–22	152	–	–
$FeBr_3 \cdot 6H_2O$	Inorganic (salt hydrate)	21	105	–	–
Polyglycol E600	Organic	17–25	127–146	0.19 (Liquid)	1126 (Liquid, 25 °C) 1232 (Solid, 4 °C)
Ethyl palmitate	Organic mixture	22.1	153	–	–
45% $Ca(NO_3)_2 \cdot 6H_2O$ +55% $Zn(NO_3)_2 \cdot 6H_2O$	Inorganic mixture	25	130	–	1930
50% $CaCl_2$+50% $MgCl_2$+$6H_2O$	Inorganic eutectic	25	95	–	–

Adapted from Cabeza et al. (2011).

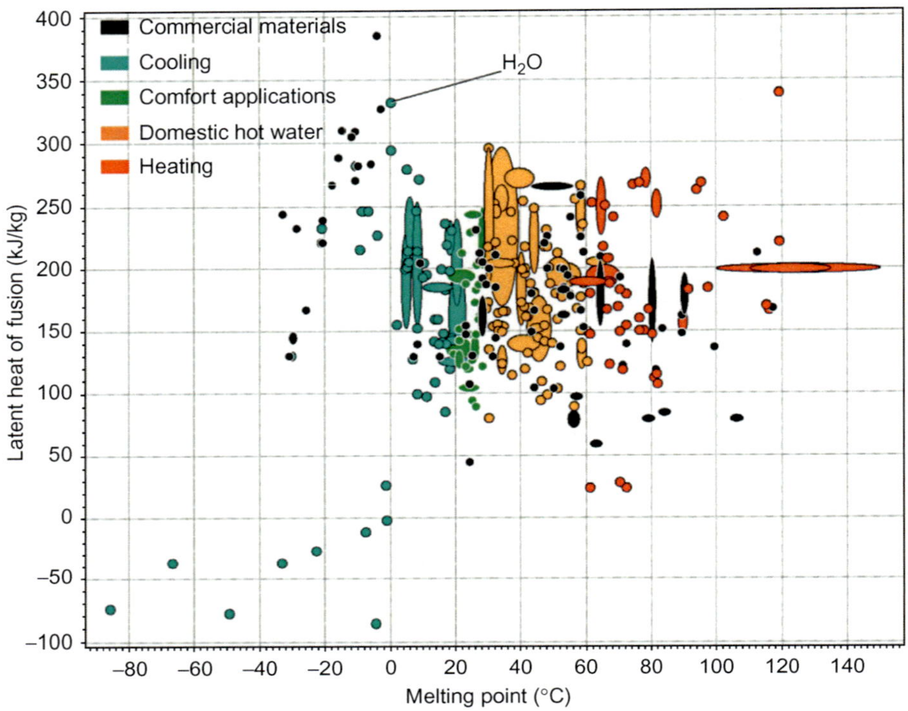

Figure 14.2 PCMs with potential applications in buildings reported by Cabeza et al. (2011) classified by Barreneche et al. (2014).

A new classification of PCM could relate to their eco-efficiency. Very broadly, this classification could be done by their nature: organic or inorganic materials. As such, PCMs can be classified as follows (Figure 14.5):

1. *Highly eco-friendly PCM*: Inorganic PCM. They have low embodied energy and the lowest footprint because they are obtained with simple chemical processes.
2. *Eco-friendly PCM*: Organics from renewable bioderivative resources, such as fatty acids and esters.
3. *Non-eco-friendly PCM*: Paraffins. These PCMs come from oil; therefore, they have high embodied energy and footprint.

14.4 Phase-change materials as a tool to mitigate building cooling demands

The use of TES in a building can smooth temperature fluctuation (Mehling and Cabeza, 2008). TES in buildings can be implemented by sensible heat (e.g., increasing and decreasing the temperature of the building envelopes) or by latent heat (with the inclusion of PCMs to increase thermal inertia). The main advantage of latent heat

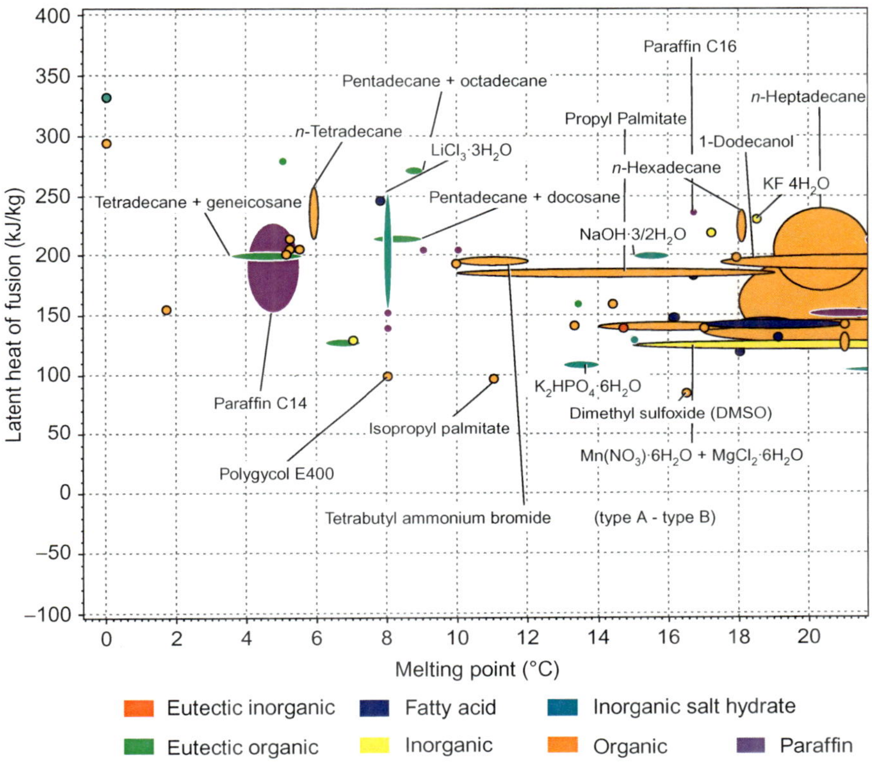

Figure 14.3 Material classification of PCMs used for cooling applications (melting temperature up to 21 °C) (Barreneche et al., 2014).

storage is the high storage density in small temperature intervals. These TES systems are known as passive systems.

PCMs have been studied for thermal storage in buildings since 1980. These systems provide a higher thermal inertia to the building that, when combined with thermal insulation, can reduce the energy consumption of the building by absorbing the heat gains and reducing the heat flow. During daytime, the PCM can absorb part of the heat through the melting process, and during the night, the heat is released by solidification of the PCM, resulting in a lower heat flow from outdoors to indoors. This can be implemented in a building as passive or active system.

14.4.1 Passive systems

In passive systems, the PCM is included in the building, usually in the building walls, and it is expected to be charged and discharged due to the outdoor temperature change. Figure 14.6 shows the effects of including PCM as passive system in the walls of a building (Cabeza et al., 2007):

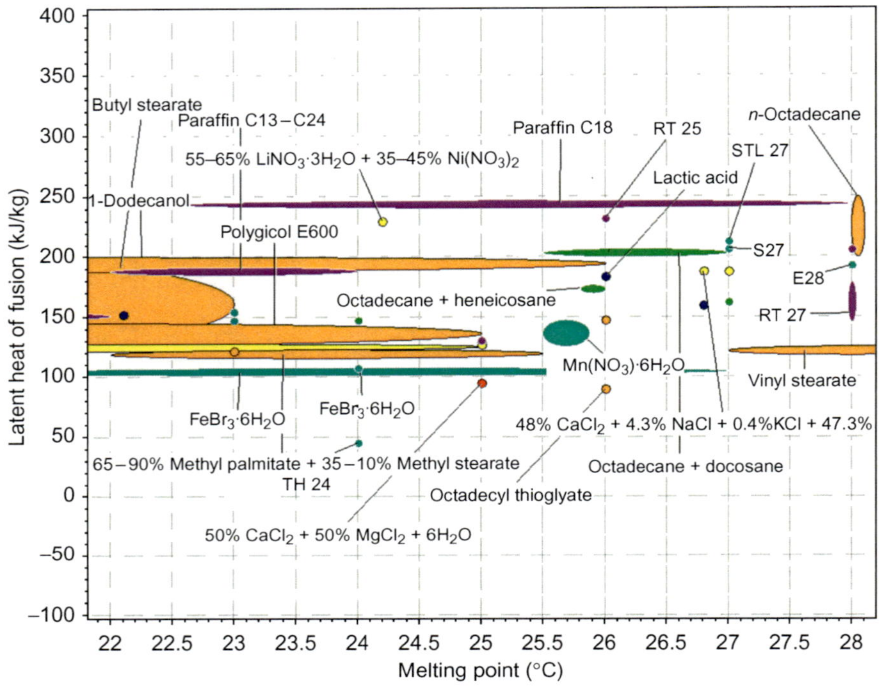

Figure 14.4 Material classification of PCMs used for comfort applications in buildings (between 20 °C and 30 °C) (Barreneche et al., 2014).

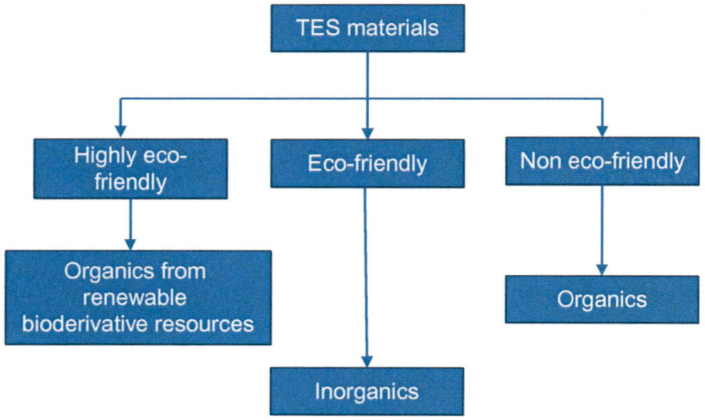

Figure 14.5 Classification of PCMs by their eco-efficiency.

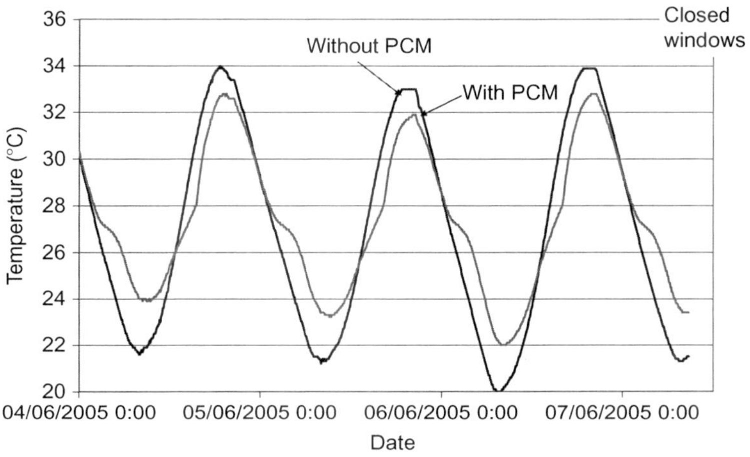

Figure 14.6 Detail of the indoor temperature of an experimental cubicle with microencapsulated PCM in its walls (Cabeza et al., 2007).

- First, the cubicle without PCM has a maximum temperature 1 °C higher than that with phase-change material and a minimum temperature 2 °C lower.
- Second, the maximum temperature in the wall with phase-change material appears about 2 hours later without PCM, that is, the thermal inertia of the wall is higher.
- Third, this thermal inertia appears again in the afternoon due to the freezing of the PCM but also earlier in the morning due to the melting of the phase-change material.

Some examples of such as system can be found in the literature (Cabeza et al., 2007; Hawes and Feldman, 1992; Schossig et al., 2002; Shapiro et al., 1987).

14.4.2 Example of passive system

The use of PCM inside building envelopes (both walls and roofs) increases the heat storage capacity of the building and might improve its energy efficiency and hence reduce the electrical energy consumption (Castell et al., 2010). Here, a detailed example of passive systems is given.

In the summer, during daytime, the sunshine and high temperatures result in a heat wave penetrating the walls of the buildings (Figure 14.7). PCM absorbs the excess heat through the melting process, delaying the heat wave inside the building and even reducing the peak. During most of the day, the room temperature remains comfortable and the cooling system consumes less energy. During nighttime, when temperatures are lower, the PCM releases the stored heat to both the internal ambient and external ambient, keeping again the room temperature comfortable and closing the cycle by solidifying.

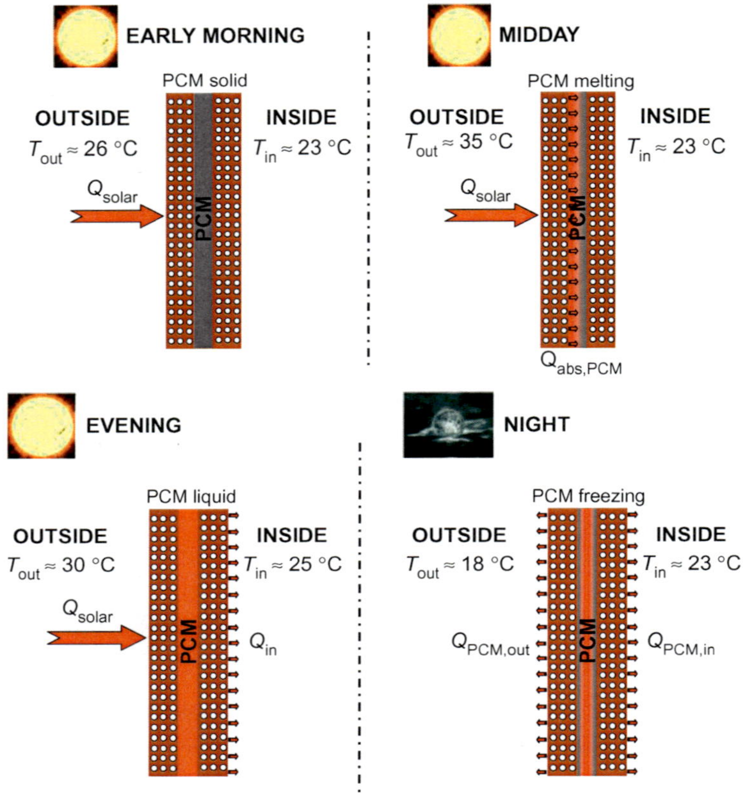

Figure 14.7 Operating principle of PCM in buildings.

Figure 14.8 Installation to test PCM in building envelopes (Puigverd de Lleida, Spain).

An experimental setup was built in Puigverd de Lleida, Spain, to evaluate passive systems for building energy efficiency (Figure 14.8). This installation consists of several cubicles, with internal dimensions of $2.4 \times 2.4 \times 2.4$ m^3, simulating real buildings and operated under real conditions. Five of those cubicles are used to test the incorporation of PCM in building envelops.

These five cubicles were built with two different types of construction (Figure 14.9):

Figure 14.9 PCM panels in a cubicle.

Table 14.3 Physical properties of PCM

Property		RT-27	SP-25-A8
Melting point (°C)		28	26
Congealing point (°C)		26	25
Heat storage capacity (kJ/kg)		179	180
Density (kg/L)	Solid	0.87	1.38
	Liquid	0.75	
Specific heat capacity (kJ/kg K)	Solid	1.8	2.5
	Liquid	2.4	
Heat conductivity (W/m K)		0.2	0.6

1. Conventional brick cubicles. Three cubicles built using conventional brick (typical Mediterranean construction) with or without insulation and with or without PCM:
   a. Reference: Built with conventional brick and air chamber (without insulation).
   b. Polyurethane: Built with conventional brick, air chamber, and 5 cm of sprayed polyurethane.
   c. Polyurethane and PCM: Built with conventional brick, air chamber, 5 cm of sprayed polyurethane, and an additional layer of macroencapsulated PCM (RT-27) (Table 14.3) on the southern and western walls and on the ceiling.
2. Alveolar brick cubicles. Two cubicles built using alveolar brick with or without macroencapsulated PCM:
   a. Reference: Built only with alveolar brick.
   b. Alveolar+PCM: Built with alveolar brick and an additional layer of macroencapsulated PCM (SP-25-A8) (Table 14.3) on the southern and western walls and on the ceiling.

Figure 14.10 presents a section of the two different constructive solutions used in the cubicles: brick and alveolar brick structure. Both sections represent the corresponding cubicle with PCM.

All the cubicles were fully instrumented and data were registered at 5 min intervals to evaluate the thermal performance.

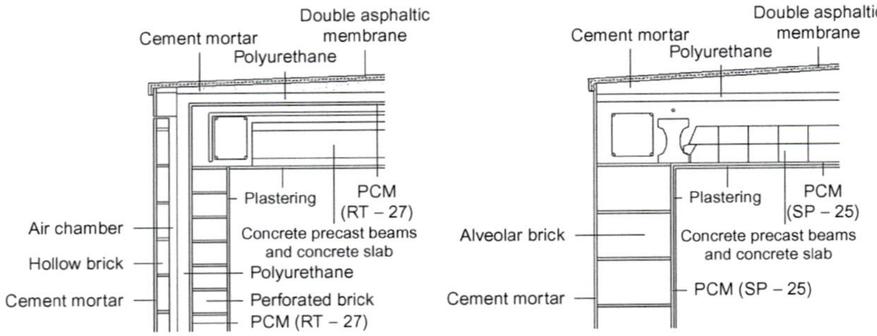

Figure 14.10 Constructive sections of the cubicles.

The main objective of the incorporation of TES is to reduce the energy consumption during summer by storing solar energy in the walls of the building. The experimental setup is designed for cooling applications, and different tests are done:

–Free-floating temperature, where no cooling system is used. The temperature conditions inside the cubicles are compared. The ones with PCM are expected to have a better behavior.
–Controlled temperature, where a heat pump is used to set a constant ambient temperature inside the cubicle. The energy consumption of the cubicles is compared. The cubicles using PCM are expected to have lower energy consumptions.

Figures 14.11–14.14 present the results for free-floating experiments in a week of August for conventional brick and alveolar brick, respectively. The results show lower peak temperatures (up to 1 °C) and more constant conditions inside the cubicles with PCM. The PCM effect is clearly visible in both constructive typologies, but especially

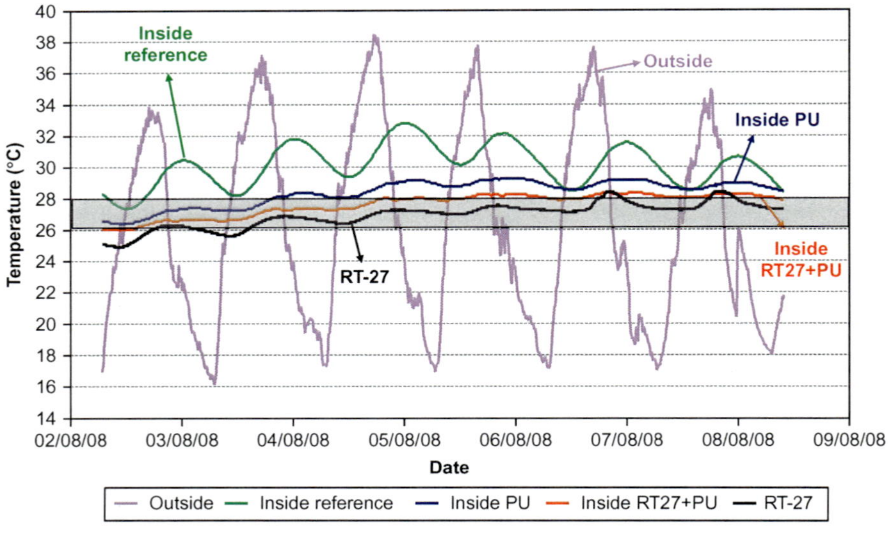

Figure 14.11 Free-floating results for the conventional brick constructive system.

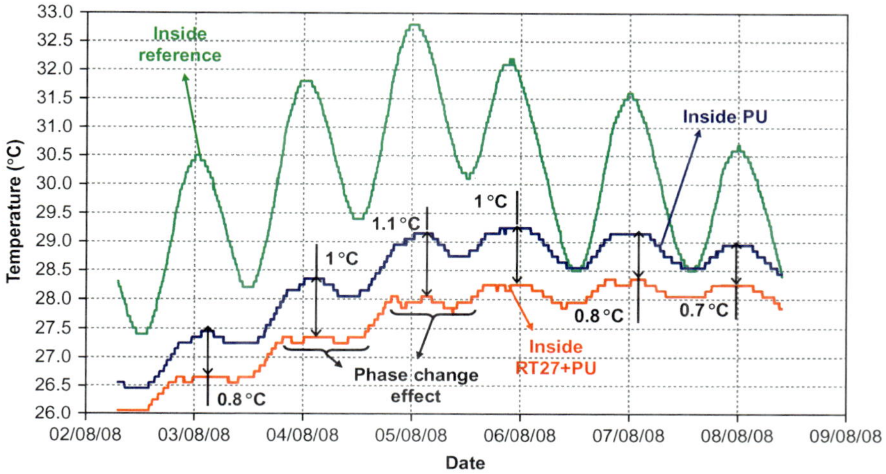

Figure 14.12 Detail of the free-floating results for the conventional brick constructive system.

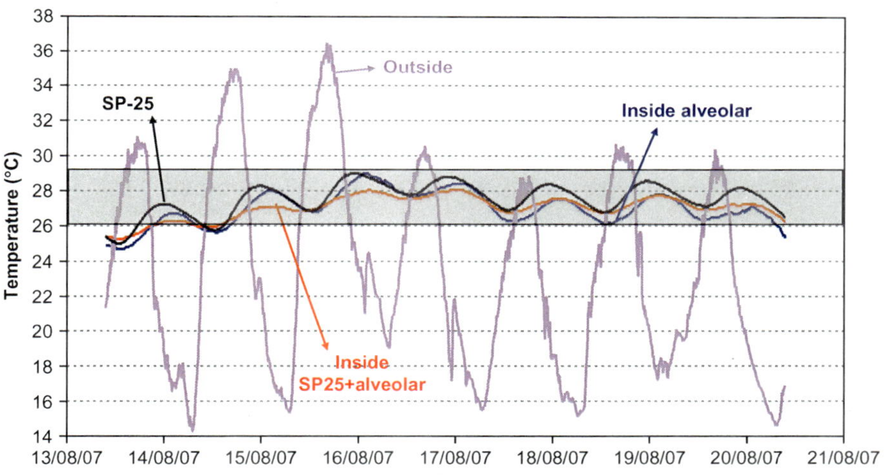

Figure 14.13 Free-floating results for the alveolar brick constructive system.

in the alveolar brick cubicle, since the thermal resistance of the wall is lower. In both cubicles, the PCMs (RT-27 and SP-25-A8) are within their melting ranges almost during all the experiment.

For the conventional brick cubicles, the reference cubicle presents high temperature fluctuations because of the lack of insulation. In the PU cubicle, those fluctuations are significantly reduced, but are still present. Finally, the inside temperature of the cubicle with PCM remains cooler (up to 1 °C) during all the experiment and the temperature fluctuations are further reduced. Moreover, for some days when the phase-change effect is maximized, the temperature fluctuations are completely prevented.

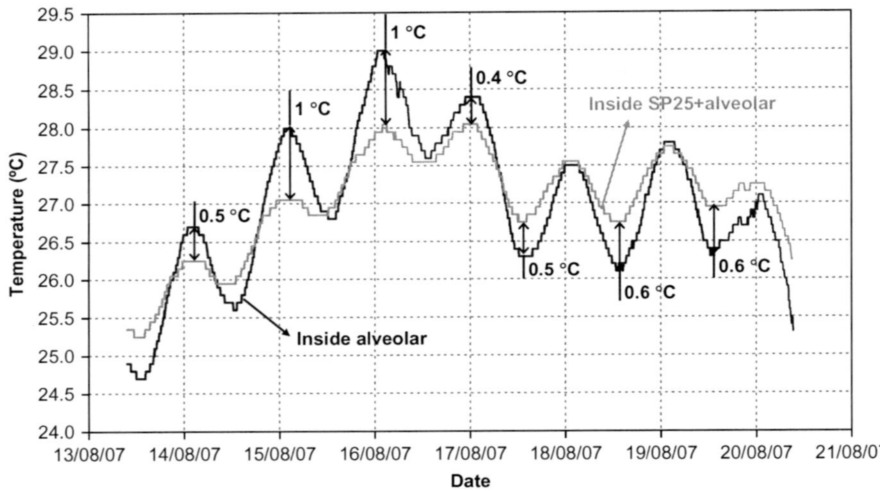

Figure 14.14 Detail of the free-floating results for the alveolar brick constructive system.

Considering the alveolar brick cubicles, when the weather is getting warmer, the PCM is melting and the maximum peak temperatures are reduced in the SP-25 + Alveolar cubicle. After some extremely warm days (outside temperatures of 36 °C), when the weather is getting cooler during the day (outside temperature of 30 °C), the cubicles with PCM present a better temperature stability, reducing the minimum peak temperatures.

The cubicles were equipped with a heat pump as a cooling system to simulate the real conditions of a house. The energy consumption of the heat pump was measured to determine the real energy savings achieved when the cubicles remain within the comfort range.

Figure 14.15 presents the results of the controlled temperature experiments using a set point of 24 °C. The accumulated energy consumption of the reference cubicle is higher than that of all the other cubicles. The RT-27 + PU cubicle is the one with the lowest energy consumption while the SP-25 + Alveolar cubicle is the second one, consuming even less energy than the PU cubicle. Finally, the Alveolar cubicle is the one that consumes more energy after the reference cubicle.

A moderate set point (like 24 °C) favors the PCM working conditions, since the inside temperature is close to the phase-change range. Both PCM cubicles reduced the energy consumption compared with the same cubicle without PCM. The RT-27 + PU cubicle achieved a reduction of 15% compared with the PU cubicle, while the SP-25 + Alveolar cubicle reached a 17% of energy savings compared with the Alveolar one (Table 14.4). Moreover, the SP-25 + Alveolar cubicle presents lower energy consumptions than the PU cubicle.

From the energy consumption of each cubicle, the CO_2 emissions to the atmosphere can be estimated. Considering the Spanish electricity production share, a CO_2 emission rate of 572.9 g/kWh is determined. Table 14.5 presents the CO_2 emissions and savings for each cubicle.

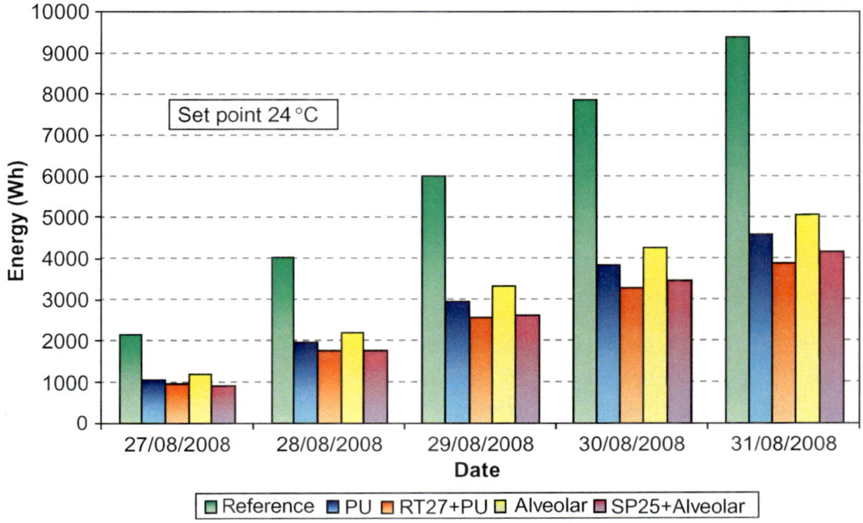

Figure 14.15 Controlled temperature results for both constructive systems.

Table 14.4 Accumulated energy consumption and savings for the different cubicles

	Energy consumption[a] (Wh)	Energy savings[b] (Wh)	Energy savings[b] (%)	Improvement[c] (%)
Reference	9376	0	0	–
PU	4583	4793	51.12	0
RT-27+PU	3907	5469	58.33	14.75
Alveolar	5053	4323	46.11	0
SP-25+Alveolar	4188	5188	55.33	17.12

[a]Set point of 24 °C during 5 days.
[b]Referred to the reference cubicle.
[c]Referred to the cubicle with analogue constructive solution and without PCM.

Table 14.5 CO_2 emissions to the atmosphere due to the energy consumption of the cubicle

	Energy consumption[a] (kWh/year/m^2)	CO_2 emissions (kg/year/m^2)	CO_2 savings[b] (kg/year/m^2)	CO_2 improvement[c] (kg/year/m^2)
Reference	29.3	16.8	0.0	
PU	14.3	8.2	8.6	0.0
RT-27+PU	12.2	7.0	9.8	1.2
Alveolar	15.8	9.1	7.7	0.0

[a]Set point of 24 °C during 90 days per year (cooling demand).
[b]Referred to the reference cubicle.
[c]Referred to the cubicle with analogue constructive solution and without PCM.

Based on these results, one can see the potential of using phase-change materials in building envelopes, especially when those ones are in contact with the external ambient. Moreover, the energy consumption of the alveolar brick constructive system with PCM is very similar to that of the conventional brick with insulation, suggesting that the thermal behavior of both systems is similar. Therefore, the thermal inertia and the dynamic behavior should be considered in the building legislation codes in order to fairly evaluate different constructive systems.

14.4.3 Active systems

In contrast to the passive system just described, active systems are mainly characterized by forced convection heat transfer into the storage material (Gil et al., 2010; Nagano et al., 2000). The storage medium itself circulates through a heat exchanger (the heat exchanger can also be a solar receiver or a steam generator). This system uses one or two tanks as storage media. Active systems are subdivided into direct and indirect systems. In a direct system, the HTF serves also as the storage medium, while in an indirect system, a second medium is used for storing the heat. Active storage systems are generally dual-medium storage systems: the HTF passes through the storage only for charging and discharging a solid material.

Free-cooling technique consisting of using outdoor coolness during the night to decrease the cooling demand of the building is implemented in many conventional cooling systems. The application of phase-change materials in free-cooling systems is suitable because of the small temperature difference between day indoors and night outdoors. However, the storage of the coolness for later cooling supply during the day is still under study. Zalba et al. (2004) developed an empirical model of a real free-cooling system concluding that this kind of system is not only technically feasible but also economically advantageous to existing cooling systems. Moreover, this kind of storage systems can be coupled to the ventilation system (Figure 14.16) or the air handling unit (Stritih and Butala, 2010).

Another type of active systems that is starting to be implemented in buildings is the thermal mass activation systems that consist of using building structure components such as walls, ceiling, or floors as a storage unit. The activation could be directly done

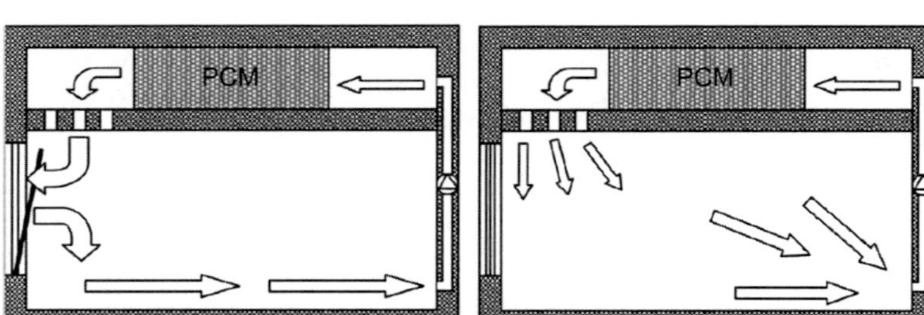

Figure 14.16 Principal function of PCM "free-cooling system": (left) cooling of PCM at night; (right) cooling of the building during the day.

inside the material by using pipes or ducts or using air cavities of the constructive systems.

Suspended ceilings have already been used for cooling purposes such as radiant cooling panels with water pipes (Roulet et al., 1999). These ceiling panels are commonly presented as an alternative of cooling systems in office buildings or building refurbishment. Hence, the design of these panels is focused on minimizing their thickness and providing good storage capacity. In order to increase the thermal inertia of these panels, Koschenz and Lehmann (2004) studied the addition of microencapsulated PCM inside a gypsum ceiling panel. The PCM in the ceiling panels melts during daytime under exposure to the inner thermal loads and solidifies at night cooled by the integrated water pipe system. The system can be coupled to renewable energies, such as solar energy or geothermal energy, for heating or cooling office buildings. Simulation study and laboratory tests carried out determined a 5 cm layer gypsum panel with 25% of PCM by weight to maintain a comfortable room temperature in standard office buildings.

A disadvantage of ceiling panel system is that the discharge period can't be controlled due to the passively process. Therefore, other systems considered active charge and discharge periods in order to provide cooling supply depending on the building demand and the weather conditions. An innovative system consisting of a ventilated facade with macroencapsulated phase-change material in its air chamber was experimentally and numerically studied by de Gracia et al. (2012). The system was designed to reduce cooling load of a building and, therefore, the energy consumption of the HVAC systems. During summer, the ventilated facade takes advantage from the low night temperatures to solidify the PCM (Figure 14.17a) and to store the cold until a cooling demand is needed (Figure 14.17b). After the cold stored in the PCM has been released to the air, the system lets the air flow from indoors to outdoors in order to prevent the overheating effect in the air channel (Figure 14.17c). Moreover, night free-cooling operation (Figure 14.16d) can also be done once the PCM-solidifying period is finished.

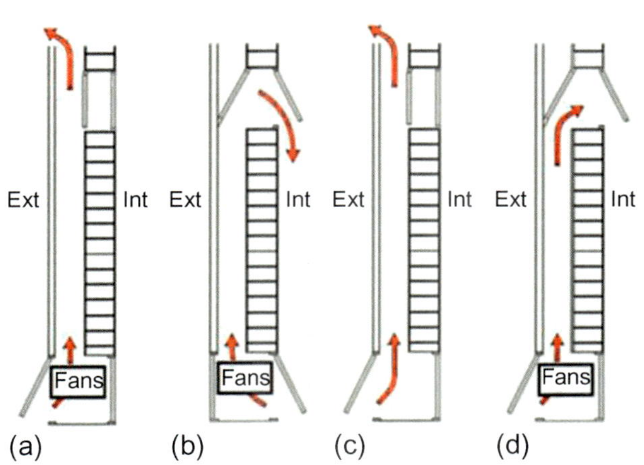

Figure 14.17 Sequence of summer operation of the ventilated facade.

Experimental measurements demonstrated the high potential of the night free-cooling effect in reducing the cooling loads of a building. On the other hand, the cold storage capacity of the system is very sensitive to the outer night temperature, since temperatures below the phase change are needed to solidify the PCM. However, no net energy savings were registered experimentally in the cooling season because of an excessive use of mechanical ventilation during the charge and discharge PCM processes in the cold storage sequence. In order to select the most appropriate schedule to solidify the PCM during nighttime, a numerical model based on control volume approach has been developed and validated experimentally (de Gracia et al., 2013).

Acknowledgments

The research leading to these results has received funding from the European Union's Seventh Framework Programme (FP7/2007-2013) under grant agreement number PIRSES-GA-2013-610692 (INNOSTORAGE). The work is partially funded by the Spanish government (ENE2011-28269-C03-02). The authors would like to thank the Catalan Government for the quality accreditation given to their research group GREA (2014 SGR 123) and research group DIOPMA (2014 SGR 1543).

References

Barreneche, C., Navarro, H., Serrano, S., Cabeza, L.F., Fernández, A.I., 2014. New database on phase change materials for thermal energy storage in buildings to help PCM selection. Energy Procedia 57, 2408–2415

Cabeza, L.F., Castellón, C., Nogués, M., Medrano, M., Leppers, R., Zubillaga, O., 2007. Use of microencapsulated PCM in concrete walls for energy savings. Energy Build. 39, 113–119.

Cabeza, L.F., Castell, A., Barreneche, C., De Gracia, A., Fernández, A.I., 2011. Materials used as PCM in thermal energy storage in buildings: a review. Renew. Sustain. Energy Rev. 15 (3), 1675–1695.

Cabeza, L.F., Barreneche, C., Miró, L., Morera, J.M., Bartolí, E., Fernández, I.A., 2013. Low carbon and low embodied energy materials in buildings: a review. Renew. Sustain. Energy Rev. 23, 536–542.

Castell, A., Martorell, I., Medrano, M., Pérez, G., Cabeza, L.F., 2010. Experimental study of using PCM in brick constructive solutions for passive cooling. Energy Build. 42 (4), 534–540.

de Gracia, A., Navarro, L., Castell, A., Ruiz-Pardo, A., Álvarez, S., Cabeza, L.F., 2012. Experimental study of a ventilated facade with PCM during winter period. Energy Build. 58, 324–332.

de Gracia, A., Navarro, L., Castell, A., Cabeza, L.F., 2013. Numerical study on the thermal performance of a ventilated facade with PCM. Appl. Therm. Eng. 61, 372–380.

Farid, M.M., Khudhair, A.M., Razack, S.A.K., Al-Hallaj, S., 2004. A review on phase change energy storage: materials and applications. Energy Convers. Manage. 45, 1597–1615.

Gil, A., Medrano, M., Martorell, I., Lázaro, A., Dolado, P., Zalba, B., Cabeza, L.F., 2010. State of the art on high temperature thermal energy storage for power generation. Part 1—concepts, materials and modellization. Renew. Sustain. Energy Rev. 14, 31–55.

Hawes, D.W., Feldman, D., 1992. Absorption of phase change materials in concrete. Sol. Energy Mater. Sol. Cells 27, 91–101.

Kenisarin, M., Mahkamov, K., 2007. Solar energy storage using phase change materials. Renew. Sustain. Energy Rev. 11, 1913–1965.

Koschenz, M., Lehmann, B., 2004. Development of a thermally activated ceiling panel with PCM for application in lightweight and retrofitted buildings. Energy Build. 36, 567–578.

Lovins, L.H., 2008. Rethinking production. State of the World 2008, p. 34.

Mehling, H., Cabeza, L.F., 2008. Heat and Cold Storage with PCM. An up to Date Introduction into Basics and Applications. Springer, Berlin.

Nagano, K., Mochida, T., Iwata, K., Hiroyoshi, H., Domanski, R., Rebow, M., 2000. Development of new PCM for TES of the cooling system. In: Benner, M., Hahne, E.W.P. (Eds.), Proceedings of the Terrastock Eighth International Conference on Thermal Energy Storage, vol. 2. Universität Stuttgart, Stuttgart, pp. 345–350.

Roulet, C.A., Rossy, J.P., Roulet, Y., 1999. Using large radiant panels for indoor climate conditioning. Energy Build. 30, 121–126.

Schossig, et al., 2002. Fraunhofer ISE. Available at www.ise.fhg.de/english.

Shapiro, M., Feldman, D., Hawes, D., Banu, D., 1987. PCM thermal storage in drywall using organic phase change material. Passive Sol. J. 4, 419–438.

Stritih, U., Butala, V., 2010. Experimental investigation of energy saving in buildings with PCM cold storage. Int. J. Refrig. 33, 1676–1683.

World Business Council for Sustainable Development, 2000, August. Eco-efficiency: Creating more with less. Retrieved from http://www.wbcsd.org/web/publications/eco_efficiency_creating_more_value.pdf.

Yadong, Y., 2013. Eco-efficiency trends in china, 1978–2010: decoupling environmental pressure from economic growth. Ecol. Indic. 24, 177–184.

Zalba, B., Marin, J.M., Cabeza, L.F., Harald, M., 2004. Free cooling of buildings with phase change materials. Int. J. Refrig. 27, 839–849.

Nanomaterial-embedded phase-change materials (PCMs) for reducing building cooling needs

15

R. Parameshwaran, S. Kalaiselvam
Anna University, Chennai, India

15.1 Introduction

In the current scenario of reducing the impacts caused by climate change and greenhouse gas (GHG) emissions, thermal energy storage (TES) technologies offer a wide opportunity for bridging the gap present between energy supply and energy demand. TES systems utilizing phase change materials (PCMs), also known as latent thermal energy storage (LTES) systems, have become increasingly attractive in recent years. The inherent thermophysical properties and phase change attributes of the PCMs enable them to be considered for the energy redistribution requirements in building cooling applications.

The PCMs incorporated either into the building fabric components (e.g., ceiling boards, wall boards, thermal mass, underfloor elements) or utilized as packed bed configurations can be highly regarded to cater the end-use energy demand through energy redistribution. Energy redistribution by TES refers to the concept of storing energy in the form of either heat or cold into the PCM during a particular duration of time, which can be retrieved from the same for later end-energy usage.

It is interesting to note that although the latent heat capacity of PCM systems seems to be lesser than that of the ice TES systems (since latent heat of fusion of ice = 333 kJ/kg), PCM systems are much preferred for their relatively high volumetric storage density. In addition, organic PCMs reveal good thermophysical properties, congruent freezing and melting processes, thermal stability, noncorrosiveness, and thermal reliability on a long-term basis, with ease of operational and maintenance aspects.

A step further, embedding of the nanomaterials into the pure PCMs would enable the thermal properties and instantaneous rate of heat transfer of the PCMs to be enhanced during the cyclic phase change processes. The terminology nanomaterials or nanostructured materials refers to a class of advanced materials in which the formation of the particles or structures occurs at the nanoscale level, with size ranging from 1 to 100 nm.

The shape and the size of the nanomaterials are equally important in the sense that the high surface-to-volume ratio of the nanomaterials facilitates their establishing excellent thermophysical property changes in the bulk materials where they have been embedded. The means through which the energy redistribution requirements can be

Eco-efficient Materials for Mitigating Building Cooling Needs. http://dx.doi.org/10.1016/B978-1-78242-380-5.00015-7

accomplished in buildings with reduced cooling loads using the PCMs embedded with nanomaterials are discussed in the forthcoming sections.

15.2 Nanomaterials for thermal energy storage

15.2.1 Nanomaterial preparation

The preparation of nanomaterials stands to be the primary approach for their being embedded into the pure PCMs dedicated for cooling TES applications in buildings. This in turn would facilitate achieving enhanced thermal storage properties of the PCMs for the long run. The broad classification of the cool TES, nanomaterials, and the global production scenario of the nanomaterials are depicted in Figures 15.1 and 15.2, and Table 15.1, respectively.

The concepts of nano-engineering being considered for producing a variety of nanostructured materials are illustrated in Figure 15.3. It can be well observed from Figure 15.3 that, in a bottom-up strategy, the desired nanostructured materials can be formed through the manipulation of the atoms. This can also be referred to as the atom-by-atom approach. The important aspects of the bottom-up strategy include

- The formation of the intricate and complex molecular structures with ultimate precision at ease
- Manufacturing cost reduction, which is made possible when compared to the required raw materials and spent energy

On the other hand, in the top-down strategy, the nanomaterials can be produced by the fragmentation of the bulk materials into smaller size/narrow size distribution ranging

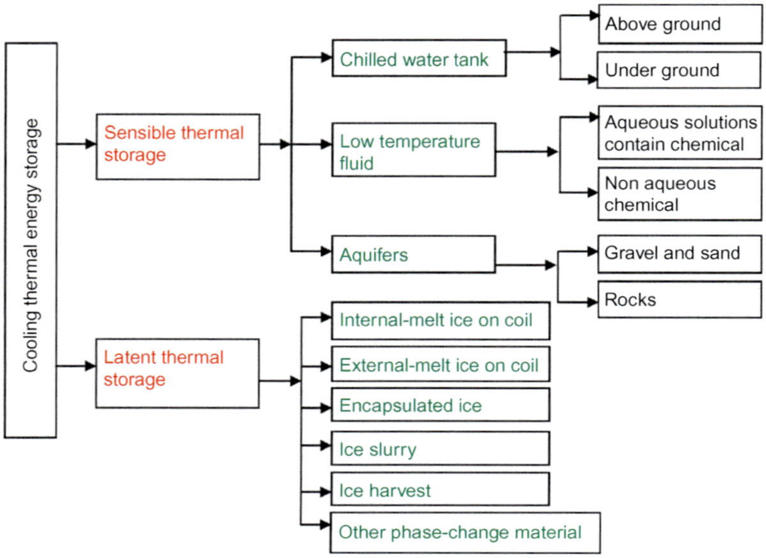

Figure 15.1 Classification of cooling thermal energy storage (Al-Abidi et al., 2012).

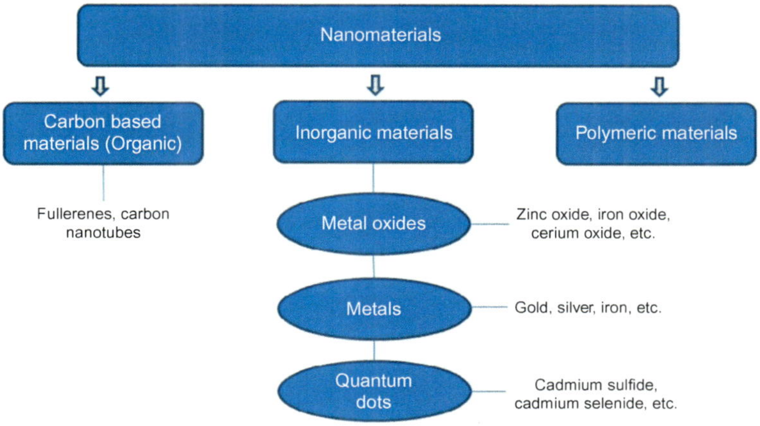

Figure 15.2 Classification of nanomaterials (Manikam et al., 2011).

Table 15.1 Global production of nanostructure materials

Application	Nanomaterial device	Estimated global production (tons/year)		
		2003/04	2010	2020
Structural application	Ceramics, catalysts, film and coating, composites, metal	10	10^3	10^4–10^5
Sink care products	Metal oxides (e.g., TiO_2 and ZnO)	10^3	10^3	10^3
Information and communication technologies	SWCNT, nanoelectronics and optoelectronics materials (excluding CMP slurries), organic light emitters	10	10^2	$>10^3$
Biotechnology	Nanocomposite, encapsulates, target drug delivery, diagnostic marker, biosensors	<1	1	10
Environmental	Nanofiltration membranes	10	10^2	10^3–10^4

(Al-Mubaddel et al., 2012; Royal Society and Royal Academy of Engineering, 2004).

from millimeter to micrometer to nanometer scale in the steps of manufacturing processes, as shown in Figure 15.3. As in the case of bottom-up strategy, the desired size or the shape of the nanomaterials being formed could not be assured at all times.

However, the top-down methodology finds its place starting from the early 1950s to the present-day scenario in producing nanostructured materials. For ease of understanding nanostructured materials, comparisons between the nanostructures on the basis of chemical composition and dimensionality are depicted in Figures 15.4 and 15.5, respectively.

It is noteworthy to say that the fundamental difference between the terms used for representing nanostructures and nanostructured materials is that the nanostructures are referred to either by form or dimensionality whereas the nanostructured materials epitomize the added composition. The classified range of dimensions (size) of different nanostructures is presented in Table 15.2. Obviously, the inner-crystallite grain

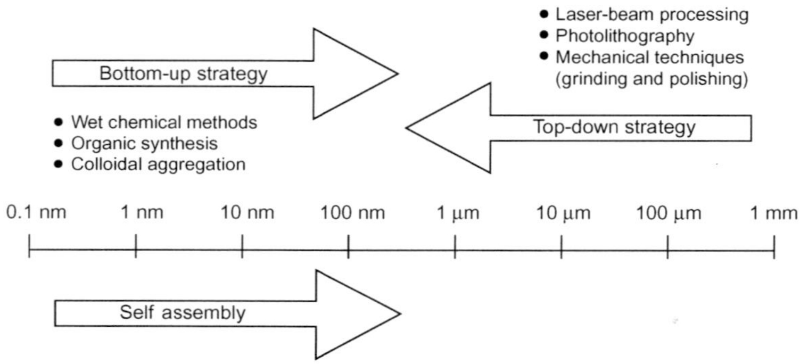

Figure 15.3 Top-down and bottom-up nanotechnology (Adams and Barbante, 2013; Ju-Nam and Lead, 2008).

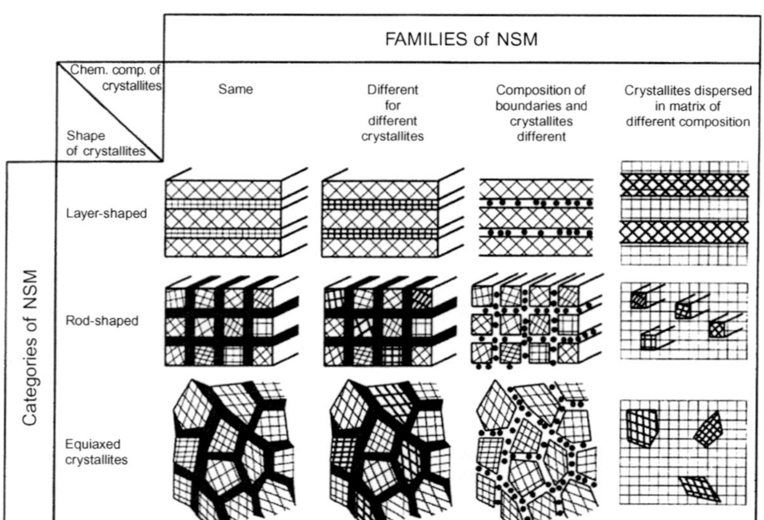

Figure 15.4 Gleiter's classification schema for nanostructured materials (NSMs) according to chemical composition and dimensionality of the crystallites forming the nanostructures (Gleiter, 2000; Manikam et al., 2011).

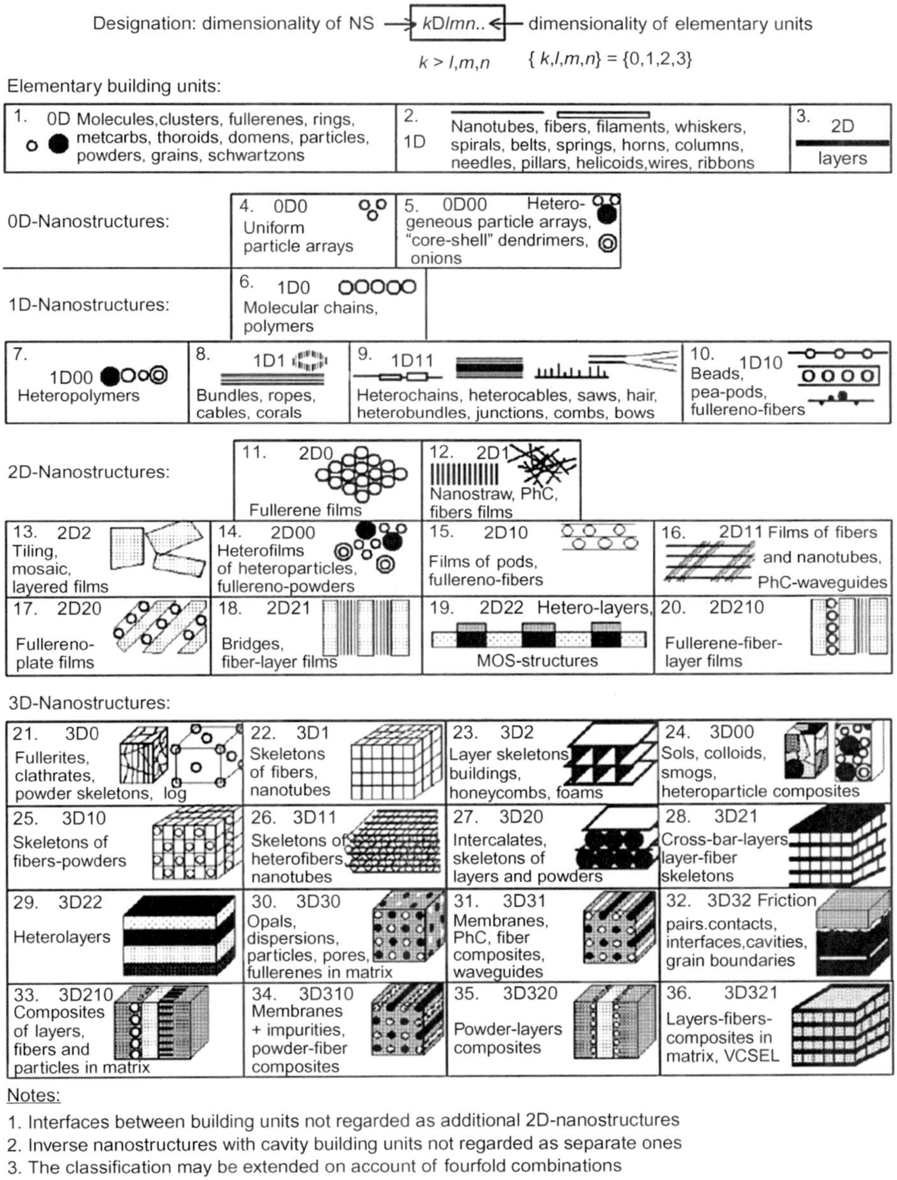

Figure 15.5 Dimensionality classification of nanostructures ($L < 100$–500 nm) (Manikam et al., 2011; Pokropivny and Skorokhod, 2007; Skorokhod et al., 2001).

Table 15.2 **Typical dimensions of nanostructures**

Nanostructure	Size (nm)	Material
Clusters, nanocrystals, quantum dots	Radius 1–10	Insulators, semi-conductors, metals magnetic materials
Other nanoparticles	Radius 1–10	Ceramic oxides
Nanobiomaterials	Radius 5–10	Membrane protein
Nanowires	Diameter 1–100	Metals, semiconductors, oxides, sulphides, nitrides
Nanotubes	Diameter 1–100	Carbon, chalcogenides
Nanobiorods	Diameter 5 nm	DNA
Surfaces and thin films	Thickness 1–1000	Insulators, semiconductors, metals, DNA
2-D arrays/nanoparticles	Area several nm^2–μm^2	Metals, semiconductors, magnetic materials
3-D superlattices/ nanoparticles	Radius several nm	Metals, semiconductors, magnetic materials

(Adams and Barbante, 2013; National Academy Press, 2008).

boundaries and the free surfaces can play a significant role in determining the essential characteristics of the nanostructures and nanostructured materials.

In fact, the grain boundaries existing in the nanostructures or nanostructured materials can produce inner classical size effects, including the enhancement of diffusion, reduction of melting point, or any changes in the lattice parameters. The preparation method for nanomaterials and nanoalloys, along with the merits and limitations, are highlighted in Table 15.3 for ready reference. The comparison provided in Table 15.2, which is nonexhaustive in nature, can be useful to scientists and engineers in selecting the suitable method for the preparation of nanomaterials.

It is interesting to note that, out of all the methods used for producing the nanomaterials, the chemical reduction method is increasingly attractive and offers good potential in terms of accomplishing physicochemical properties, better handling, equipment configuration, and the associated cost factors involved in other processes. The most interesting part of the chemical method is that, by utilizing suitable precursors (raw materials), reducing agents, and stabilizing agents (usually chemicals), the nanomaterials of relatively smaller size/narrow size distribution and desired shape can be effectively synthesized. Moreover, the stability of the nanomaterials being dispersed into the base material (PCM) can be established very well without experiencing agglomeration (or settlement) of the particles.

Besides, the shape of the nanomaterials (either spherical, rod-like, tube bundles, or wires) can also influence the overall thermophysical property changes exhibited by the base material (PCM). The characterization of the nanomaterials has a significant role to play, which can help to ensure whether the desired physicochemical and thermophysical property requirements have been achieved by both the nanomaterials and the base material (PCM) being embedded in them. Detailed information on the analytical modalities for characterizing nanomaterials can be found in Lin et al. (2013).

Table 15.3 **Mainstream methods for synthesizing nanoparticles and nanoalloys (Manikam et al., 2011)**

Method	Description of method	Advantages	Disadvantages	References
Molecular beams	Beams are directed towards specific metal targets using a variety of methods; laser vaporization, pulsed arc, ion and magnetron sputtering. This creates clusters of metallic nanoparticles including nanoalloys	(1) Any type of nanoparticle or nanoalloy can be created from metallic/alloy targets. (2) Nanoparticle synthesis is quick and not lengthy	(1) Process is expensive, and requires equipment setup in most cases	Alonso et al. (2009), Ferrando et al. (2008)
Chemical reduction	Use of precursor salts, reducing agents and stabilizer to synthesize nanoparticles. In most cases a catalyst and some heating is used.	(1) Can readily produce bulk quantities of nanoparticles and nanoalloys. (2) Process can be easily scaled up to meet mass manufacturing needs. (3) Process enables synthesis of particles close to 1 nm and this can easily be controlled. (4) Process is relatively cheaper compared to other synthesis methods since the technology is quite standard.	(1) Mass use of chemicals and some may be harmful to the environment. (2) Processing is time consuming and depends on many parameters.	Ferrando et al. (2008), Hu and Easterly (2009), Khaydarov et al. (2009), Jian et al. (2005), Starowicz et al. (2006), Ritson (2008), Liu et al. (2001), Salkar et al. (1999), Tunc et al. (2008), Guzman et al. (2008), Zhang et al. (2007)

Continued

Table 15.3 **Continued**

Method	Description of method	Advantages	Disadvantages	References
Thermal decomposition of metals	Thermal decomposition of metal or metal complexes (for nanoalloys) is produced using high temperature mediums or solvents.	(1) Nanoparticles can be created at relatively low temperatures. (2) Process can create nanoparticles in a wide range of sizes.	(1) Requires use of chemicals and solvents which may be harmful to the environment.	Ferrando et al. (2008)
Ion implantation	This method is used to create nanoalloys by implanting two or more metal ions into a specific matrix.	(1) Metallic ions can be implanted into exact positions in a matrix. (2) Various combinations of ions can be used to yield nanoalloy clusters.	(1) Requires equipment setup which is relatively expensive compared to chemical reduction.	Ferrando et al. (2008)
Electro-chemical synthesis	Using an electrolysis cell and two electrodes of metallic elements, bimetallic alloys/ nanoalloys can be created in solution. Core–shell structures have also been created via this method.	(1) Various nanoalloy combinations can be synthesized. (2) Cell setup is rather easy and does not need extensive equipment.	(1) Use of chemicals as electrolytes which may yield harmful/toxic gases as by products from the process.	Ferrando et al. (2008), Hu and Easterly (2009), Khaydarov et al. (2009), Jian et al. (2005), Starowicz et al. (2006)

Radiolysis	Radiolysis of an aqueous solution of metal ions to produce nanoparticles. This method has also been used to create nanoalloys/bimetallic particles.	(1) Irradiation of molecules is able to create nanoparticles with a wide range of sizes, as well as very narrow sizes.	(1) Requires expensive equipment setup. (2) Radiation is harmful to the health of living organisms, including humans. (3) The use of this method requires extensive clearance from concerned authorities.	Ferrando et al. (2008)
Sonochemical synthesis	Irradiation of metal salt solutions using ultrasound to create nanoparticles and nanoalloys.	(1) Sonic wave irradiation is able to create narrow particle sizes.		Ferrando et al. (2008), Ritson (2008), Liu et al. (2001), Salkar et al. (1999)
Biosynthesis	Biological means are used to synthesize nanoparticles and nanoalloys using microorganisms or plants. Synthesis of nanoparticles using these mediums can make the nanoparticles more bio-compatible.	(1) This method is cheap and uses sources from nature. (2) The method does not produce waste detrimental to human beings. (3) Does not require extensive equipment setup.	(1) This method is slow and takes time.	Ferrando et al. (2008), Niemeyer (2001), Parameshwaran et al. (2013c)

15.2.2 Apposite nanomaterials for PCM thermal storage

In the spectrum of nanomaterials being produced for a variety of scientific and engineering applications, it has always been a fact that each nanomaterial has its own inherent physicochemical and thermophysical properties, distinct from those of another. The property changes of the particles observed at the nanometer level also depend on their morphology, crystal structure, thermal conductivity, stability, surface charge, surface-to-volume ratio, and so on. Thus, it becomes a prerequisite to identify the quality of the properties exhibited by the nanomaterials that would suit the intended specific application.

The potential opportunities available for enhancing the thermal storage properties of the PCM can be effectually accomplished by embedding the nanomaterials into the pure (or base) PCM. In the case of LTES systems, based on past literature reports, it is observed that the majority of PCMs (especially for building applications) utilize metallic and metal-oxide nanomaterials. The reason behind this could be the high thermal conductivity of the metallic and metal-oxide nanoparticles, and in some instances it is also decided by the cost factors involved. In general, the well-established nanomaterials being considered for the LTES applications can be categorized as given below:

- Metallic nanomaterials
  - Silver (Ag)
  - Copper (Cu)
  - Aluminum (Al)
- Metallic-oxide nanomaterials
  - Copper oxide (CuO)
  - Aluminum oxide or alumina (Al_2O_3)
  - Silicon dioxide (SiO_2)
  - Titanium dioxide or titania (TiO_2)
- Non-metallic nanomaterials
  - Multi-walled carbon nanotubes (CNT/MWCNT)
  - Graphite/expanded graphite
  - Graphene/graphene oxide

15.2.3 Hybrid nanomaterials

The research efforts pertaining to the development of hybrid nanocomposite (HyNC) materials are gaining momentum in recent years, wherein the metal nanoparticles are adsorbed over the surface of the metal-oxide nanoparticles (Parameshwaran and Kalaiselvam, 2013a). The interesting features of the nanomaterials of different morphologies and functional properties have created a quest towards the development of hybrid nanomaterials.

The term "hybrid nanomaterial" signifies the amalgamation of one or more distinct materials at nanometer scale into a singular and multifunctional material, which would eventually exhibit synergetic properties that the constituent (parent) materials do not have in them when they are existing individually. The preparation of the hybrid nanomaterials depends on many parameters, including:

- Composition of individual elements (or parent materials)
- Shape and size (morphology)
- Surface effects (Gibbs free energy criteria)
- Density factors
- Heterogeneous nucleation
- Growth kinetics
- Coordinated organic linkages
- Precursor-to-seed or reducing agent ratio during synthesis

The generic growth models for the synthesis of the hybrid nanostructures of different configurations are presented in Figure 15.6. For instance, the metal-oxide nanoparticles serve as both the substrate and nucleation centers for the metal nanoparticles to be formed over them during the synthesis. The HyNC materials being prepared in recent times have been embedded into the pure PCM and real-time tested, which facilitates their being considered potential candidates for cool TES applications in buildings (Parameshwaran and Kalaiselvam, 2013a).

15.3 Enhanced thermophysical property attributes

The prime objective behind the incorporation of the nanomaterials into the pure PCM is to achieve better thermal storage properties of the PCM compared to the same one existing at its bulk state. Embedding the nanomaterials into the pure PCMs can have a significant role to play in terms of:

- Improved thermal conductivity of the PCMs (especially organic PCMs)
- Swift phase change characteristics (freezing and melting)
- Enhanced thermal stability of the PCMs
- Performance improvement on heat storage capability

Many research studies have been reported in recent years demonstrating the potential of infusing the nanomaterials into the pure PCM for attaining improved thermal storage properties. From this viewpoint, the metallic nanoparticles (Copper), which are added to the PCM in varying molar proportions, yielded reduction in the freezing rate of the PCM (Fang et al., 2008). This enables the LTES system to consume less energy per unit mass of PCM during the freezing process.

Nanomaterials can also be used as fillers and nucleating agents for the enhancement of thermal storage properties, especially in improving thermal conductivity and heterogeneous nucleation (Khodadadi and Hosseinizadeh, 2007; Meng et al., 2013; Pacheco-Torgal, 2014; Pacheco-Torgal and Jalali, 2011; Pielichowska and Pielichowski, 2010; Wang et al., 2010a; Zeng et al., 2009). Due to the surface properties, the adsorption established between the nanomaterial and the PCM facilitates creating a network of densely packed thermally conductive interfaces in the PCM matrix layers. This in turn enables the nanomaterial-embedded PCM to exhibit

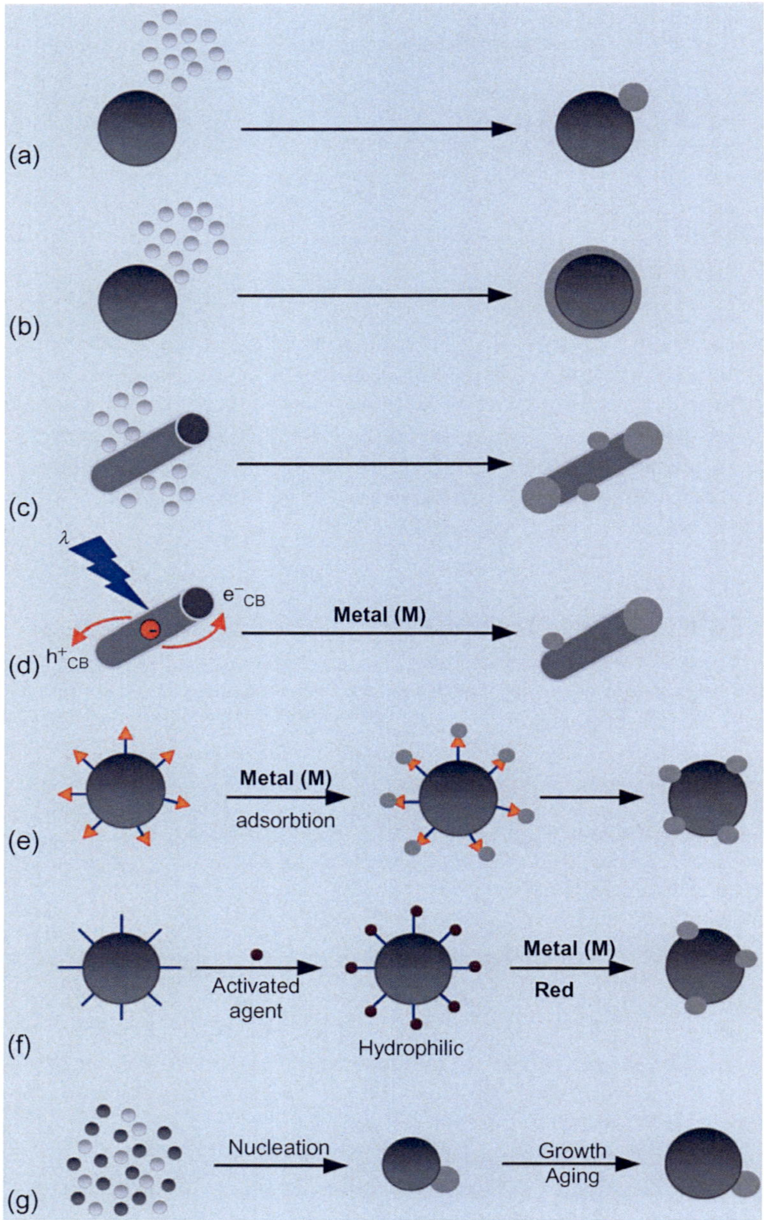

Figure 15.6 A general sketch of the reaction mechanisms for the formation of the hybrid nanostructures: (a–c) heterogeneous nucleation and growth of the secondary precursors on the preformed seeds; (d) secondary precursors adsorbed on the opposite-charged surfaced supports by photo-irradiation; (e) reduced precipitation of the secondary precursors on the hydrophilic-surfaced supports; (f) secondary precursor growth on the activated-surfaced supports; (g) "one-pot" self-controlled nucleation-growth (Khodadadi et al., 2013).

better thermal conductivity and nucleation properties compared to the PCM in its purest form.

The heat storage performance of a variety of organic PCMs has been analyzed using the metal-oxide nanomaterials, including zirconium oxide, alumina, and aluminum nanoparticles. It is apparent that thermal conductivity, heterogeneous nucleation, and thermal stability have been improved through the addition of these nanomaterials into the pure PCM. Furthermore, the PCMs embedded with these nanomaterials have been investigated by both analytical and experimental approaches (Ai et al., 2010; Kalaiselvam et al., 2012; Wang et al., 2010b) as well. It is noteworthy that the enhanced thermal storage properties of the PCMs are directly coupled with the mole concentrations/weight proportions of the nanomaterials being embedded into them.

As mentioned earlier, beyond metallic and metal-oxide nanomaterials, nonmetallic nanostructured materials can also influence the thermophysical and physicochemical properties of the PCMs. It has been found that the thermal conductivity, freezing, and melting characteristics of the organic PCM (paraffin-based) have been improved significantly by embedding the graphite nanostructures into the PCM (Sanusi et al., 2011; Yavari et al., 2011).

15.4 Thermal energy storage properties of nanomaterial-embedded PCMs

15.4.1 Thermal conductivity

As far as TES application in buildings is concerned, organic PCMs are highly preferred over salt hydrate PCMs, because when subjected to several thermal cycles, the latter may get dissociated (destabilized) over time. Moreover, supercooling effects and slow nucleation rates are inherently present in salt hydrate PCMs, which limits them from being considered for real-time active or passive building cooling applications at times. Though the thermal conductivity of salt hydrate PCMs is very high compared to organic PCMs, the former is less considered in building cooling applications for the aforementioned reasons.

Thermal conductivity of the PCM has a significant role in acquiring the desired thermal storage capability for reducing the building cooling load. As such, thermal conductivity has no direct influence on reducing building cooling needs, but it has an interdependent effect on the rate of charging and discharging the PCM. In general, PCMs with low thermal conductivity may establish congruent freezing and melting, provided the rate at which the freezing and melting takes place is under question.

This is because, in building cooling applications, the PCM undergoes a charging process (herein referred to as the freezing process) during part load conditions (evening/nighttime periods). By this, the cold energy can be stored in the PCM (through either passive or active mode of operation) and the same can be retrieved (herein referred to as the melting process) during on-peak load conditions (daytime periods).

In this case, if the PCM is not completely solidified even after the charging periods, the cold energy contained by the PCM could not be retrieved entirely when the demand arises. This is the point where the thermal conductivity of the PCM can be expected to influence the heat transfer rate of the PCM with the surrounding heat transfer medium (air or fluid). Thus, the PCM with high thermal conductivity can be helpful in achieving swift freezing and melting rates, depending upon the cooling load requirements in buildings.

In this context, since the majority of organic PCMs have inherently low thermal conductivity relative to that of the surrounding heat transfer medium, it becomes an essential task to improve thermal conductivity of such PCMs using thermal enhancers. In practice, there are several thermal enhancement techniques available for this purpose, including finned capsulation, metallic rings, and brushes. It is certain that the thermal conductivity of the PCMs can be enhanced using these methods; however, the process of enhancement could be at the macroscopic level.

Having said this, a step ahead in technology—the incorporation of nanomaterials into pure PCMs—can really establish micro-/nano-scale heat transfer characteristics with the molecules of the PCM, thereby accomplishing faster charging and discharging characteristics. By virtue of embedding the nanomaterials into the PCM, the network of thermal interfaces being created can help serve as a thermal energy carrier for enabling effectual transfer of heat energy. The inherent Brownian motion, diffusion, surface effects, and agglomeration of the nanomaterials can play a vital role in achieving enhancement of thermal conductivity of the PCMs.

This is even more evident from the studies conducted on the PCMs embedded with nanographite and silver–titania HyNC (Li, 2013; Parameshwaran et al., 2014). The appearance of the nanographite and its amalgamation with the paraffin PCM are shown in Figures 15.7 and 15.8, respectively. It can be observed from the micrographs of the nanographite-incorporated paraffin PCM that the PCM gets occupied into the layers of the nanographite, thereby establishing a more tightly packed composite network. This enables the nanographite/PCM to exhibit improved thermal conductivity. The enhancement of thermal conductivity for this composite PCM is about 2.89–7.41 times, compared to the pure paraffin PCM for 1% and 10% mass fraction of nanographite addition (Figure 15.9).

Figure 15.7 Appearance of the nanographite (Li, 2013).

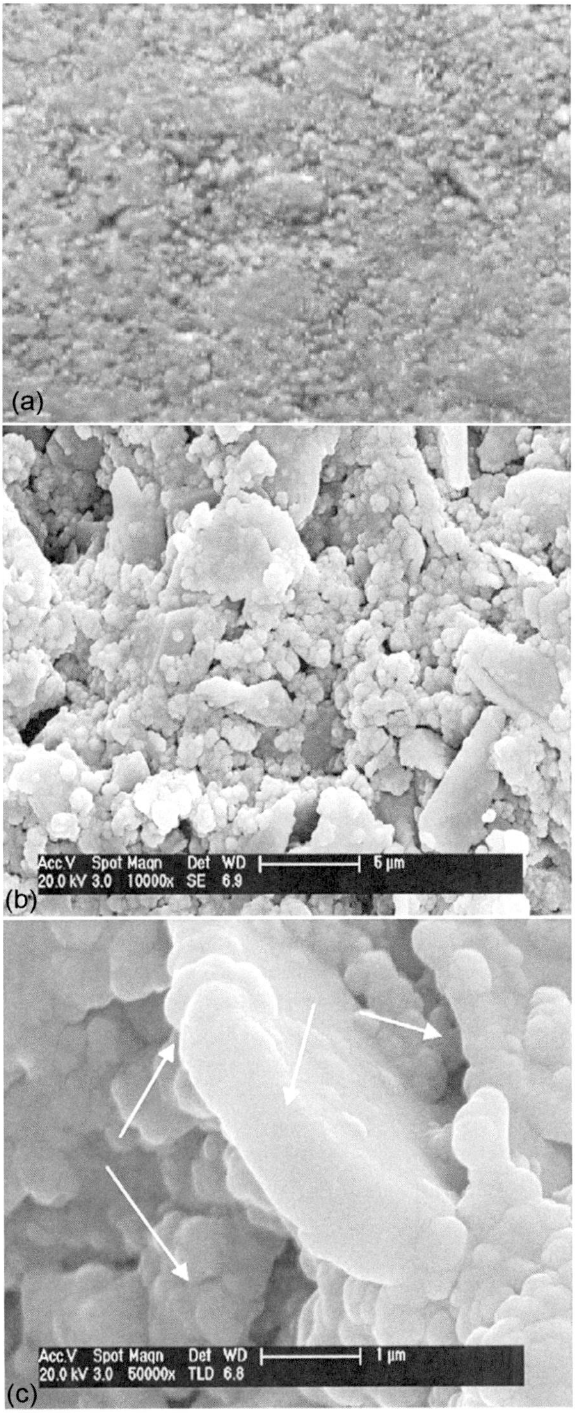

Figure 15.8 The microstructure of the paraffin/nanographite composite: (a) macromorphology; (b) 10,000 magnification; (c) 50,000 magnification (Li, 2013).

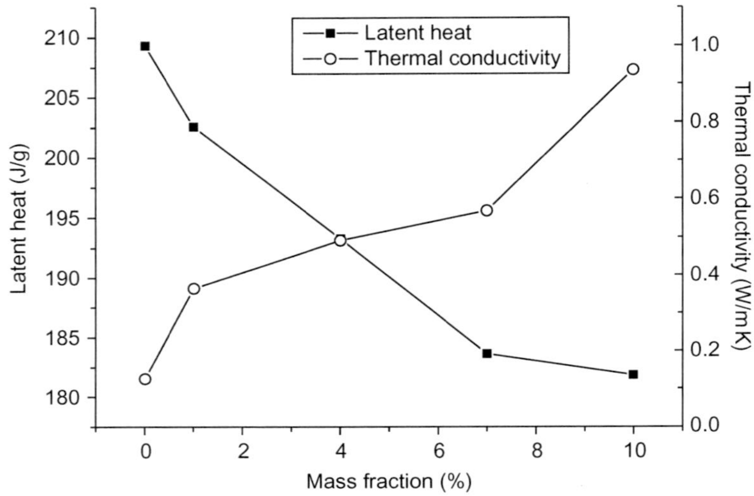

Figure 15.9 The latent heat and thermal conductivity of the composite as a function of nanographite mass fraction (Li, 2013).

The enhancement of thermal conductivity can also be viable in case of the HyNC particles being embedded into the PCM. The formation (morphology) of the silver–titania HyNC and its dispersion into the organic PCM are depicted in Figure 15.10. The morphology of the silver–titania (Ag–TiO_2) HyNC, as shown in Figure 15.10, clearly reveals the arrangement of the silver nanoparticles on the surface of the titania nanoparticles.

It is pertinent to note that, during the formation of the HyNC, the silver nanoparticles were adsorbed on the surface of the titania nanoparticles. The silver nanoparticles present on the peripheral surface of the titania particles resemble the extended heat transfer-like surfaces (fins-like structure) at the nanoscale level.

The variations of latent heat enthalpy and thermal conductivity of the HyNC-embedded PCM with respect to the HyNC concentration are shown in Figure 15.11. For the HyNC mass concentration of 0.1% through 1.5%, thermal conductivity of the PCM increased from 0.286 W/m K to 0.538 W/m K respectively, when compared to its base value (0.257 W/m K). The nonlinear increase of thermal conductivity, with respect to the increased mass concentration of the HyNC, could be ascribed to the inherent Brownian motion, interparticulate diffusion (or phonon interaction), surface morphology, heat exchange potential, and clustering of the HyNC, as suggested by Kim et al. (2009) and Gao et al. (2009).

Even though there could be a condition where the latent heat of enthalpy might decrease at the cost of embedding the nanomaterials, the enhancement of thermal conductivity of the PCM can be accomplished substantially. The dispersed nanomaterials in turn facilitate the transfer of thermal energy to and from the PCM during the charging and discharging processes for the reasons stated above. The variety of nanomaterials (e.g., metallic, metal-oxide, and non-metallic types) being utilized as thermal conductivity enhancement materials for the PCMs (both cooling and heating applications) are summarized in Table 15.4.

Figure 15.10 Morphology of (a–b) Ag–TiO_2 HyNC using field emission scanning electron microscopy (FESEM), (c) Ag–TiO_2 HyNC dispersed in the organic ester PCM using TEM (Parameshwaran et al., 2013a).

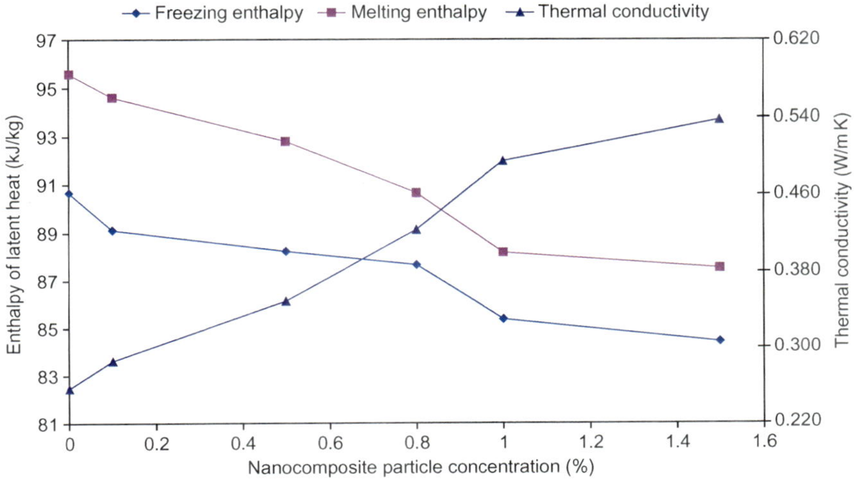

Figure 15.11 Variation of latent heat enthalpy and thermal conductivity of the HyNC PCM with different mass proportions of the HyNC (Parameshwaran et al., 2014).

Table 15.4 Summary of utilized PCM and nano-structured thermal conductivity enhancers (Khodadadi et al., 2013)

Authors (year)	PCM	Nano-structured materials		
	Materials and properties	Materials and properties	Dimension etc.	Fractions of enhancers
Siegel (1977)	Molten salts, k: 0.4–4 W/mK	Stainless steel, iron (Fe), aluminum (Al), and copper (Cu) particles, k: 15, 60, 204, and 386 W/mK	N/A	0–100 vol%[a]
Seeniraj et al. (2002)	Molten salts, k: 0.4–4 W/mK	Stainless steel, and Cu particles, k: 50 and 380 W/mK	N/A	0–100 vol%
Elgafy and Lafdi (2005)	Paraffin wax, T_m : ~67 °C, k: ~0.24 W/mK, α: 1.61×10^{-7} m^2/s	Carbon nanofibers (CNF), ρ: ~1600 kg/m^3	Outer diameter: ~100 nm, length: ~20 μm	1, 2, 3, and 4 wt%
Khodadadi and Hosseinizadeh (2007)	Water, ρ: 997.1 kg/m^3, C_p: 4179 J/kg K, k: 0.6 W/mK, μ: 8.9×10^{-4} kg/ms, L: ~335,000 J/kg	Cu nanoparticles, ρ: 8954 kg/m^3, C_p: 383 J/kg K, k: 400 W/mK	10 nm	10 and 20 vol%
Hong et al. (2007a,b)	PAC[b], EG solutions (with 50 vol% water), T_f: −35.6 °C	Single-walled carbon nanotubes (SWCNT), alumina (Al_2O_3), MgO nanoparticles	N/A	0.05, 0.1, and 0.2 wt%
Zeng et al. (2007)	1-Tetradecanol ($C_{14}H_{30}O$), T_m: ~38 °C, L: ~ 230,000 J/kg	Silver (Ag) nanoparticles	500 nm	2–94 wt%
Xie et al. (2008)	Deionized (DI) water	Al_2O_3, titanium oxide (TiO_2) nanoparticles	~8 nm (Al_2O_3); ~3.8 nm (TiO_2)	1.54–6.19 wt% (Al_2O_3); 2.616–7.85 wt% (TiO_2)

Weinstein et al. (2008)	Paraffin wax, T_m: 56 °C, C_p: 2100 J/kg K, *k:* 0.25 W/mK, *L:* ~234,000 J/kg	Graphite nanofibers (GNF) of three types: ribbon, platelet and herringbone	Diameter: 4–10 nm, length: ~1 μm	0.25, 0.5, 1 and 5 wt%
Zeng et al. (2008)	$C_{14}H_{30}O$ and $C_{14}H_{30}O$/PANI[c], T_m: ~35 °C, *L:* 221, 250 J/kg and 119,140 J/kg	Multi-walled carbon nanotubes (MWCNT)	Outer diameter: 10–30 nm, length: 5–15 μm	0.5, 1, 2.5, and 5 wt%
Shaikh et al. (2008)	Paraffin wax (shell wax 100), *L:* 156,300 J/kg	SWCNT, MWCNT, and CNF	Diameter: 1, 10, and 100 nm	0.1, 0.4, 0.7, and 1 vol%
Wang et al. (2008)	Palmitic acid (PA), purity 98%, T_m: 62.5–64 °C, ρ: 853 kg/m^3, *k:* 0.24 W/mK, *L:* 207,800 J/kg	MWCNT, surface-oxidized by a mixed acid with 1:3 of concentrated nitric and sulfuric acids	Diameter: 30 nm, length: 50 μm, specific surface area: 60 m^2/g	0.5, 1, 2, and 5 wt%
Kim and Drzal (2009)	n-Docosane, T_m: 53–57 °C, *k:* 0.26 W/mK, *L:* 157,300 J/kg	Exfoliated graphite nanoplatelets (xGnP)	Diameter: 15 μm, thickness: <10 nm, surface area: 30 m^2/g	1, 2, 3, 5, and 7 wt%
Zeng et al. (2009)	PA, T_m: 59.48 °C, T_f: 58.78 °C, *k:* 0.318 W/mK, *L:* ~201,000 J/kg	Long and short pristine MWCNT; surface-oxidized MWCNT (by two acids)	Outer diameter: 10–30 nm, length: 5–15 μm (long), 1–2 μm (short)	0.099–4.76 wt% (without surfactants), and 0.095–4.5 wt% (with surfactants)
Liu et al. (2009)	Saturated barium chloride ($BaCl_2$) aqueous solution, T_f: −8 °C, pH: 8	TiO_2 nanoparticles	20 nm	0.167, 0.283, 0.565, and 1.13 vol%
Wu et al. (2009)	Distilled water, *k:* 0.6008 W/mK	Al_2O_3 nanoparticles	20 nm	0.05, 0.1, and 0.2 wt%
Ho and Gao (2009), Gao (2008)	n-Octadecane ($C_{18}H_{38}$), T_m: 26.5 °C, T_f: 25.1 °C, *L:* ~243,100 J/kg	Al_2O_3 nanoparticles, ρ: 3600 kg/m	33 nm (159.6 and 196.0 nm insuspensions)	5 and 10 wt%

Continued

Table 15.4 **Continued**

Authors (year)	PCM	Nano-structured materials		
	Materials and properties	Materials and properties	Dimension etc.	Fractions of enhancers
Wang et al. (2009)	Paraffin wax, T_m: 53 °C, *L*: ~165,300 J/kg	MWCNT, treated by ball milling	Diameter: 30 nm, length: 50 μm, specific surface area: 60 m^2/g	0.2, 0.5, 1, and 2 wt%
Zeng et al. (2010)	$C_{14}H_{30}O$, *k*: 0.32 W/mK, *L*: ~220,000 J/kg	Ag nanowires	N/A	9.09–62.73 wt%
Wang et al. (2010a)	PA, purity 98%, T_m: 62.4 °C, ρ: 853 kg/m^3, *k*: 0.22 W/mK (solid), 0.16 W/mK (liquid), *L*: 208,000 J/kg	MWCNT, treated by mechano-chemical reaction with potassium hydroxide/ball milling	Diameter: 30 nm, length: 50 μm, specific surface area: 60 m^2/g	0.2, 0.5, and 1 wt %
Wang et al. (2010b)	Paraffin wax, T_m: ~48.1 °C, *L*: 142,200 J/kg	γ-Al_2O_3 nanoparticles ρ: 3900 kg/m^3	20 nm	1, 2, and 5 wt%
Wang et al. (2010c)	PA, purity 98%, T_m: 62.5–64 °C, *k*: 0.223 W/mK (solid), 0.154 W/mK (liquid)	MWCNT, treated by surface oxidation, mechano-chemical reaction, ball milling, and grafting following acid oxidation	Diameter: 30 nm, length: 50 μm, specific surface area: 60 m^2/g	0.2, 0.5, and 1 wt %
Wang et al. (2010d)	Paraffin wax, purity 99.99%, T_m: 58–60 °C	Cu nanoparticles, purity: 99.9%	Diameter: 25 nm, specific surface area: 30–50 m^2/g	0.1, 0.5, 1, and 2 wt%
Wu et al. (2010)	Paraffin, T_m: 58–60 °C, *k*: 0.2699 W/mK (solid), 0.1687 W/mK (liquid), *L*: 204,000 J/kg	Cu, Al, and C/Cu nanoparticles, purity: 99.9%	Diameter: 25 nm	0.1 wt% for all three kinds of nanoparticles, 0.1, 0.5, 1, and 2 wt% for Cu nanoparticles

Mo et al. (2011)	DI water	MWCNT	Diameters: 10–30, 40–60, and 60–100 nm, length: 5–15 μm	0.1 wt%
Cui et al. (2011)	Paraffin and soy wax, T_m: 52–54 °C	CNF, MWCNT (purity: 95%)	Outer diameter: 200 nm (CNF); diameter: 30 nm, length: 50 μm, specific surface area: 60 m^2/g (MWCNT)	1, 2, 5, and 10 wt%
Xiang and Drzal (2011)	n-Docosane, T_m: 53–57 °C	Exfoliated graphite nanoplatelets (xGnP); xGnP with ball milling treatment	Diameter: 15 μm, thickness: 10 nm, surface area: 20–40 m^2/g, and diameter: 1 μm, surface area: 100–130 m^2/g after ball milling	1, 2, 4, 6, 8, and 10 wt%
Yavari et al. (2011)	1-Octadecanol, T_m: ~66 °C, ρ: 812 kg/m^3, *k*: 0.38 W/mK, *L*: ~250,000 J/kg	Graphene flakes	N/A	0.2, 0.5, 1, 2, and 4 wt%
Fan and Khodadadi (2012)	Cyclohexane, T_m: 6.5 °C, ρ: 850 kg/m^3 (solid), 779 kg/m^3 (liquid), *L*: 32,557 J/kg	Copper oxide (CuO) nanoparticles	Diameter: 5–15 nm	1, 2, and 4 wt%
Fan and Khodadadi (2011)	Eicosane, T_m: ~37 °C	Copper oxide (CuO) nanoparticles	Diameter: 5–15 nm	1, 2, 5, and 10 wt%

[a]Vol% and wt% stand for volumetric and mass fractions, respectively.
[b]PAC and EG stand for prediluted antifreeze coolant and ethylene glycol, respectively.
[c]PANI denotes polyaniline.

15.4.2 Phase change temperature and latent heat of enthalpy

In LTES applications, phase change temperature and latent heat of enthalpy are considered the two decisive parameters, which actually reveals the viability of PCMs for specific applications (either cooling or heating) in buildings. It was also mentioned that water as a PCM possesses better latent heat enthalpy (latent heat of fusion) than organic and salt hydrate PCMs. This is because water, when utilized as a PCM, may undergo incongruent phase transition and supercooling, which limits its application for reducing building cooling requirements. Besides, for the same cooling load requirements, the TES systems utilizing water or ice require huge storage tanks compared to the PCM-based TES systems.

Supercooling is the state attained by the PCM wherein even below the phase transition temperature the PCM still remains in liquid form without freezing. Further cooling of the PCM can lead to the formation of ice crystals inside the PCM, and therefore it starts to freeze completely. However, if the PCM has a high degree of supercooling, it is a direct measure for higher energy consumption during the cooling process. That is, during the charging process (nighttime), most of the cooling energy would be consumed for initiating the freezing process to take place in the PCM. It is therefore a prime task to reduce the supercooling effects in the PCM.

In recent years, there have been a variety of research studies performed on reducing supercooling in PCMs, of which incorporation of the nanomaterials has gained impetus in this regard. The swift thermal response exhibited by PCMs embedded with nanomaterials has paved the way for accomplishing reduced cooling energy needs in buildings. The infusion of nanomaterials into the PCM at suitable mass proportions can help attain the desirable thermophysical properties for storage of cool thermal energy.

In this context, the fatty acid-based PCM (capric acid–palmitic acid; CA–PA) supported by the graphite interconnected nanosheets has been experimentally and numerically studied for energy efficient building construction (Sayyar et al., 2014). This nano-PCM has been incorporated into the gypsum wallboard to form a sandwich structure, as shown in Figure 15.12.

The as-prepared nano-PCM gypsum wallboard is then subjected to testing under the simulated day and night temperature change mode with reference to the conventional drywall panels. The experimental arrangements with the configuration of the nano-PCM sandwich structure are depicted in Figure 15.13. The micrographs of the graphite nanosheets and the nanosheets filled with the PCM are presented in Figure 15.14.

Based on the scanning electron microscope image (Figure 15.14a), it can be visualized that the graphite interconnected nanosheets are comprised of mesopores and

Figure 15.12 Sandwiched structure of gypsum wallboard (Sayyar et al., 2014).

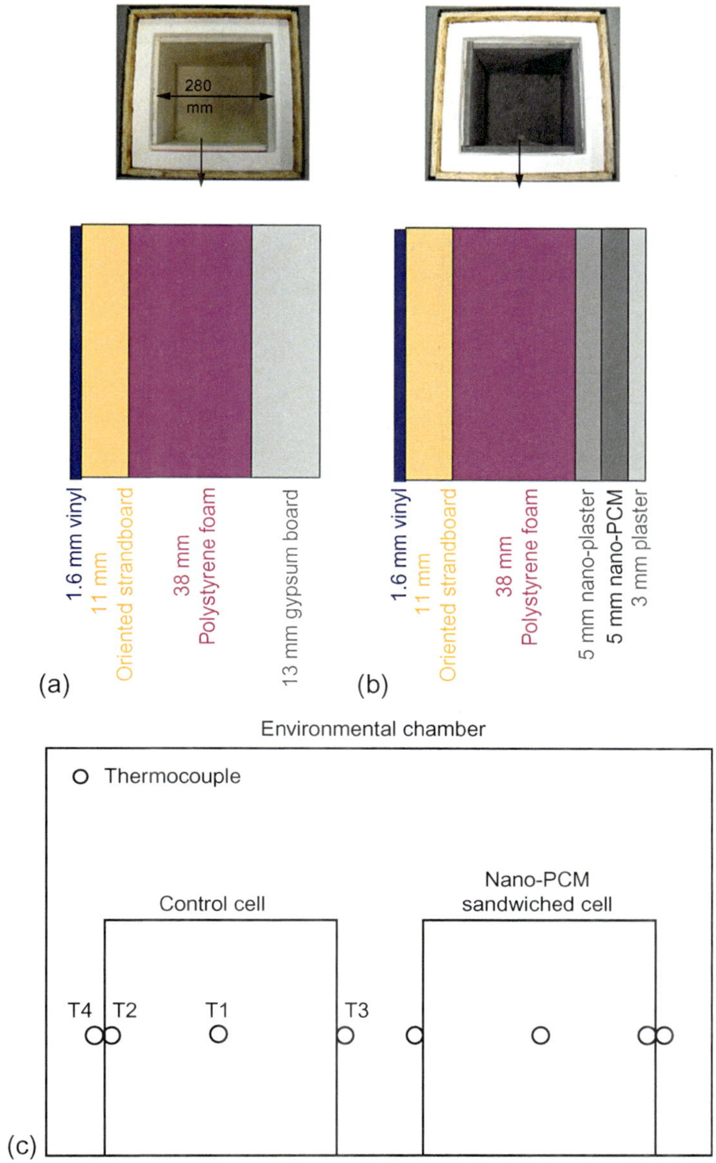

Figure 15.13 Test cells with (a) control envelope panels and (b) sandwich envelope panels incorporating nano-PCM (c) for their thermal performance evaluation in an environmental chamber (T1–T4, four thermocouples) (Sayyar et al., 2014).

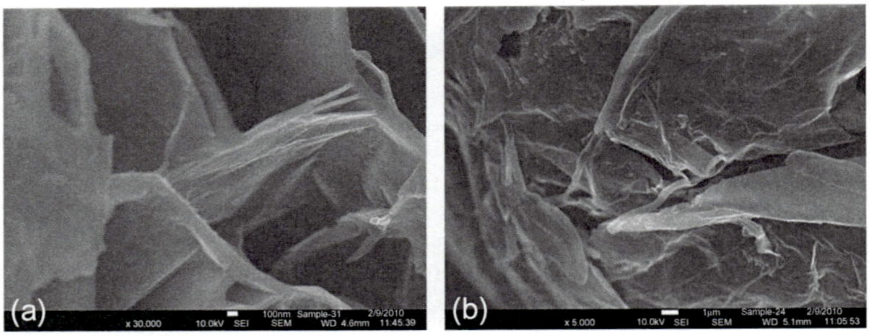

Figure 15.14 SEM micrographs of graphite interconnected nanosheets (a) and CA–PA nanocomposite (b) (Sayyar et al., 2014).

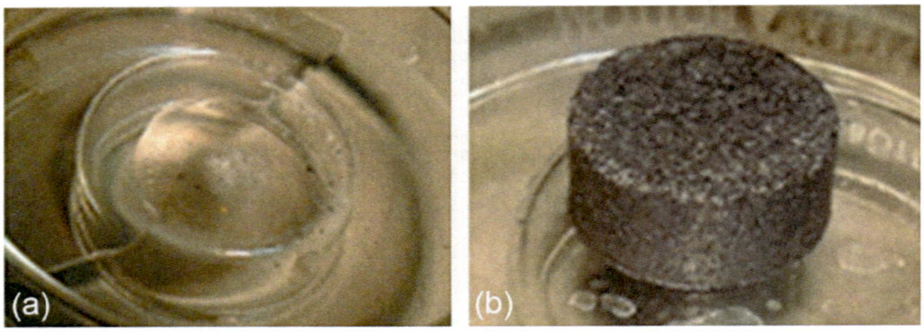

Figure 15.15 Appearance at elevated temperatures (overnight at 35 °C) of control PCM (a) versus corresponding nano-PCM (b) (Sayyar et al., 2014).

macropores, which indicate their high liquid sorption capacity. The graphite nanosheets have a surface area of $\sim$100 m^2/g, which permits higher absorbability. In Figure 15.14b, it can be seen that the CA–PA PCM gets filled in the pores of the graphite interconnected nanosheets.

It is interesting to note that because of the filling of the PCM into the graphite nanosheets, the shape of the nano-PCM is unaltered (shape stabilized) during the phase change processes. The appearance of the control PCM and the nano-PCM at elevated temperatures (overnight at 35 °C) are shown in Figure 15.15. This could be due to the immobilization of the molten PCM due to the capillary action, surface tension, and high absorbability of the graphite interconnected nanosheets.

The analysis on the control wall and the nano-PCM wall being performed infers that, by incorporating the graphite interconnected nanosheets-based PCM, there exist differences in the heat fluxes (heating/cooling) energy needed for maintaining the indoor temperature within the comfort range. The differences in the heat fluxes are illustrated by the colored regions in Figure 15.16.

It can be clearly observed that the heating/cooling energy demand in the building space has been reduced considerably for the nano-based PCM walls compared to the control walls. The heating/cooling energy demand for the indoor temperature control

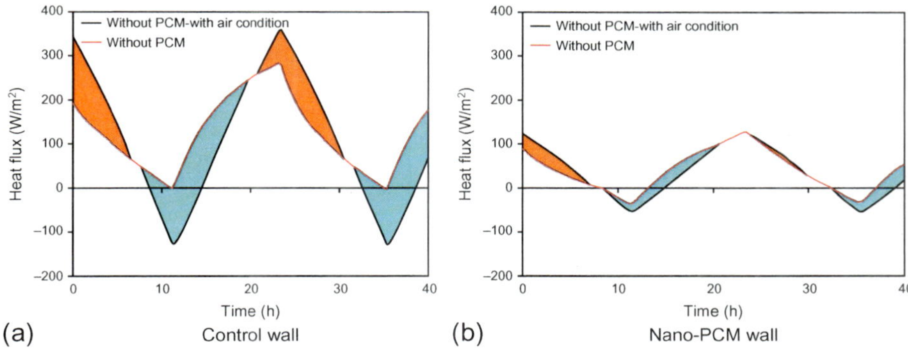

Figure 15.16 Calculated heat fluxes and heating/cooling energy demands for cases with control (a) and nano-PCM (b) walls (Sayyar et al., 2014).

Table 15.5 Energy demands for control and nano-PCM walls (size: 1.0 m × 1.0 m) (Sayyar et al., 2014)

	Energy demand (J)		
	Cooling	**Heating**	**Total**
Nano-PCM wall	544	332	877
Control wall	2224	1967	4191

over a 24-hour period, as presented in Table 15.5, reveals that the incorporation of the nano-PCM walls exhibited better energy savings capability than that of the control walls. Overall, about 79% of the total energy savings has been achieved by the nano-PCM walls, which accounts for the total reduction in the heating/cooling energy demand from 4191 J to 877 J.

The development of nanoencapsulated PCMs for TES applications in buildings is also gaining momentum in recent times. It is reported that nanocapsules containing the PCM as the core, with the shell made up of copolymer, can exhibit better TES properties for building cooling applications. Typically, the organic PCM (*n*-octadecane) as the core material is encapsulated with the St (styrene)—MMA (methylmethacrylate) copolymer shell using the miniemulsion *in situ* polymerisation method (Tumirah et al., 2014).

Different types of samples have been prepared for estimating the thermal storage properties of the nanoencapsulated PCMs. The transmission electron microscope (TEM) images for the as-prepared nanoencapsulated PCMs are shown in Figure 15.17. The micrographs clearly show that the organic PCM (dark spherical particle) has been encapsulated by the copolymer shell (outer ring). The tested nanocapsules actually exhibited increased stability (thermal and chemical) with compact surface, smoothness, and perfect spherical morphology. Besides, the average diameter of the nanocapsules has been measured to be 102 nm. The mass ratio of the St/MMA

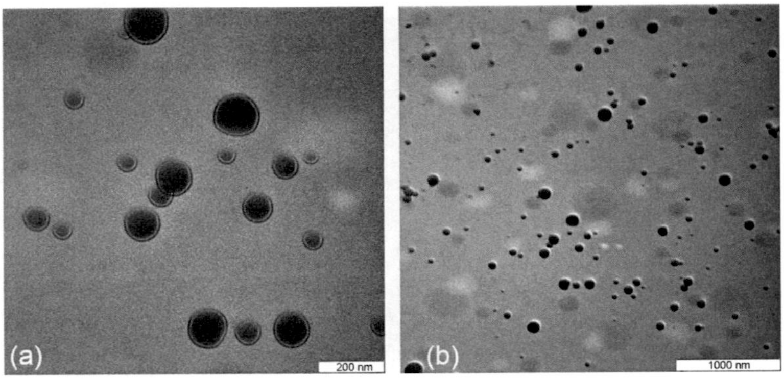

Figure 15.17 TEM image of the *n*-octadecane/St-MMA nanocapsules of Sample A4 at two different TEM setting views: (a) magnification 100k (b) magnification 30k (Tumirah et al., 2014).

has a vital role to play in the encapsulation efficiency and the morphology of the nanocapsules as well.

Though the thermal conductivity of the nanocapsules varies with the encapsulation efficiency due to the smaller size of the materials, the surface-to-volume ratio is high, which results in achieving swift heat transfer during the charging and discharging processes. As evident from the differential scanning calorimetry measurements (Figure 15.18), the nanocapsules exhibited better phase change temperature (freezing: 24.6 °C; melting: 29.5 °C) and a latent heat of enthalpy (freezing: 104.9 J/g; melting: 107.9 J/g), which is highly suitable for building applications.

Similar to the above study, the development of the nanoencapsulated PCM (NEPCM) for cold energy storage applications utilizing air conditioning in buildings has been reported (Fang et al., 2013). Herein, the *n*-tetradecane (Tet) acts as the core material and the polystyrene (PS) serves as the shell material, and the NEPCM has been prepared through the ultrasonic-assistant miniemulsion in situ polymerisation technique.

The morphology of the NEPCM obtained using the TEM and the DSC curves of the Tet and Tet/PS nanocapsule is shown in Figures 15.19 and 15.20. It is thus suggested that the NEPCM, which exhibits phase transition temperature (freezing: −3.43 °C; melting: 4.04 °C) and latent heat of enthalpy (freezing: 91.27 J/g; melting: 98.71 J/g), can be a suitable candidate for cold TES applications in buildings. The nanocapsules that are subjected to more than 40 rounds of the thermal cycles (freezing-melting) show no delamination and are complete and stable.

15.4.3 Nucleation and thermal energy storage potential

One of the prime reasons for embedding nanomaterials into the PCM could be the particle–particle interaction taking place during the phase transition processes. Nanoparticles, whose size is comparatively larger than the pure (base) fluid

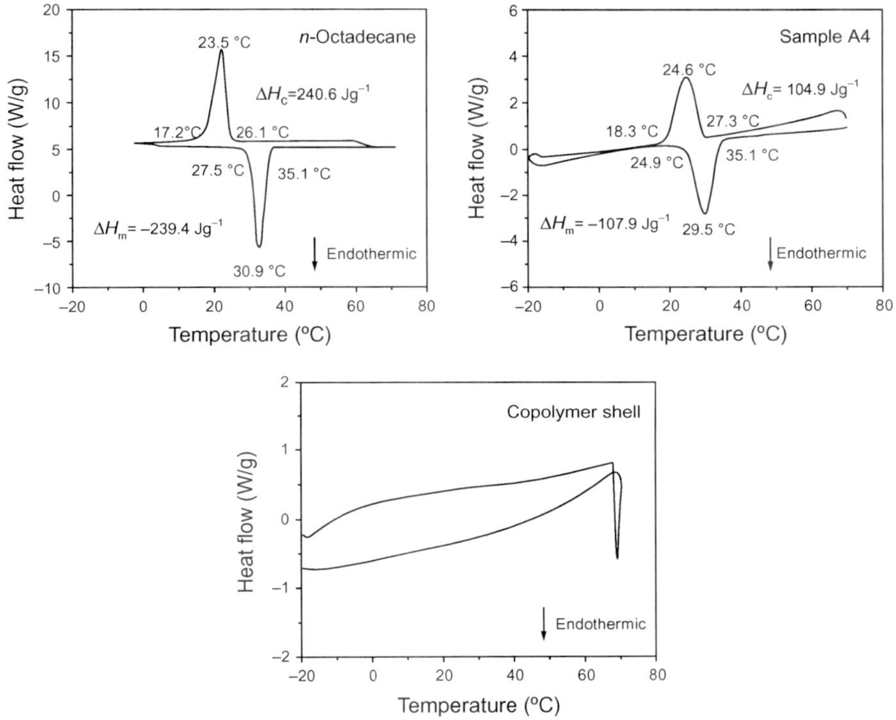

Figure 15.18 DSC thermograms of *n*-octadecane, copolymer shell and *n*-octadecane/St-MMA nanocapsules of Sample A4 (Tumirah et al., 2014).

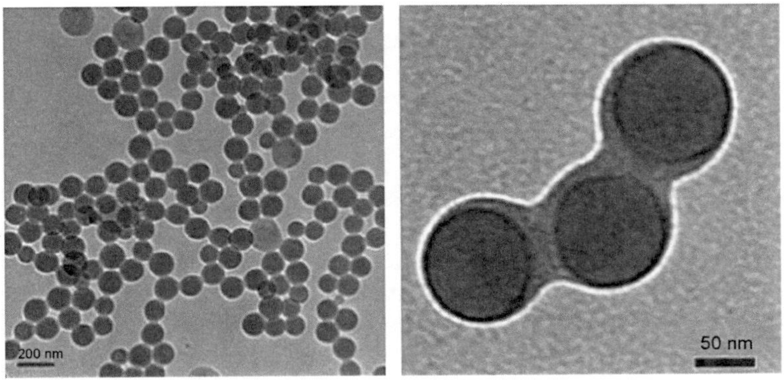

Figure 15.19 TEM images of Tet/PS nanocapsule (Fang et al., 2013).

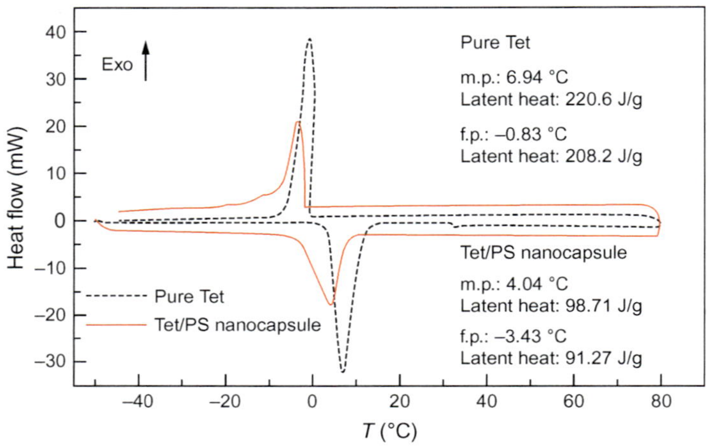

Figure 15.20 DSC curves of Tet and Tet/PS nanocapsule (Fang et al., 2013).

molecules, can participate in initiating the freezing and melting processes through the nucleation phenomenon. The presence of the nanoparticles (either in individual or cluster form) in the pure PCM can create potential nucleation sites for the formation and subsequent growth of ice-like crystals during the freezing (charging) process (Parameshwaran and Kalaiselvam, 2013b,c; Parameshwaran et al., 2013b).

Because nanomaterials can act as a foreign material to the PCM, heterogeneous nucleation is said to occur in the PCM for it to change its phase from liquid to solid. The cold energy being supplied by the external heat transfer medium (air or fluid) can be swiftly transferred by means of the conductive thermal network being created by the nanoparticles within the PCM matrix layers. The activation energy, which is actually required to initiate the phase transition, is supplied to the PCM molecules through the nanoparticles. By this, the PCM embedded with the nanomaterials can be made to freeze at a faster rate, thereby reducing cooling energy needs during the charging process.

Likewise, in the melting process (discharging), the formation of the immediate liquid front can be accomplished by means of the nanoparticles present in the pure PCM, thereby initiating the ice-crystals of the PCM to melt at a faster rate (for the aforesaid reasons). The increase of entropy that is occurring at the molecular level within the PCM due to the presence of nanoparticles enables the rigid and complex-structured ice-crystals to break, which paves way for the formation of the solid–liquid interface boundary.

Furthermore, the gain in entropy over the internal bonding enthalpy losses can also favor the spatial randomization of the nanoparticles and the PCM molecules, leading to the melting of the PCM at a faster rate. The solid–liquid interface or the moving boundary layer actually tends to transfer the heat energy to the adjacent molecules of the PCM. The sensible heat gain of the interface layer can be ascribed to promoting the conductive-convective-like heat transfer within the PCM, thereby augmenting the melting process. It is pertinent to note that the solid–liquid interface layer is also expected to contain the nanoparticles, provided they are in the well-dispersed state within the PCM.

The pictorial representations of the organic ester PCM embedded with the silver–titania HyNC undergoing the freezing and melting processes are shown in Figure 15.21 (Parameshwaran et al., 2013a). During the freezing process, the effective nucleation (or nucleation sites) being formed in the PCM by means of the HyNC particles can be very well observed, as highlighted by green spots in Figure 15.21. The formation of ice-crystals through the heterogeneous nucleation and subsequent movement of the solid–liquid interface boundary can also be clearly visualized in Figure 15.21.

The interesting part of embedding the nanomaterials into the PCM is that, because they are well dispersed within the PCM matrix layers, the heterogeneous nucleation takes place from every potential site of the PCM. In the conventional approach, as the PCM starts to freeze, the movement of the solid–liquid interface boundary would be from the periphery to the center of the PCM containment. In the case of doping of nanomaterials, in addition to the aforementioned process, the nanomaterials present in between the center and the peripheral locations can also be expected to promote the heterogeneous nucleation (Parameshwaran and Kalaiselvam, 2013b).

The case is similar for the melting process encountered in the PCM infused with nanomaterials. However, due to the infusion of the nanomaterials into the pure PCM, there may be some possibilities for having shifts in the latent heat of enthalpy and the phase transition temperature during the freezing and melting processes.

The reason this happens may be the surface tension forces or the physical amalgamation of the nanomaterials with the pure PCM during the preparation stages. Keeping the shifts in these parameters at a minimum level through the optimum doping concentration of the nanomaterials would help facilitate the enhanced TES capabilities of the PCMs dedicated to cooling applications in buildings.

The simulation and the experimental freezing and melting characteristics of the silver–titania HyNC-embedded organic ester PCM are depicted in Figure 15.22. It is evident that, for the reasons stated above and for the specific weight proportions, the HyNC-embedded PCM exhibited reduced freezing and melting times compared to the PCM at its purest form. Factually, the energy spent charging the HyNC-embedded PCM is comparatively less than that of the pure PCM. This results in having reduced cooling requirements in buildings without compromising the energy redistribution requirements between on-peak and part load conditions.

In order to substantiate the positive effects of embedding nanomaterials into the PCM for reducing building cooling needs, a chiller equipped with the LTES system containing the HyNC PCM has been experimentally determined for an office building application and compared with that of the pure PCM (Parameshwaran and Kalaiselvam, 2013a). The schematic representation of the HyNC PCM incorporated LTES air conditioning system (HiTES system) is shown in Figure 15.23. In principle, the hydronic (chilled water) circuit can be divided into two folds: (1) a portion of the chilled water can be routed to the HyNC PCM TES tank, and (2) the remaining portion

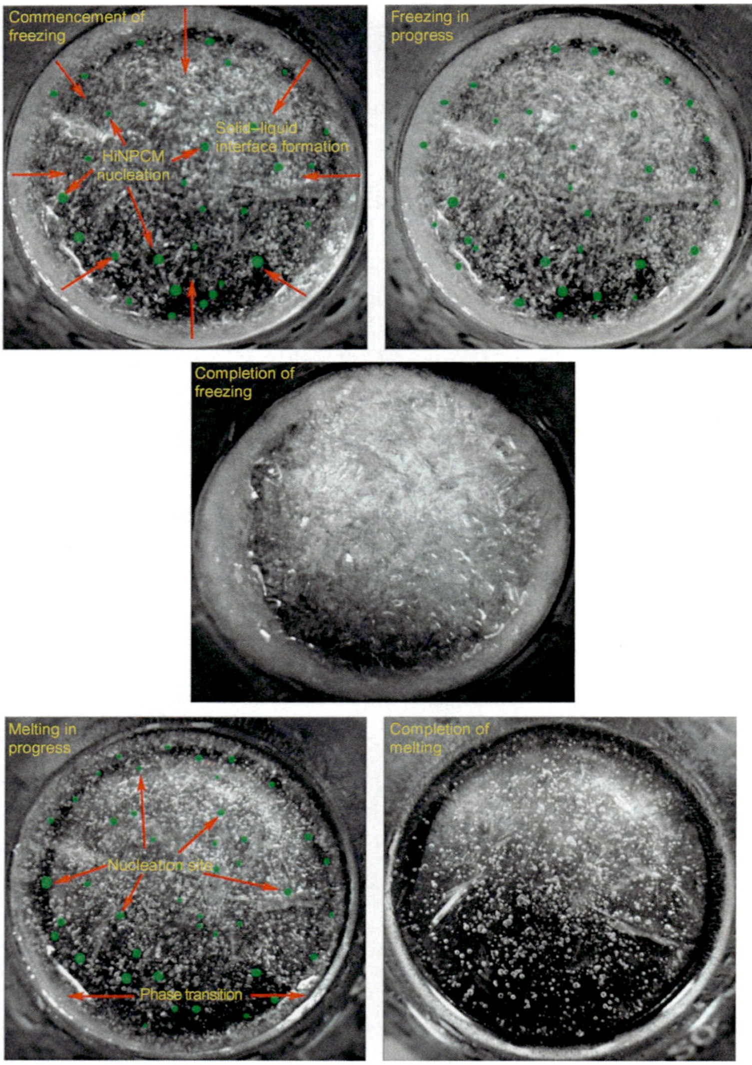

Figure 15.21 Pictorial views of the Ag–TiO_2 HyNC-embedded PCM undergoing freezing and melting processes for a specific concentration of the HyNC (Parameshwaran et al., 2013a).

of the chilled water can be diverted to the energy transfer station (or air handling cooling coil unit) at the building side.

During part load conditions (charging process—nighttime), the chilled water is supplied to both the TES tank and the building cooling coil unit. Since the temperature of the chilled water is below the fusion temperature of the HyNC PCM, charging of the same takes place at a much faster rate. During on-peak load conditions (discharging

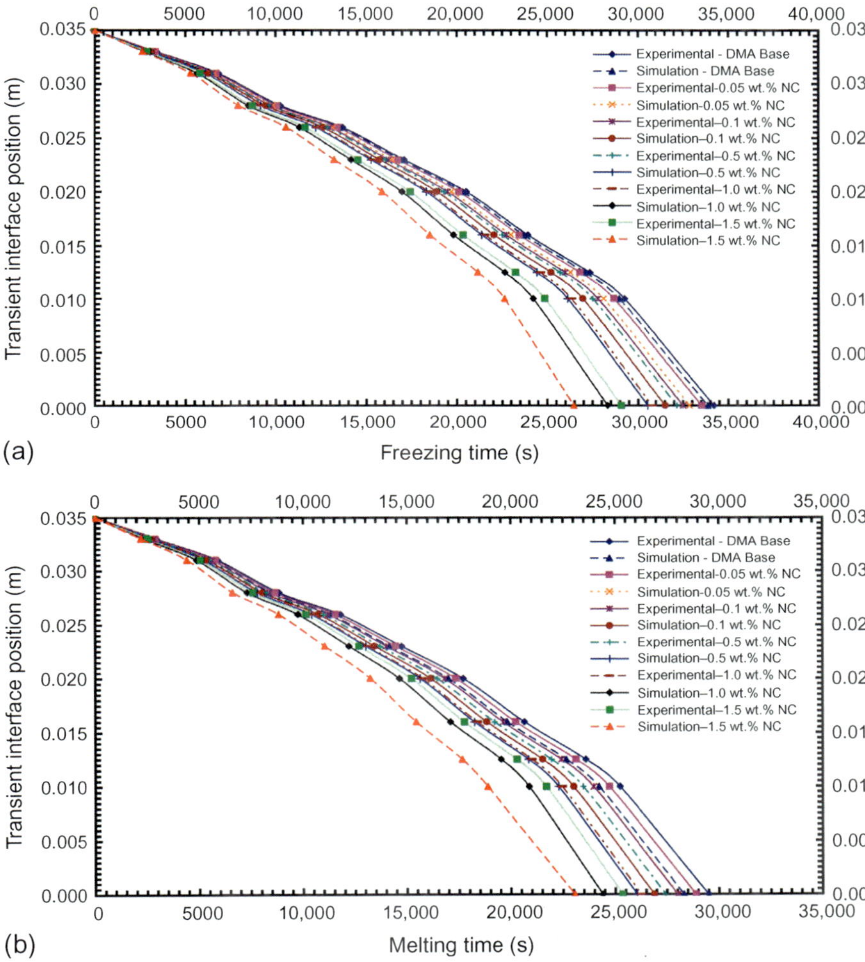

Figure 15.22 Comparison of simulation and experimental results characterizing HyNC PCM in (a) charging (freezing) and (b) discharging (melting) (Parameshwaran and Kalaiselvam, 2013a).

process—daytime), the warm chilled water at a temperature higher than the fusion temperature discharges the stored cooled energy from the HyNC PCM. Through this strategy, the energy redistribution requirements in buildings can be effectually achieved in addition to the reduction in cooling energy needs.

Based on the test inferences as summarized in Table 15.6, it is suggested that the chiller equipped with the LTES system containing the HyNC PCM established a reduction in the nominal cooling capacity in both summer and winter design conditions. This means that the energy consumed by the chiller for meeting the cooling demand in buildings can be reduced considerably using the HyNC PCM. It is therefore

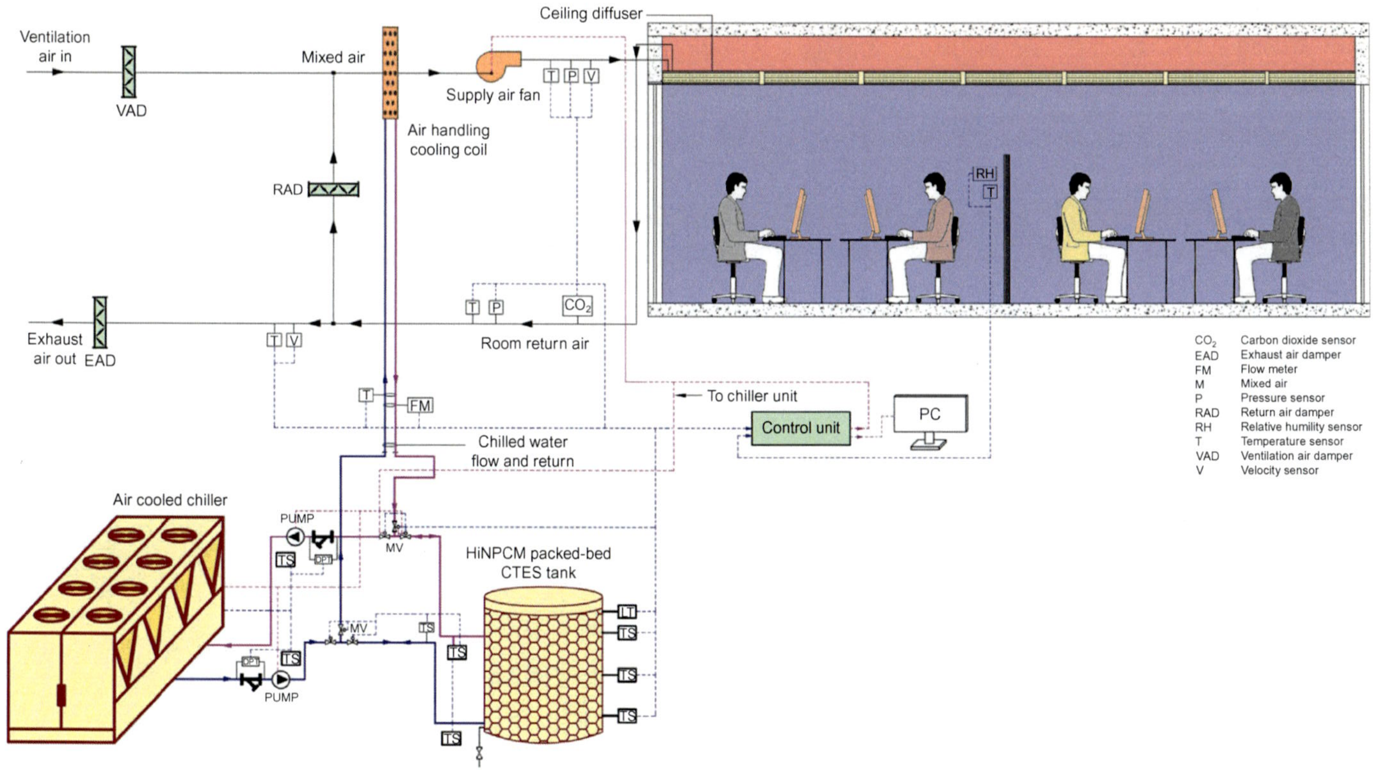

Figure 15.23 Schematic representation of the proposed HiTES air conditioning system (Parameshwaran and Kalaiselvam, 2013a).

Table 15.6 Cooling capacity of conventional and HiTES A/C systems (Parameshwaran and Kalaiselvam, 2013a)

Office space description	Floor area (m^2)	Cooling capacity (W/m^2)	Nominal cooling capacity in summer (kW)					Nominal cooling capacity in Winter (kW)				
			Existing chiller	Chiller with 0.05 HiTES		Chiller with 1.5 HiTES		Existing chiller	Chiller with 0.05 HiTES		Chiller with 1.5 HiTES	
				On-peak	Part load	On-peak	Part load		On-peak	Part load	On-peak	Part load
Zone 1	14	185	17.4	11.0	9.8	10.5	9.35	15.7	9.9	8.8	9.5	8.42
Zone 2	18	183										
Zone 3	15	181										
Zone 4	17	189										
Zone 5	14	187										
Zone 6	16	186										

suggested that, on the basis of the inferences obtained from these studies, the PCMs embedded with nanomaterials possessing improved thermal storage properties can be considered potential candidates for cool TES applications in buildings.

15.4.4 Nanomaterial-embedded PCMs—energetic aspects

The energy redistribution aspects in terms of the cooling capacity of the demonstrated HyNC-embedded PCM LTES comfort cooling system can be devised based on the following relations (Parameshwaran and Kalaiselvam, 2013a):

The nominal cooling capacity of the chiller is given by

$$\text{Nominal cooling capacity of the chiller} = \left(\frac{\text{Total cooling load of the thermal zone in the building}}{\begin{array}{c}(\text{Total charging hours} * \text{efficiency factor}) + \\ (\text{Total discharging hours} * \text{efficiency factor})\end{array}} \right) \tag{15.1}$$

The design cooling load sharing capacity of the nanomaterial-embedded PCM-LTES cooling system can be expressed by

$$\begin{aligned}&(\text{Cooling capacity shared by the nano-based PCM-LTES system})_{\text{on peak}} \\ &= (\text{Total cooling load of the zone at the specific hour}) \\ &-(\text{Nominal cooling capacity of the chiller at the same hour} * \text{efficiency factor})\end{aligned} \tag{15.2}$$

The cooling capacity balance of the nanomaterial-embedded PCM-LTES cooling system (under charging mode of the chiller) at the completion of part load conditions is given by

$$\begin{aligned}&\text{Cooling capacity of nano-based PCM-LTES} \\ &= (\text{Total cooling load-}((\text{Nominal cooling capacity of the chiller} \\ &* \text{peak load operating hours} * \text{efficiency factor}) \\ &+ \sum \text{ Cooling capacity of chiller during charging at part load conditions}))\end{aligned} \tag{15.3}$$

The cooling capacity balance remaining in the storage tank at on-peak load conditions can be determined by

$$\begin{aligned}&\text{Cooling capacity shared by the nano-based PCM-LTES} \\ &= (\text{Cooling capacity of nano-based PCM LTES at the completion of part load})) \\ &-(\text{Total cooling load-Chiller cooling capacity during direct cooling})\end{aligned} \tag{15.4}$$

15.5 Scope for future research

In the pathway of research oriented towards the development of nanomaterial-embedded PCMs, it is evident that huge potential is available for exploring innovative thoughts and means for producing energy efficient and cost effective LTES systems for cooling applications. Some suggestions for future research are presented below:

- The investigation of the self-diffusion mechanism of eco-friendly nanomaterial-embedded PCMs using molecular dynamics simulation with subsequent experimental validation
- Performance assessment of various combinations of nanomaterial-embedded PCMs with other sensible heat storage methods to reduce the overall cooling energy needs in buildings
- The implementation of energy conservative ventilation techniques, which would collectively merit towards achieving energy redistribution requirements and energy efficiency for the development of a sustainable built environment

Acknowledgments

The authors gratefully acknowledge DST, New Delhi for providing financial support to carry out this research work under the PURSE scheme and UGC Major Research Project (F. No. 42-894/2013 (SR)).

References

Adams, F.C., Barbante, C., 2013. Nanoscience, nanotechnology and spectrometry. Spectrochim. Acta, Part B 86, 3–13.

Ai, D., Su, L., Gao, Z., Deng, C., Dai, X., 2010. Study of ZrO_2 nanopowders based stearic acid phase change materials. Particuology 8, 394–397.

Al-Abidi, A.A., Mat, S.B., Sopian, K., Sulaiman, M.Y., Lim, C.H., Abdulrahman, T., 2012. Review of thermal energy storage for air conditioning systems. Renew. Sustain. Energy Rev. 16, 5802–5819.

Al-Mubaddel, F.S., Haider, S., Al-Masry, W.A., et al., 2012. Engineered nanostructures: a review of their synthesis, characterization and toxic hazard considerations. Arabian J. Chem.. http://dx.doi.org/10.1016/j.arabjc.2012.09.010.

Alonso, J.C., Diamant, R., Castillo, P., Acosta-García, M.C., Batina, N., Haro-Poniatowski, E., 2009. Thin films of silver nanoparticles deposited in vacuum by pulsed laser ablation using a YAG:Nd laser. Appl. Surf. Sci. 255, 4933–4937.

Cui, Y., Liu, C., Hu, S., Yu, X., 2011. The experimental exploration of carbon nanofiber and carbon nanotube additives on thermal behavior of phase change materials. Sol. Energy Mater. Sol. Cells 95, 1208–1212.

Elgafy, A., Lafdi, K., 2005. Effect of carbon nanofiber additives on thermal behavior of phase change materials. Carbon 43, 3067–3074.

Fan, L., Khodadadi, J.M., 2011. Temperature-dependent thermal conductivity of eicosane-based phase change materials with copper oxide nanoparticles. In: International Symposium on Thermal and Materials Nanoscience and Nanotechnology, Antalya, Turkey.

Fan, L., Khodadadi, J.M., 2012. An experimental investigation of enhanced thermal conductivity and expedited unidirectional freezing of cyclohexane-based nanoparticle suspensions utilized as nano-enhanced phase change materials (NePCM). Int. J. Therm. Sci. 62, 120–126.

Fang, Y., Kuang, S., Gao, X., Zhang, Z., 2008. Preparation and characterization of novel nanoencapsulated phase change materials. Energy Convers. Manage. 49, 3704–3707.

Fang, Y., Yu, H., WWan, W., Gao, X., Zhang, Z., 2013. Preparation and thermal performance of polystyrene/*n*-tetradecane composite nanoencapsulated cold energy storage phase change materials. Energy Convers. Manage. 76, 430–436.

Ferrando, R., Jellinek, J., Johnston, R.L., 2008. Nanoalloys: from theory to applications of alloy clusters and nanoparticles. Chem. Rev. 108, 845–910.

Gao, J.Y., 2008. An Experimental Study on Melting Heat Transfer Behavior of a Phasechange-Material Containing Al_2o_3 Nanoparticles in a Vertical Rectangular Enclosure. M.S. Thesis, National Cheng Kung University, Taiwan. Available online at: http://ethesys.lib.ncku.edu.tw/ETD-db/ETD-search/view_etd?URN=etd-0825108-153106.

Gao, J.W., Zheng, R.T., Ohtani, H., Zhu, D.S., Chen, G., 2009. Experimental investigation of heat conduction mechanisms in nanofluids. Clue on clustering. Nano Lett. 9, 4128–4132.

Gleiter, H., 2000. Nanostructured materials: basic concepts and microstructure. Acta Mater. 48, 1–29.

Guzman, M.G., Dille, J., Godet, S., 2008. Synthesis of silver nanoparticles by chemical reduction method and their antibacterial activity. World Acad. Sci. Eng. Technol. 43, 357–364.

Ho, C.J., Gao, J.Y., 2009. Preparation and thermophysical properties of nanoparticle-inparaffin emulsion as phase change material. Int. Commun. Heat Mass Transfer 36, 467–470.

Hong, H., Wensel, J., Peterson, S., Roy, W., 2007a. Efficiently lowering the freezing point in heat transfer coolants using carbon nanotubes. J. Thermophys. Heat Transfer 21, 446–448.

Hong, H., Zheng, Y., Roy, W., 2007b. Nanomaterials for efficiently lowering the freezing point of anti-freeze coolants. J. Nanosci. Nanotechnol. 7, 1–5.

Hu, M.Z., Easterly, C.E., 2009. A novel thermal electrochemical synthesis method for production of stable colloids of "naked" metal (Ag) nanocrystals. Mater. Sci. Eng. C 29, 726–736.

Jian, Z., Xian, Z., Yongchang, W., 2005. Electrochemical synthesis and fluorescence spectrum properties of silver nanospheres. Microelectron. Eng. 77, 58–62.

Ju-Nam, Y., Lead, J.R., 2008. Manufactured nanoparticles: an overview of their chemistry, interactions and potential environmental implications. Sci. Total Environ. 400, 396–414.

Kalaiselvam, S., Parameshwaran, R., Harikrishnan, S., 2012. Analytical and experimental investigations of nanoparticles embedded phase change materials for cooling application in modern buildings. Renew. Energy 39, 375–387.

Khaydarov, R.A., Khaydarov, R.R., Gapurova, O., Estrin, Y., Scheper, T., 2009. Electrochemical method for the synthesis of silver nanoparticles. J. Nanopart. Res. 11, 1193–1200.

Khodadadi, J.M., Hosseinizadeh, S.F., 2007. Nanoparticle-enhanced phase change materials (NEPCM) with great potential for improved thermal energy storage. Int. J. Heat Mass Transfer 34, 534–543.

Khodadadi, J.M., Fan, L., Babaei, H., 2013. Thermal conductivity enhancement of nanostructure-based colloidal suspensions utilized as phase change materials for thermal energy storage: a review. Renew. Sustain. Energy Rev. 24, 418–444.

Kim, S., Drzal, L.T., 2009. High latent heat storage and high thermal conductive phase change materials using exfoliated graphite nanoplatelets. Sol. Energy Mater. Sol. Cells 93, 136–142.

Kim, D., Kwon, Y., Cho, Y., Li, C., Cheong, S., Hwang, Y., et al., 2009. Convective heat transfer characteristics of nanofluids under laminar and turbulent flow conditions. Curr. Appl. Phys. 9, 119–123.

Li, M., 2013. A nano-graphite/paraffin phase change material with high thermal conductivity. Appl. Energy 106, 25–30.

Lin, P.-C., Lin, S., Wang, P.C., Sridhar, R., 2013. Techniques for physicochemical characterization of nanomaterials. Biotechnol. Adv. 32, 711–726. http://dx.doi.org/10.1016/j.biotechadv.2013.11.006.

Liu, S., Huang, W., Chen, S., Avivi, S., Gedanken, A., 2001. Synthesis of X-ray amorphous silver nanoparticles by the pulse sonoelectrochemical method. J. Non-Cryst. Solids 283, 231–236.

Liu, Y.-D., Zhou, Y.-G., Tong, M.-W., Zhou, X.-S., 2009. Experimental study of thermal conductivity and phase change performance of nanofluids PCMs. Microfluid. Nanofluid. 7, 579–584.

Manikam, V.R., Cheong, K.Y., Razak, K.A., 2011. Chemical reduction methods for synthesizing Ag and Al nanoparticles and their respective nanoalloys. Mater. Sci. Eng. B 176, 187–203.

Meng, X., Zhang, H., Sun, L., Xu, F., Jiao, Q., Zhao, Z., Zhang, J., Zhou, H., Sawada, Y., Liu, Y., 2013. Preparation and thermal properties of fatty acids/CNTs composite as shape-stabilized phase change materials. J. Therm. Anal. Calorim. 111, 377–384.

Mo, S., Chen, Y., Yang, J., Luo, X., 2011. Experimental study on solidification behavior of carbon nanotube nanofluid. Adv. Mater. Res. 171–172, 333–336.

Nguyen, T.-D., 2013. Portraits of colloidal hybrid nanostructures: controlled synthesis and potential applications. Colloids Surf. B 103, 326–344.

Niemeyer, C.M., 2001. Nanoparticles, proteins, and nucleic acids: biotechnology meets materials science. Angew. Chem. Int. Ed. 40, 4128–4158.

Pacheco-Torgal, F., 2014. Eco-efficient construction and building materials research under the EU Framework Programme Horizon 2020. Constr. Build. Mater. 51, 151–162.

Pacheco-Torgal, F., Jalali, S., 2011. Eco-efficient Construction and Building Materials, first ed. Springer, London.

Parameshwaran, R., Kalaiselvam, S., 2013a. Energy efficient hybrid nanocomposite-based cool thermal storage air conditioning system for sustainable buildings. Energy 59, 194–214.

Parameshwaran, R., Kalaiselvam, S., 2013b. Effect of aggregation on thermal conductivity and heat transfer in hybrid nanocomposite phase change colloidal suspensions. Appl. Phys. Lett. 103, 1–7.

Parameshwaran, R., Kalaiselvam, S., 2013c. Thermal Energy Storage Technologies, Nearly Zero Energy Building Refurbishment: A Multidisciplinary Approach. Springer Publications, London, pp. 483–536.

Parameshwaran, R., Dhamodharan, P., Kalaiselvam, S., 2013a. Study on thermal storage properties of hybrid nanocomposite-dibasic ester as phase change material. Thermochim. Acta 573, 106–120.

Parameshwaran, R., Jayavel, R., Kalaiselvam, S., 2013b. Study on thermal properties of organic ester phase-change material embedded with silver nanoparticles. J. Therm. Anal. Calorim. 114, 845–858.

Parameshwaran, R., Kalaiselvam, S., Jayavel, R., 2013c. Green synthesis of silver nanoparticles using Beta vulgaris: role of process conditions on size distribution and surface structure. Mater. Chem. Phys. 140, 135–147.

Parameshwaran, R., Deepak, K., Saravanan, R., Kalaiselvam, S., 2014. Preparation, thermal and rheological properties of hybrid nanocomposite phase change material for thermal energy storage. Appl. Energy 115, 320–330.

Pielichowska, K., Pielichowski, K., 2010. Crystallization behaviour of PEO with carbon-based nanonucleants for thermal energy storage. Thermochim. Acta 510, 173–184.

Pokropivny, V.V., Skorokhod, V.V., 2007. Classification of nanostructures by dimensionality and concept of surface forms engineering in nanomaterial science. Mater. Sci. Eng. C 27, 990–993.

Ritson, 2008. The sonoelectrochemical synthesis of silver nanoparticles their applications to SERS", PhD dissertation, University of Liverpool.

Royal Society and Royal Academy of Engineering, 2004. Nanoscience and Nanotechnologies: Opportunities and Uncertainties. The Royal Society, London.

Salkar, R.A., Jeevanandam, P., Aruna, T., Koltypin, Y., Gedanken, A., 1999. The sonochemical preparation of amorphous silver nanoparticles. J. Mater. Chem. 9, 1333–1335.

Sanusi, O., Warzoha, R., Fleischer, A.S., 2011. Energy storage and solidification of paraffin phase change material embedded with graphite nanofibers. Int. J. Heat Mass Transfer 54, 4429–4436.

Sayyar, M., Weerasiri, R.R., Soroushian, P., Lu, J., 2014. Experimental and numerical study of shape-stable phase-change nanocomposite toward energy-efficient building constructions. Energy Build. 75, 249–255.

Seeniraj, R.V., Velraj, R., Narasimhan, N.L., 2002. Heat transfer enhancement study of a LHTS unit containing dispersed high conductivity particles. J. Sol. Energy Eng. 124, 243–249.

Shaikh, S., Lafdi, K., Hallinan, K., 2008. Carbon nanoadditives to enhance latent energy storage of phase change materials. J. Appl. Phys. 103, 6.

Siegel, R., 1977. Solidification of low conductivity material containing dispersed high conductivity particles. Int. J. Heat Mass Transfer 20, 1087–1089.

Skorokhod, V., Ragulya, A., Uvarova, I., 2001. Physico-chemical kinetics in nanostructured systems. Kyiv: Academperiodica, 180.

Starowicz, M., Stypuła, B., Banas, J., 2006. Electrochemical synthesis of silver nanoparticles. Electrochem. Commun. 8, 227–230.

Tumirah, K., Hussein, M.Z., Zulkarnain, Z., Rafeadah, R., 2014. Nano-encapsulated organic phase change material based on copolymer nanocomposites for thermal energy storage. Energy 66, 881–890.

Tunc, I., Guvenc, H.O., Sezen, H., Suzer, S., Correa-Duarte, M.A., Liz-Marzán, L.M., 2008. Optical response of Ag–Au bimetallic nanoparticles to electron storage in aqueous medium. Nanosci. Nanotechnol. 8, 3003–3007.

Wang, J., Xie, H., Xin, Z., 2008. Thermal properties of heat storage composites containing multiwalled carbon nanotubes. J. Appl. Phys. 104, 5.

Wang, J., Xie, H., Xin, Z., 2009. Thermal properties of paraffin based composites containing multi-walled carbon nanotubes. Thermochim. Acta 488, 39–42.

Wang, J., Xie, H., Li, Y., Xin, Z., 2010a. PW based phase change nanocomposites containing γ-Al2O3. J. Therm. Anal. Calorim. 102, 709–713.

Wang, J., Xie, H., Xin, Z., Li, Y., 2010b. Increasing the thermal conductivity of palmitic acid by the addition of carbon nanotubes. Carbon 48, 3979–3986.

Wang, J., Xie, H., Xin, Z., Li, Y., Chen, L., 2010c. Enhancing thermal conductivity of palmitic acid based phase change materials with carbon nanotubes as fillers. Sol. Energy 84, 339–344.

Wang, N., Yang, S., Zhu, D., Ju, X., 2010d. Preparation and heat transfer behavior of paraffin based composites containing nano-copper particles. In: Proceedings of the Seventh International Conference on Multiphase Flow, Tampa, FL.

Weinstein, R.D., Kopec, T.C., Fleischer, A.S., D′Addio, E., Bessel, C.A., 2008. The experimental exploration of embedding phase change materials with graphite nanofibers for the thermal management of electronics. J. Heat Transfer 130, 8.

Wu, S., Zhu, D., Li, X., Li, H., Lei, J., 2009. Thermal energy storage behavior of Al_2O_3–H_2O nanofluids. Thermochim. Acta 483, 73–77.

Wu, S., Zhu, D., Zhang, X., Huang, J., 2010. Preparation and melting/freezing characteristics of Cu/paraffin nanofluid as phase-change material (PCM). Energy Fuel 24, 1894–1898.

Xiang, J., Drzal, L.T., 2011. Investigation of exfoliated graphite nanoplatelets (xGnP) in improving thermal conductivity of paraffin wax-based phase change material. Sol. Energy Mater. Sol. Cells 95, 1811–1818.

Xie, H., Wan, J., Chen, L., 2008. Effects on the phase transformation temperature of nanofluids by the nanoparticles. J. Mater. Sci. Technol. 25, 742–744.

Yavari, F., Raeisi Fard, H., Pashayi, K., Rafiee, M.A., Zamiri, A., Yu, Z., et al., 2011. Enhanced thermal conductivity in a nanostructured phase change composite due to low concentration graphene additives. J. Phys. Chem. C 115, 8753–8758.

Zeng, J.L., Sun, L.X., Xu, F., Tan, Z.C., Zhang, Z.H., Zhang, J., et al., 2007. Study of a PCM based energy storage system containing Ag nanoparticles. J. Therm. Anal. Calorim. 87, 369–373.

Zeng, J.L., Liu, Y.Y., Cao, Z.X., Zhang, J., Zhang, Z.H., Sun, X.L., et al., 2008. Thermal conductivityenhancement of MWNTS on the PANI/tetradecanol form-stable PCM. J. Therm. Anal. Calorim. 91, 443–446.

Zeng, J.L., Cao, Z., Yang, D.W., Xu, F., Sun, L.X., Zhang, X.F., et al., 2009. Effects of MWNTS on phase change enthalpy and thermal conductivity of a solid–liquid organic PCM. J. Therm. Anal. Calorim. 95, 507–512.

Zeng, J.L., Cao, Z., Yang, D.W., Sun, L.X., Zhang, L., 2010. Thermal conductivity enhancement of Ag nanowires on an organic phase change material. J. Therm. Anal. Calorim. 101, 385–389.

Zhang, W., Qiao, X., Chen, J., 2007. Synthesis of silver nanoparticles—effects of concerned parameters in water/oil microemulsion. J. Mater. Sci. Eng. B 142, 1–15.

Fenestration for reducing building cooling needs: an introduction to spectral selectivity, thermochromics, and electrochromics

16

C.G. Granqvist
Uppsala University, Uppsala, Sweden

16.1 Introduction

Surface coatings are used on a huge and rapidly increasing scale in order to enhance the performance of windows and glass facades—referred to jointly as "glazings"— and make the buildings where they are used more eco-efficient so that less energy is needed for cooling and heating purposes (Smith and Granqvist, 2010). We first observe that the flat glass market is huge by any standards and is predicted to reach a stunning 9.2 billion square meters per year in 2016 (Freedonia Group, 2013). Most of the flat glass is produced by the float process, has high quality, and is suitable for surface coating and implementation in glazings.

The objects of this chapter are to (i) explain why coatings can lead to very significant improvements of the glazings' eco-efficiency, (ii) introduce the basics of today's technologies, and (iii) outline forthcoming technologies that already exist as prototypes or are under development in research laboratories. In particular, the focus is on thermochromic and electrochromic technologies, which allow the throughput of solar energy and visible light to be varied depending on temperature and on the application of a low electrical voltage, respectively. Coating technology as such is not discussed but is covered elsewhere with a view towards eco-efficiency by Granqvist (2012a, 2013a) and with a more general scope in books by Pulker (1999), Gläser (2000), Ohring (2002), Mattox (2010), and Piegari and Flory (2013).

16.2 Light, solar energy, thermal radiation, and more

16.2.1 Spectral selectivity of ambient radiation

Energy-efficient fenestration is conveniently introduced by first considering the electromagnetic radiation that surrounds us (Granqvist, 1981; Smith and Granqvist, 2010). This radiation is reported in Figure 16.1, which illustrates a number of characteristics on a common logarithmic wavelength scale.

Eco-efficient Materials for Mitigating Building Cooling Needs. http://dx.doi.org/10.1016/B978-1-78242-380-5.00016-9

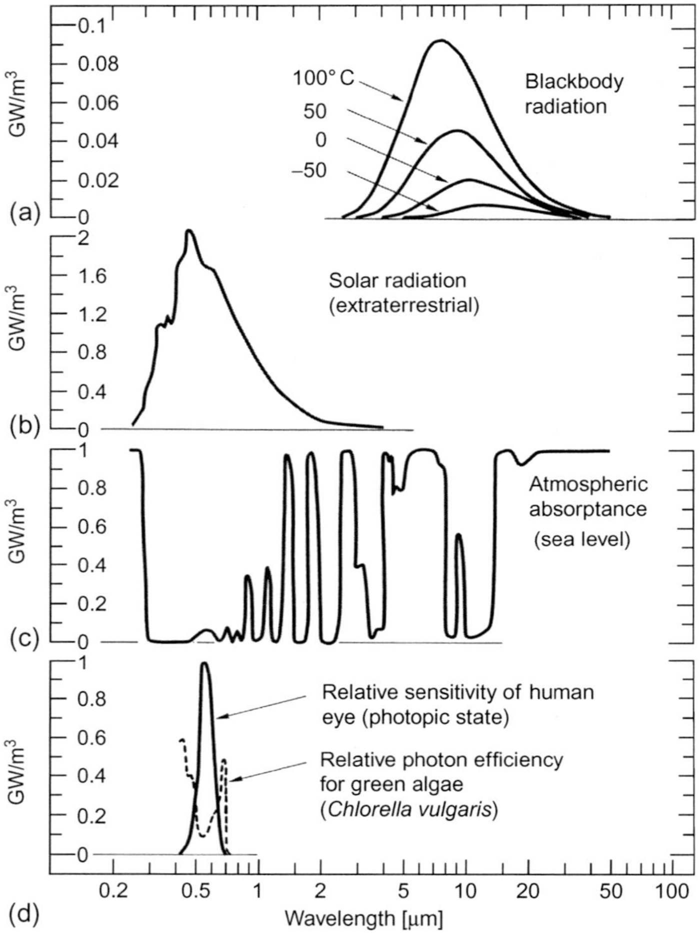

Figure 16.1 Spectra for radiation in our natural surroundings, showing (a) blackbody radiation from surfaces at four "normal" temperatures, (b) the solar spectrum outside the Earth's atmosphere, (c) absorption in the atmospheric envelope during clear weather, and (d) spectral responses for the human eye and for photosynthesis in green plants.
From Granqvist (1981).

Everything around us emits electromagnetic radiation and receives radiation. This radiation can be understood from Planck's radiation law, which is a cornerstone in modern quantum physics. The Planck spectrum is governed by the temperature and the surface properties of the radiating object. Figure 16.1(a) shows four bell-shaped graphs that represent the Planck spectra for a number of temperatures of practical relevance. The actual emitted radiation is then obtained as the Planck spectrum multiplied by an emittance, which is a material-related property that in general is wavelength-dependent. Most objects in our surroundings—including glass—have an emittance of 0.85–0.90, which means that the curves at the top of Figure 16.1 are good descriptions of emitted *thermal radiation*. The only important cases of materials with low emittance are clean metal surfaces, whose emittance can be as low as a few percent. However, no matter the magnitude of the emittance, thermal radiation at common temperatures lies in the $2 < \lambda < 50$ μm interval, where λ denotes wavelength.

Solar radiation, which is considered next, arrives at wavelengths in the $0.25 < \lambda < 3$ μm range, as shown in Figure 16.1(b). This curve has a bell shape that makes it resemble a Planck spectrum, and it is well described by such a spectrum for a temperature of about 5500 °C. This temperature is then defining the Sun's surface temperature, which is very much smaller than the temperature of some 15 million degrees in the interior of the Sun and is a consequence of nuclear reactions powering the Sun. The wavelength-integrated area below the solar spectrum gives the "solar constant," which is 1353 ± 21 W m^{-2} and represents the power density on a surface positioned at right angles to the incident solar radiation and placed just outside the Earth's atmosphere. It is now possible to make an essential observation: the wavelength regions for incident solar radiation and for emitted thermal radiation are almost nonoverlapping, which means that surfaces can have completely different properties for solar radiation and for thermal radiation.

The solar radiation that falls onto a glazing has passed through the atmosphere, and the radiation that is emitted from a glazing goes into air. Clearly, *atmospheric absorptance* is of much importance, and Figure 16.1(c) shows the typical absorptance—from zero to unity—vertically through the full atmosphere at sea level and during clear weather. It is evident that most of the solar irradiation is able to pass through the atmosphere, and the maximum power density at ground level is about 1000 Wm^{-2}. Strong atmospheric absorption occurs at the shortest wavelengths and is due to ozone. This absorption is very important since intense short-wavelength radiation is dangerous and may cause melanoma. At long wavelengths—beyond approximately one micrometer—one observes several intense absorption bands. They emanate from water vapor, carbon dioxide, and ozone, whereas the main constituents of the atmosphere, nitrogen and oxygen, play a very minor role for the absorption. There is a very notable exception with regard to the atmospheric absorption, however, and the atmosphere can be highly transparent in the $8 < \lambda < 13$ μm interval, which is called the "atmospheric window." This is a "window" that is open day and night during clear weather conditions and is positioned with regard to wavelengths where thermal radiation is at its maximum. The "atmospheric window" is essential for the Earth's thermal balance, and it serves as a major channel for energy dissipation. If the amounts of carbon dioxide and/or water vapor are altered in the atmosphere, which is the case if large quantities of carbon dioxide are injected as the result of burning of hydrocarbon-based fuels, the "window" gets narrower, the energy loss becomes reduced, and global warming ensues. We return briefly to the influence of this "window" in Section 16.6 below.

Figure 16.1(d), finally, illustrates two very significant biological properties. The solid, bell-shaped curve is the *eye's relative sensitivity*, which lies at $0.4 < \lambda < 0.7$ μm and has a maximum at 0.55 μm. The dashed curve in the same panel signifies *photosynthetic efficiency*, which typically occurs for about the same wavelengths as those that our eyes can perceive. It is now possible to draw another important conclusion: only part of the solar radiation is useful for human vision and for (most) plants. Solar radiation at $\lambda < 0.4$ μm, that is, in the ultraviolet, contains some 5% of the solar energy, while radiation at $\lambda > 0.7$ μm, which is in the infrared, holds about 50% of the energy. The remaining 45% of the solar radiation is what our eyes can make use of. These percentages are of great importance for energy-efficient glazings, as will be apparent shortly.

16.2.2 Nature's "rules" and how they can be used

We now summarize the main lessons learned from Figure 16.1: well-defined and sometimes nonoverlapping wavelength intervals characterize the electromagnetic radiation surrounding us. This property is known as "*wavelength selectivity*" and gives many different possibilities to devise eco-efficient glazing solutions. Another key feature for the ambient radiation is its *angular dependence*, which refers to the known fact that radiation is often incident from, or leaving, at a well-defined angle. Direct solar energy, for example, falls onto a glazing from an element of solid angle high up on the vault of heaven for most of the day, while visual indoor–outdoor contact normally is of main concern along a more or less horizontal line of sight. It is then evident that one can minimize the solar energy ingress into a room while preserving a good view to the surroundings by a judicious use of overhangs, awnings, blinds, or nanostructured materials. Finally, it is also very well known that the weather normally changes during the day and season; sometimes, it may be warm and sometimes cold. This *temporal variability* may demand energy-guzzling cooling or heating of buildings unless one employs carefully devised materials with properties that can be altered through some external stimulus, for example, to modify the solar energy throughput of windows in an appropriate manner. Such materials are commonly called "*chromogenic*"; the term was coined some decades ago by Lampert and Granqvist (1990) and Granqvist (1990). Chromogenics is discussed further below in Sections 16.4 and 16.5.

Bringing it all together, nature provides us with a set of "rules"—which involve wavelength selectivity, angular dependence, and temporal variability—and energy-efficient fenestration and eco-efficient buildings in general must work in harmony with these "rules."

We now consider what the observations about the radiation around us lead to for the optical properties of a glazing, namely, its absorptance A, reflectance R, transmittance T, and emittance E. Here, A, T, and R signify the fractions of the radiation that are absorbed, transmitted, and reflected, respectively, while E is the fraction of the black-body radiation that is emitted. Energy conservation dictates that

$$A(\lambda) + R(\lambda) + T(\lambda) = 1 \tag{16.1}$$

and

$$E(\lambda) = A(\lambda) \tag{16.2}$$

at each wavelength. Equation (16.2) is commonly known as Kirchhoff's law and pertains to thermodynamic equilibrium.

16.3 Eco-efficient glazings with static properties

16.3.1 The importance of low thermal emittance and solar control

The outdoor temperature is usually not the same as the one desired indoors, and hence, it is necessary to heat or cool—by one technique or another—in order to create a comfortable climate inside a building. Good thermal insulation is demanded so that excessive amounts of energy are not wasted for this heating and cooling. Therefore, we now contemplate the thermal properties of glazings.

Heat transfer occurs via three mechanisms—radiation, conduction, and convection—and can be measured and computed. Figure 16.2 illustrates calculated results for glazings embodying one, two, and three vertical glass panes separated by air gaps (Granqvist, 1991); the glass surfaces are numbered consecutively with the outermost surface denoted 1. One of

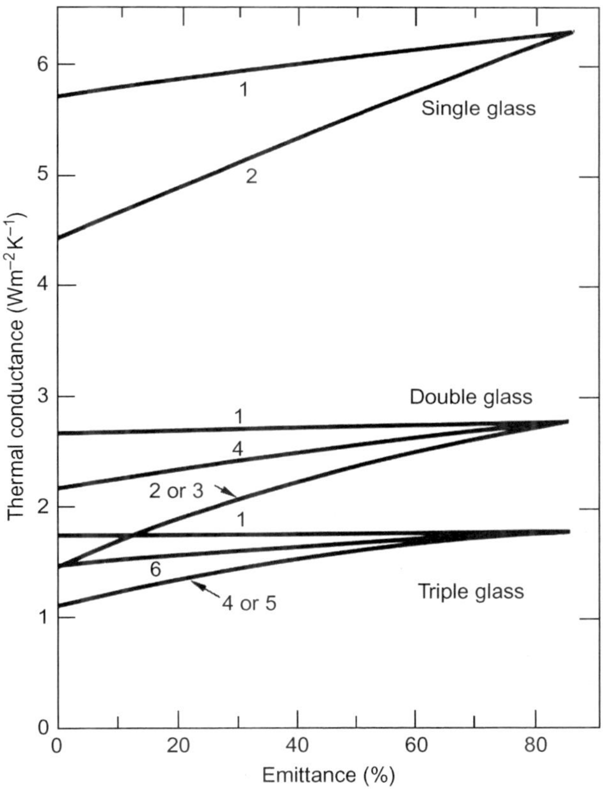

Figure 16.2 Computed thermal transfer for different glazings with one surface having low thermal emittance. The glass surfaces are numbered consecutively with the outermost surface labeled 1, and the panes are separated by air gaps of 12.7 mm (typical for IGUs). The calculations assume a low outdoor temperature and some wind and are not described in detail. From Granqvist (1991).

the surfaces has an emittance in the range between zero and 0.85, where the upper value applies to normal float glass, so that the calculations show the effect of a lowered surface emittance. The upper two curves in Figure 16.2 pertain to a single glass pane and show that the heat transfer is diminished somewhat from about 6 $Wm^{-2}\,K^{-1}$ when the emittance drops. In the case of double or triple panes, the relative influence of emittance changes is much larger, however, and the heat transfer then goes from 2.8 to 1.4 $Wm^{-2}\,K^{-1}$ and from 1.8 to 1.2 $Wm^{-2}\,K^{-1}$, respectively, when the emittance approaches zero. It is evident that low emittance should be on one of the surfaces that bound an air gap. One observes that the heat transfer in a double-pane construction can be almost halved when one of the surfaces displays low emittance. (For completeness, we note that modern glazings seldom have air in the spaces between the glass panes; instead, the glazing comprises an "insulated glass unit" (IGU) typically with argon between the panes, which yields an additional decrease of the heat transfer though not as large as the drop governed by a low emittance.)

The obvious question is now how it is possible to create the desired low emittance for a glass surface. To answer this question, it is necessary to consider the properties of window glass, which obviously is transparent for visible light and usually also is fairly transparent for solar radiation, whereas it is opaque for thermal radiation. A low value of E means that A is low, and Equation (16.1) dictates that the corresponding R must be high. This property can be obtained by coating the glass with a thin layer (usually referred to as a "thin film") with the following ideal property:

$$T(\lambda) = 1 \quad \text{for } 0.4 < \lambda < 3\ \mu\text{m} \tag{16.3a}$$

$$R(\lambda) = 1 \quad \text{for } 3 < \lambda < 50\ \mu\text{m} \tag{16.3b}$$

A practically useful glazing should be transparent for visible light and have low heat transfer, but what should the transparency be for solar energy? We now recall that Figure 16.1 indicated that about 50% of the solar energy lies in the infrared, so the question above is whether this infrared radiation should be transmitted or not. If the building where the glazing is to be used needs heating (more exactly, if the room where the glazing is placed requires heating), the solar energy ingress should be large so that that the properties described by Equations (16.3a) and (16.3b) are desired. However, many buildings do not need solar heating but have a cooling requirement instead, which is so for almost all modern commercial buildings, irrespectively of their geographic location, and also for residential buildings in warm climates. In this latter case, the wanted properties can be expressed as

$$T(\lambda) = 1 \quad \text{for } 0.4 < \lambda < 0.7\ \mu\text{m} \tag{16.4a}$$

$$R(\lambda) = 1 \quad \text{for } 0.7 < \lambda < 50\ \mu\text{m} \tag{16.4b}$$

Coatings giving optical properties approximated by Equations (16.3a), (16.3b) and (16.4a), (16.4b) are called "low-E coatings" and "solar control coatings," respectively. They are electrically conducting and often alternatively referred to as "transparent conductors"; they can be used as electrodes, for example, in electrochromics-based glazings, as discussed in Section 16.5, and in solar cells, light-emitting diodes, etc.

16.3.2 Coatings for "low E" and "solar control": metal-based options

Many different coating materials can be employed to achieve low thermal emittance and solar control. The most common coatings for energy-efficient glazings make use of extremely thin metal layers, in particular silver-based, and are treated first. In Section 16.3.3, we then look at semiconductor-based coatings and some other options.

Consider first what happens when a metal is deposited onto glass or plastic by a common coating technique such as vacuum evaporation or sputter deposition! The deposit on the surface does not evolve uniformly, but the growth resembles water vapor condensing on an oily surface, albeit with the important difference that the "drops" of metal have sizes of nanometers to micrometers. Figure 16.3 shows some scanning electron micrographs for gold films prepared by sputter deposition onto glass from the work by Lansåker et al. (2009); the film thicknesses are "equivalent" and represent hypothetical values for uniform deposits. It is found that the initial deposition creates tiny metallic nuclei on the substrate and that continued deposition makes these nuclei grow, which is expected to take place by diffusion of gold atoms over the glass surface and by direct impingement of such atoms. Continued deposition makes some of the "islands" touch and rearrange to form larger and more irregular entities; this phenomenon is referred to as "coalescence growth." Subsequently, the growing film exhibits "large-scale coalescence" (also referred to as "percolation"), implying that a macroscopic metallic network is formed. Metallic conduction can now be measured along the film, and there is a concurrent onset infrared reflectance. Continued deposition decreases the voids between the metallic paths, and a uniform metal film may form eventually.

A thin film of silver or gold, with a thickness of the order of 10 nm, can display some optical transparency, which is mainly limited by reflectance at the two surfaces of the film (Smith and Granqvist, 2010). If this thin film is embedded between dielectric layers with properly adjusted thicknesses, its reflection can be very much diminished as a result of destructive optical interference, and there is a corresponding increase in the film's transmittance. Figure 16.4 shows some representative optical

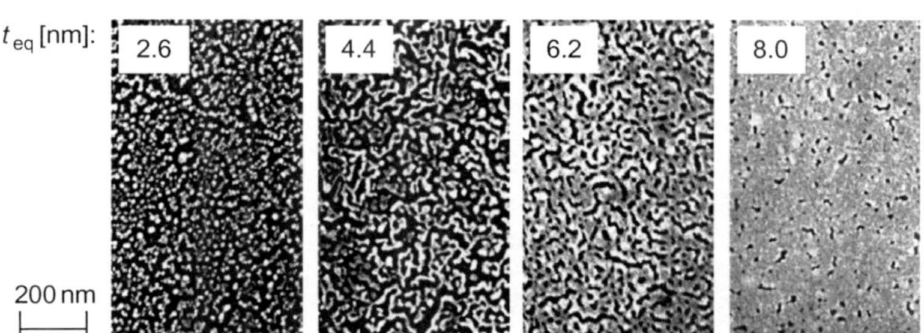

Figure 16.3 Scanning electron micrographs for gold films prepared by sputter deposition onto glass to the shown equivalent thicknesses t_{eq}. Gold appears bright and the uncoated parts of the substrate are dark.
From Lansåker et al. (2009).

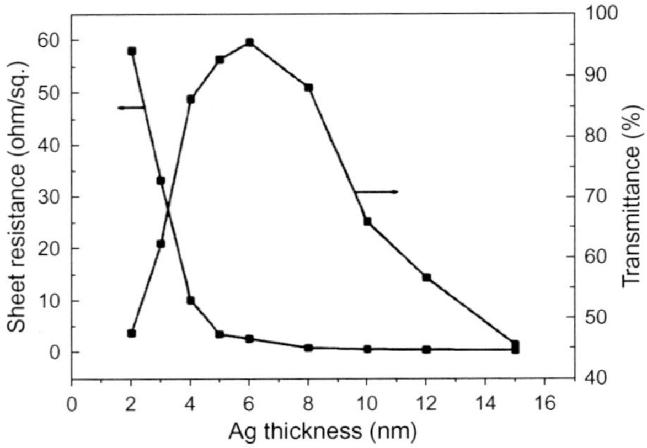

Figure 16.4 Electrical resistance and maximum optical transmittance as a function of Ag thickness in a three-layer structure consisting of ZnO (20 nm)/Ag/ZnO (20 nm).
From Sahu et al. (2009).

and electrical data for the specific case of ZnO/Ag/ZnO three-layer structures, with ZnO thicknesses of 20 nm and varying silver thicknesses, according to Sahu et al. (2006). The maximum optical transmittance exceeds 90%. The electrical resistance is high for the lowest silver thicknesses because very thin metal films are not fully contiguous, but the resistance falls off in proportion to the thickness of the silver film when it is larger than about 6 nm.

It follows that well-designed silver-based coatings can have high transmittance at short wavelengths and high reflectance at long wavelengths. The onset of high reflectance occurs at a wavelength that depends on the thicknesses of the dielectric/silver/dielectric stack (and in analogous more elaborated multilayer structures that are used in practice). Multilayer structures of this kind have widespread applications both for "low-E" and "solar control" applications as surveyed by Jelle (2013). The silver will tarnish rapidly if exposed to air, so it has to be employed in IGUs with inert gas between the glass panes. The electrical resistivity is low and may be of the order of 10^{-5} to 10^{-6} Ω cm. The left-hand panel in Figure 16.5 illustrates the performance of a commercial silver-based "solar control" coating, which transmits 80% of visible light and 41% of solar energy (International Glazing Database, 2009).

16.3.3 Coatings for "low E" and "solar control": semiconductor-based options and some alternatives

Several different oxides can be used as alternatives to the metal-based coatings, especially oxides that combine a wide optical band gap (visible transparency) with good electrical conductivity and hence infrared reflectivity. Indium oxide-based films may be the superior example. In_2O_3 is transparent, and replacing a few percent of the In atoms with Sn makes the material electrically conducting (Hamberg and Granqvist, 1986). The density of the free electrons depends on the Sn content and leads to metal-like properties for wavelengths larger than the "plasma wavelength," denoted λ_p. It is possible to make indium tin oxide coatings with $\lambda_p \approx 2$ μm, which means that such a coating can exhibit a low thermal emittance, but not as tiny as for a silver-based

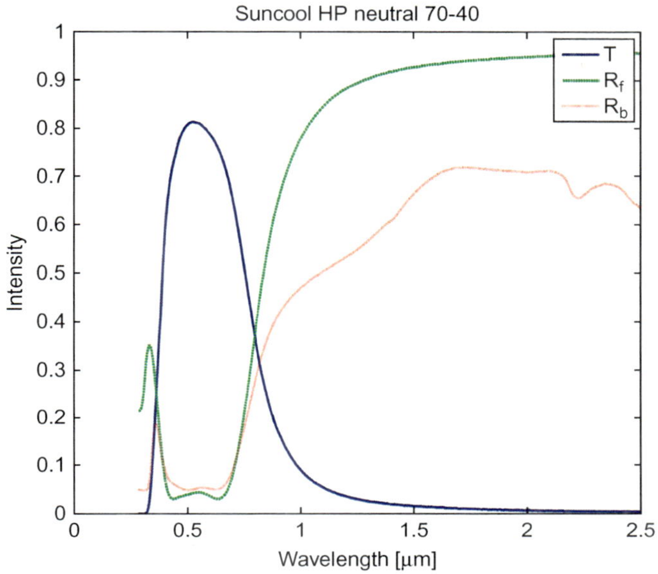

Figure 16.5 Spectral transmittance T and reflectance from the front (coated) side (R_f) and the back side (R_b) for a commercial silver-based coating on glass and optimized for "solar control." From the International Glazing Database (2009).

coating, while the transmittance for solar energy is large. Such a coating therefore can function for "low-E" applications but not for "solar control" applications. Alternative coatings are Sn oxide containing some fluorine, Zn oxide with some Al or Ga, Ti oxide with some Nb, and others, as discussed in detail by Granqvist (2007, 2014). Tin oxide is particularly hard and corrosion-resistant and can be applied to glass surfaces exposed to air. The electrical resistivity may be as low as 10^{-4} Ω cm, and typical film thicknesses are a few hundred nanometers (which calls for good thickness control in order to avoid visibly disturbing features such as iridescence). The resistivity is rather independent of thickness. Figure 16.6 shows data for a commercial window coating based on tin oxide; its transmittance is 83% for visible light and 71% for solar energy (International Glazing Database, 2009).

There are several emerging alternatives to the metal-based and semiconductor-based coatings mentioned above. Some of them use carbon, which can form an infinite number of nanostructures some of which are depicted in Figure 16.7, which is reproduced from the work by Geim and Novoselov (2007). Graphite consists of weakly bound layers as shown to the right, and graphene is formed if these layers are separated ("exfoliated"). Carbon is also able to form nanotubes, as shown in the middle part of the figure, and nanoparticles such as 60-atom units (C_{60}, called a "fullerene molecule" or "buckyball"). Single-walled carbon nanotubes can be as long as several centimeters and consist of one layer of graphene rolled to form a seamless cylinder with a radius up to a few nanometers. There are two types of such

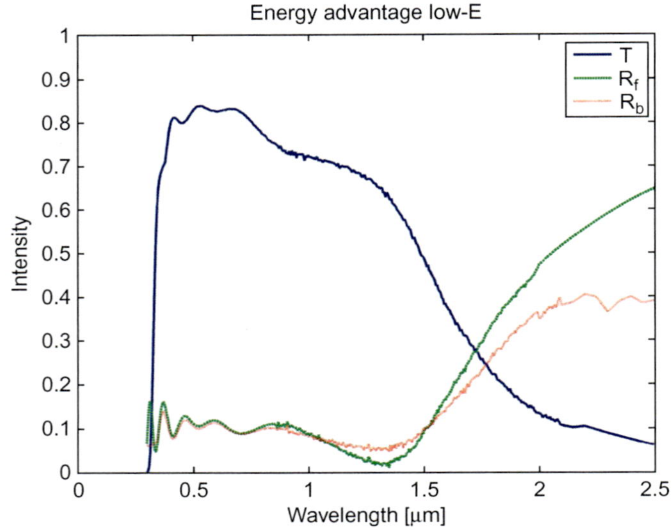

Figure 16.6 Spectral transmittance T and reflectance from the front (coated) side (R_f) and the back side (R_b) for a commercial tin oxide-based coating on glass and optimized for "low-E." From the International Glazing Database (2009).

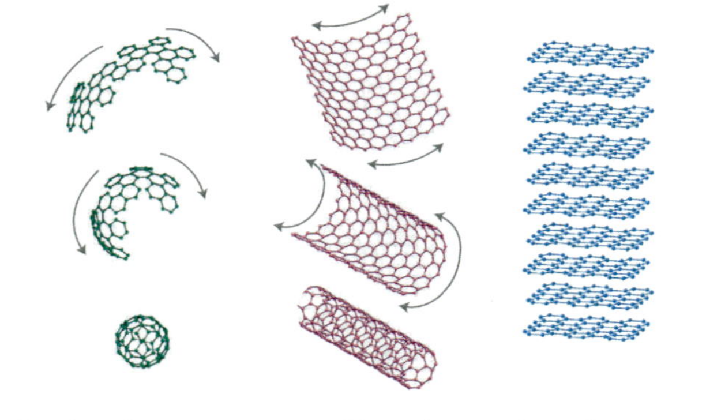

Figure 16.7 Carbon-based nanostructures and how they are formed. From Geim and Novoselov (2007).

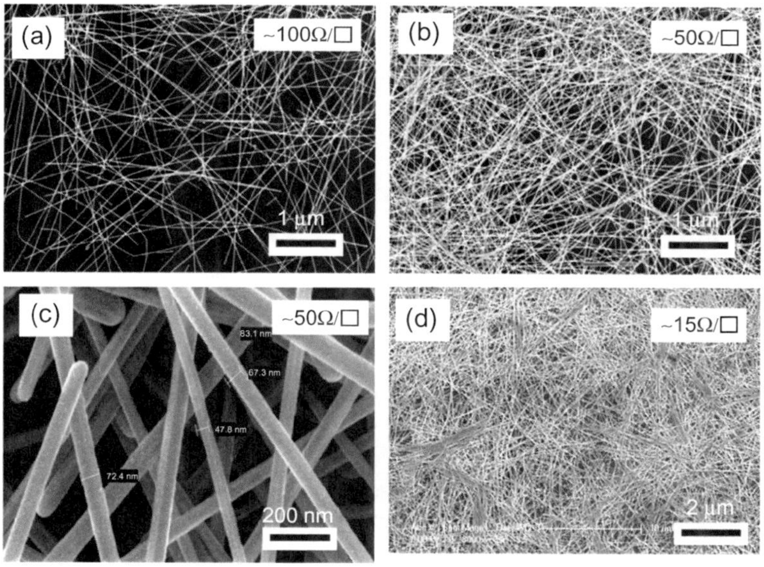

Figure 16.8 Scanning electron micrographs of silver nanowire meshes with the shown resistance.
From Hu et al. (2010).

nanowires: with metallic and semiconducting properties and having electrical conductivities comparable to those of Cu and Si, respectively. Meshes of such nanowires can serve as transparent conductors and have properties approaching—but as yet not exceeding—those of the oxides mentioned above. Graphene is a particularly interesting transparent conductor, which may be possible to mass manufacture (Bae et al., 2010).

Meshes composed of nanorods of silver or other metals are another possibility to make transparent conductors (Du et al., 2014; Hu et al., 2010). They are somewhat optically scattering (Preston et al., 2013), which may limit their usefulness for normal glazings (while scattering can be beneficial for other applications, e.g., in solar cells and light-emitting devices). Large quantities of nanowires, such as those shown in Figure 16.8, can be produced by inexpensive reduction of silver nitrate.

16.4 Chromogenic glazings: thermochromics

The spectrally selective coatings introduced above have been developed during decades and have reached near perfection with regard to optical and thermal performance. However, fenestration with superior eco-efficiency can be accomplished by contemplating other "rules" for our ambient radiation—introduced in Section 16.2.2—specifically that this radiation normally varies during the day and season so that the demand for solar energy ingress through glazings needs to be altered in response to varying demands. This variation can be obtained by traditional techniques based on overhangs, blinds, awnings, etc., but these options are often undesirable since they tend to obstruct the view and may require expensive maintenance. Principally better solutions can be obtained by coatings that not only are spectrally selective but also are able to vary the throughput of visible light and solar energy.

We therefore turn to "*chromogenic*" materials, which are characterized by their ability to have variable optical properties (Granqvist, 1990; Lampert and Granqvist, 1990) and are applied in glazings that frequently are referred to as "switchable," "smart," or "intelligent." Two types of such "smart" glazings are of main interest for eco-efficient buildings: we first consider the *thermochromic* option, which uses materials that respond to temperature changes, and in the following section, we discuss *electrochromic* devices that use electricity.

16.4.1 Basics of vanadium dioxide

Thermochromic materials are conveniently introduced by looking at their electrical properties. Figure 16.9 shows conductivity as a function of temperature for a number of oxides and a few other compounds, as reproduced from Jorgenson and Lee (1986). Abrupt changes take place at well-defined temperatures and signal reversible changes in the crystal structure. Vanadium dioxide, VO_2, is of particular interest, and it is seen that the conductivity varies by several orders of magnitude at a temperature that is not too far from room temperature. Specifically, there is switching between a low-temperature phase with semiconducting properties and large infrared transmittance and a high-temperature phase with metallic-like conductivity and high infrared reflectance; this switching takes place at a temperature around 68 °C.

Vanadium dioxide is the only one of today's known materials that display optical properties that give good prospects to serve as a basis for developing practically useful thermochromic glazings, and it has been attracting attention for many years, as discussed by Babulanam et al. (1987), Gao et al. (2012), Li et al. (2012), Taylor et al. (2013), Hoffmann et al. (2014), and Warwick and Binions (2014). Figure 16.10 illustrates the most salient properties of VO_2-based materials. Left-hand panels show spectral transmittance (upper) and reflectance (lower) for 50 nm thick VO_2 films in the wavelength range appropriate for solar radiation, and it is seen that $T(\lambda)$ is much larger in the semiconducting state than in the metallic state—that is, below and above the phase change temperature of about 68 °C—for $\lambda > 1$ μm. The difference in $T(\lambda)$ between low temperature and high temperature grows for increasing wavelength. This type of variation in $T(\lambda)$ is clearly the desired one, in principle, and a glazing with a thermochromic VO_2 film lets in more energy at low temperature than at high temperature (Saeli et al., 2010a,b). Figure 16.10 also demonstrates that the corresponding curve for $R(\lambda)$ goes up monotonically towards long wavelengths for the metallic state, which is the expected behavior. The right-hand panels in Figure 16.10 refer to nanoparticles of VO_2 and are discussed later.

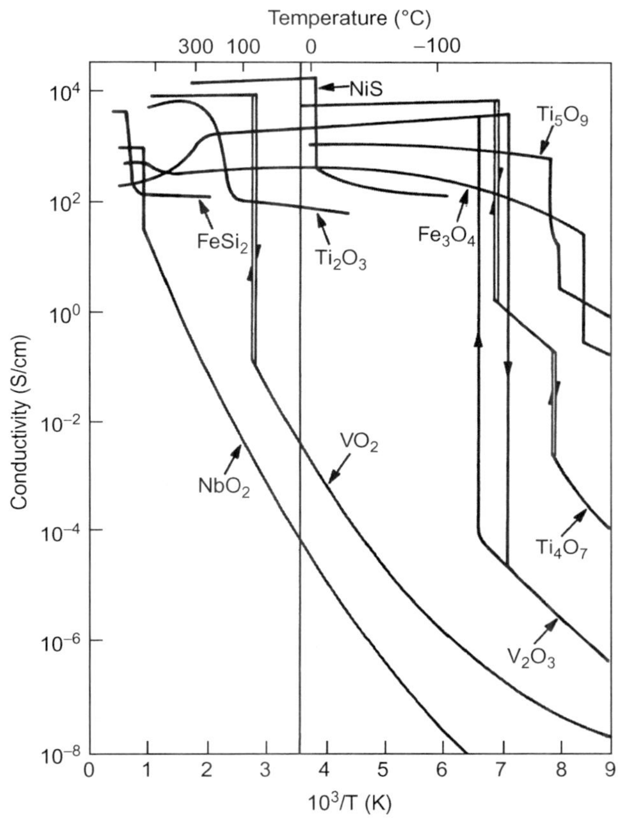

Figure 16.9 Electrical conductivity as a function of reciprocal temperature (lower horizontal axis) and temperature (upper horizontal axis) for several metal-based compounds.
From Jorgenson and Lee (1986).

16.4.2 Towards practically useful vanadium dioxide-based coatings for glazings

Let us now have a closer look at $T(\lambda)$ for the VO_2 film, as illustrated in Figure 16.10. We can identify three challenges for practical implementation of these films on glazings (Li et al., 2012):

(i) The shift between the semiconducting and metallic states takes place at about 68 °C, which obviously is too high for building-related applications.
(ii) The visible transmittance is only about 40%, which is too low for most practical glazings; the film of course would be more transparent if it is thinner than 50 nm, but such films have a smaller modulation of the solar energy transmittance between the semiconducting and metallic states.
(iii) The transmittance modulation is strong only for wavelengths at which the solar irradiation is rather weak (cf. Figure 16.1); this limits the solar energy modulation to roughly 10%, which is too small to be of much practical interest.

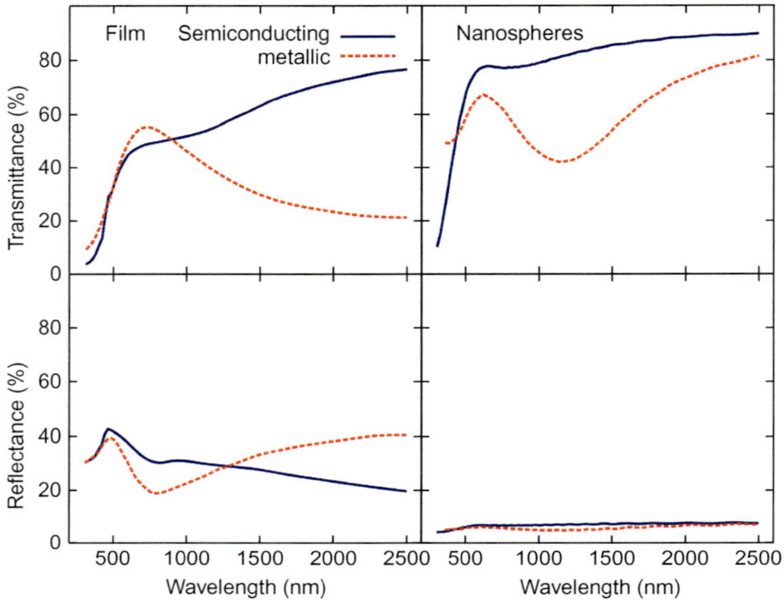

Figure 16.10 Spectral transmittance (upper panels) and reflectance (lower panels) for a 50 nm thick thermochromic VO_2 film (left-hand panels) and for a layer consisting of a dilute dispersion of VO_2 nanospheres, with an equivalent VO_2 thickness of 50 nm, in a medium mimicking typical transparent glass and polymers.

How can these challenges be met? Issue (i) is the easiest, and replacing some percent of the vanadium atoms with tungsten to make $W_xV_{1-x}O_2$ films can decrease the transition between the semiconducting and metallic states to room temperature as illustrated in Figure 16.11, which is based on the work by Mlyuka et al. (2009a), without significantly deteriorating the optical performance.

Issue (ii) is associated with an undesirable optical absorption at short wavelengths, which is clearly seen in Figure 16.10. This absorption is due to the fact that VO_2 has a small optical band gap, and an at least partial solution of this problem can be reached by adding a band gap-widening agent, specifically magnesium. Figure 16.12 depicts $T(\lambda)$ for 50 nm thick films of $Mg_yV_{1-y}O_2$ in semiconducting and metallic states and shows that the addition of Mg not only displaces the high-transmittance region towards shorter wavelengths but also partially erodes the transmittance change between the semiconducting and metallic states at long wavelengths (Mlyuka et al., 2009a). A similar effect was seen recently in sputter-deposited $Zn_yV_{1-y}O_2$ films (Jiang et al., 2014), and transmittance increase can be accomplished also by fluorine incorporation (Burkhardt et al., 2002; Khan and Granqvist, 1989; Kiri et al., 2011). The shift at short wavelengths is indeed due to a widening of the optical band gap, as clarified recently for $Mg_yV_{1-y}O_2$ (Li et al., 2013a). The semiconductor-metal phase

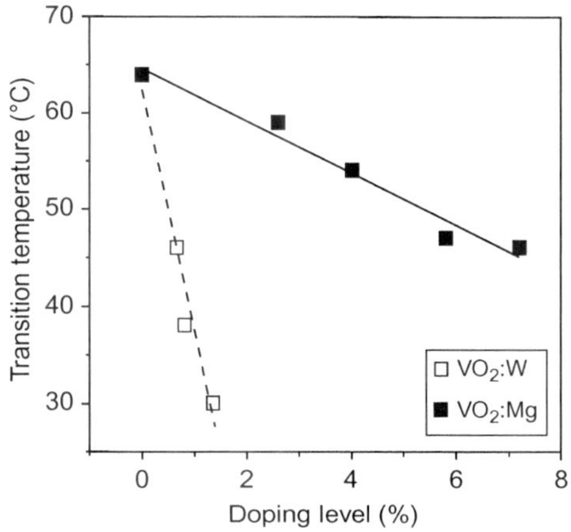

Figure 16.11 Temperature for the transition between semiconducting and metallic states in thermochromic thin films of $W_xV_{1-x}O_2$ and $Mg_yV_{1-y}O_2$, where x and y are referred to as "doping level" and lines were drawn for convenience.
From Mlyuka et al. (2009a).

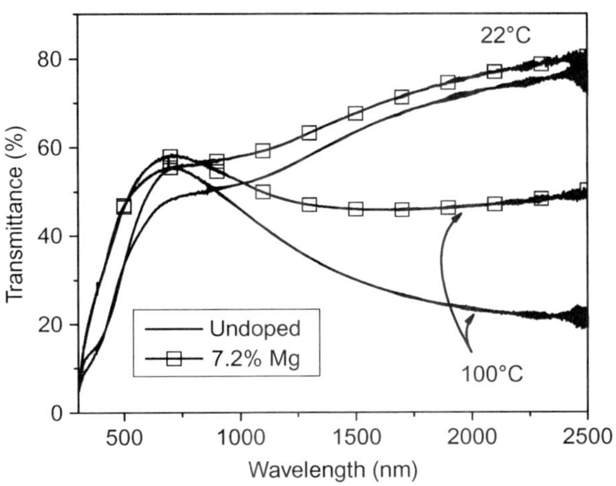

Figure 16.12 Spectral transmittance of 50 nm thick films of VO_2 and $Mg_{0.072}V_{0.928}O_2$ in semiconducting (at 22 °C) and metallic (at 100 °C) states.
From Mlyuka et al. (2009a).

transition temperature drops in proportion with the Mg content, as reported in Figure 16.11, though not as abruptly as for tungsten addition. Figure 16.13 shows luminous transmittance—obtained by averaging the spectral data over the sensitivity curve for the human eye (cf. Figure 16.1)—and shows that this property goes from 39% to 51% when the Mg content is increased from zero to 7.2% (Mlyuka et al., 2009a).

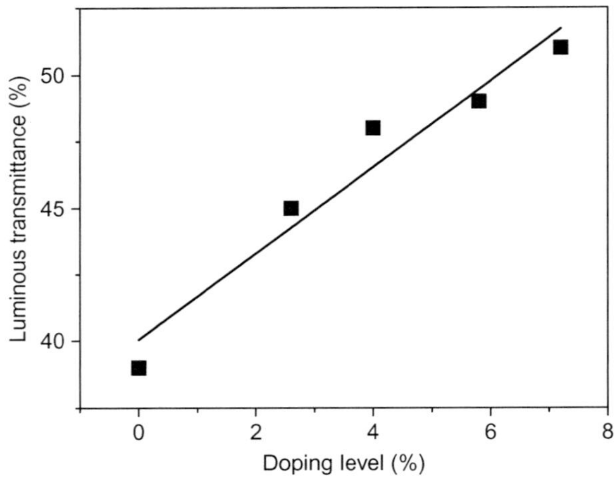

Figure 16.13 Luminous transmittance as a function of Mg content ("doping level") in 50 nm thick films of $Mg_yV_{1-y}O_2$.
From Mlyuka et al. (2009a).

16.4.3 Nanothermochromism in composite materials with VO_2-based nanoparticles

The limited modulation of the solar energy transmittance between the semiconducting and metallic states in VO_2 has been the hardest nut to crack in order to approach practically interesting thermochromic glazings. However, it was recently realized by Li et al. (2010, 2011) that a layer containing VO_2 nanoparticles, rather than a continuous VO_2 film, can lead to a radical improvement. Calculations have shown that the particle size should be of the order of 20 nm or less in order not to produce optical scattering ("haze") (Laaksonen et al., 2014). Such "nanothermochromism" was introduced in the right-hand panels in Figure 16.10, which report $T(\lambda)$ and $R(\lambda)$ for a 5 μm thick layer of a transparent medium, with properties representative of typical glasses and polymers, containing 1 vol.% of well-dispersed VO_2 nanospheres with diameters much less than 20 nm. This choice of parameters gives an equivalent VO_2 thickness of 50 nm, which makes the data for the nanospheres and the film straightforward to compare. A number of important features distinguish the two sets of data:

(i) The material containing nanospheres is much more transparent than the corresponding film.
(ii) The metallic nanoparticles have a strong transmittance minimum in the $0.7 < \lambda < 1.5$ μm range; this significantly limits the solar transmittance, which is strong at these wavelengths, while the optical properties are hardly affected for visible light.
(iii) The nanospheres absorb rather than reflect.

The striking transmittance minimum (absorption maximum) at infrared wavelengths is due to plasmon resonance at the wavelength $\lambda_p \approx 1.2$ μm (cf. Section 16.3.3), which is almost ideally located for solar control (Lopez et al., 2002). VO_2-based nanoparticles and nanorods can be prepared by numerous chemical and physical techniques. Mg-containing VO_2 nanoparticles were demonstrated recently and displayed enhanced visible transmittance (Zhou et al., 2013). Figure 16.14 illustrates nanorods prepared by Li et al. (2013b) through sputter deposition onto heated glass substrates

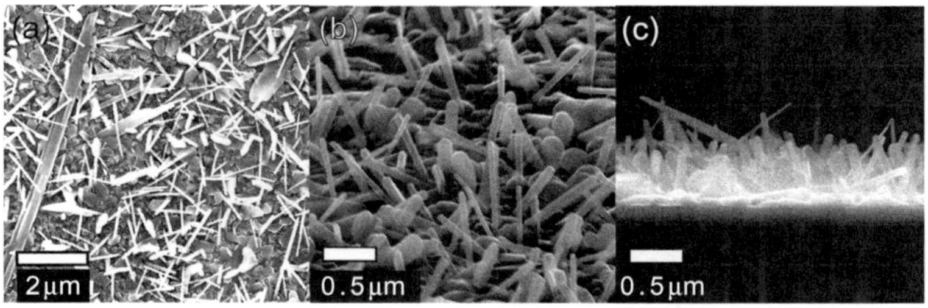

Figure 16.14 Scanning electron micrographs for a VO_2 layer deposited onto glass heated to ~460 °C and having an equivalent thickness of 102 nm. (a) Top view, (b) imaging at 70° between electron beam and sample normal, and (c) cross-sectional view. Note that the magnifications differ among the images.
From Li et al. (2013b).

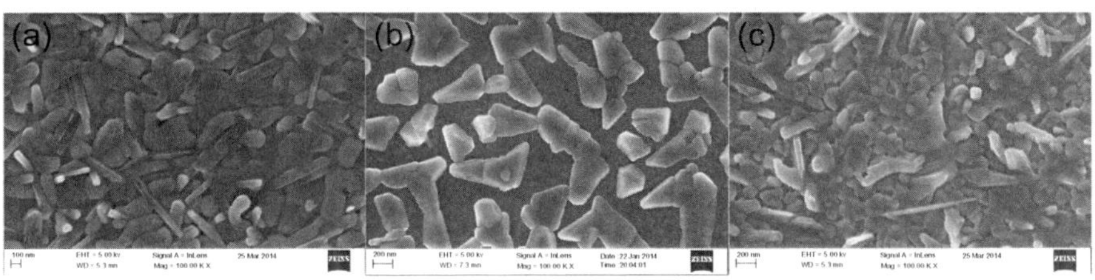

Figure 16.15 Scanning electron micrographs for VO_2 layers deposited onto glass heated to ~390 °C (a) and analogously to glass precoated with films of In oxide (b) and Sn oxide (c).
From Montero (2014).

under judiciously controlled conditions. Many parameters affect the nanofeatures, one of them being the substrate material, and Figure 16.15 depicts VO_2 nanoparticles prepared by sputtering onto surfaces of glass and transparent In oxide and Sn oxide under otherwise identical conditions (Montero, 2014).

16.4.4 Approximate performance limits for thermochromic glazings

It should be apparent from the discussion so far that various combinations of luminous transmittance and solar energy modulation can be accomplished in different VO_2-based materials. Figure 16.16, from Li et al. (2014), presents an overview over the current state of the art. Pure VO_2 films give the lowest values on both parameters, but the situation can be improved by antireflection (AR) coatings, analogously to what was discussed in Section 16.3.2 for metal-based coatings. Particularly good properties were obtained with multilayers of TiO_2 and VO_2 (Mlyuka et al., 2009b;

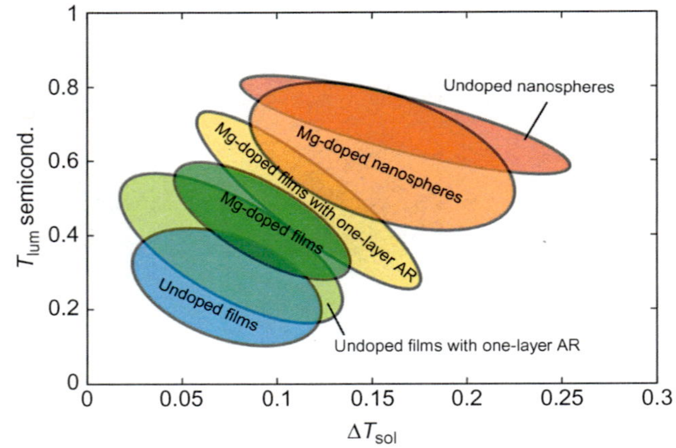

Figure 16.16 Schematic illustration of performance limits of the shown VO_2-based materials with regard to thermochromic glazings. Specifically shown are luminous transmittance T_{lum} for semiconducting films and particles and solar energy transmittance modulation ΔT_{sol} for thin films—including antireflection (AR) coated films—and dilute nanoparticle composites. From Li et al. (2014).

Chen et al., 2011). Cone-shaped "bio-inspired" surfaces offer another possibility to enhance the luminous transmittance (Taylor et al., 2013). Mg-containing VO_2 films have superior properties, which can be boosted by antireflection. However, the champion performance is achieved with VO_2 nanoparticles that, for example, can display ~60% luminous transmittance combined with more than 20% solar energy modulation. Interestingly, Mg-containing nanoparticles do not yield better performance than nanoparticles of pure VO_2. The reason for this seems to be that the Mg partly removes the solar modulation, as shown in Figure 16.12, to an extent that is not compensated by the better transmittance at shorter wavelengths. Minor improvements of the performance are conceivable by selecting nanoparticles with optimized shape and orientation (Li et al., 2010) and by having core-shell configurations (Li et al., 2011).

Concerning practical implementation, VO_2-based thin films can be applied in IGUs in a manner similar to that for metal-based coatings with static properties. Nanoparticles might be dispersed in polymers and used for glass lamination, as sketched in Figure 16.17.

16.4.5 On the durability of vanadium dioxide-based materials

Any coating that is devised for glazings must be durable enough to last for many years without losing its performance more than marginally. This requirement needs careful consideration for thermochromic VO_2-based materials since vanadium pentoxide, that is, V_2O_5, is the thermodynamically stable oxide.

Figure 16.18 shows results from some recent studies of 80 nm thick VO_2 films subjected to heating in dry air at 300 °C for one hour in the work by Ji et al. (2014).

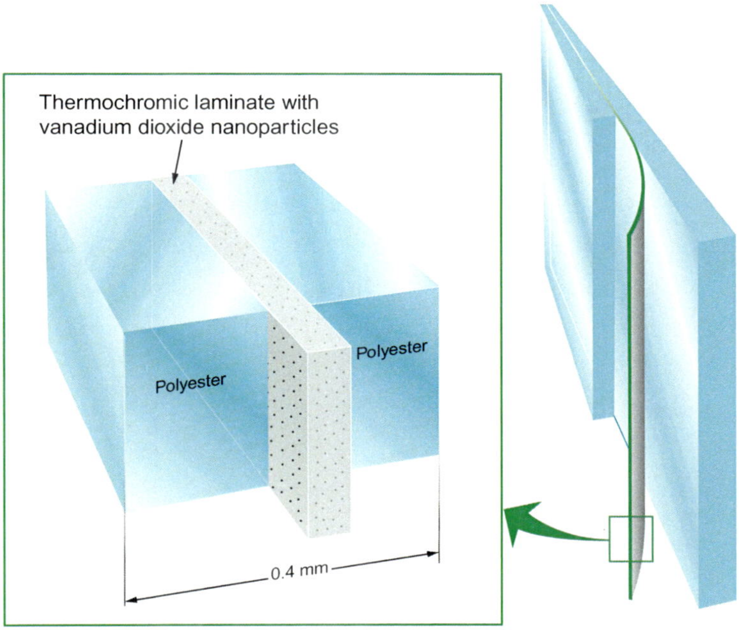

Figure 16.17 Construction principle for a foil incorporating thermochromic nanoparticles.

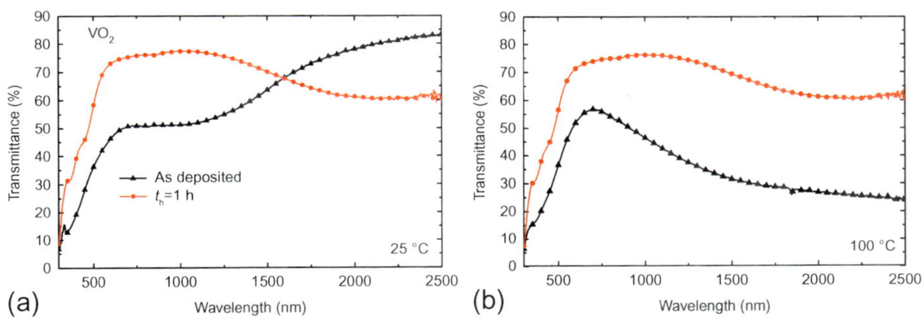

Figure 16.18 Spectral transmittance for an 80 nm thick VO_2 film in as-deposited state and after heating at 300 °C for one hour. Data were taken for films in semiconducting (at 25 °C) and metallic (at 100 °C) states.
From Ji et al. (2014).

The as-deposited film displays the expected thermochromic performance in its spectral transmittance (cf. Figures 16.10 and 16.12). However, the heat-treated film exhibits entirely different data on $T(\lambda)$, and an examination of the spectral features reveals that it has been converted to nonthermochromic V_2O_5 (Lykissa et al., 2014; Talledo and Granqvist, 1995). This transformation would be very much slower at the temperatures

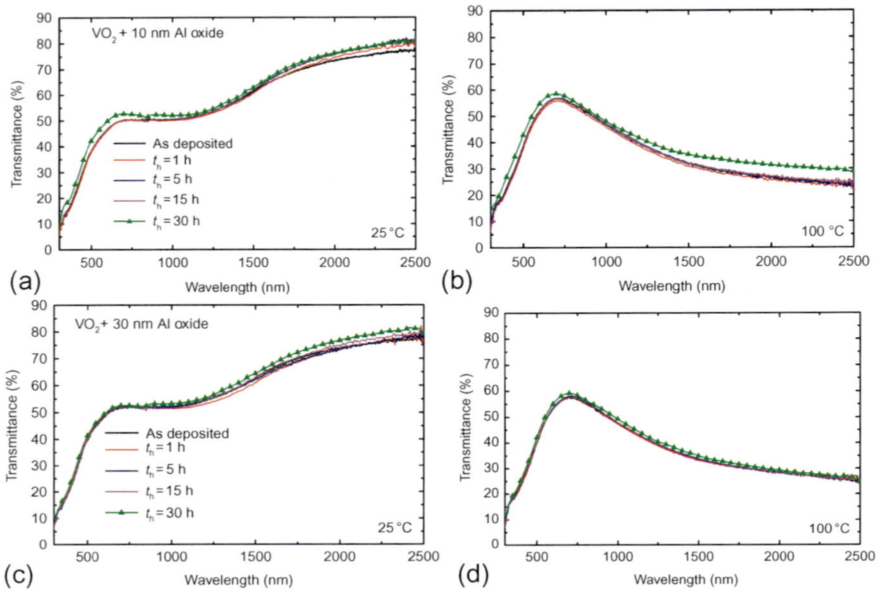

Figure 16.19 Spectral transmittance for 80 nm thick VO_2 films, coated with 10 nm (panels a and b) and 30 nm (panels c and d) of Al oxide, in as-deposited state and after heating at 300 °C for the shown durations t_h. Data were taken for films in semiconducting (at 25 °C) and metallic (at 100 °C) states.
From Ji et al. (2014).

of concern for practical IGUs, but the data in Figure 16.18 nevertheless give clear signs that the VO_2-based materials must be protected against further oxidation.

Figure 16.19 contains spectral data for analogous VO_2 films with top coats of 10 and 30 nm of aluminum oxide and heat-treated at 300 °C for durations in the range $1 < t_h < 30$ h. It is apparent that the top coat provides excellent protection, and only the thinnest Al oxide and the longest heating period gave some noticeable, though very small, decay of the thermochromic performance. Multilayer coatings are readily produced by several deposition techniques, and core-shell structures with cores of VO_2 and shells of SiO_2 or TiO_2 have been demonstrated several times (Li et al., 2013c,d), which means that the data in Figure 16.19 give good prospects for long-term applications of VO_2-based materials in glazings.

16.5 Chromogenic glazings: electrochromics

Electrochromic glazings are presently (2014) making their appearance on a limited scale and are expected to become more common in the future as discussed by Baetens et al. (2010), Granqvist (2012b, 2014), and Jelle et al. (2012). These glazings are able to provide energy efficiency together with indoor comfort, as shown in several studies (Azens and Granqvist, 2003; Clear et al., 2006; Lee et al., 2006a,b, 2012; Loonen et al., 2014; Tavares et al., 2014) and discussed in some detail in Chapter 18. The comfort issue is multifaceted and includes the maintenance of a narrow temperature interval—which tends to be energy-intensive (Arens et al., 2010; Roaf et al., 2010)—and glare suppression.

16.5.1 General device design

The principle of an electrochromic device is illustrated in Figure 16.20 (Bayrak Pehlivan et al., 2014; Granqvist, 1995). A transparent electrolyte, for example, an ion-conducting polymer, is located in the center and joins two nanoporous oxide films, for example, of tungsten oxide and nickel oxide (Niklasson and Granqvist, 2007). This three-layer stack is positioned between transparent electrical conductors; any one of the alternatives in Section 16.3 above can be used, but an oxide-based coating may be preferable since it is rugged. The application of a voltage between the transparent conductors—typically a few volt DC—shuttles charge between the tungsten oxide and nickel oxide layers. The electrical requirements are suitable for solar cell powering (Lampert, 2003). The function of the electrochromic device is similar to that of an

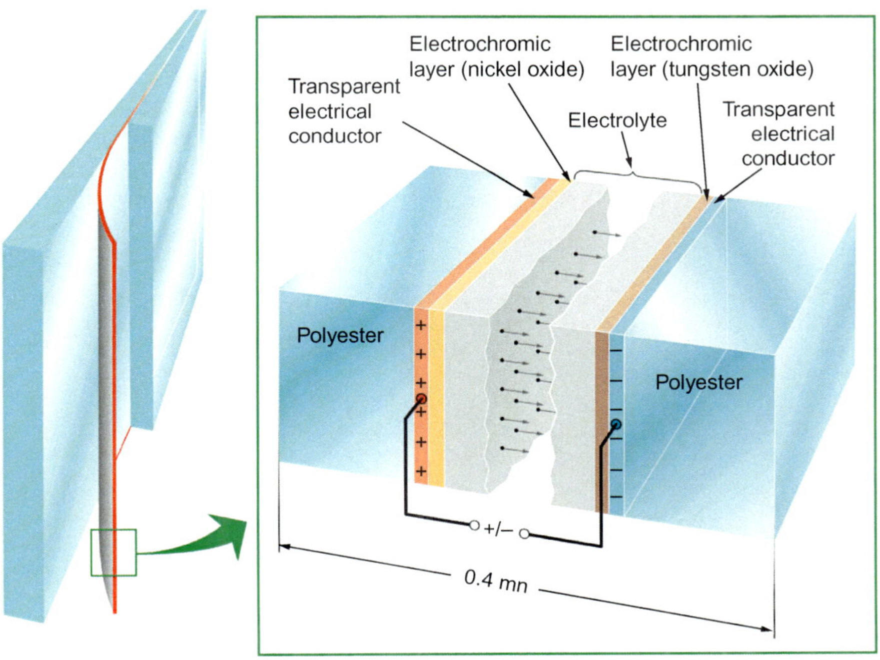

Figure 16.20 Construction principle for an electrochromic foil-type device implemented as a laminate between glass panes.
From Bayrak Pehlivan et al. (2014).

electrical battery, with tungsten oxide and nickel oxide serving as cathode and anode, respectively, and it is instructive to regard the electrochromic device as an electrical battery with optical absorption depending on the charging state. The device absorbs across the solar spectrum when the mobile charge resides in the tungsten oxide and is transparent when this charge is in the nickel oxide. The optical properties can remain unchanged—dark, bleached, or intermediary—for days under open-circuit conditions. The optical changes are gradual and occur at a rate that depends on the size of the device; the response time can be tens of minutes for a large glazing (whereas an area of a few square centimeters may darken and bleach in seconds). Time constants of the order of minutes are convenient since they allow the eye to light-adapt, and the optical modulation of the glazing therefore may be barely noticeable.

16.5.2 How does it work? The underlying physics and electrochemistry

Electrochromism occurs in some oxides that can serve as mixed conductors for electrons and ions (Granqvist, 1995). Tungsten oxide is a particularly well-known electrochromic material whose atomic arrangement is shown in Figure 16.21. This oxide consists of corner-sharing octahedral building blocks, each with a centrally positioned tungsten atom surrounded by six oxygen atoms so that the overall composition is WO_3. The electrochromic oxides may be oxygen-deficient, which means that some portions of the octahedra are edge-sharing, as also depicted in Figure 16.21. The shown atomic arrangement leads to an extended three-dimensional "tunnel structure" between the octahedra, and these "tunnels" are wide enough to serve as conduits for mobile ions in an electrical field. The most appropriate mobile species are hydrogen ions (H^+; "protons") and lithium ions (Li^+). The nanoporous nature of the oxide, and therefore its

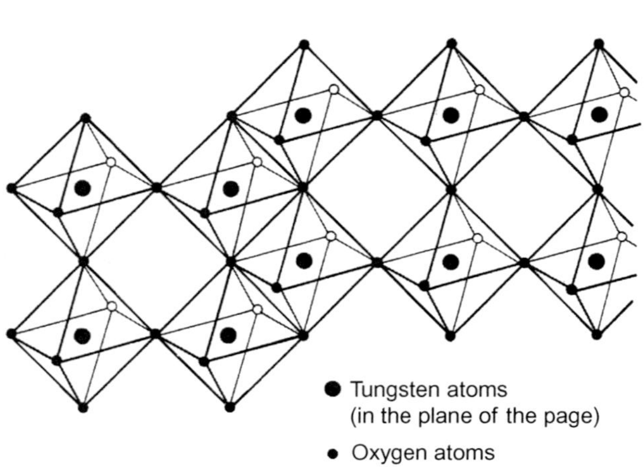

Figure 16.21 Schematic of corner-sharing and edge-sharing octahedra in tungsten oxide. From Granqvist (1995).

ability to sustain ion transport, can be boosted if it is prepared under conditions rendering it highly disordered on a length scale that is larger than the one of the octahedra.

The ion insertion into tungsten oxide (WO_3) is accompanied by inflow of charge-balancing electrons from the transparent electrodes in the electrochromic device. These electrons change the valency of the tungsten atoms, and some of them are transformed from W^{6+} to W^{5+}. A photon that impinges on a tungsten oxide layer can have sufficient energy to transfer an electron between two neighboring tungsten ions (on sites *i* and *j*, say). This process can be written, schematically, as

$$W_i^5 + W_j^6 + \text{photon} \rightarrow W_i^6 + W_j^{5+},$$

which means that the photon's energy is used to transfer an electron so that optical absorption takes place locally. The intensity of the overall absorption in the tungsten oxide is then governed by the density of the inserted electrons, which in its turn is proportional to the content of inserted ions, that is, the charging state.

Tungsten oxide is one of the oxides that become optically absorbing under charge *insertion*, and oxides of molybdenum, titanium, and niobium are alternatives (albeit with somewhat lower optical efficiency). There are also oxides that darken under charge *extraction*, such as nickel oxide and iridium oxide. It is therefore advantageous to have electrochromic devices that combine oxides of the two types, for example, tungsten oxide and nickel oxide, since charge transfer from nickel oxide into tungsten oxide darkens both of them, while charge transfer in the opposite direction bleaches them both. Such a device design was shown in Figure 16.20 above.

16.5.3 Practical device types and some data on a foil-based option

Figure 16.20 also illustrates how electrochromic glazings can be devised in practice: one can start with two flexible plastic foils, for example, of polyester, and coat them with two thin films each; this is feasible by inexpensive roll-to-roll deposition (Bishop, 2011). The thin oxide films, which serve as anode and cathode, are then joined by the polymer electrolyte, and the entire electrochromic device is used as a laminate connecting two glass panes, as indicated in the left-hand part of Figure 16.20. However, this is not the only way to construct an electrochromic device, and another "monolithic" alternative that is currently being commercialized by some major glass producers is to apply the full five-layer stack shown in Figure 16.20 directly onto the surface of a single glass pane by sequential deposition technology. This means that the electrolyte layer in the "monolithic" devices has to be a thin film that guarantees that the anodes and cathodes are reliably separated. The five-layer coating is a delicate object and must be protected against mechanical and chemical damage, most conveniently by integrating the coated glass in an IGU immediately after the thin-film stack has been produced.

The laminated and the all-thin-film "monolithic" approaches differ in several regards. Thus, the laminated construction allows the production of a generic electrochromic foil that can be cut to size and adapted, in principle, to any glazing construction of any shape. It thus separates the electrochromic functionality from the window manufacturing, which may be advantageous and leads to entirely different business models for implementation of electrochromic glazings produced according to the two approaches. Lamination may be desirable also for other reasons and can yield spall shielding, burglar protection, acoustic damping, and more. Even infrared-absorbing or thermochromic nanoparticles may be incorporated in the laminate.

Figure 16.22 illustrates the transmittance modulation of a small electrochromic device and shows that the midluminous transmittance goes between 17% and 68%. This operation can be extended to tens of thousands of cycles. Figure 16.23 shows a full-size ($0.8 \times 1.8\ m^2$) electrochromic window in an office environment.

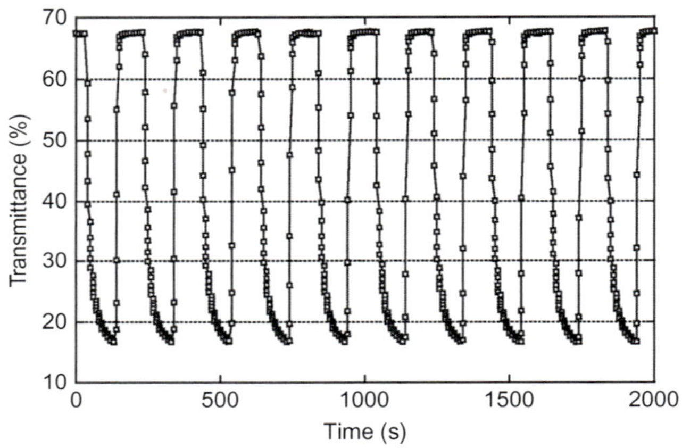

Figure 16.22 Modulation of the transmittance of visible light during repeated charging–discharging of a small electrochromic device of the kind shown in Figure 16.20.

Figure 16.23 Part of an office equipped with laminated electrochromic glazing, with one window being transparent and two windows in their dark state.

16.6 Comments and conclusions

This chapter covered a number of ways to improve glazings and make them applicable to eco-efficient fenestration. In particular, the discussion encompassed coatings with static optical properties devised for solar control and low thermal emittance and coatings and nanoparticle composites with dynamic optical properties and thermochromic or electrochromic features.

The technological advances that were presented above are not the only ones of interest for eco-efficient buildings. Other recent developments concern photocatalytic coatings, usually based on titanium dioxide. Ultraviolet solar irradiation activates these surfaces and allows them to decompose organic pollutants and cause almost complete wetting (Fujishima et al., 2008; Smith and Granqvist, 2010). Both of these effects contribute to making the surface easy to clean. The same type of coatings is able to kill microorganisms and have antiseptic features.

"Low-E" coatings facing the space between glass panes in IGUs improve their thermal insulation. This space should be filled with a gas, such as argon or krypton, which reduces heat conduction and convection. Even better thermal insulation can be achieved with vacuum between the panes, which requires spacers to prevent implosion. Such spacers can be made from tiny glass pillars or from silica aerogel (Baetens et al., 2011; Buratti and Moretti, 2013). The latter material consists of interconnected SiO_2 nanoparticles with sizes down to one nanometer; the overall porosity can be larger than 99%, so the weight is extremely small.

Improved thermal insulation is an asset, but it may lead to some (perhaps) unexpected side effects. Thus, if a highly insulated glazing is exposed towards the clear night sky, it can be cooled as a consequence of the imbalance between outgoing radiation from the glass and incoming radiation from the sky. This imbalance ensues from the "atmospheric window" at the $8<\lambda<13$ μm wavelength interval, noted in Figure 16.1, which can produce a continuous heat loss as large as about 100 W m^{-2} at ambient temperature and yield nighttime temperature drops as large as ~10 °C for surfaces that are fully exposed to the sky (Granqvist and Hjortsberg, 1981). The glazing can reach temperatures below the dew point of the air, implying that water condensation takes place. This effect is well known for buildings in cold regions and usually occurs before sunrise. A good remedy for the dew formation is to put a "low-E" coating on the outermost surface of the glazing, as was demonstrated for car windscreens already in the 1980s (Hamberg et al., 1987). Another solution is to employ a super-hydrophilic surface on the glazing, which serves to make liquid water invisible. Figure 16.24 shows the effect of a super-hydrophilic titanium dioxide coating on the outermost surface of a window (Smith and Granqvist, 2010).

Finally, we note that the building sector consumes more energy than any other sector in society, which largely is an effect of poor—but well-entrenched—building practices. There are many possibilities to radically cut the energy expenditure, and building envelopes that function in harmony with the nature's normal energy flows are a key (Richter et al., 2008; Smith and Granqvist, 2010). Such a strategy involves spectrally selective surfaces on glazings and other surfaces that reflect, transmit,

Figure 16.24 Photograph of a window in a private house in Uppsala, Sweden. Both parts of the window are triple-glazed and have a center-of-glass thermal conductance as small as 0.65 $Wm^{-2}K^{-1}$ (cf. Figure 16.2). The window is in a northeast-facing facade with an open view towards the sky. The left-hand part of the window has an outer glass with a super-hydrophilic TiO_2-based coating, and the right-hand part has standard uncoated glass.
From Smith and Granqvist (2010).

absorb, and emit controlled amounts of energy in well-defined wavelength ranges (and possibly directions), and further improvements can be obtained with chromogenic materials, which are able to modulate the properties of the building envelope to compensate for varying temperatures, solar radiation flows, and illumination levels. It is important to note that the new technologies can be implemented while the indoor comfort is improved.

The developments of building envelopes that presently are in progress may have far-reaching implications and involve a paradigmatic shift: thus, future buildings can serve as dynamic entities that adapt and adjust to varying demands for cooling and heating rather than being today's static blocks of concrete, glass, wood, etc. Another development line can lead to lightweight climate shells surrounding buildings and blurring the zone between indoors and outdoors. However, none of these changes will be rapid, and the timescale for introducing new and benign building technologies concerns decades or more.

Acknowledgment

Financial support during the writing of this chapter was received from the European Research Council under the European Community's Seventh Framework Program (FP7/2007-2013)/ERC grant agreement no. 267234 (GRINDOOR).

References

Arens, A., Humphreys, M.A., de Dear, R., Zhang, H., 2010. Are 'class A' temperature requirements realistic or desirable? Build. Environ. 45, 4–10.

Azens, A., Granqvist, C.G., 2003. Electrochromic smart windows: energy efficiency and device aspects. J. Solid State Electrochem. 7, 64–68.

Babulanam, S.M., Eriksson, T.S., Niklasson, G.A., Granqvist, C.G., 1987. Thermochromic VO_2 films for energy-efficient windows. Sol. Energy Mater. Sol. Cells 16, 347–363.

Bae, S., Kim, H., Lee, Y., Xu, X., Park, J.-S., Zheng, Y., Balakrishnan, J., Lei, T., Kim, H.R., Song, Y.I., Kim, Y.-J., Kim, K.S., Özyilmaz, B., Ahn, J.-H., Hong, B.H., Iijima, S., 2010. Roll-to-roll production of 30-inch graphene films for transparent electrodes. Nat. Nanotechnol. 5, 574–578.

Baetens, R., Jelle, B.P., Gustavsen, A., 2010. Properties, requirements and possibilities of smart windows for dynamic daylight and solar energy control in buildings: a state-of-the-art review. Sol. Energy Mater. Sol. Cells 94, 87–105.

Baetens, R., Jelle, B.P., Gustavsen, A., 2011. Aerogel insulation for building application: a state-of-the-art review. Energy Build. 43, 761–769.

Bayrak Pehlivan, İ., Marsal, R., Pehlivan, E., Runnerstrom, E.L., Milliron, D.J., Granqvist, C. G., Niklasson, G.A., 2014. Electrochromic devices with polymer electrolytes functionalized by SiO_2 and In_2O_3:Sn nanoparticles: rapid coloration/bleaching dynamics and strong near-infrared absorption. Sol. Energy Mater. Sol. Cells 126, 241–247.

Bishop, C.A., 2011. Vacuum Deposition onto Webs, Films and Foils, second ed. William Andrew, Waltham, MA, USA.

Buratti, C., Moretti, E., 2013. Silica nanogel for energy-efficient windows. In: Pacheco-Torgal, F., Diamanti, M.V., Nazari, A., Granqvist, C.G. (Eds.), Nanotechnology in Eco-Efficient Construction. Woodhead, Cambridge, UK, pp. 207–235.

Burkhardt, W., Christmann, T., Franke, S., Kriegseis, W., Meister, D., Meyer, B.K., Niessner, W., Schalch, D., Scharmann, A., 2002. Tungsten and fluorine co-doping of VO_2 films. Thin Solid Films 402, 226–231.

Chen, Z., Gao, Y., Kang, L., Du, J., Zhang, Z., Luo, H., Miao, H., Tan, G., 2011. VO_2-based double-layered films for smart windows: optical design, all-solution preparation and improved properties. Sol. Energy Mater. Sol. Cells 95, 2677–2684.

Clear, R.D., Inkarojrit, V., Lee, E.S., 2006. Subject responses to electrochromic windows. Energy Build. 38, 758–779.

Du, J., Pei, S., Ma, L., Cheng, H.-M., 2014. Carbon nanotube- and graphene-based transparent conductive films for optoelectronic devices. Adv. Mater. 26, 195–1991.

Freedonia Group, 2013. World Flat Glass to 2016: Industry Market Research, Market Share, Market Size, Sales, Demand Forecast, Market Leaders, Company Profiles, Industry Trends. The Freedonia Group, Cleveland, OH, USA. http://www.freedoniagroup.com/World-Flat-Glass.html.

Fujishima, A., Zhang, X., Tryk, D.A., 2008. TiO_2 photocatalysis and related phenomena. Surf. Sci. Rep. 63, 515–582.

Gao, Y., Luo, H., Zhang, Z., Kang, L., Chen, Z., Du, J., Kanehira, M., Cao, C., 2012. Nanoceramic VO_2 thermochromic smart glass: a review on progress in solution processing. Nano Energy 1, 221–246.

Geim, A.K., Novoselov, K.S., 2007. The rise of graphene. Nat. Mater. 6, 183–191.

Gläser, H.J., 2000. Large Area Glass Coating. von Ardenne Anlagentechnik, Dresden, Germany.

Granqvist, C.G., 1981. Radiative heating and cooling with spectrally selective surfaces. Appl. Optics 20, 2606–2615.

Granqvist, C.G., 1990. Chromogenic materials for energy control of large-area windows. Crit. Rev. Solid State Mater. Sci. 16, 291–308.

Granqvist, C.G., 1991. Energy-efficient windows: present and forthcoming technology. In: Granqvist, C.G. (Ed.), Materials Science for Solar Energy Conversion Systems. Pergamon, Oxford, UK, pp. 106–167.

Granqvist, C.G., 1995. Handbook of Inorganic Electrochromic Materials. Elsevier, Amsterdam, The Netherlands.

Granqvist, C.G., 2007. Transparent conductors as solar energy materials: a panoramic review. Sol. Energy Mater. Sol. Cells 91, 1529–1598.

Granqvist, C.G., 2012a. Preparation of thin films and nanostructured coatings for clean tech applications: a primer. Sol. Energy Mater. Sol. Cells 99, 166–175.

Granqvist, C.G., 2012b. Oxide electrochromics: an introduction to devices and materials. Sol. Energy Mater. Sol. Cells 99, 1–13.

Granqvist, C.G., 2013. Thin films and nanostructured coatings for eco-efficient buildings. In: Pacheco-Torgal, F., Diamanti, M.V., Nazari, A., Granqvist, C.G. (Eds.), Nanotechnology in Eco-Efficient Construction. Woodhead, Cambridge, UK, pp. 161–187.

Granqvist, C.G., 2014. Electrochromics for smart windows: oxide-based thin films and devices. Thin Solid Films 564, 1–38.

Granqvist, C.G., Hjortsberg, A., 1981. Radiative cooling to low temperatures: general considerations and application to selectively emitting SiO films. J. Appl. Phys. 52, 4205–4220.

Hamberg, I., Granqvist, C.G., 1986. Evaporated Sn-doped In_2O_3 films: basic optical properties and applications to energy-efficient windows. J. Appl. Phys. 60, R123–R159.

Hamberg, I., Svensson, J.S.E.M., Eriksson, T.S., Granqvist, C.G., Arrenius, P., Norin, F., 1987. Radiative cooling and frost formation on surfaces with different thermal emittance: theoretical analysis and practical experience. Appl. Optics 26, 2131–2136.

Hoffmann, S., Lee, E.S., Clavero, C., 2014. Examination of the technical potential of near-infrared switching thermochromic windows for commercial building applications. Sol. Energy Mater. Sol. Cells 123, 65–80.

Hu, L., Kim, H.S., Lee, J.-Y., Peumans, P., Cui, Y., 2010. Scalable coating and properties of transparent, flexible, silver nanowire electrodes. ACS Nano 4, 2955–2963.

International Glazing Database, 2009. Lawrence Berkeley National Laboratory, Berkeley, CA, USA; http://windows.lbl.gov/software/window/window.html.

Jelle, B.P., 2013. Solar radiation glazing factors for window panes, glass structures and electrochromic windows in buildings: measurement and calculation. Sol. Energy Mater. Sol. Cells 116, 291–323.

Jelle, B.P., Hynd, A., Gustavsen, A., Arasteh, D., Goudey, H., Hart, R., 2012. Fenestration of today and tomorrow: a state-of-the-art review and future research opportunities. Sol. Energy Mater. Sol. Cells 96, 1–28.

Ji, Y.-X., Li, S.-Y., Niklasson, G.A., Granqvist, C.G., 2014. Durability of thermochromic VO_2 thin films under heating and humidity: effect of Al oxide top coatings. Thin Solid Films 562, 568–573.

Jiang, M., Bao, S., Cao, X., Li, Y., Li, S., Zhou, H., Luo, H., Jin, P., 2014. Improved luminous transmittance and diminished yellow color in VO_2 energy efficient smart thin films by Zn doping. Ceram. Int. 40, 6331–6334.

Jorgenson, G.V., Lee, J.C., 1986. Doped vanadium oxide for optical switching films. Sol. Energy Mater. 14, 205–214.

Khan, K.A., Granqvist, C.G., 1989. Thermochromic sputter-deposited vanadium oxyfluoride coatings with low luminous absorptance. Appl. Phys. Lett. 55, 4–6.

Kiri, P., Warwick, M.E.A., Ridley, I., Binions, R., 2011. Fluorine doped vanadium dioxide thin films for smart windows. Thin Solid Films 520, 1363–1366.

Laaksonen, K., Li, S.-Y., Puisto, S.R., Rostedt, N.K.J., Ala-Nissila, T., Granqvist, C.G., Nieminen, R.M., Niklasson, G.A., 2014. Nanoparticles of TiO_2 and VO_2 in dielectric media: conditions for low optical scattering, and comparison between effective medium and four-flux theories. Sol. Energy Mater. Sol. Cells 130, 132–137.

Lampert, C.M., 2003. Large-area smart glass and integrated photovoltaics. Sol. Energy Mater. Sol. Cells 76, 489–499.

Lampert, C.M., Granqvist, C.G. (Eds.), 1990. Large-area chromogenics: materials and devices for transmittance control. In: SPIE Institutes for Advanced Optical Technologies, vol. 4, SPIE Optical Engineering Press. Bellingham, WA, USA.

Lansåker, P.C., Backholm, J., Niklasson, G.A., Granqvist, C.G., 2009. TiO_2/Au/TiO_2 multilayer thin films: novel metal-based transparent conductors for electrochromic devices. Thin Solid Films 518, 1225–1229.

Lee, E.S., DiBartolomeo, D.L., Selkowitz, S.E., 2006a. Daylighting control performance of a thin-film ceramic electrochromic window: field study results. Energy Build. 38, 30–44.

Lee, E.S., Selkowitz, S.E., Clear, R.D., DiBartolomeo, D.L., Klems, J.H., Fernandes, L.L., Ward, G.J., Inkarojrit, V., Yazdanian, M., 2006b. Advancement of Electrochromic Windows, California Energy Commission, PIER, Publication number CEC-500–2006–052.

Lee, E.S., Claybaugh, E.S., LaFrance, M., 2012. End user impacts of automated electrochromic windows in a pilot retrofit application. Energy Build. 47, 267–284.

Li, S.-Y., Niklasson, G.A., Granqvist, C.G., 2010. Nanothermochromics: calculations for VO_2 nanoparticles in dielectric hosts show much improved luminous transmittance and solar energy transmittance modulation. J. Appl. Phys. 108, 063525/1–063525/8.

Li, S.-Y., Niklasson, G.A., Granqvist, C.G., 2011. Nanothermochromics with VO_2-based core–shell structures: calculated luminous and solar optical properties. J. Appl. Phys. 109, 113515/1–113515/5.

Li, S.-Y., Niklasson, G.A., Granqvist, C.G., 2012. Thermochromic fenestration with VO_2-based materials: three challenges and how they can be met. Thin Solid Films 520, 3823–3828.

Li, S.-Y., Mlyuka, N.R., Primetzhofer, D., Hallén, A., Possnert, G., Niklasson, G.A., Granqvist, C.G., 2013a. Bandgap widening in thermochromic Mg-doped thin films: quantitative data based on optical absorption. Appl. Phys. Lett. 103, 161907/1–161907/4.

Li, S.-Y., Namura, K., Suzuki, M., Niklasson, G.A., Granqvist, C.G., 2013b. Thermochromic VO_2 nanorods made by sputter deposition: growth conditions and optical modeling. J. Appl. Phys. 114, 033516/1–033516/11.

Li, S., Li, Y., Jiang, M., Ji, S., Luo, H., Gao, Y., Jin, P., 2013c. Preparation and characterization of self-supporting thermochromic films composed of VO_2 (M) @SiO_2 nanofibers. ACS Appl. Mater. Interfaces 5, 6453–6457.

Li, Y., Ji, S., Gao, Y., Luo, H., Jin, P., 2013d. Modification of Mott phase transition characteristics in VO_2@TiO_2 core/shell nanostructures by misfit-strained heteroepitaxy. ACS Appl. Mater. Interfaces 5, 6603–6614.

Li, S.-Y., Niklasson, G.A., Granqvist, C.G., 2014. Thermochromic undoped and Mg-doped VO_2 thin films and nanoparticles: optical properties and performance limits for energy efficient windows. J. Appl. Phys. 115, 053513/1–052513/10.

Loonen, R.C.G.M., Singaravel, S., Trčka, M., Cóstola, D., Hensen, J.L.M., 2014. Simulation-based support for product development of innovative building envelope components. Autom. Constr. 45, 86–95.

Lopez, R., Haynes, T.E., Boatner, L.A., Feldman, L.C., Haglund Jr., R.F., 2002. Temperature-controlled surface plasmon resonance in VO_2 nanorods. Opt. Lett. 27, 1327–1329.

Lykissa, I., Li, S.-Y., Ramzan, M., Chakraborty, S., Ahuja, R., Granqvist, C.G., Niklasson, G.A., 2014. Electronic density-of-states of amorphous vanadium pentoxide films: electrochemical data and density functional theory calculations. J. Appl. Phys. 115, 183701/1–183701/5.

Mattox, D.M., 2010. Handbook of Physical Vapor Deposition (PVD) Processing, second ed. Elsevier/William Andrew, Norwich, NY, USA.

Mlyuka, N.R., Niklasson, G.A., Granqvist, C.G., 2009a. Mg doping of thermochromic VO_2 films enhances the optical transmittance and decreases the metal–insulator transition temperature. Appl. Phys. Lett. 94, 171909/1–171909/3.

Mlyuka, N.R., Niklasson, G.A., Granqvist, C.G., 2009b. Thermochromic VO_2-based multilayer films with enhanced luminous transmittance and solar modulation. Phys. Status Solidi A 206, 2155–2160.

Montero, J. 2014. Private communication.

Niklasson, G.A., Granqvist, C.G., 2007. Electrochromics for smart windows: thin films of tungsten oxide and nickel oxide, and devices based on these. J. Mater. Chem. 17, 127–156.

Ohring, M., 2002. The Materials Science of Thin Films: Deposition and Structure, second ed. Academic, New York, NY, USA.

Piegari, A., Flory, F., 2013. Optical Thin Films and Coatings: From Materials to Applications. Woodhead, Cambridge, UK.

Preston, C., Xu, Y., Munday, J.N., Hu, L., 2013. Optical haze of transparent and conductive silver nanowire films. Nano Res. 6, 461–468.

Pulker, H., 1999. Coatings on Glass, second ed. Elsevier, Amsterdam, The Netherlands.

Richter, B., Goldston, D., Crabtree, G., Glicksman, L., Goldstein, D., Greene, D., Kammen, D., Levine, M., Lubell, M., Savitz, M., Sperling, D., Schlachter, F., Scofield, J., Dawson, J., 2008. How America can look within to achieve energy security and reduce global warming. Rev. Mod. Phys. 80, S1–S107.

Roaf, S., Nicol, F., Humphreys, M., Tuohy, P., Boerstra, A., 2010. Twentieth century standards for thermal comfort: promoting high energy standards. Archit. Sci. Rev. 53, 65–77.

Saeli, M., Piccirillo, C., Parkin, I.P., Binions, R., Ridley, I., 2010a. Energy modeling studies of thermochromic glazing. Energy Build. 42, 1666–1673.

Saeli, M., Piccirillo, C., Parkin, I.P., Ridley, I., Binions, R., 2010b. Nano-composite thermochromic thin films and their application in energy-efficient glazing. Sol. Energy Mater. Sol. Cells 94, 141–151.

Sahu, D.R., Lin, S.-Y., Huang, J.-L., 2006. ZnO/Ag/ZnO multilayer films for the application of a very low transparent electrode. Appl. Surf. Sci. 252, 7509–7514.

Smith, G.B., Granqvist, C.G., 2010. Green Nanotechnology: Solutions for Sustainability and Energy in the Built Environment. CRC Press, Boca Raton, FL, USA.

Talledo, A., Granqvist, C.G., 1995. Electrochromic vanadium-pentoxide-based films: structural electrochemical and optical properties. J. Appl. Phys. 77, 4655–4666.

Tavares, P.F., Gaspar, A.R., Martins, A.G., Frontini, F., 2014. Evaluation od electrochromic windows impact in the energy performance of buildings in Mediterranean climates. Energy Policy 67, 68–81.

Taylor, A., Parkin, I., Noor, N., Tummeltshammer, C., Brown, M.S., Papakonstantinou, I., 2013. A bioinspired solution for spectrally selective thermochromic VO_2 coated intelligent glazing. Opt. Express 21, A750–A764.

Warwick, M.E.A., Binions, R., 2014. Advances in thermochromic vanadium dioxide films. J. Mater. Chem. A 2, 3275–3292.

Zhou, J., Gao, Y., Liu, X., Chen, Z., Dai, L., Cao, C., Luo, H., Kanahira, M., Sun, C., Yan, L., 2013. Mg-doped VO_2 nanoparticles: hydrothermal synthesis, enhanced visible transmittance and decreased metal–insulator transition temperature. Phys. Chem. Chem. Phys. 15, 7505–7511.

Electrochromic glazing and walls for reducing building cooling needs

17

M. Pittaluga
Cagliari University, Cagliari, Italy

17.1 Introduction

Innovation is never the end of a path, only the application of a new idea. Innovation is the starting point for countless scenarios in which outcomes cannot be predicted. This chapter tells the story of an innovation, namely, the use of electrochromic glass in architecture. But this innovation is the starting point for the study and research of new applications and new solutions—future scenarios for a better life.

In the second section of the chapter, the need for a dynamic envelope, which is able to modify its behavior according to daily and seasonal climate changes, is briefly described. The most important examples of dynamic envelopes are listed, and some architectonic theories are discussed.

The third section describes the most important characteristics of electrochromic materials and their uses in the building sector for the improvement of energy efficiency. The major producers of electrochromic glazing are listed, and the first and most important examples of use of electrochromic materials in the history of architecture are analyzed and their energy behavior is discussed. The following section analyzes the most important research projects in the specific field with their outcomes, deliverables, and energy-saving data. The last section describes how the electrochromic glazings are the starting point toward other uses and innovative solutions. The possible use of electrochromic glazing in the opaque partition of the architectonic envelope is also discussed (Pittaluga, 2013).

17.2 The building envelope as a dynamic organism

If we imagine having to choose just one outfit to use for the whole year, what would we choose? If we choose a pair of shorts and a shirt, we will be fine during summer, during the day, but we will feel very cold during winter and sometimes even at night, during summer. If we choose a very heavy outfit, during summer days we will suffer a lot because of the heat. Our choice would probably also be a consequence of the place where we live.

In warmer areas of our planet, such as the deserts of Africa, people cover themselves, for example, with heavy woolen clothes, even during the day, in order to create

Eco-efficient Materials for Mitigating Building Cooling Needs. http://dx.doi.org/10.1016/B978-1-78242-380-5.00017-0

a microclimate that regulates itself on body temperature (around 37 °C), much lower than the outside temperature that significantly exceeds 40 °C. Fortunately, we do not have to make the choice of a single outfit. We can change our clothing, our second skin, depending on the climatic conditions in which we live.

Our third skin, namely, the envelope of the buildings in which we live and work, on the other hand, is always the same. When a building is first designed and then built, the designer assigns to it a certain standoff to external actions and the capability to isolate itself from the surrounding climate. An outfit is chosen and it can no longer be changed. Also in this case, there is the problem of which outfit we have to choose. In most cases, regulations come to our aid, providing us guidance on which fabric to choose, on how many layers to use, on which is the best sleeve length, and so on. As end users, we have only the ability to adjust the opening or closing of the windows. And this has happened for a very long time. For centuries, architecture has been characterized by low constraints (structural and technical ones) and by the large availability of raw materials and manpower, by little attention to the housing needs of the poorest social classes, by lack of interest in housing comfort, and, above all, by the awareness of being a static entity.

A milestone in history is the advent of modern architecture. Modern architecture studies the use of new materials such as glass, steel, and reinforced concrete; it analyzes new forms and typologies. For the first time, it addresses some problems that in the past had never been taken into account; these problems have become fundamental in the art of building.

At the beginning of the 1920s, after the defeat in World War I, the Weimar Republic planned an emergency program for housing. In Frankfurt am Main, about 15,000 dwellings were built with the supervision of the German architect Ernst May, thanks to organization and efficiency of both the design phase and the building phase. This approach inevitably led to the formulation of spatial standards known as *Existenzminimum* (subsistence dwelling, in terms of acceptable floor space, density, fresh air, green spaces, and other resident issues), in contrast to the idealistic appeal of the Swiss architect Le Corbusier to an *Existenzmaximum*.

The minimal standards of May translated into general use of ingenious wardrobes built into the wall and folding beds and especially into the development of an ultra efficient kitchen, like a laboratory, the *Frankfurter Kuche*, designed by the Austrian architect G. Schutte Lihotzky; it is the kitchen concept that we use nowadays.

From the manic exploitation of the little available space came the idea to use in different ways the little space available, so as to make it flexible, adaptable, and therefore dynamic. The most groundbreaking example is the 1924 Gerrit Rietveld's Schröder house in a suburb of Utrecht. Upstairs, there is a single dynamic living space that has the option of being left open or subdivided by a system of sliding and revolving panels. When it is entirely closed, the floor accommodates three bedrooms, a bathroom, and a living room (see Figure 17.1). Hence came the concept of open space, adaptable according to the needs of large office buildings of the last century. But the concept of adaptability and flexibility was still limited to interior spaces.

Figure 17.1 Sketch of the first floor of the Schröder house in Utrecht, 1924.

In the first decades of the twentieth century, attempts to make the envelope of buildings dynamic were characterized in two different ways: the media paradigm and the organic paradigm.

The building envelope has always been a means of communication, for example: the paintings on the walls of Egyptian tombs and temples of the twelfth century BC, the bas reliefs of Greek temple facades, the Byzantine mosaics, the Gothic polychrome glazed facades, and the Art Nouveau ceramics (Schittisch, 2003). In the early decades of the 1900s, thanks to the advent of electricity, the facades of buildings became a dynamic means of communication. In 1928, in Times Square in New York, the first large electric moving display was mounted in the headquarters of The Times newspaper. It consisted of a tape made of 14,800 light bulbs, 16 m long, which told the breaking news. As shown in Figure 17.2, today, thanks to liquid crystal displays (LCDs), light-emitting diodes (LEDs), and organic light-emitting diodes (OLEDs), larger screens can be realized on building facades (Pittaluga, 2005).

In 1916, Le Corbusier suggested for the first time that the behavior of the human skin could be copied in architectural organisms. The thermoregulatory system of the human body makes possible, depending on the case, the production of heat (thermogenesis) and its elimination (thermolysis), adapting itself to the outer temperature and humidity, through sweating.

In 1931, Le Corbusier, in a project for the Citè de Refuge, in Paris, planned a double-glazed envelope, in the main facade, with a mechanical ventilation system (*respiration exacte*) (Banham, 1995; Wigginton, 1996). *Mur neutralisant* would have had to keep the air temperature constant at 18 °C inside the building. During colder periods, high-temperature air would have been blown inside the cavity, while in the warmer periods, cold air would have been introduced. The system, high-tech for that time, was studied, thanks to the aid of the French glass manufacturer company Saint-Gobain, even in some test rooms. Due to the reduction of the budget, the facade was

Figure 17.2 The LCD facade of the Nasdaq building in Times Square, New York, 2000.

created without the *mur neutralisant*. The end result was obviously disappointing from the point of view of air conditioning; in fact, in 1933, some openable partitions were placed in the glass facade.

The American architecture historian Reyner Banham firmly criticized, Le Corbusier's attempt to solve the environmental problems of the building with a completely primitive ventilation system, which had already been used by Joseph Paxton, English architect and botanist, in the house for Victoria Regia water lilies, at Chatsworth in 1850. Actually, the system designed by Paxton did not foresee a double wall; as reported by Michel Conan (2002), "thirty openings between the piers of the basement wall allowed for low-level ventilation, and opening lights in the roof *'made to open by simple machinery'* gave additional ventilation when required." According to Wigginton (1996), it was not until the creation of Lloyd's headquarters, in London, by Richard Rogers Partnership, that it was possible to see Le Corbusier's idea correctly applied.

The same Reyner Banham (1995), however, recognized Le Corbusier for the innovation of external solar shading, the *brise-soleil*: *it is without doubts one of his most masterful inventions, one of the few architectural inventions happened lately useful for environmental monitoring.*

The attempt to make the architectural envelope dynamic, over the past 70 years, has focused on the double-skin glazed facades, with systems of forced and/or natural ventilation in the cavities, and on the external solar shadings. *Brise-soleil*, or *louver*, especially in contemporary curtain-walls office buildings, assumes considerable importance, thanks to the possibility of being mechanically controlled as a function of the path of the sun and of the outer weather conditions (Compagno, 2002; Schittisch, 2003; Wigginton and Harris, 2002). Very often, such shielding systems of solar radiation are positioned within the gaps between the two sides of a double-skin glazed envelope.

The possibility of an automatic control of some parts of the buildings, as if they were provided with a complex nervous system (Mc Luhan, 1967), represents a further opportunity in research toward the creation of a dynamic envelope. Between 1981 and 1987, the French architect Jean Nouvel, in the South facade of the Arab World Institute in Paris, reinterpreted traditional Arab perforated screens, the *moucharabieh*, creating 30,000 sensitive diaphragms that were designed in order to regulate the penetration of light to the interior of the building (Figure 17.3). Like human eyes, these attractive steel diaphragms, placed between two sheets of glass, open and close themselves, set up one by one by a computer, depending on the incidence of solar radiation. The overall design is a fascinating game of geometries, where there are many symbolisms, which vary from hour to hour.

Today, the possibility of creating a dynamic and organic envelope is a concrete reality, which puts into practice the utopian vision of a wall for all seasons, as articulated by the British researcher Mike Davies (Davies, 1981): *What is needed is an environmental diode, a progressive thermal and spectral switching device, a dynamic interactive multi-capability processor acting as a building skin.*

The protagonists of the polyvalent and dynamic wall are chromogenic, or photodynamic, materials; they are able to vary their characteristics of transparency to solar radiation as a function of the variation of certain physical parameters. Today, chromogenic materials are the decisive means by which the third skin, the envelope of the

Figure 17.3 South facade of the Arab World Institute in Paris, Jean Nouvel.

buildings in which we live, may allow adaptation to sudden and violent changes of climate and, especially, a reduction in energy consumption for HVAC. Among the chromogenic materials, the electrochromic ones are most interesting.

17.3 Electrochromic materials for the building envelope

Electrochromic materials are able to vary their coloration and transparency to solar radiation, in a reversible manner, when they are subjected to a small electric field (1–5 V). The main materials with electrochromic properties are metal oxides of transition, in particular WO_3, MoO_3, IrO_2, NiO, and V_2O_5.

In the early 1980s, the electrochromic (EC) materials were the subject of great interest due to their possible use in displays and watches. Due to durability issues and manufacturing problems, another type of chromogenic material, liquid crystals, was chosen for displays. Furthermore, liquid crystals were able to switch faster from a transparent state to an opaque one, within fractions of a second. The interest in EC materials had a new impetus in the early 1990s, particularly in the aerospace and aviation sectors (Dornier Aerospace) and automotive sector (Toyota).

EC materials, produced in the form of thin layers, are normally inserted within a panel of insulating glass or laminated glass making up the so-called smart windows or switchable windows. In this way, the electrochromic materials are studied in view of their possible use in architecture.

As mentioned above, what makes EC glazing particularly attractive is its ability to vary the transparency to solar radiation. While other types of chromogenic materials, such as photochromics, liquid crystals, and suspended particle devices (SPDs), allow a variation of the transmission of solar radiation belonging mostly to the visible field (i.e., in the range 380–780 nm), thermochromics, electrochromics, and gasochromics (GC) can also shield the radiation belonging to the infrared range (i.e., in the range 780–2500 nm). In this way, EC glazings can prevent the greenhouse effect in the interior of the buildings. This effect allows the overheating of the buildings, especially during summer and in the hottest climates, and it requires a high energy consumption for cooling. The EC glazing transmission of solar radiation in the infrared range, on average, varies between 50% (in the clear or "off" state) and 10% (in the colored or "on" state).

In 1988, the Japanese company Asahi Glass Co., specializing in the production of automotive glass, developed an electrochromic glass that was used in the Seto Bridge museum, in Kojima (Lampert, 1991). Since the 1970s, the German company FLABEG has become specialized in solar glazing; in 1980, the company was taken over by the multinational Pilkington. A few years later, the first electrochromic glazing was produced: the FLABEG EControl.

EC glazing had a high cost (about € $1000/m^2$); their effectiveness was still uncertain and therefore they were rarely used. The few examples of the use of EC glazing, in those years, are listed below (Pittaluga et al., 2010).

In 1999, the Sparkasse headquarter in Dresden was completed, by a project of Bauer and Keller, with a large glass area of 150 m^2 on the main facade, made of FLABEG EControl electrochromic glazing (Figure 17.4). According to the energy consumption data, presented at the time by the same manufacturer (Figure 17.5), the use of EC glazing allowed an energy saving of about 18% to 24% if compared with thermal insulation glazing. The consumption with EC glazing was, more or less, the same as in the case of use of conventional glazing and external sun shadings. In 2000, FLABEG broke away from the Pilkington Group. In May 2000, the OstseeSparkasse, in Rostock, was completed by a project of Heynsen and Meisterlin; FLABEG EC glazings were used in the roof of the main hall. In 2002, the Fraunhofer Gymnasium in Cham was completed; Architekturburo Berthold used FLABEG EC glazing in the main facade.

One of the most important examples of the use of EC glazing is the Regional Library of Saxony, in Dresden, designed by German architects Ortner and Ortner; it was completed in April 2001. The roof of the central reading room was made of FLABEG EC glazing. During a correspondence with the designers, they

Figure 17.4 The main facade of the Sparkasse headquarters in Dresden.

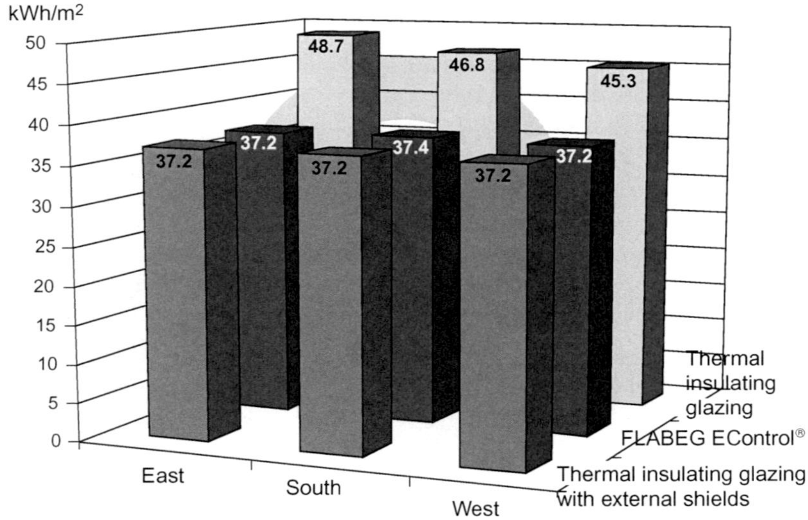

Figure 17.5 Energy consumption for HVAC of Stadtsparkasse headquarter.

indicated that "for sun protection of the glazed roofs they used EC glasses but, unfortunately, in the meantime it became clear that they do not function; so the client is forced to change them and to replace them with a conventional protection against the sun."

In 2007, after several years of development in the Pilkington/FLABEG group, a new company, EControl-Glas, was established and it continued the production of EC glazing.

On the other side of the world, since 1989, SAGE Electrochromics has been studying, producing, and commercializing EC glazing in the United States. Over the last two decades, SAGE Electrochromics was involved in several research projects in the United States and in Europe. The company developed a technology that coats float glass with thin films of metal oxides using sputter deposition to form the active layer. In 2010, the French glass giant Saint-Gobain made an important strategic investment in SAGE, and since 2012, the American company, located in Faribault, Minnesota, has been a wholly owned subsidiary of Saint-Gobain.

In the last years, other companies started the production of EC glazing. The German company Gesimat, since 1989, has been developing several different switchable glass products, such as thermochromic, electrochromic, and electrotropic laminated glass.

In the United States, in 2006, the Soladigm company, originally named Echromics, was founded. In 2012, Soladigm announced a name change to View, Inc. In the same year, the first switchable electrochromic window and skylight products from Soladigm reached the American market.

17.4 Research in the field of electrochromic glazing

The first study on EC glazing for the architectural envelope was carried out in 1999 by the Lawrence Berkeley National Laboratory (LBNL) of California (Lee and DiBartolomeo, 2000). FLABEG EControl glazings (with a Pilkington Optitherm S low-e layer) were studied, and their performance was monitored for four months, during winter, by calculating the energy savings on lighting. The EC glazing used had a light transmission value in the visible range (TV) variable between 0.11 and 0.38; they were mounted on the southeast facade on the fifth floor of a federal building in Oakland, California.

The EC glazings were mounted on two adjacent rooms; the reference room was hypothesized in three different conditions: the first one simulated a traditional glass (TV = 38%), the second one simulated a solar glass (TV = 11%), and the third one suggested two different regulations: The upper range of glass was clear (38%) and the lower one was colored (TV = 11%). In the test room, EC glazings were activated when the luminance was greater than 540–700 lux. Measurement of luminance was carried out every 5–8 minutes.

The energy saving was between +6% and +24% (test room compared with the reference room in the configuration completely dark), between −13% and +3% (test room compared with the reference room in the confiuration completely clear), and between +3% and 14% (test room compared with the reference room 11–38%). The obtained results were not very positive, as expected.

The three-year SWIFT (Switchable Facade Technology) Project, funded by the European Union, within the Fifth Framework Programme, started in 2000 and finished in April 2003.

Three pilot facades were studied (SWIFT, 2003): at the Eindhoven University of Technology, in Rome (at the ENEA headquarter in Casaccia), and in Freiburg (at ISE). The EC glazings were monitored and their performance was evaluated through the administration of questionnaires to a sample of users of the various structures. This activity was focused on the optimization of the interaction of the three components that can simultaneously allow energy saving and home comfort: photodynamic glazing system, lighting system, and shading systems. Useful guidelines to users, but also to designers and builders, were also defined during the project.

The electrochromic glazing used during the project had the following stratifications: an external laminated float glass, equipped with an electrochromic layer in the inner face, with a thickness of 9 mm, a gap of 16 mm containing gas argon, and a second pane of glass, equipped with a low-e coating, with a thickness of 4 mm, for a total of 29 mm. The transmittance was $U = 1.6–1.1\ W/m^2\ K$. The sound insulation Rw = 35 dB. The transmission in the visible range (TV) varied between 0.52 and 0.16, while the solar heat gain coefficient (SHGC) value ranged between 0.40 and 0.16.

The test at the Eindhoven University of Technology (Tenner and Zonneveldt, 2002) had as its objective the identification of the optimal conformation of the electrochromic facade, for glare reduction and optimization of lighting. The test chamber,

3.6×6.2, with a height of 2.6 m, included a glazed area of about 3.6×1.7. The optimal conformation consisted of a vertical split into three bands. Horizontally, instead, three panels with reduced size appeared at the top and at the bottom; they were openable and equipped with a low-e glazing. The central parts of the facade foresaw in the center, at the left and at the right, 2+2 electrochromic panels.

The testing of the Swift Project in Rome aimed to evaluate the response of users and the level of satisfaction with reference to the photodynamic technology. Therefore, the electrochromic glazings were installed in one office. Thirty users, who had the ability to independently adjust the system, participated in the test. The experimentation was conducted between August and October 2001. It generally started in the afternoon, when the facing west window was affected by solar irradiation, and it was necessary to decrease the brightness; at the end of the afternoon, a questionnaire was administered.

In summary, 84% of interviewed people said they were excited about the use of the new technology and only 10% deemed it not particularly useful. Some users would also have preferred a faster activation system (activation time, then as now, was a few minutes). The alteration of the external environment, following the activation of the EC glazing, was perceived but accepted. Even the possibility of manual control of the glazing was welcomed.

The third pilot of the SWIFT Project (SWIFT, 2003) was held at the offices of the Fraunhofer Institute for Solar Energy Systems (ISE) in Freiburg. The objective of the testing in Freiburg was to demonstrate that it was possible to integrate the photodynamic technology in existing buildings and the acquisition of experience in photodynamic glazing control systems.

A further phase of the SWIFT Project involved the simulation, using special software, of the lighting conditions and energy consumption in workplaces as a result of the use of photodynamic glazing. The simulation was conducted with the collaboration of the research group of the IEA (International Energy Agency), Tasks 27 and 31. Two test rooms were defined with the hypothesis of three different locations: Stockholm, Brussels, and Rome.

The energy consumption for artificial lighting, the glare and the level of illumination, and the energy consumption for heating and cooling were measured through the simulation. The following software was used: Radiance, Daysim (for calculation of luminance and energy consumption related to the artificial lighting), Exp-r52, and TRNSYS, for energy simulations.

Thanks to the use of the TRNSYS software, energy consumption for cooling was calculated taking into account four different kinds of glazing: electrochromic glazing (EC; $U = 1.1$ W/m^2 K and SHGC = 40–15%), gasochromic glazing (GC; $U = 0.9$ W/m^2 K and SHGC = 48–18%), double glazing with low-e coating (SC; $U = 1.1$ W/m^2 K and SHGC = 33%), and double glazing or heat mirror (HM; $U = 1.3$ W/m^2 K and SHGC = 62%). A facade with a window wall ratio (WWR) of 30% was firstly assumed.

Paradoxically, it was found that the percentage reduction in energy consumption for cooling was greater in northern Europe than in southern Europe (Table 17.1): in Brussels, EC glazing allowed consumption, for cooling, of 81% of what was

Table 17.1 SWIFT project simulation; HVAC energy consumption in Rome, Brussels, and Stockholm with WWR 30%

WWR: 30%		SC		HM	HM:SC (%)		EC	EC:SC (%)		GC	GC:SC (%)
Rome	Heating energy (kWh/m^2 a)	5.5	–	3.7	−32.73	–	5.8	5.45	–	4.9	−10.91
Rome	Cooling energy (kWh/m^2 a)	24.2	–	45.5	88.02	–	14.1	−41.74	–	15.2	−37.19
Rome	Total energy (kWh/m^2 a)	29.7	–	49.2	65.66	–	19.9	−33.00	–	20.1	−32.32
Brussels	Heating energy (kWh/m^2 a)	20.3	–	16.6	−18.23	–	19.9	19.88	–	17.2	3.61
Brussels	Cooling energy (kWh/m^2 a)	6.8	–	16.3	139.71	–	3	−81.60	–	3.4	−79.14
Brussels	Total energy (kWh/m^2 a)	27.1	–	32.9	21.40	–	22.9	−15.50	–	20.6	−23.99
Stockholm	Heating energy (kWh/m^2 a)	39.4	–	33.8	−14.21	–	37.9	−3.81	–	33.1	−15.99
Stockholm	Cooling energy (kWh/m^2 a)	7.3	–	18.8	157.53	–	2.6	−64.38	–	3.1	−57.53
Stockholm	Total energy (kWh/m^2 a)	46.7	–	52.6	12.63	–	40.5	−13.28	–	36.2	−22.48

SC, double glazing with low e-coating; HM, double glazing or heat mirror; EC, electrochromic glazing; GC, gasochromic glazing.

required if double glazing with low-e coating (SC) was used. In Stockholm, this value was 64%, while in Rome, only 41%.

Taking into account a facade with a greater transparent surface (WWR 60%), it could be noticed that the greatest saving for cooling, thanks to the use of EC glasses, was obtained in Stockholm and that, in general, the GC glazing allowed a greater saving (Table 17.2).

In 2002, thanks to the support of the DOE (US Department of Energy), of the California Energy Commission and the US Department of Housing and Urban Development, LBNL started a new three-year research project (PIER, Public Interest Energy Research) on EC glazing (Lee et al., 2006). The research project involved the study and experimentation of an electrochromic glass prototype, with the size of 0.46×0.89, provided by the American company SAGE Electrochromics Inc., characterized by the presence of a tungsten trioxide layer, as well as all the other EC glasses mentioned in the chapter.

While the test in Oakland had foreseen the use of German glasses, based on a polymeric coating, in this case, glasses with a base coating on a thin ceramic film were used; they were produced more quickly and at lower cost. The stratification of the glazing consisted of an outer sheet of 4 mm tempered glass, with an EC layer in the inner face, a 15 mm cavity, and a tempered glass panel of 6 mm.

LBNL, during this project, carried out the first studies in relation to the control system of the EC glazing. SAGE Glass provided a rudimentary controller that had appeared, by now, inadequate, as it had only two operating states (on and off), and it was not able to activate all the panel at the same time. Thanks to the collaboration between LBNL and SAGE, a more powerful controller was created.

The main objective of this project, as well as the most interesting, in this context, was to quantify the benefits, costs, and risks of EC technology in real conditions, in order to stimulate the interest of the market and facilitate the development of such technology.

The test lasted for about two and a half years; three test rooms were created: the first (test case) was with active EC glazing, and the second and third (reference cases) were kept constantly and, alternatively, in the clear state and in the colored state.

Daily energy consumption for cooling, saving relative to the peaks of consumption for cooling, daily energy consumption for lighting, and consumption peaks for lighting were monitored; measurements were taken every minute between 06:00 and 18:00.

Starting from the data collected *in situ*, and considering the requirements of ASHRAE 90.1 2005 regulation, which defines U and SHGC values, thanks to the simulation through the DOE-2.1 software, it was noticed that the use of EC glazing allowed a reduction of the annual demand of primary energy for an office building of a quantity between 15% and 23% (WWR = 0.30) and of a quantity between 10% and 24% in the case of WWR = 0.60.

Using the DOE-2.1 software, LBNL conducted, during the same period, another interesting study, which aimed to quantify the potential impact of electrochromic glazing for saving of primary energy in commercial buildings in the United States, in 16 different climate zones in California, and in 6 different climates in the United States

Table 17.2 SWIFT project simulation; cooling energy consumption in Rome, Brussels, and Stockholm with WWR 60%

WWR: 60%		SC		HM	HM:SC		EC	EC:SC (%)		GC	GC:SC (%)
Rome	Cooling energy (kWh/m^2 a)	110	–			–	63	−42.73	–	60	−45.45
Brussels	Cooling energy (kWh/m^2 a)	44	–			–	23	−47.73	–	22	−50.00
Stockholm	Cooling energy (kWh/m^2 a)	63	–			–	32	−49.21	–	30	−52.38

(Lee et al., 2004). The simulation was conducted having as a building type the prototype developed by the LBNL for some ASHRAE standards in force in the United States. The building had a square plan and stood on three levels aboveground. Each floor had an area of 139 m^2; each side had 10 offices (3.04×4.6 m). The internal height was 2.74 m and each facade was oriented in the direction of one of the cardinal points. Both the exterior walls and the roof conformed to the insulation values prescribed by the ASHRAE American regulations (2005). It was assumed to use different kinds of glass, in single or multiple sheets, and different types of frameworks, with and without thermal cutting, depending on the requirements of the regulations in relation to the transmittance and solar factor.

The EC panel consisted of two plates of 6 mm (the outer one with EC layer) and an air gap of 1.27 cm. The "daylight control" strategy was used for the activation of EC glass: the glass was activated when, in a specific point of the room (0.76 m from the floor and 3.05 m from the window), the value of illuminance of 538 lux was exceeded. Obviously, a lighting system and an HVAC system, in order to guarantee the comfort conditions, were provided.

In summary, EC glazing, without internal shielding and without a lighting control system, allowed energy saving and a reduction of consumption peaks in all the monitored situations: for small windows (WWR = 0.30), there was a reduction of energy consumption that was variable between 15% and 28% in all climates, while consumption peaks were reduced by an amount between 10% and 25%; for large windows (WWR = 0.60), there was a reduction of energy consumption variable between 10% and 24% in most climates, while consumption peaks were reduced by an amount between 10% and 15%; in a few cases, there was an increase in peaks.

The summary table (Table 17.3) shows the energy saving that EC glazing allowed, if compared to the glass panes that were, at the time, the most efficient ones (double evergreen tint low-e glazing, TV = 0.43, SHGC = 0.27, and $U = 2.61$ W/m^2 K).

A study conducted at the DICAAR (Department of Civil Engineering, Environment and Architecture) of the Cagliari University (Loddo et al., 2009), between 2009 and 2010, consisted of the simulation, through software operating in dynamic regime, of the performance of electrochromic glazing in an office building with a transparent curtain-wall envelope, located in Cagliari (Figure 17.6).

It was thought to replace the glazing of the facades using two possible alternative solutions. The first one involved the use of double-glazed units, representing at that time the best option available on the market; the second one foresaw the use of electrochromic glazing produced by the American company SAGE Electrochromics Inc. The stratifications of EC glazing were the following: the internal sheet of 6 mm, a cavity with gas argon with a thickness of 12.7 mm, the electrochromic layer, and finally the outer sheet of 6 mm, for a total of about 25 mm. The study aimed to quantify the energy consumption required for heating and cooling in the two situations and the savings resulting from the use of EC glazing. A cost-benefit analysis was also conducted, in order to understand the time of return of the investments: it was too long, due to the high cost of electrochromic glass. Table 17.4 summarizes the economic analysis.

Table 17.3 Synoptic table of the most important research projects conducted on electrochromic glazing in recent years

Project/ coordinator	Period	EC glazing	Thickness (mm)	Glazing dimension (cm)	TV	U (W/ m^2 K)	SHGC	Mode	Location	Facade orientation	WWR	Terms for comparison (TfC)	EC HVAC consumption (kWh/m^2)	TfC HVAC consumption (kWh/m^2)	Saving (%)
LBNL	1999–2000	FLABEG E-control		62.1 × 43.2	0.11–0.38		0.12–0.36	Sim	Oakland, California	SE					
SWIFT—IEA	2000–2003	FLABEG E-control	29		0.16–0.52	1.1	0.15–0.40	Sim	Rome	S, N	0.3	Double low-e glazing	19.9	29.7	33.00
								Sim	Brussels	S, N	0.3	U = 1.1, W/m^2 K,	22.9	27.1	15.50
								Sim	Stockolm	S, N	0.3	SHGC = 0.33	40.5	46.7	13.20
													Only cooling		
								Sim	Rome	S, N	0.6		63	110	42.73
								Sim	Brussels	S, N	0.6		23	44	47.73
								Sim	Stockolm	S, N	0.6		32	63	49.21
LBNL	2002–2005	SAGE Glass	25	46 × 89	0.05–0.60	1.87	0.09–0.42	Test	Berkeley, California	S	0.3 0.6	Double low-e glazing TV = 0.42, SHGC = 0.219, U = 1.41 W/m^2 K			15–23 10–24
LBNL	2002–2004	SAGE Glass	24		0.02–0.45	0.49	0.09–0.42	Sim	Houston	S	0.6	Double evergreen tint	304.79	346.74	12.10
								Sim	Phoenix	S	0.6	low-e glazing	293.08	338.83	13.50
								Sim	Atlanta	S	0.6	TV = 0.43,	196.69	233.60	15.80
								Sim	Minneapolis	S	0.6	SHGC = 0.27,	339.28	395.43	14.20
								Sim	Chicago	S	0.6	U = 2.61 W/m^2 K	283.73	332.63	14.70
								Sim	Washington	S	0.6		232.92	289.71	19.60
Unica/ DICAAR	2009–2010	SAGE Glass	25	150 × 100	0.03–0.65	1.58	0.09–0.43	Sim	UIL Building, Cagliari (Italy)	N, S, W, E	1	Double low-e glazing—TV = 0.61, SHGC = 0.68, U = 1.93 W/m^2 K	19.21	67.54	71.00
RAS/ Laborvetro/ DICAAR	2010–2012	SAGE Glass	25	150 × 100	0.03–0.62	1.49	0.08–0.48	Test and sim	Cagliari (Italy)	S-E, S-W	0.1	Double low-e glazing—Planitherm TV = 0.78, SHGC = 0.58, U = 1.30 W/m^2 K	114.15	129.55	11.88
DICAAR	2012	SAGE Glass	25		0.03–0.62	1.32	0.08–0.48	Sim	Cagliari (Italy)	S-E, S-W	0.3	Single glass TV = 0.88, SHGC = 0.81, U = 6.12 W/m^2 K	128.05	136.21	5.99

Figure 17.6 The glazed tower of the UIL building in Cagliari.

Table 17.4 Economic analysis for the two hypotheses of requalification of the UIL building in Cagliari

Items	Conventional glazing	Electrochromic glazing
A—cost of curtain-wall realization (€)	363,513	631,052
Heating load (kWh/year)	13,120	21,380
Cooling load (kWh/year)	128,780	18,980
Total energy to be supplied (kWh/year)	141,900	40,360
Energy consumption for HVAC (kWh/year)	47,300	13,450
G—cost for HVAC (€/year)	5868	1668
Cost for 627 m^2 of external shielding (€)	21,945	–
Cost for activation (1533 h/year) of EC glazing (€)	–	205
Total costs (A+G)	385,458	631,052
Higher cost for the realization of EC facades	€245,594	
Annual saving for HVAC thanks to EC glazing	€3995	
Time of return (inflation 3% yearly)	34.7 years	

The use of conventional glazing, as shown in Table 17.3, involved a total energy consumption of 67.54 kWh/year per m^2, while the EC glazing required an energy consumption of 19.21 kWh/year per m^2. The reduction of energy consumption in the case of use of EC glazing was therefore 71%.

Table 17.3 is very interesting due to the fact that, for the first time, it summarizes the results, in terms of energy saving, of all the most important research projects carried out on electrochromic glazing in recent years. Apart from being a point of reference and easily readable, it allows for some important comparisons.

First of all, it can be noted that with the increase of the WWR value, we obtain higher values of energy saving. Through the use of electrochromic glazing, energy saving can be achieved especially in warmer weather, in relation to consumption for cooling. Finally, it is necessary to consider the different terms of comparison that are taken into consideration: the glazings, used as terms of comparison, vary depending on the values of transmittance and SHGC. The same electrochromic glazings have evolved over time in relation to their physical characteristics, to the possibility to vary beyond the two states of on and off, and in relation to the control system.

The analysis of the described research projects highlights, in any case, two needs: First is that live testing, in real weather conditions, is necessary; second is that further simulations, carried out on different types of buildings, in terms of function, the period of construction, and building techniques, are needed in order to identify the optimal size of the electrochromic glazing to obtain an economic convenience. These are the research topics that, in the last years, the DICAAR of Cagliari University have faced.

Thanks to a grant of the Autonomous Region of Sardinia (European Social Fund, 2007–2013), a new research project started in 2010 in partnership between DICAAR, the Italian company Laborvetro Ltd., and SAGE Electrochromics Inc. (Pittaluga et al., 2013). The project involved the construction of two test rooms (Figure 17.7), the first one with EC glazing and the second one with conventional glazing, and the monitoring, using appropriate survey instruments, of the performance of the different kinds of glazing over a period that included winter, spring, and summer. A preliminary phase involved the simulation, through DesignBuilder software, of the different performances of the two test rooms.

The test rooms were built in Cagliari with the balloon frame technique. They had plan dimensions of $4 \times 4\ m^2$. In the S-E facade of each room, there is a small window of 120×80 cm; in the S-W facade, there is a larger window of $150 \times 100\ cm^2$; the other facades were blind.

The EC room was equipped with SAGE Electrochromic double glazing (single panel of 6 mm; gap filled with argon, 13 mm; and tempered glass panel, 6 mm, for a total of 25 mm). The U value was 1.49 W/m^2 K. The solar factor SHGC varied between 0.48 and 0.08. TV varied between 0.62 and 0.035.

The LE room was equipped with a double glazing with Saint-Gobain Planitherm ultra low-e coating, 4 mm; gap filled with argon, 13 mm; and double-plate Saint-Gobain Planilux with PVB film, 6 mm, for a total of 23 mm. The U value was 1.30 W/m^2 K, SHGC = 0.58, and TV was 0.78. The transmittance of the blind parts of the envelope was $U = 0.172\ W/m^2$ K.

The most important objective of the project was to quantify the different performances of EC glazing compared with conventional glazing. The measurements started in December 2011 and ended in August 2012. For all the days of the testing, for each room, the following parameters were measured: the average relative humidity, the average internal temperature (in °C), the maximum and minimum internal

Figure 17.7 The room with EC glazing, above, and the room with low-e glazing, below.

temperature (in °C) and the related time, the average temperature of the globe thermometer (in °C), the average solar radiation (expressed in W/m^2), the maximum solar radiation and the relative time, the total incoming solar radiation during 24 hours (expressed in Wh/m^2), and the difference of the total solar radiation between the two rooms. For each day of the survey, a brief comment on the data was made and the discomfort index was calculated.

In summary, it can be argued that, consistently, throughout the duration of the testing, in the EC room, lower temperatures and a lower amount of incoming solar radiation were recorded, if they were compared to the LE room ones. This resulted in an almost constant better state of comfort recorded in the EC room if compared to the LE one.

Paradoxically, the best performances of EC glazing were found in winter and spring months. In these months, the reduced height of the sun implied a greater input of solar direct radiation inside the rooms. In particular, at certain times of the day, in the LE room, temperatures more than 14 °C higher (compared with EC room ones) were recorded (Figure 17.8). In those moments of the day, one in the morning and one in the afternoon, corresponding to times when the solar radiation was almost

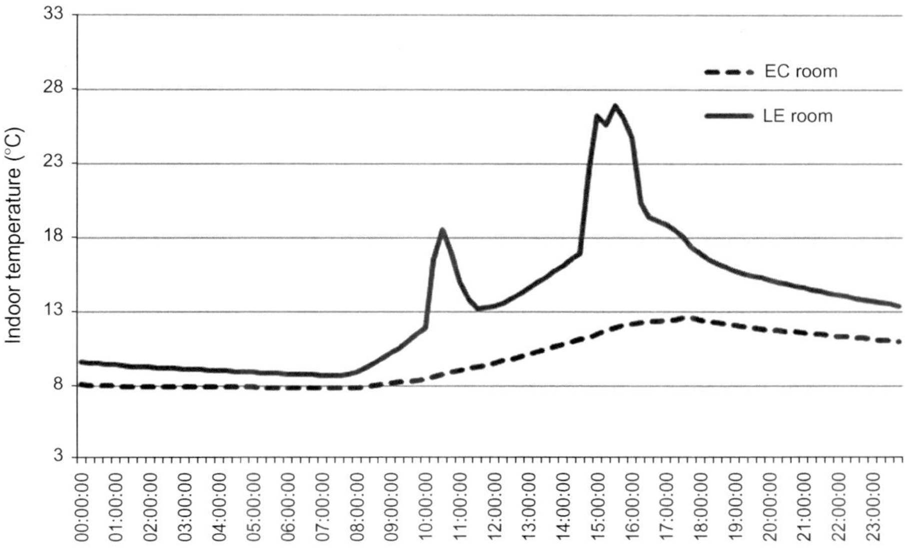

Figure 17.8 Comparison of indoor temperature between EC room and LE room on 16th of February.

perpendicular, incident on windows, in the LE room, there was significant discomfort (hot), whereas in the EC room, these unpleasant effects due to solar radiation were canceled thanks to electrochromic glazing. In warmer periods in both rooms, it was necessary to use air-conditioning systems, but the energy required by the EC room was definitely lower.

It is interesting to analyze the energy consumption for cooling, calculated during the simulation, in the months from May to August in the two rooms; in these months in fact, there was a great consistency between the simulation results and testing ones. As can be seen in Table 17.3 and Figure 17.9, the savings obtained thanks to the use of electrochromic glass, for summer cooling, was approximately 22.5%.

In recent years, as previously mentioned, the research group of DICAAR has been carrying out further simulations on buildings of different types and from different periods of construction, in order to study the savings arising from the use of EC glazing and to define the optimal surfaces (the so-called WWR) beyond which there is an economic advantage in the use of electrochromic glazing.

In order to verify the energetic performance, thanks to the use of EC glazing, in envelopes with design features not conforming to current standards, a study was carried out in 2012, having as a subject the City Hall Building of Cagliari (Ludoni et al., 2012). The Art Nouveau construction is characterized, on all the facades, by large glazed surfaces.

The simulation was conducted using the DesignBuilder software with the EnergyPlus module for the calculation of thermophysical parameters. The simulation was

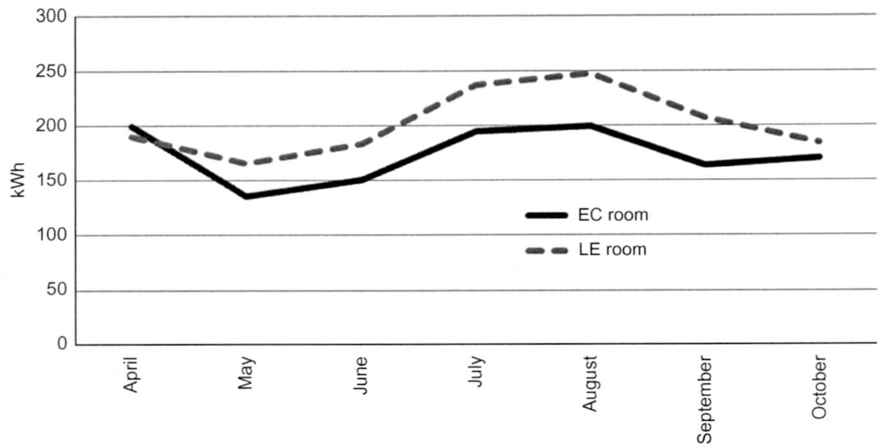

Figure 17.9 Comparison of total energy consumption of the EC room and the LE room.

set in a time interval of a year, with monthly trend. The heating system consisted of a centralized system and two radiators for each room; the cooling one instead was represented by a VRV inverter machine, which could simultaneously supply more rooms. The most important outputs of the simulation were thermal comfort, energy consumptions, internal intakes, and CO_2 emissions.

In summary, with reference to the thermal comfort, the monthly and daily averages were calculated; on average, there were 78.60 monthly discomfort hours and 2.62 daily ones in the actual situation and, respectively, 59.88 and 1.99 discomfort hours for the case with EC glazing. The reduction of discomfort hours, thanks to the use of EC glazing, was about 24%.

With reference to the HVAC system, energy consumption for cooling and heating were calculated. The use of EC glazing reduced the value of energy consumption for cooling by 27%, if compared to the case of single glazing. The energy saving for heating was less evident, because the gain, due to the contribution of solar radiant energy, was greater in the case of single glazing. During winter, thanks to the use of double EC glazing, there was, on the other hand, a lower dispersion of heat flow, due to the good performance of double glazing itself.

With reference to the monthly average of CO_2 emissions, an average value of 114,537 kg for single glazing and an average value of 106,313 kg for the EC glazing were recorded, with a reduction of about 7%.

The study also showed the convenience of using EC glazing in historic buildings. In this case, the hypothesis of refurbishment consisted only of the replacement of glazing. It was minimally invasive, a very important factor in the case of refurbishment of buildings that have almost the vocation of monuments.

17.5 Future trends and innovative applications

The technological evolution, in the field of electrochromic materials and glazing, proceeds in several directions. On the one hand, from the viewpoint of materials science, new nanocomposites, consisting of nanocrystalline indium oxide, embedded in matrices based on niobium oxide have been studied (Llordes et al., 2013). From the production point of view of electrochromic glazing, in order to improve the performance and durability of the electrochromic layers, stratification of the panels has been modified. In the past, the electrochromic layer was placed on the inner face of the outer pane of a double glazing. Today, the electrochromic layer is inserted between two glass panes that constitute the outer part of the double-glazed unit (DGU) (SAGE and EControl). The glazing thickness remains almost unchanged, between 25 and 29 mm.

From the energy saving point of view, it is useful, as noted above, to continue studying simulations on both existing and new buildings. In the case of existing buildings, it is necessary to understand when replacing windows with electrochromic panels can allow an economic return in the short term. The payback time obviously depends on the building type, geographic location, and especially percentage of transparent surfaces (WWR). In the case of new buildings, it is useful to identify some guidelines that may assist in the design phase, to better understand, as appropriate, the orientation of electrochromic glass surfaces and their extension. Another interesting field of research is the study of coupling electrochromic glazing and photovoltaic layers, which could produce, at least, the energy required for the operation of the glasses themselves.

Finally, from the architectural technology point of view, it is interesting to study new ways to use EC glazing. One hypothesis involves the use of electrochromic glass not in the transparent parts of the facade but in the opaque ones.

A recent work (Pittaluga, 2013) analyzes the hypothesis of fusion between traditional technologies, ventilated facades and Trombe walls, and electrochromic materials. Ventilated facades are usually composed of five layers: the wall of the building, the substructure for attaching the cladding, the thermal insulation, a ventilated cavity, and the external cladding (normally made of opaque materials). Due to the generation of the chimney effect inside the cavity, especially in summer, the overheating of the inner wall is reduced.

On the other hand, the Trombe wall is a passive system, usually oriented to south, for the accumulation of heat. It is usually composed of three layers: a solid wall, which is the thermal mass heat storage, with ventilation holes at the bottom and at the top; a cavity; and the cladding (normally made of transparent materials). Due to the generation of the greenhouse effect, the solar radiation is trapped inside the cavity and converted into heat. During winter, the hot air of the cavity is directed toward the interior of the building, whereas during summer, it is directed outside, thanks to dedicated holes.

In hottest climates, during summer days, the air inside the cavity reaches very high temperatures and a large amount of heat could be transmitted through the wall into the interior of the building. In order to avoid these problems, the wall can be shielded with exterior blinds or opaque sliding elements. From these considerations rises the idea of using electrochromic glazing as the cladding of the Trombe wall. The new system, the electrochromic wall, during winter, with EC glazings in their "off" state, will work as a Trombe wall; during summer, with EC glazings in their "on" state, it will work as a ventilated facade.

The simulation of the electrochromic wall behavior was carried out using the DesignBuilder software. A three-dimensional test room model was prepared and it was subjected to a dynamic simulation, for a period of 12 months. The test room was assumed initially without interior partitions.

The south facade of the test room was studied in three different configurations: the base case (NW, normal wall), the south facade with a Trombe wall (TW), and the south facade with an electrochromic wall (EW). The physical parameters characterizing the three systems and their performance were measured and compared.

Thanks to the TW solution, higher temperatures can be reached for the whole year (Figure 17.10). The indoor air temperature in the NW solution is always lower than the TW configuration; the difference varies between 2.66 °C (in October) and 0.5 °C (in December). The Trombe wall solution is more effective in winter and the NW one is more effective in summer. The electrochromic solution (continuous line) guarantees the lowest temperatures in summer, when the EC glazings are switched "on" (from May to September), and temperatures very similar to the TW solution in months from October to April, when EC glazings are in the "off" state.

If we analyze the relative humidity (Figure 17.11), we can see that thanks to the NW solution, higher values can be reached for the whole year. The relative humidity in the TW solution is always lower than that in the NW configuration. The EW solution (continuous line) guarantees the lowest values in summer, when the EC glazings are switched "on," and humidity values very similar to the TW solution in the other months, when EC glazings are in the "off" state.

In order to identify the solution that achieves the best hygrothermal comfort, the Fanger method (Fanger, 1967 and 1970), which defines the predicted mean vote

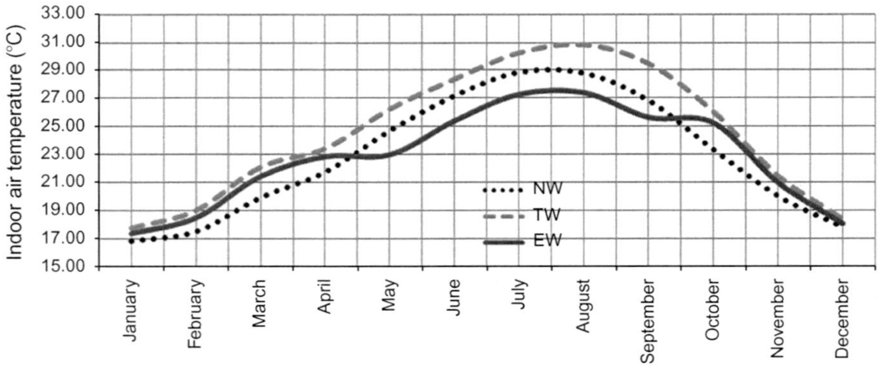

Figure 17.10 Comparison of indoor air temperature in the three models (normal wall, Trombe wall, and electrochromic wall).

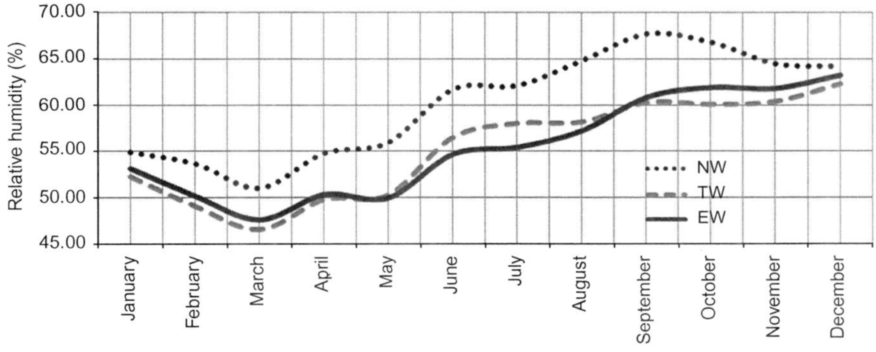

Figure 17.11 Comparison of indoor relative humidity in the three models (normal wall, Trombe wall, and electrochromic wall).

(PMV), was used. Two groups of parameters were analyzed: from the environmental point of view, air temperature, relative humidity, operating temperature, surface temperature of the floor, vertical difference of air temperature, radiant temperature, and air velocity and, from a human point of view, clothing, activities, and also on. The PMV index, according to the ASHRAE thermal sensation scale, predicts the mean response of a larger group of people: +3 hot, +2 warm, +1 slightly warm, 0 neutral, −1 slightly cool, −2 cool, and −3 cold.

Looking at the graph of the Fanger index (Figure 17.12), it is clear that the closest configuration to the optimum value of 0 is the one with the electrochromic wall. The monthly average PMV value is in fact 1.05 for the EW solution, 1.14 for the TW, and 1.15 for the NW.

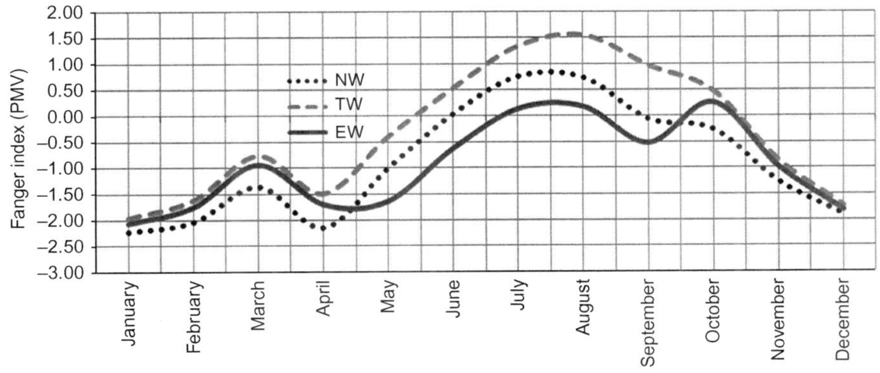

Figure 17.12 Comparison of Fanger Index (PMV) in the three models (normal wall, Trombe wall, and electrochromic wall).

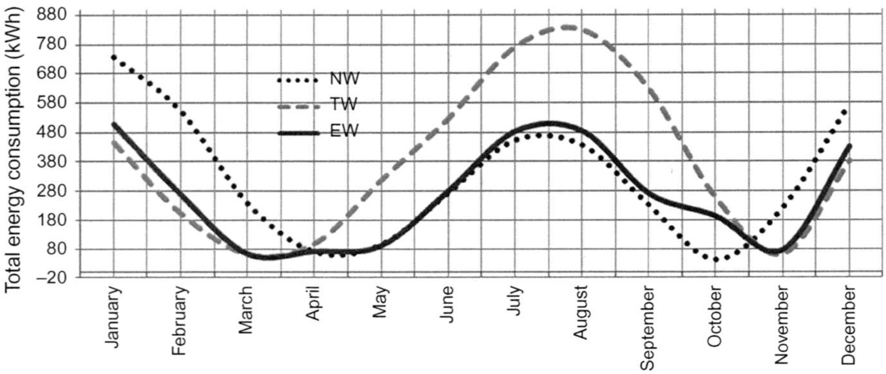

Figure 17.13 Comparison of total energy consumption in the three models (normal wall, Trombe wall, and electrochromic wall).

During the simulation, an HVAC system was foreseen: winter heating was provided by a natural gas-fired boiler and radiators; summer cooling system consisted of heat pumps powered by electricity. The total energy consumption diagram (Figure 17.13) shows that during the colder months, from October to April, the NW solution requires the highest energy amount; but this solution is the most convenient in the other months: during the hottest months, in fact, the TW solution requires the highest energy consumption.

The NW configuration has a consumption, in a year, of 3921 kWh, while the TW configuration has a consumption of 4584 kWh. On the other hand, the EW configuration is the best, with a consumption of 3229 kWh. The energy consumption values of the EW solution are in fact similar to the TW values during the coldest months and similar to the NW values in the hottest months. If we compare the EW solution to the NW one, we notice an energy saving of 17.6%; on the other hand, when we compare EW to TW, the saving is 29.5%.

The objective of the study was to understand whether a technological innovation could lead to an energy saving and may therefore be considered a useful option for the future. The results obtained are encouraging from a quantitative point of view, but we have to bear in mind that currently, codes for calculating the energy performances of buildings are still not 100% reliable. Uncertainty and imprecision of the results increase in particular when complex systems such as ventilated facades or Trombe walls are simulated. Further studies, live testings, and experiments are required in order to measure the real energy performance through appropriate instruments of survey and in order to better define the geometry of the system, the best size of the ventilated cavity, the position and size of the ventilation holes, the size of the glazed surfaces, and so on.

In conclusion, according to the thought of Mies van der Rohe, one of the greatest architects in history, we must remember that everything depends on how we use a material, not on the material itself. New materials are not necessarily superior. Each material is only what we make it.

References

Banham, R., 1995. Ambiente e tecnica nell. Architettura Moderna, Roma Laterza.

Compagno, A., 2002. Intelligent Glass Facades. Birkhauser, Basel.

Conan, M., 2002. Bourgeois and Aristocratic Cultural Encounters in Garden Art. Dumbarton Oaks, Washington, DC, 1550–1850.

Davies, M., 1981. A wall for all Seasons. Riba J. 88 (2), 55–57.

Fanger, P.O., 1967. Calculation of thermal comfort: introduction of a basic comfort equation. ASHRAE Trans. 73 (2), III4.1–III4.20.

Fanger, P.O., 1970. Thermal Comfort. Danish Technical Press, Copenhagen.

Lampert, C.M., 1991. Large area chromogenic for smart windows. IEEE Circuits Devices Magazine 8, 19–26, LBL 27720

Lee, E.S., DiBartolomeo, D.L., 2000. Application issues for large electrochromic windows in commercial building. Solar Energy Materials and Solar Cells 71, 465–491, LBNL-45841.

Lee, E.S., Yazdanian, M., Selkowitz, S.E., 2004. The Energy-Savings Potential of Electrochromic Windows in the US Commercial Buildings Sector, Building Technologies Program Environmental Energy Technologies Division, Lawrence Berkeley National Laboratory, 1 Cyclotron Road, Berkeley, CA 94720, USA, Mailstop 90–3111.

Lee, E.S., Selkowitz, S.E., Clear, R.D., DiBartolomeo, D.L., Klems, J.H., Fernandes, L.L., Ward, G.J., Inkarojrit, V., Yazdanian, M., 2006. Advancement of Electrochromic Windows, California Energy Commission, PIER. Publication number CEC-500-2006-052.

Llordes, A., Garcia, G., Gazquez, J., Milliron, D.J., 2013. Tunable near-infrared and visible-light transmittance in nanocrystal-in-glass composites. Nature 500, 323–326.

Loddo, G., Mandas, N., Pittaluga, M., Cossu, G., Fenu, L., 2009. Study addressing the performance of an electrochromic envelope in an office building in the Mediterranean climate. In: Proceedings of Glass Performance Days, Tampere.

Ludoni, D., Loddo, G., Cossu, G.P., Pittaluga, M., 2012. Technologies of the future for the re-use of the past: the electrochromic glass. In: Proceedings of Heritage 2012, 3rd International Conference on Heritage and Sustainable Development. Greenlines Institute, Porto.

McLuhan, M., 1967. Understanding Media, The Extensions of Man. The MIT Press, Cambridge.

Pittaluga, M., 2005. Media City: ai confini della realtà. In: Proceedings of International Conference on Theory and Practice of Construction: Knowledge, Means and Models. Ravenna, Moderna, pp. 1269–1279.

Pittaluga, M., 2013. The electrochromic wall. Energy Build. 66, 49–56.

Pittaluga, M., Loddo, G., Cossu, G.P., Ludoni, D., 2010. Electrochromic glazing in contemporary architecture: examples of use and opportunities for the future. In: Proceedings of XXXVII IAHS World Congress on Housing, Design, Technology, Refurbishment and Management of Building. Santander, Spain.

Pittaluga, M., Loddo, G., Cossu, G.P., Ludoni, D., 2013. Performance and properties testing of electrochromic glazing in Mediterranean climate. In: Proceedings of 39th IAHS Congress on Housing Sciences, Milano.

Schittisch, C., 2003. Building Skins. Birkhauser, Basel.

SWIFT, 2003. 031011-Architectural and technical guidelines.

Tenner, A.D., Zonneveldt, L., 2002. Switchable facades and visual comfort. Right Light 5, 29–31.

Wigginton, M., 1996. Glass in Architecture. Phaidon, New York.

Wigginton, M., Harris, J., 2002. Intelligent Skins. Oxford Elsevier, UK.

The impact of electrochromic windows on the energy performance of buildings in Mediterranean climates: a case study

18

P.F. Tavares[1,2,3], *A.R. Gaspar*[3], *A.G. Martins*[2,3], *F. Frontini*[4]
[1]Polytechnic Institute of Coimbra, Coimbra, Portugal; [2]Institute for Systems Engineering and Computers, Coimbra, Portugal; [3]University of Coimbra, Coimbra, Portugal; [4]University for Applied Sciences and Arts of Southern Switzerland (SUPSI), Canobbio, Switzerland

18.1 Introduction

The components of a building envelope that are more permeable to energy exchange with the outside are usually the fenestrations (Assimakopoulos et al., 2004; Roos et al., 2005). They are the most problematic elements of the building façade, because they may allow high amounts of solar energy to enter the building in the summer, imposing high energy costs for space cooling. This problem of overheating is particularly critical in modern commercial buildings, heavily influenced by internal gains from thermal energy released by computers and other office equipment, and also the latent and sensible energy released by occupants (Azens and Granqvist, 2003).

On the other hand, they are the components of the façade with higher U-values and are therefore responsible for relevant energy losses in winter, contributing to energy consumption for space heating.

The measures normally taken to avoid overheating due to excessive solar gain from glazing are the installation of blinds and awnings with manual action that, in practice, stay actuated for long periods. Thus, the use of glass with controllable properties, with responses that can be made automatically, may be a strategy to overcome this limitation. The glass that allows this control are designated chromogenic (Granqvist, 2006a,b).

When applied to high performance glazing, the term "chromogenic" is used to indicate that the optical properties can be changed in response to an external stimulus (Granqvist et al., 2010). In this context, the main chromogenic technologies are electrochromic (EC), in response to a voltage or electric charge stimulus; thermochromic (TC), in response to a temperature change; photochromic, in response to the variation of the luminous radiation, typically in the ultraviolet range; and gasochromic, which depends on exposure to reducing or oxidizing gases (Carmody et al., 2004; Granqvist, 2006a; Granqvist et al., 2009, 2010).

Eco-efficient Materials for Mitigating Building Cooling Needs. http://dx.doi.org/10.1016/B978-1-78242-380-5.00018-2

The chromogenic technologies are also grouped into two categories, depending on the type of activation (electrical or nonelectrical) (Vergaz et al., 2007). In the first category fall the electrochromic and the gasochromic windows, glasses with suspended particles and glasses of liquid crystals dispersed in polymers. In the second category are the photochromic and the thermochromic glasses.

EC glass can be controlled in an infinite number of intermediate states between the clear and the colored extremes. The technology, the advantages, the performance analysis of EC glass, and the main companies that commercialize this type of glass may be found in Chapter 17 and in this chapter. Complete reviews of this technology can be found in Cronin et al. (1999), Granqvist (2000), Granqvist et al. (2003), Lampert (2003), Azens et al. (2005), Avendaño et al. (2006), Niklasson and Granqvist (2007), Granqvist (2007), Georg and Georg (2009), Baetens et al. (2010), Granqvist (2012), Jelle et al. (2012), and Jelle (2013).

Building rehabilitation is a major task in the building industry, with a strong impact on global sustainability (Voss, 2000), and energy saving measures should be part of the goals of rehabilitation. Thus, it is significantly more attractive to consider the combination of renewal and improvement of energy efficiency from both a technological and an economic perspective than to consider these issues separately (Spiegelhalter, 2009; Voss, 2000).

In this context, the present chapter describes a study of building energy analysis using EC windows in rehabilitation.

The chapter is divided into four sections. After this introduction, Section 18.2 presents the methodology used to assess the energy performance of the EC glass in a building that will be refurbished. The adopted simulator is presented, with the description of the module used to simulate EC window controls. Some considerations based on previous works are undertaken to justify the procedure considered in this case study.

The third section is dedicated to the analysis of the case study, a school building constructed in the middle of the last century. In this section, the building model is described and the results are presented and discussed.

The last section presents the main conclusions of this chapter, highlighting the main achievements regarding the impact of using EC glass in the energy performance of buildings to be rehabilitated.

18.2 Methodology for electrochromic (EC) energy performance assessment

There are several energy building simulation programs used by the scientific community that present results with accuracy. Crawley et al. (2008) present an interesting survey about the performance and main characteristics of most of the simulation building programs available.

Most of the building simulation programs have the capacity to simulate EC glass in binary mode (activated and deactivated state). However, a detailed analysis of the effect of EC glass on building performance assessment requires the modulation of multiple variables and characteristics of EC glass in the intermediate states.

From our research we find a module in the ESP-r simulator (Clarke, 2001), as it is one of the building energy analysis programs that the research team has used that has implemented an advanced module—Black Box Model (BBM)—to simulate controllable shutters and complex transparent façades (Frontini et al., 2009). Therefore, the authors adapted the algorithm implemented in the BBM to permit the simulation of the EC glass control.

For the case study developed in Section 18.3, the ESP-r version 11.10 for Linux™ (Crawley et al., 2005a,b, 2008) was used to perform energy performance simulations. The ESP-r is an energy simulation program able to model phenomena involving exchanges of energy and mass inside buildings as well as other auxiliary systems. It also allows for the simulation of real conditions in a building, analyzing the building's performance in terms of energy consumption, occupant comfort indices, evolution of temperature, air quality, control systems, etc. The ESP-r is a computer program (simulator) available at no cost under an open source license (Open Source), developed by the "Energy Systems Research Unit" (ESRU), University of Strathclyde, Scotland.

Simulator fundamentals (e.g., modules, mathematical description, models) in general and in particular for ESP-r can be found in Clarke (2001). The simulation of shading devices in ESP-r is based on the methodologies presented in Strachan (1990) and Haugaard (2003).

The equations of this model were first presented in Kuhn (2006a) with the objective to build a general method for realistic performance evaluation of solar control properties of façades for façades with sun-shading or other solar control systems. The accuracy of this method is described in Kuhn (2006b). In 2009 this method was implemented in ESP-r (Frontini et al., 2009) and given the name "Black Box Model" (BBM). The main advantage of the new method is that it only uses measurable quantities of the transparent or translucent part of the façade as a whole (Kuhn et al., 2011).

The objective behind the BBM model is to describe every complex façade with a two layer model (e.g., two panes of glass). The glazing system has defined bidirectional transmissivity and reflectivity; each of the two virtual layers has effective solar absorption, also with a bidirectional definition. All the optical properties are a function of both solar azimuth and altitude. Each data set "i" is valid for different directions of the incident solar radiation and different settings of a control parameter "β_k," which, for a venetian blind, is the tilt angle of the slats, or, in the case of façades with switchable properties, a parameter field that characterizes the glass state and solar radiation incidence angle.

The new interface for the BBM needs as input the bidirectional angle-dependent total energy transmittance (g-value) and direct solar transmittance (τ_{solar}), and the U-value of the total façade unit under known laboratory conditions. If available, the solar reflectance ρ_{solar} (for solar radiation reaching the outer surface of the façade unit) can also be used as input, which will improve the accuracy of the external surface temperature modeling. It is emphasized that it is not necessary to measure g-value, τ_{solar}, and ρ_{solar} directly, but these properties can also be calculated with mathematical models.

There are two control options implemented. The first involves specifying set points of a user defined parameter, which are tested at each timestep in the simulation to determine which dataset to use. The second is to use a temporal definitions file to specify which datasets are to be used at each timestep in the simulation. The validation and the accuracy of this BBM are shown in Frontini (2011) and Kuhn et al. (2011).

The parameterisation of the EC glass in the BBM consists of the creation of a file with the optical properties of the glass. Although the control of the state of the EC glass can be done continuously, in this modulation it is only possible to define a set of six possible states of colouration, including the clear state (state 1) and full-color (condition 6). These six states were defined with the list of dimensionless properties listed in Figure 18.1 (solar factor: g; solar absorptance: $\alpha_{\text{solar;}}$ solar transmittance: τ_{solar}; and solar reflectivity: ρ_{solar}).

A text file for each of the geographical orientations studied with EC (east, south, and west), with the optical properties for each of the glass states, was prepared in a format recognized by the simulator (Frontini et al., 2009). The formulation used (see Equations 18.1–18.4) is based on the equations from Rosenfeld (1996), Roos et al. (2000), Roos et al. (2001), Kuhn (2006a), Kuhn et al. (2011), and validated by Platzer (2000), Rosenfeld et al. (2000), Karlsson and Roos (2000), Roos et al. (2000), and Karlsson et al. (2001). Based on the parameters a_{roos}, b_{roos}, c_{roos}, α_{roos}, β_{roos}, and γ_{roos} presented in Table 18.1, the values of τ_{solar}, g, and ρ_{solar}, for the solar azimuth range of [−90°, 90°] and solar height range of [−90°, 90°], were computed with increments of 5°.

Each text file has the values of τ_{solar}, g, and ρ_{solar} for solar incidence range in each of the six discrete states defined for EC glass. During the simulation process, the state of the EC glass is determined according to the value of the control variable. Based on the instant height and solar azimuth, in each step of the simulation, the optical properties of the EC glass were obtained from the text file.

Equation (18.1) for calculating the solar factor of the glass (dependent on the angle of incidence: α_{in}):

$$g_{\text{glass}}[\alpha_{\text{in}}] = g_{\text{glass}}[0^{\circ}]\left(1 - a_{\text{roos}}\left(\frac{\alpha_{\text{in}}}{90^{\circ}}\right)^{\alpha_{\text{roos}}} - b_{\text{roos}}\left(\frac{\alpha_{\text{in}}}{90^{\circ}}\right)^{\beta_{\text{roos}}} - c_{\text{roos}}\left(\frac{\alpha_{\text{in}}}{90^{\circ}}\right)^{\gamma_{\text{roos}}}\right) \tag{18.1}$$

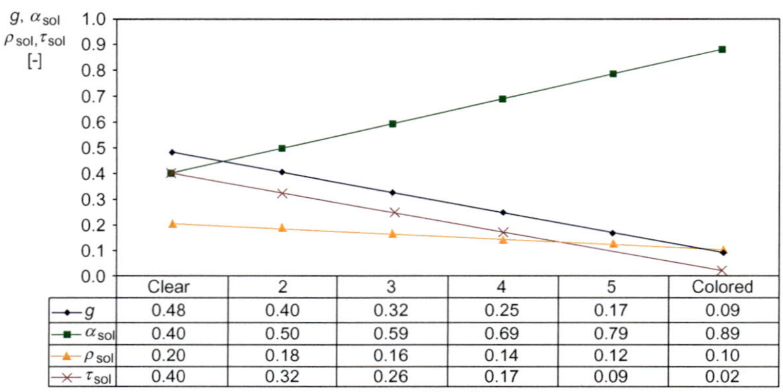

	Clear	2	3	4	5	Colored
g	0.48	0.40	0.32	0.25	0.17	0.09
α_{sol}	0.40	0.50	0.59	0.69	0.79	0.89
ρ_{sol}	0.20	0.18	0.16	0.14	0.12	0.10
τ_{sol}	0.40	0.32	0.26	0.17	0.09	0.02

States of the EC glass

Figure 18.1 Optical properties, for normal incidence, of the six states of the modeled EC glass.

Table 18.1 Parameters used to calculate (τ_{sol}–g–ρ_{sol})

Parameters	Value	Notes
p_{roos}	2.00	Number of glass panes separated by an air layer
q_{roos}	1.00	Fixed parameter (for EC glasses)
a_{roos}	8.00	Fixed parameter
$b_{roos}=0.25/q_{roos}$	0.25	Fixed parameter
$c_{roos}=1-a_{roos}-b_{roos}$	−7.25	Fixed parameter
$\alpha_{roos}=5.2+0.7\times q_{roos}$	5.90	Fixed parameter
$\beta_{roos}=2$	2.00	Fixed parameter
$\gamma_{roos}=(5.26+0.06\times p_{roos})+(0.73+0.04\times p_{roos})\times q_{roos}$	6.19	Fixed parameter
$g\ (\alpha=0^o)$	0.48	Variable parameter
$\tau_{sol}\ (\alpha=0^o)$	0.40	Variable parameter
$\rho_{sol}\ (\alpha=0^o)$	0.20	Variable parameter
$\alpha_{sol}\ (\alpha_{in}=0^o)$	0.40	Fixed parameter (α_{in}—solar incidence angle)
ϕ	0.00	Orientation of the façade (0° = south, positive west)
ϕ_s	–	Variable parameter—solar azimuth angle (0° = south, west positive)
ϕ_f	–	Variable parameter—$\phi_f=\phi_s-\phi$ (façade azimuth angle, 0° parallel to façade normal)
α_s	–	Variable parameter—solar altitude angle

Equation (18.2) for calculating the solar transmittance of the glass (also dependent on the angle of incidence: α_{in}):

$$\tau_{sol,glass}[\alpha_{in}] \approx \tau_{sol,glass}[0^o]\left(1-a_{roos}\left(\frac{\alpha_{in}}{90^o}\right)^{\alpha_{roos}}-b_{roos}\left(\frac{\alpha_{in}}{90^o}\right)^{\beta_{roos}}-c_{roos}\left(\frac{\alpha_{in}}{90^o}\right)^{\gamma_{roos}}\right) \tag{18.2}$$

Equation (18.3) for calculating the solar reflectance of the glass (dependent on the angle of incidence: α_{in}):

$$\rho_{sol,glass}(\alpha_{in}) \approx \begin{cases} 1-\tau_{sol,glass}(\alpha_{in})-\underbrace{\left[1-\rho_{sol,glass}(\alpha_{in}=0^o)-\tau_{sol,glass}(\alpha_{in}=0^o)\right]}_{\text{absortivity for normal incidence } \alpha(\alpha_{in}=0^o)} & \text{if } \alpha_{in}\leq 5^o \\ 1-\tau_{sol,glass}(\alpha_{in})-\alpha(\alpha_{in}=0^o)\dfrac{\alpha_{in}-90^o}{15^o} & \text{if } \alpha_{in}>75^o \end{cases} \tag{18.3}$$

Equation (18.4) for calculating the angle of incidence (dependent on the solar altitude angle: α_s and on the façade azimuth: ϕ_f):

$$\alpha_{in}(\alpha_s,\phi_f)=\arccos(\cos(\alpha_s)\cos(\phi_f)) \tag{18.4}$$

The BBM has five possible glazing control options (Frontini et al., 2009): zone air temperature; outside air temperature; incident radiation on an outside surface; level of zone illuminance; and other zone air temperature.

Following the criterion of Roos et al. (2005), an hourly timestep was adopted for the dynamic simulations, which is considered adequate for operation and control conditions of EC windows.

The transition between states of EC glass can be made according to several variables. In previous works, Tavares et al. (2011, 2012, 2014) analyzed the EC control by the indoor temperature, the outdoor temperature, and the solar radiation incident on the glass. It was concluded that in Mediterranean climates, the best way to control EC glass is by incident radiation. It was also found that for east and west façades, the best criteria to control EC glass in the heating season consists of keeping EC glass always in the clear state, while during cooling season maintenance in the colorful state is the best option. In terms of annual energy balance, the best results for the east or west façade were obtained with a value of solar radiation incident on the glass of 150 W/m^2 to impose a total colored state. However, for this control strategy the results are almost insensitive to the selected ranges of transition between states. For south façades the application of EC glass presents no advantage.

18.3 Case study

18.3.1 The building model

18.3.1.1 Introduction

The building selected as the case study houses the Faculty of Arts and Humanities of the University of Coimbra (FAHUC). This building was selected because in the near future it will need to be rehabilitated. Moreover, given the shape and orientation of the building façades of FAHUC, this building is particularly interesting because it allows the performing of multiple studies in the four main geographical orientations, with glazing subjected to different levels of solar incidence or degree of shading. Finally, there are limitations/restrictions regarding this building as historical, and by the fact of its integration in the World Heritage of University of Coimbra, Alta and Sofia, recognized by the United Nations Organization for Education, Science and Culture (UNESCO). The electrochromic glass technology, which receives special dedication in this work, could present itself as a solution with good potential for application in the context of rehabilitation of this building.

18.3.1.2 Building description

The FAHUC building (see Figures 18.2 and 18.3) was finished in 1951. It consists of seven floors, with the hall of honor (main entrance) located on the fourth floor, at the level of "Largo da Porta Férrea." The other entrance is located on the opposite façade of the building, in front of "Machado de Castro" Museum, and allows direct and easy communication with the central library located on the second floor.

Although several floors show identical geometry and area, there are differences between them because some are partial underground spaces, with different types of partitioning and uses.

The information from energy bills and an energy audit conducted to list the equipment, luminaries, and respective power installed by space; determine the load

Figure 18.2 Main façade of the FAHUC building (adapted from FLUC, 2010), south oriented.

Figure 18.3 Aerial view of FAHUC building (Google, 2014).

diagrams; and break down the consumption per floor was used to parameterise the building model and validate the simulation for the base condition, from which all other studies were performed.

18.3.1.3 Building model

The geometric model was defined based on the drawings kindly provided by the University of Coimbra and several walkthrough visits in order to better understand the use of the building and adopt adequate simplifications for the building model, particularly in spaces grouping.

The zoning of the building model was carried out, taking into account the current use of the spaces and the orientation of the room façades, resulting in the definition of 73 thermal zones (spaces). In Figure 18.4, an example of a floor plan level zoning is shown.

According to the building project and the walkthrough observations, it was considered a height of 3.9 m for the first floor, and 4 m for the remaining floors (floors 2–7). Following the building model geometric simplification (zoning), the fenestration area was also grouped for each thermal zone.

Figure 18.4 Zoning considered on floor 6.

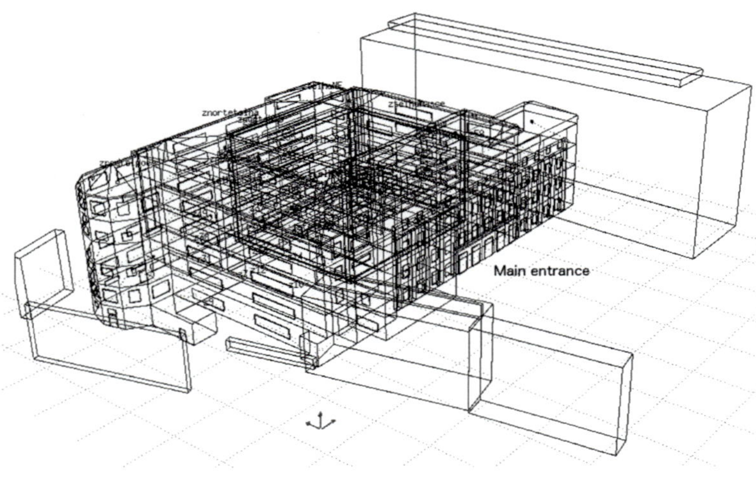

Figure 18.5 View of the west/south façade model in ESP-r.

Figure 18.5 shows an outlook of the building model as it was parameterised in the program, and some blocks representing adjacent buildings that shadow the building under analysis.

The parameterisation of the model in terms of physical properties of the construction elements (walls, floors, ceilings, windows) was reported to the existing specifications of the FAHUC building design dating from the 1940s. In the absence of information, typical data for the expected materials were adopted.

18.3.1.4 Simulation validation

To validate the model, a comparison was made between the real building consumption and the simulation results. With respect to electricity, monthly energy consumption (provided through the monthly bills of the university campus) was considered. Regarding fuel oil, billing information was limited to a filling of 10 tons during the reference year, insufficient information to conclude about the real fuel oil consumption.

The information from an energy audit made in 2006 was used to define the profile diagrams of lighting, equipment, and occupancy, and to determine the installed power per different uses.

From the electrical bills, the energy consumption in the reference year was 147,789 kWh, and from the building energy simulation, the value of 135,902 kWh was obtained. The difference between real and simulation energy consumption is 8%, a value below the reference 10% usually accepted as the maximum deviation for this kind of energy evaluation. The difference corresponds to some loads that are quite impossible to quantify with accuracy, namely related loads of two old elevators, some movable heaters, and some recent split air conditioning units. Therefore, the results of the simulation in real conditions give enough confidence about the adequacy of the ESP-r building model parameterisation and representativeness of the energy performance of the building under analysis. The energy breakdown per final use for electricity is presented in Figure 18.6.

18.3.2 Simulation planning

In Portugal and south European regions, the north façades only receive diffuse radiation during wintertime, and during the cooling period they receive direct solar radiation during a short period in the early morning and late afternoon. Therefore, the north façades were not considered for the study with EC glass, because their control does not result in significant energy savings.

The base situation corresponds to the current situation of the FAHUC building, which has windows with single glass. For this situation with single glass, the percentage of shading of each glazing defined in the model was initially determined in order to consider this parameter in the analysis of the influence of EC glass in the thermal performance of the building.

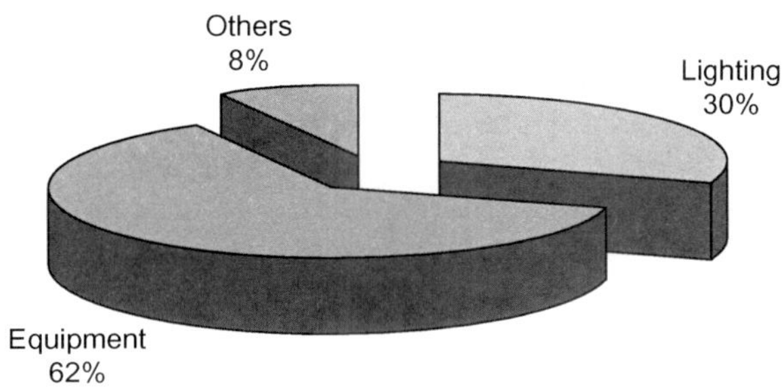

Figure 18.6 Electricity use breakdown obtained from the simulation results.

After that, the free behavior of the building (without air-conditioning system) and with indoor control temperature were determined for subsequently determining the differences in energy needs with the use of double glazing and double EC glazing. With the results of the energy requirements for air conditioning, studying the feasibility of applying EC glazing in buildings was a simple economic analysis.

The simulator does not allow the use of the BBM in more than one zone model. To overcome this limitation, the simulations with EC glass solution were done considering a limited number of zones. Thus, the representative simulated zones with EC glass were selected taking into account the diversity of the use type, the power density, and also the shading degree of the glazing.

The use of EC glass from SAGE™ was only taken into account in the windows of the east and west façades. Thus, four zones were considered in each of these façades, corresponding to a total of eight different simulations. The choice of each one of the zones took the shading degree of the glazing area into account to ensure that this shading degree was identical to the real glazing, since in the model the grouping of glazing per thermal zone was considered. In the selection, it was considered that the selected zones correspond to a diversification in the shading degree and the zones had glazing in no more than a single façade. When it was not possible to apply this criterion, we considered the EC glass in the façade to be analyzed and conventional double glazing in the remaining façades.

With these eight simulation results and the help of a linear regression, the application of EC glass in the entire building was verified. After the performance analysis was done façade by façade, an overall simulation was made in order to simplify the economic analysis.

The properties of the three glass types studied are listed in Table 18.2. The option was to study three types of commercially available glass, to perform simulation with glass properties obtained from the manufacturers and, in this way, to obtain more realistic results. The properties for the double EC glass from SAGE™ in the completely clear to colored states are presented in Table 18.2, and the properties in intermediate states are presented in Figure 18.1.

Table 18.2 Physical properties of the three types of glass analyzed

	Single	Double	Double EC SAGE™ (4 mm/12 mm/4 mm)	
Properties	**Clear 3 mm**	**Clear (4 mm/12 mm/4 mm)**	**Clear**	**Colored**
U [W/(m^2 K)]	5.79	2.80	1.60	
$g_{\perp}$	0.86	0.75	0.48	0.090
$\rho_{solar,\ \perp}$	0.07	0.12	0.20	0.100
$\tau_{solar,\ \perp}$	0.82	0.67	0.40	0.015
$\alpha_{solar,\ \perp}$	0.11	0.21	0.40	0.885
$\tau_{v,\ \perp}$	0.89	0.80	0.62	0.035

18.3.3 Energy performance analysis

18.3.3.1 Façade-oriented analysis

The thermal loads from the internal gains and solar gains of the studied zones (zones with different annual shading values and with different uses of the interior spaces) are presented in Table 18.3 for the heating and cooling seasons considering windows with simple glass. These values were calculated monthly, and a month was considered in the heating or the cooling season if the calculated energy needs for heating and cooling were higher, respectively. The presented values of internal gains were obtained using the simulator and calculated from the maximum hourly average power for lighting, occupancy, and equipment, multiplied by the respective daily profiles and summed for a given period.

All data were processed and analyzed in relation to the reference building, the situation with single glass, which corresponds to the present building case study and represents the most common solution for existing buildings to be renovated in the near future in Portugal.

Figure 18.7 shows the results obtained for passive operation of selected zones, which means results of simulation without control of the indoor thermal conditions; therefore, no energy is determined for space heating and cooling. The graphical representations show, for the zones in east and west façades, the percentage of the occupancy period in which the indoor ambient is considered uncomfortable. In the heating season this corresponds to indoor temperature below 20 °C (large white bars), and in the cooling season to indoor temperature above 25 °C (large black bars). The comparison is made with single glass and EC glass solutions. The small bars represent the percentage variation obtained by the replacement of single glass with double EC glass, and must be read in the right axis. These bars are represented just for the zones where EC glass was considered. The pattern colors of the small bars have identical meaning to the base bars (black with white dots for temperature above 25 °C and white with black dots for temperature below 20 °C).

The results show that the use of EC glass improves indoor thermal comfort in the spaces of both façades during the cooling season (1% to 7% of occupancy time when the temperature is above 25 °C). However, during winter the use of EC glass increases thermal discomfort (a variation from 1% to 6% of occupancy time when the temperature is below 20 °C). The positive effect of EC glass increases with the increase of the cooling needs of the space.

The results considering the utilization of air-conditioning are presented in Table 18.4 in terms of the energy needs of the building case study with single glass and with EC glass from SAGE™.

The results of Table 18.4 show that EC glass has different impacts during the heating and cooling seasons. The heating needs increase due to the reduction of solar gain, since the solar factor of EC glass is smaller even in the clear position (0.48) than the one of the single glass (0.86). This reduction is not compensated by the reduction of the thermal losses achieved by the change from single glazing to double glazing with a low solar factor. The advantages of EC glass are more visible in the reduction of energy needs for cooling, following the general conclusions found in the literature.

Table 18.3 **Internal and solar gains of the selected zones on the east and west façades for the base case (single glass)**

		Internal gains during				Solar gains during				Annual gains				
		Heating		Cooling		Heating		Cooling		Internal		Solar		
Zone	Simulated area (m^2)	(kWh)	(kWh/m^2)	(kWh)	(kWh/m^2)	(kWh)	(kWh/m^2)	(kWh)	(kWh/m^2)	(kWh)	(kWh/m^2)	(kWh)	(kWh/m^2)	Annual shading (%)
East façade														
z5	301	11490.0	38.2	0.0	0.0	1332.5	4.4	0.0	0.0	11490.0	38.2	1332.5	4.4	91
z24	269	8492.0	31.6	4317.6	16.1	4622.7	17.2	4057.2	15.1	12809.6	47.6	8679.9	32.3	85
z33	140	7360.1	52.6	5360.8	38.3	550.7	3.9	556.9	4.0	12720.9	90.9	1107.6	7.9	63
z54	269	6335.5	23.6	4544.1	16.9	6384.6	23.7	8719.3	32.4	10879.6	40.4	15103.9	56.1	74
West façade														
z9	438	15518.9	35.4	5244.7	12.0	22525.3	51.4	12061.3	27.5	20763.7	47.4	34586.5	79.0	66
z17	138	1735.4	12.6	0.0	0.0	5413.9	39.2	0.0	0.0	1735.4	12.6	5413.9	39.2	87
z38	431	7647.3	17.7	3888.7	9.0	17772.2	41.2	14605.8	33.9	11535.9	26.8	32378.0	75.1	56
z49	431	9193.3	21.3	9269.0	21.5	16708.4	38.8	27145.9	63.0	18462.2	42.8	43854.3	101.8	56

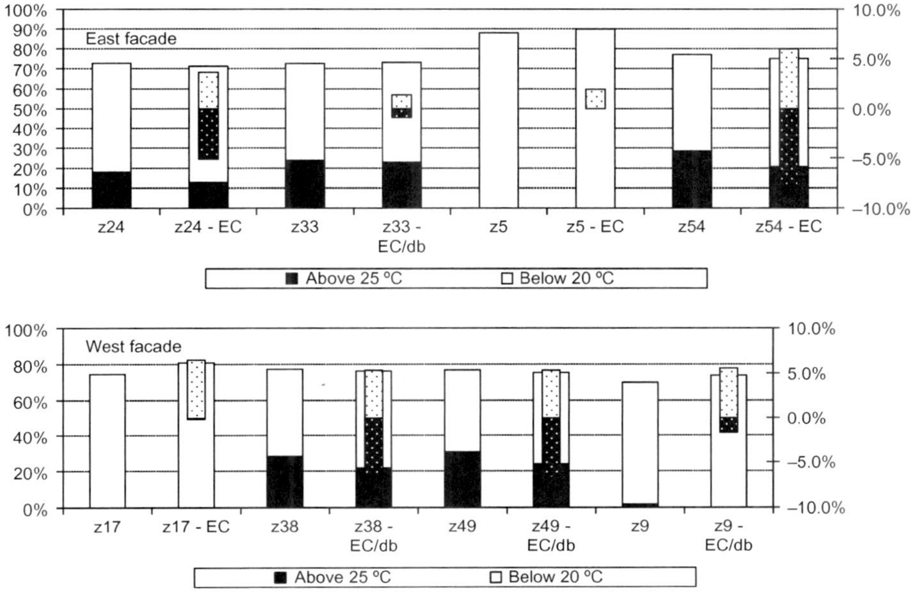

Figure 18.7 Percentages of occupancy time with the indoor air temperature above 25 °C and below 20 °C and corresponding variations by the use of EC glass.

For the spaces located in the east façade, only zone 5 has no advantage in having EC windows, mainly because it is a zone without cooling needs (partially underground) and the window has a high level of shading (91%). On the west façade, zones 9 and 17 also do not have advantages in the adopting of EC glass, mainly because they present low energy needs for cooling due to high levels of shading (66% and 87%, respectively). Thus, the positive effect in the cooling period is not compensated by the negative effect in the heating period.

The presented results validate the findings of Warner et al. (1992) that EC windows can be controlled in response to solar gains in office buildings, leading to the reduction of energy required for cooling. However, in cold climates, the use of EC windows could increase energy needs for heating. The end result must be well balanced between gains and losses for cooling and heating energy needs.

In Figure 18.8, the values of maximum thermal load for heating and cooling are presented, with single and EC glazing applied in the simulated zones. The upper limit of each of the vertical lines corresponds to the maximum heating load required for zone heating, and the lower limit the maximum load required for zone cooling.

Figure 18.8 shows the advantages of using EC glass to decrease the thermal load to control the temperature of the space. The positive effects occur both for heating and cooling, however, the effects are more significant for the west façade (heating: from 3.9% to 15.3%; cooling: from 33% to 100%—zone 9 (z9) no longer has energy needs for cooling) than for the east façade (heating: from 1.1% to 8.7%; cooling: from 0% to 32.2%).

Table 18.4 Energy requirements for space conditioning with single and EC glass and calculated relative energy differences

Zone	Simulated area (m^2)	Single glass		EC glass		Relative difference					
		Annual energy needs for									
		Heating	Cooling	Heating	Cooling	Heating		Cooling		Annual	
		(kWh/m^2)	(kWh/m^2)	(kWh/m^2)	(kWh/m^2)	(kWh/m^2)	(%)	(kWh/m^2)	(%)	(kWh/m^2)	(%)
East façade											
z5	301	42.7	0.0	43.3	0.0	0.5	1.3	0.0	0.0	0.5	1.3
z24	269	22.9	−6.0	−24.0	−3.3	1.1	4.9	−2.7	−45.2	−1.6	−5.5
z33	140	12.1	−10.5	12.0	−9.3	−0.2	−1.4	−1.2	−11.1	−1.3	−5.9
z54	269	20.3	−13.0	22.2	−6.2	1.9	9.2	−6.9	−52.6	−5.0	−15.0
West façade											
z9	438	29.1	−0.4	29.6	0.0	0.5	1.7	−0.4	100.0	0.1	0.4
z17	138	56.8	0.0	60.3	0.0	3.5	6.1	0.0	0.0	3.5	6.1
z38	431	21.2	−11.0	22.6	−5.4	1.3	6.3	−5.5	−50.5	−4.2	−13.0
z49	431	17.9	−15.3	17.1	−8.2	−0.8	−4.5	−7.1	−46.5	−7.9	−23.9

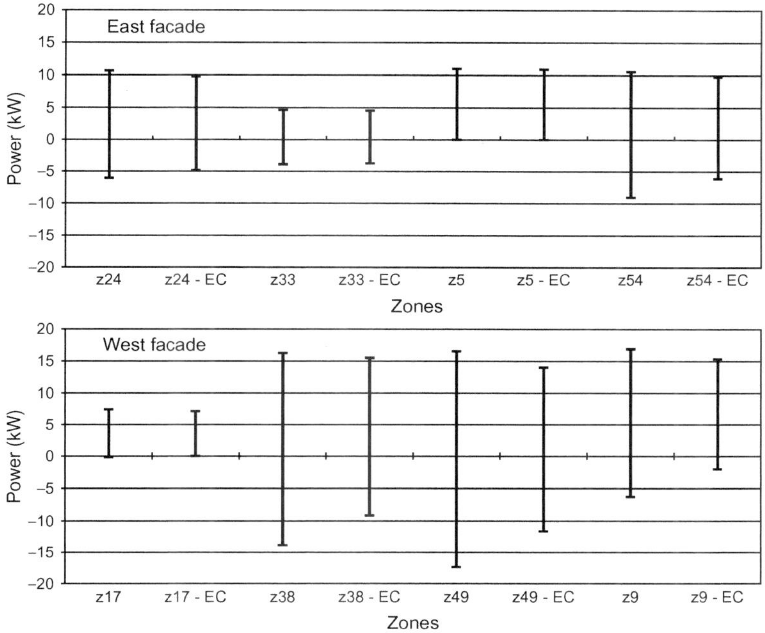

Figure 18.8 Maximum heating and cooling load for the situations of single and EC glass.

18.3.3.2 Global energy assessment

For the global energy assessment considering the application of EC glass all over the present analyzed case study, the previous energy results obtained from selected representative spaces were used to determine multiple regressions, considering as independent variables the internal gains and the solar gains obtained for the solution with single glass. For the regression generation MATLAB® (R2010b version) was used—«least_square_polyfit_xyz.m». The obtained function is of the type $f(x, y) = a + bx + cx^2 + dxy + y + fy^2$, where f is the difference in energy consumption between the use of EC glass and use of single glass, x the internal gains, and y the solar gains obtained for the solution with single glass.

The coefficients a, b, c, d, e, and f were determined based on the results obtained by the simulation of the eight initial zones. Six regressions were also established: three regressions for the east façade and three regressions for the west (considering the energy needs for heating, cooling and annual energy needs—see Table 18.5). The results showed a lower mean quadratic deviation considering the heating and cooling individually. Thus, the seasonal regressions were used instead of the annual regressions.

One of the initial criteria for the selection of the zones to be simulated was based on shading level diversity. Therefore, the remaining zones to be analyzed have a shading level within the range of application of the determined regressions.

Table 18.5 **Regression coefficients and respective mean quadratic deviation**

Regression coefficients	East façade			West façade		
	Heating	Cooling	Annual	Heating	Cooling	Annual
a	−1.2289	0.0000	−0.2471	1.8414	0.0000	9.6499
b	0.1769	0.3594	0.0811	−0.7304	0.1287	−0.7276
c	−0.0029	−0.0082	−0.0012	0.0166	0.0545	0.0117
d	0.0077	−0.0103	0.0052	−0.0079	−0.0132	0.0014
e	−0.5423	−0.3441	−0.3398	0.1738	−0.1341	0.1044
f	0.0161	0.0059	0.0003	0.0034	−0.0022	−0.0024
Mean quadratic deviation (kWh/m^2 year)	0.0961	0.0568	0.5132	0.1837	0.0465	0.3243

The results obtained for all the studied building thermal zones with EC glass are shown in Table 18.6. It should be noted that all areas of the east and west façades where the use of EC glass resulted in an unfavorable annual energy performance (compared to single glazing) should have been considered as to have windows with conventional double glazing (presented in Table 18.2), and therefore were removed from this table.

From the results of Table 18.6, it is concluded that it is advantageous to replace the single glass in the east and west façades with EC glass. Although in most zones there is an increase of energy for heating, the reduction of energy requirements for cooling is higher. For the east façade, annual energy reductions range from 3.7% to 24.4%, and on the west façade from 5.3% to 23.9%.

Although for most spaces the application of EC glass results in an increase of energy for heating, there are some zones with a heating demand reduction, including zones z33 (bar on the fifth floor that has east EC glass windows and double glazing to the north), z68 (corridor on the third floor that has a small east-oriented EC window and is mainly an interior space), and zpoentehab (cabinet zone on the sixth floor, near the roof, which has east-oriented EC glass windows). The results show that when the energy needs for cooling using single glass are low both in east and west façades (e.g., in zones z5, z17, and z9—see Table 18.5), the application of EC glass leads to an increase in annual energy needs in relation to the situation of single glazing, due to the decreased solar gains in the heating season. The zones where the use of EC glass leads to increased annual energy needs in relation to the use of single glass were grouped with the other south and north zones where it is recommended to replace the single glass with conventional double glazing without improved characteristics, as shown by the results in Table 18.7.

For the 19 zones identified in Table 18.6, the annual energy reduction for thermal correction is 14.1%. The reduction of energy per unit area for heating is 0.2 kWh/m^2 year and 4.1 kWh/m^2 year for cooling, or a yearly reduction of 4.3 kWh/m^2 year.

Table 18.7 shows the results of the remaining zones of the building where the use of EC windows is disadvantageous in terms of energy savings. This set corresponds to all zones in the north and south façades; zones z4, z5, and z20 in the east façade; zones z9, z12, and z17 in the west front; underground zones z8 and z10; and an auditorium/theatre "Paulo Quintela" (z19).

Table 18.6 Annual differences in energy needs for heating and cooling by using EC glass. Results obtained using multiple regressions for heating and cooling

				Single glass		Electrochromic			
				Annual energy needs for					
				Heating	Cooling	Heating	Cooling		
Zone	Façade(s)	Typology	Simulated area (m^2)	(kWh/m^2 year)	(kWh/m^2 year)	(kWh/m^2 year)	(kWh/m^2 year)	Differences (kWh/m^2 year)	Variation (%)
z24	E	Classroom	269	22.88	−5.96	24.00	−3.27	−1.57	−5.5
z44	E	Classroom	269	20.74	−10.08	21.16	−4.15	−5.51	−17.9
z54	E	Classroom	269	20.32	−13.04	22.19	−6.18	−4.99	−15.0
z14	E (courtyard)	Cabinets	165	24.36	−8.15	24.65	−4.37	−3.50	−10.8
z30	E (courtyard)	Rooms	165	22.57	−10.21	21.59	−5.00	−6.20	−18.9
z36	E (courtyard)	Cabinets	165	18.26	−14.51	19.28	−6.61	−6.89	−21.0
z50	E (courtyard)	Cabinets	165	21.75	−12.41	21.85	−6.39	−5.92	−17.3
zpoentehab	E (courtyard)	Cabinets	149	28.52	−27.41	28.03	−25.85	−2.05	−3.7
z33	E (courtyard)	Bar	140	12.13	−10.48	11.97	−9.31	−1.33	−5.9
z68	E and interior	Corridor	229	12.03	−0.71	10.22	−0.69	−1.83	−14.4
z27	E and interior	Corridor	229	10.59	−2.92	9.06	−2.73	−1.73	−12.8
z42	E and interior	Corridor	229	9.57	−4.52	7.44	−3.61	−3.04	−21.6
z51	E and interior	Corridor	229	8.35	−5.94	6.18	−4.61	−3.49	−24.4
z31	W	Rooms	431	23.54	−9.15	22.15	−6.45	−4.09	−12.5
z38	W	Rooms	431	21.21	−10.96	22.55	−5.42	−4.19	−13.0
z49	W	Classroom	431	17.87	−15.31	17.07	−8.20	−7.92	−23.9
z25	W (courtyard)	Corridor	138	43.09	−2.17	42.85	0.00	−2.42	−5.3
z40	W (courtyard)	Corridor	138	38.67	−5.85	38.94	0.00	−5.58	−12.5
z52	W (courtyard)	Corridor	138	35.60	−9.39	39.36	−0.57	−5.05	−11.2
		Total:	4379	20.54	−9.53	20.38	−5.46	−4.23	−14.1

Table 18.7 **Annual differences in energy requirements for air conditioning by the use of double glazing in zones where the use of EC windows is a disadvantage**

Simulated area (m^2)	Single glass		Double glass		Differences for heating (kWh)	Differences for cooling (kWh)	Annual differences (kWh)
	Annual energy needs for						
	Heating (kWh)	Cooling (kWh)	Heating (kWh)	Cooling (kWh)			
9854	258,768 kWh/m^2 year 26.26	−49,547 kWh/m^2 year −5.03	215,709 kWh/m^2 year 21.89	−67,831 kWh/m^2 year −6.88	−43,059 kWh/m^2 year −4.37 −16.6%	18,284 kWh/m^2 year 1.86 36.9%	−24,775 kWh/m^2 year −2.51 −8.0%

The results in Table 18.7 show that the use of double glazing in the heating season is an advantage due to the reduction of heat transfer loss, but a disadvantage in the cooling season. Globally for these zones there is an annual energy savings of 8% for heating and cooling, increasing the thermal energy for heating in 4.37 kWh/m^2 year and decreasing the thermal energy for cooling in 1.86 kWh/m^2 year, or an annual increase of 2.51 kWh/m^2 year.

18.3.4 Discussion of economic aspects

As mentioned in this chapter, the cost of EC glass is higher than the cost of current glass technology, leading to a small EC glass market. In this context, and to complement the results of the detailed thermal behavior of the building, a simplified economic study using data from Gesimat and SAGE™ products was developed from the perspective of the costs that make the technology viable.

This simplified economic analysis was based on the data relating to EC glass mentioned in the introduction to this chapter. In addition, a lifetime of the EC glass technology reported in the literature periods of 20–30 years was considered (Baetens et al, 2010), along with the SAGE™, E-Control®, and Gesimat companies' present product warranty of 10 years (Baetens et al, 2010; Ritchie, 2010).

Considering that EC glass has a "memory effect," i.e., it is only necessary to provide energy to stimulate a transition from one state to another and not to maintain the state, and that the energy required for this stimulus is negligible, the supplied electrical energy is not considered in this analysis.

Investment decisions on technology replacements depend strongly on the discount rate assumed by each decision maker, which translates directly into the time span for full investment recovery. In order to facilitate the assessment of the EC option, admissible extra cost values of the EC option as compared to double glass units were calculated for various scenarios. Considering only the building energy operational cost and assuming a unitary coefficient of performance (COP) for heating and an energy efficiency ratio (ERR) of 2 for cooling, Table 18.8 presents the exercise for the application of EC glass in this case study.

Beyond this operational cost analysis, a more detailed analysis could discuss other savings components, namely the reduction of the installed air-conditioning system capacity; no need for additional blinds; increased thermal and visual comfort (no glare); increased productivity; possible energy savings in the artificial lighting system; decrease of the installed electrical power of the building; and maximum load to the electrical grid (Sanders, 2010; Sanders and Podbelski, 2009, 2010).

The achieved energy savings can contribute to mitigating the high cost of current commercially available EC glass (€1000/m^2), thus increasing the use on a large scale and reducing production cost.

Several authors reported results of annual energy savings above 100 kWh/m^2_glass year by the use of EC glass. For example, admitting a solar radiation incidence on a south-oriented window of 1000 kWh/m^2 year, Granqvist (2001), Azens and Granqvist (2003), Azens et al. (2005), Niklasson and Granqvist (2007), Granqvist (2007), and Granqvist (2008) reported annual savings of 170 kWh/m^2_glass year.

Table 18.8 Simplified economic analysis of possible additional costs for EC glass

Heating energy saved from the use of EC glass		700.64	kWh
Cooling energy saved from the use of EC glass		8911.27	kWh
Annual energy savings resulting from the use of EC glass		9611.91	kWh
EC glass area		402.46	m^2
Annual energy saved per unit area		23.88	kWh/m^2 ECglass year
kWh price (without considering power rate and taxes)		€0.14	
Maximum additional allowable costs (€/m^2) to recover the investment in EC windows in:	1 year	€3.34	
	5 years	€16.72	
	10 years	€33.44	
	15 years	€50.15	
	20 years	€66.87	

Syrrakou et al. (2005) refer typical values for energy savings by the use of EC glass in the order of 120–340 kWh/m^2_glass year. Karlsson (2001) analyzed three different climates (Stockholm, Sweden; Denver, USA; and Miami, USA) and obtained annual energy savings between 0 and 150 kWh/m^2_glass year, indicating an average value for annual energy savings of 100 kWh/m^2_glass year.

From the present study the operational energy savings are in the range of 23.88 kWh/m^2_glass year (see Table 18.7), showing less favorable results for the use of EC windows. The first reason for the lesser impact of EC glass obtained in the present study has to do with the climate considered. In fact, better results were obtained for warmer climates (with energy requirements for heating near zero–Miami in the USA, for example, Karlsson (2001)). Moreover, some studies referred to in the first section of this chapter do not mention the shading of the façades (e.g., Lee et al., 2004), which the present study proved to be an important factor in the viability analysis of EC window use.

Several studies (Jonsson and Roos, 2010a,b; Karlsson, 2001; Lee et al., 2004) also show the feasibility of using EC windows on the south façade. However, according to the results and the discussion in Section 18.2, the use of EC glass is only interesting on the east and west façades. The discrepancy in the EC energy performance applied on the south façade could be related to the EC physical characteristics used and the period of analysis. Actually, the abovementioned works do not use the physical characteristics of a commercially available EC glass, and they calculate energy savings only for the cooling season.

18.4 Conclusions

The use of a general building energy simulation tool (ESP-r) that can be adapted to accommodate this technology (EC windows) is a solution of great importance in helping building designers (for new or refurbished constructions), to easily calculate the savings resulting from the use of EC windows. It was thus possible to mobilize a general simulation tool to address the need for analysis of an emerging technology.

The analysis of a case study building was performed in five distinct phases. In the first stage, passive behavior was addressed and it was found that, in general, the performance of the building in the heating season is less satisfactory than in the cooling season. In the second phase, the energy needs for heating and cooling were evaluated in the baseline situation, considering the current situation of single glazing. For both studies it was found that the exterior shading of windows, the use profile, and the internal gains of the spaces are variables with non-negligible influence on the results.

The third phase consisted of studying the application of EC glazing windows in east and west façades and calculating the savings resulting from this application. It was found that it is not always advantageous to use EC glass for east and west façades, with the results depending on the level of shading of the windows and on the internal gains of spaces. The annual savings from the building as a whole were estimated, considering double EC glass where advantageous in the east and west façades, and conventional double glass in the other building windows. A difference in annual energy needs with the replacement of single glazed by conventional double glazing was obtained in the order of 8%, and savings of about 14% in annual energy needs resulting from single glass replacement with double EC glass controlled by solar radiation.

A simplified economic analysis of the feasibility of using EC windows was developed, and it was found that for energy savings per m^2 glass area of about 24 kWh/year, there is an economic attractiveness of using this type of glass (whereas EC glass is used in east and west façades with different levels of shading). For this energy savings, the maximum permissible additional cost per m^2 of EC glass for a simple recovery period of 10 years is €33.44.

For both studies, passive behavior and the use of an air conditioning system, exterior shading of windows, the type of use, and the internal gains of the spaces are variables that influence the results. The lower the levels of shading of the east and west façades, the greater the savings achieved by the use of EC windows.

If the exterior shading is less than 91% on the east façade and 66% on the west façade, the replacement of single glass by EC glass should be considered. In the remaining zones, only the replacement of single glass by conventional double glazing should be considered.

In the studied building, the replacement of single glass by conventional double glazing to all considered zones (north, south, and those where there is no feasibility of using EC glass in east and west façades) resulted in savings of 2.5 kWh/m^2 year, or a reduction of 8%.

Savings resulting from single glass replacement for double EC glazing controlled by incident solar radiation to the set of zones considered (in east and west) are 4.2 kWh/m^2 year, which corresponds to a 14% reduction of the annual energy needs compared to the base case.

The results presented here are on an annual basis, taking into account the heating and cooling seasons. This annual approach and the presence of exterior shading could be the main justification for the less favorable results obtained.

EC windows are an efficient solution in terms of energy (technically and economically), being advantageous for use in the rehabilitation of buildings, on the east and west façades, where it is difficult to control solar gains.

Acknowledgments

This work is being developed in the context of the "Energy for Sustainability" initiative of the University of Coimbra and is supported by the Portuguese "Fundação para a Ciência e a Tecnologia" (FCT—Foundation for Science and Technology) under grant SFRH/BD/35910/2007. This work has also been partially supported by the Portuguese Foundation for Science and Technology under project grant PEst-OE/EEI/UI308/2014.

References

Assimakopoulos, M.N., Tsangrassoulis, A., Guarracino, G., Santamouris, M., 2004. Integrated energetic approach for a controlable electrochromic device. Energy Build. 36, 415–422. http://dx.doi.org/10.1016/j.enbuild.2004.01.040.

Avendaño, E., Berggren, L., Niklasson, G.A., Granqvist, C.G., Azens, A., 2006. Electrochromic materials and devices: brief survey and new data on optical absorption in tungsten oxide and nickel oxide films. Thin Solid Films 496, 30–36. http://dx.doi.org/10.1016/j.tsf.2005.08.183.

Azens, A., Granqvist, C.G., 2003. Electrochromic smart windows: energy efficiency and device aspects. J. Solid State Electrochem. 7, 64–68. http://dx.doi.org/10.1007/s10008-002-0313-4.

Azens, A., Avendaño, E., Backholm, J., Berggren, L., Gustavsson, G., Karmhag, R., Niklasson, G.A., Roos, A., Granqvist, C.G., 2005. Flexible foils with electrochromic coatings: science, technology and applications. Mater. Sci. Eng., B 119, 214–223. http://dx.doi.org/10.1016/j.mseb.2004.12.085.

Baetens, R., Jelle, B.P., Gustavsen, A., 2010. Properties, requirements and possibilities of smart windows for dynamic daylight and solar energy control in buildings: a state-of-the-art review. Sol. Energy Mater. Sol. Cells 94, 87–105. http://dx.doi.org/10.1016/j.solmat.2009.08.021.

Carmody, J., Selkowitz, S., Lee, E.S., Arasteh, D., Willmert, T., 2004. Window Systems for High-Performance Buildings. W. W. Norton & Company, Inc, New York ISBN: 0-393-73121-9.

Clarke, J.A., 2001. Energy Simulation in Building Design, second ed. Butterworth-Heinemann, Oxford ISBN: 0-7506-5082-6.

Crawley, D.B., Hand, J.W., Kummert, M., Griffith, B.T., 2005a. Contrasting the Capabilities of Building Energy Performance Simulation Programs. Report, version 1.0. Available at: http://apps1.eere.energy.gov/buildings/tools_directory/pdfs/contrasting_the_capabilities_of_building_energy_performance_simulation_programs_v1.0.pdf.

Crawley, D.B., Hand, J.W., Kummert, M., Griffith, B.T., 2005b. Contrasting the Capabilities of Building Energy Performance Simulation Programs. In: Proceedings of "Building Simulation 2005, Ninth International IBPSA Conference", Montreal, Canada, 2005 15–18 August.

Crawley, D.B., Hand, J.W., Kummert, M., Griffith, B.T., 2008. Contrasting the capabilities of building energy performance simulation programs. Build. Environ. 43, 661–673. http://dx.doi.org/10.1016/j.buildenv.2006.10.027.

Cronin, J.P., Gudgel, T.J., Kennedy, S.R., Agrawal, A., Uhlmann, D.R., 1999. Electrochromic glazing. Mater. Res. 2 (1), 1–9.

FLUC, 2010. "Faculdade de Letras—Universidade de Coimbra". Web page accessed on 12 August 2011, <http://ezequielbarelfilho.blogspot.pt/p/galeria-de-fotos.html > Photograph authored by Ezequiel Barel Filho (<http://www.ezequielbarelfilho.blogspot.com>).

Frontini, F., 2011. Daylight and Solar Control in Building. Fraunhofer Verlag, Stuttgard—DEU (ISBN 978-3-8396-0238-6).

Frontini, F., Kuhn, T.E., Herkel, S., Strachan, P., Kokogiannakis, G., 2009. Implementation and Applications of a New Bi-Directional Solar Modelling Method for Complex Facades within the ESP-r Building Simulation Program. In: Proceedings of "Eleventh International IBPSA Conference (Building Simulation 2009)", Glasgow (Scotland), 27–30 July 2009.

Georg, A., Georg, A., 2009. Electrochromic device with a redox electrolyte. Sol. Energy Mater. Sol. Cells 93, 1329–1337. http://dx.doi.org/10.1016/j.solmat.2009.02.009.

Google, 2014. 3D image from Google Earth. Web page accessed 19.08.14: http://maps.google.pt/.

Granqvist, C.G., 2000. Electrochromic tungsten oxide films: review of progress 1993–1998. Sol. Energy Mater. Sol. Cells 60, 201–262.

Granqvist, C.G., 2001. Electrochromic smart windows: toward an energy efficient architecture. Electrochem. Soc. Interface 3, 18–19, Fall.

Granqvist, C.G., 2006a. Electrochromic materials—out of a niche. Nat. Mater. 5, 89–90.

Granqvist, C.G., 2006b. Electrochromics: finally a technology for large-scale applications. SPIE Newsroom. http://dx.doi.org/10.1117/2.1200602.0140. Published online: 1 April 2006. http://spie.org/x8814.xml (accessed 15.04.09).

Granqvist, C.G., 2007. Transparent conductors as solar energy materials: a panoramic review. Sol. Energy Mater. Sol. Cells 91, 1529–1598. http://dx.doi.org/10.1016/j.solmat.2007.04.031.

Granqvist, C.G., 2008. Oxide electrochromics: why, how, and whither. Sol. Energy Mater. Sol. Cells 92, 203–208. http://dx.doi.org/10.1016/j.solmat.2006.10.027.

Granqvist, C.G., 2012. Oxide electrochromics: an introduction to devices and materials. Sol. Energy Mater. Sol. Cells 99, 1–13. http://dx.doi.org/10.1016/j.solmat.2011.08.021.

Granqvist, C.G., Avendaño, E., Azens, A., 2003. Electrochromic coatings and devices: survey of some recent advances. Thin Solid Films 442, 201–211. http://dx.doi.org/10.1016/S0040-6090(03)00983-0.

Granqvist, C.G., Lansaker, P.C., Mlyuka, N.R., Niklasson, G.A., Avendaño, E., 2009. Progress in chromogenics: new results for electrochromic and thermochromic materials and devices. Sol. Energy Mater. Sol. Cells 93, 2032–2039. http://dx.doi.org/10.1016/j.solmat.2009.02.026.

Granqvist, C.G., Green, S., Niklasson, G.A., Mlyuka, N.R., von Kræmer, S., Georén, P., 2010. Advances in chromogenic materials and devices. Thin Solid Films 518, 3046–3053. http://dx.doi.org/10.1016/j.tsf.2009.08.058.

Haugaard, P., 2003. Investigation and implementation of building simulation programmes—especially ESP-r". MS ThesisTechnical University of Denmark, Denmark.

Jelle, B.P., 2013. Solar radiation glazing factors for window panes, glass structures and electrochromic windows in buildings—measurement and calculation. Sol. Energy Mater. Sol. Cells 116, 291–323. http://dx.doi.org/10.1016/j.solmat.2013.04.032.

Jelle, B.P., Hynd, A., Gustavsen, A., Arasteh, D., Goudey, H., Hart, R., 2012. Fenestration of today and tomorrow: a state-of-the-art review and future research opportunities. Sol. Energy Mater. Sol. Cells 96, 1–28. http://dx.doi.org/10.1016/j.solmat.2011.08.010.

Jonsson, A., Roos, A., 2010a. Evaluation of control strategies for different smart window combinations using computer simulations. Sol. Energy 84, 1–9. http://dx.doi.org/10.1016/j.solener.2009.10.021.

Jonsson, A., Roos, A., 2010b. Visual and energy performance of switchable windows with antireflection coatings. Sol. Energy 84, 1370–1375. http://dx.doi.org/10.1016/j.solener.2010.04.016.

Karlsson, J., 2001. Control System and Energy Saving Potential for Switchable Windows. In: Proceedings of "7th International IBPSA Conference", Rio de Janeiro, Brazil, 13–15 August 2001.

Karlsson, J., Roos, A., 2000. Modelling the angular behaviour of the total solar energy transmittance of windows. Sol. Energy 69 (4), 321–329.

Karlsson, J., Rubin, M., Roos, A., 2001. Evaluation of predictive models for the angle-dependent total solar energy transmittance of glazing materials. Sol. Energy 71, 23–31.

Kuhn, T.E., 2006a. Solar control: a general evaluation method for facades with venetian blinds or other solar control systems. Energy Build. 38, 648–660. http://dx.doi.org/10.1016/j.enbuild.2005.10.002.

Kuhn, T.E., 2006b. Solar control comparison of two new systems with the state of the art on the basis of a new general evaluation method for facades with venetian blinds or other solar control systems. Energy Build. 38, 661–672. http://dx.doi.org/10.1016/j.enbuild.2005.10.001.

Kuhn, T., Herkel, E.S., Frontini, F., Strachan, P., Kokogiannakis, G., 2011. Solar control: a general method for modelling of solar gains through complex facades in building simulation programs. Energy Build. 43, 19–27. http://dx.doi.org/10.1016/j.enbuild.2010.07.015.

Lampert, C.M., 2003. Large-area smart glass and integrated photovoltaics. Sol. Energy Mater. Sol. Cells 76, 489–499.

Lee, E.S., DiBartolomeo, D.L., Rubinstein, F.M., Selkowitz, S.E., 2004. Low-cost networking for dynamic window systems. Energy Build. 36, 503–513. http://dx.doi.org/10.1016/j.enbuild.2003.12.008.

Niklasson, G.A., Granqvist, C.G., 2007. Electrochromics for smart windows: thin films of tungsten oxide and nickel oxide, and devices based on these. J. Mater. Chem. 17, 127–156. http://dx.doi.org/10.1039/b612174h.

Platzer, W.J., 2000. The ALTSET Project: Measurement of Angular Properties for Complex Glazing. In: Atas de "EuroSun 2000 Conference", Copenhagen, Dinamarca, 19–22 June 2000.

Ritchie, J., 2010. "Energy efficiency through electrochromism and gasochromism of tungsten oxide films" (EECE 580, 17 March 2010). Web page accessed on 15 September 2011, http://pt.scribd.com/doc/53333679/Electrochromics-for-Smart-Windows-for-class.

Roos, A., Polato, P., Nijnatten, P.A.V., Hutchins, M.G., Olive, F., Anderson, C., 2000. Angular-dependent optical properties of low-e and solar control windows—simulations versus measurements. Sol. Energy 69 (Suppl. 6), 15–26, 1–6.

Roos, A., Karisson, J., van Nijnatten, P.A., Hutchins, M.G., Polato, P., Nichelatti, E., Montecchi, M., Olive, F., Anderson, C., 2001. Angular dependent optical properties of coated glazings—validation of two predictive algorithms. In: Lampert, C.M.,

Granqvist, C.-G., Lewis, K.L. (Eds.), Solar and Switching Materials. In: Proceedings of SPIE, 4458, pp. 19–28.

Roos, A., Persson, M.-L., Platzer, W., Köhl, M., 2005. Energy Efficiency of Switchable Glazing in Office Buildings. In: Proceedings of Glass Processing Days 2005—9th International Glass Conference, Tampere, Finland, 17–20 June 2005.

Rosenfeld, J.L.J., 1996. On the Calculation of the Total Solar Energy Transmittance of Complex Glazings. In: Proceedings of 8th International Meeting on Transparent Insulation Material, Freiburg, Germany.

Rosenfeld, J.L.J., Platzer, W.J., Dijk, H.v., Maccari, A., 2000. Modelling the Optical and Thermal Properties of Advanced Glazing: Overview of Recent Developments. In: Proceedings of EuroSun 2000 Conference, Copenhagen, Denmark, 19–22 June 2000.

Sanders, H., 2010. Getting to net zero energy buildings with electronically tintable glass. Verre 16 (4), 42–47.

Sanders, H., Podbelski, L., 2009. Electronically Tintable Glass: The Future of High Performance Facades is Here. In: Proceedings of Glass Performance Days 2009 (Architectural Challenges & Solutions), Tampere, Finland, 12–15 June 2009.

Sanders, H.E., Podbelski, L., 2010. Electronically Tintable Glass: The Future of High Performance Façades is Here. In: Proceedings of "Engineered Transparency—International Conference at glasstec, Düsseldorf, Germany, 29–30 September 2010.

Spiegelhalter, T., 2009. Zero-fossil-energy powered infrastructure and buildings on Catalina Island in Southern California. Ecosyst. Sustainable Dev. VII 122, 293–304.

Strachan, P., 1990. "Adition of blinds/shutter control to transparent multi-layer constructions and other improvements to the solar routines of ESPsim". Web page accessed on 15 April 2009, <http://www.esru.strath.ac.uk/Documents/90/strachan_solar_mods.pdf>.

Syrrakou, E., Papaefthimiou, S., Skarpentzos, N., Yianoulis, P., 2005. Electrochromic windows: physical characteristics and environmental profile. Ionics 11, 281–288.

Tavares, P.F., Martins, A.G., Gaspar, A.R., 2011. "Reduction Building Energy Needs by the Use of Automation and Control of Electro-Chromic Windows". Proceedings/Lectures Book (ISBN:978-84-95010-41-4), edited by ATECYR—Associación Técnica Española de Climatización y Refrigeración, (and pen drive) of international conference: CLIMAMED 2011—VI Mediterranean Congress of HVAC, 2–3rd June, 2011, Madrid, Spain, pp. 375–394.

Tavares, Paulo Filipe de A. F. Tavares, 2012. "Estratégias de Aumento do Potencial Bioclimático de Edifícios com Recurso à Automação—A Opção Pelos Vidros EC". University of Coimbra, Ph.D. Thesis (in Portuguese). Available at <https://estudogeral.sib.uc.pt/>.

Tavares, P.F., Gaspar, A.R., Martins, A.G., Frontini, F., 2014. Evaluation of electrochromic windows impact in the energy performance of buildings in Mediterranean climates. Energy Policy 67, 68–81. http://dx.doi.org/10.1016/j.enpol.2013.07.038.

Vergaz, R., Pena, J.M.S., Barrios, D., Pérez, I., Torres, J.C., 2007. Electrooptical behaviour and control of a suspended particle device. Opto-Electron. Rev. 15 (3), 154–158. http://dx.doi.org/10.2478/s11772-007-0013-9.

Voss, K., 2000. Solar energy in building renovation—results and experience of international demonstration buildings. Energy Build. 32, 291–302.

Warner, J.L., Reilly, M.S., Selkowitz, S.E., Arasteh, D.K., 1992. Utility and Economic Benefits of Electrochromic Smart Windows. In: Proceedings of "ACEEE 1992 Summer Study on Energy Efficiency in Buildings, Pacific Grove, California, USA, 30 August to 5 September 1992.

Appendix: nomenclature

U	thermal transmission coefficient $W/(m^2\,K)$
g	solar factor
α_{solar}	solar absorptance
τ_{solar}	solar transmittance
τ_v	visual transmittance
ρ_{solar}	solar reflectivity
p_{roos}	coefficients from Roos et al. (2000)
q_{roos}	Roos et al. (2001)
a_{roos}	Kuhn (2006a)
b_{roos}	Kuhn et al. (2011)
c_{roos}	validated by Platzer (2000)
α_{roos}	Rosenfeld et al. (2000)
β_{roos}	Karlsson and Roos (2000)
γ_{roos}	Roos et al. (2000) and Karlsson et al. (2001)
α_{in}	solar incidence angle (°)
ϕ	orientation of the façade (°)
ϕ_s	solar azimuth angle (°)
ϕ_f	$\phi_f = \phi_s - \phi$ (°)
α_s	solar altitude angle (°)

Index

Note: Page numbers followed by *f* indicate figures and *t* indicate tables.

C

D

E

T

U

V

W

Z

Printed in the United States
By Bookmasters